T0387992

Optoelectronic Organic–Inorganic Semiconductor Heterojunctions

Optoelectronic Organic–Inorganic Semiconductor Heterojunctions

Ye Zhou

CRC Press is an imprint of the
Taylor & Francis Group, an **informa** business

First edition published 2021
by CRC Press
6000 Broken Sound Parkway NW, Suite 300, Boca Raton, FL 33487-2742

and by CRC Press
2 Park Square, Milton Park, Abingdon, Oxon, OX14 4RN

Library of Congress Cataloging-in-Publication Data
Names: Zhou, Ye (Semiconductor engineer), editor.
Title: Optoelectronic organic-inorganic semiconductor heterojunctions /
[edited] by Ye Zhou. Description: First edition. | Boca Raton, FL : CRC Press/Taylor & Francis
Group, LLC, 2021. | Includes bibliographical references and index. |
Summary: "The book summarizes advances in development of
organic-inorganic semiconductor heterojunctions, challenges and possible
solutions for material/device design, and prospects for commercial
applications. It introduces the concept and basic mechanism of
semiconductor heterojunctions. It describes a series of
organic-inorganic semiconductor heterojunctions with desirable electrical and optical properties for
optoelectronic devices. Typical
devices such as solar cells, photo-detectors and optoelectronic memories
are discussed. Materials, device challenges, and strategies are discussed to promote the commercial
translation of semiconductor
heterojunction based optoelectronic devices" -- Provided by publisher.
Identifiers: LCCN 2020043631 (print) | LCCN 2020043632 (ebook) | ISBN
9780367342128 (hardback) | ISBN 9780367348175 (ebook)
Subjects: LCSH: Heterojunctions. | Optoelectronic devices. | Organic
semiconductors.
Classification: LCC TK7874.53 .O68 2021 (print) | LCC TK7874.53 (ebook) |
DDC 621.3815/2--dc23
LC record available at https://lccn.loc.gov/2020043631
LC ebook record available at https://lccn.loc.gov/2020043632

ISBN: 978-0-367-34212-8 (hbk)
ISBN: 978-0-367-34817-5 (ebk)

Typeset in Times
by SPi Global, India

Contents

Preface

A semiconductor heterojunction is the interface between two layers or regions of different semiconductor materials. When combining organic semiconductor with inorganic semiconductor, a hybrid heterojunction can be formed, either in planar structure or bulk structure. Organic–inorganic heterojunctions are promising candidates for functional electronic devices owing to the expected co-activation between organic and inorganic components. The hybrid system is beneficial from both the advantages of organic semiconductors and inorganic counterpart.

Optoelectronic semiconductor organic–inorganic heterojunctions have attracted intense research interests in the past decade. If the two semiconductors are combined through a reasonable interface design, the strengths can be enhanced. It is expected to obtain better photoelectric performance with proper design of the heterojunction materials. These kinds of heterojunctions have great potential to be applied in transistors, light-emitting diodes, sensors, solar cells, and memories. Scientists have further explored the one-dimensional and zero-dimensional electronic behaviors in heterogeneous structures. It is expected that new phenomena will be discovered in the future, and more novel heterogeneous structural components will emerge.

In this book, we have introduced the concept and basic mechanism of semiconductor heterojunctions. We investigate a series of organic–inorganic semiconductor heterojunctions with desirable electrical and optical properties for optoelectronic devices. The typical devices such as solar cells, photo-detectors, and optoelectronic memories are discussed. Materials and device challenges as well as possible strategies are also discussed to promote the commercial translation of these semiconductor heterojunctions-based optoelectronic devices.

I would like to acknowledge all the fellow authors who have contributed in this book. I also want to express my gratitude to Gabrielle Vernachio, Allison Shatkin, Lara S. Loes, and Camilla Michael at CRC Press/Taylor & Francis and Karthik Thiruvengadam at SPi Global, for all the help during the book editorial process, and for the excellent experience of working with them. Specially, I want to thank all the readers for their interest in our book. Our aim is to give a comprehensive, critical, and up-to-date book here. I hope that this book can be useful as a reference guide for researchers and students who work in the field of organic–inorganic heterojunctions and related electronic devices.

Ye Zhou

Editor Biography

Prof. Ye Zhou is an IAS Fellow and group leader in the Institute for Advanced Study, Shenzhen University. He received his B.S. (2008) in Electronic Science and Engineering from Nanjing University, M.S. (2009) in Electronic Engineering from Hong Kong University of Science and Technology, and Ph.D. (2013) in Physics and Materials Science from City University of Hong Kong. His research interests include flexible and printed electronics, nano-materials, nano-composite materials, and nano-scale devices for technological applications such as logic circuits, memories, photonics and sensors. He has edited 4 books, 3 USA patents, 11 China patents, and over 120 SCI papers in journals such as *Science, Nature Electronics, Nature Communications, Chemical Reviews, Advanced Materials*, etc. More than 40 of his works have been highlighted as cover pages or frontispieces. These published papers have been extensively accessed and cited by prestigious journals. He is the Associate Editor of *STAM, Applied Nanoscience and IEEE Access*, and sits at Editorial/ Community Board of *Materials Horizons, Multifunctional Materials, Chemistry, PLOS ONE, and Chemistry Proceedings*.

Contibutors

Dayan Ban
Department of Electrical and Computer Engineering
Waterloo Institute for Nanotechnology, University of Waterloo
Waterloo, Ontario, Canada

Sylke Blumstengel
Institut für Physik, Institut für Chemie & IRIS Adlershof
Humboldt-Universität zu Berlin
Berlin, Germany

J. Bruckbauer
Department of Physics
SUPA, University of Strathclyde
Glasgow, Scotland, UK

Shengnan Duan
Key Laboratory of Physics and Technology for Advanced Batteries (Ministry of Education)
College of Physics, Jilin University
Changchun, P. R. China

N. J. Findlay
WestCHEM, School of Chemistry
University of Glasgow
Glasgow, Scotland, UK

Shuang Gao
CAS Key Laboratory of Magnetic Materials and Devices & Zhejiang Province Key Laboratory of Magnetic Materials and Application Technology
Ningbo Institute of Materials Technology and Engineering, Chinese Academy of Sciences
Ningbo, Zhejiang, China

Jie Guan
School of Chemistry and Chemical Engineering
Yangzhou University
Yangzhou, Jiangsu, China

Yanbing Guo
Central China Normal University
Wuhan, Hubei, China

Zhengyuan Jin
College of Physics and Optoelectronic Engineering
Shenzhen University
Shenzhen, China

Xin Jin
Key Laboratory of Physics and Technology for Advanced Batteries (Ministry of Education)
College of Physics, Jilin University
Changchun, P. R. China

Sadaf Bashir Khan
Institute for Advanced Study
Shenzhen University
Shenzhen, Guangdong, China

Asif Abdullah Khan
Department of Electrical and Computer Engineering
Waterloo Institute for Nanotechnology, University of Waterloo
Waterloo, Ontario, Canada

Norbert Koch
Institut für Physik & IRIS Adlershof
Humboldt-Universität zu Berlin
Berlin, Germany

Helmholtz-Zentrum Berlin für Materialien und Energie
Berlin, Germany

Shern Long Lee
Institute for Advanced Study
Shenzhen University
Shenzhen, Guangdong, China

Run-Wei Li
CAS Key Laboratory of Magnetic Materials and Devices & Zhejiang Province Key Laboratory of Magnetic Materials and Application Technology
Ningbo Institute of Materials Technology and Engineering, Chinese Academy of Sciences
Ningbo, Zhejiang, China

Xiaoguang Li
Institute for Advanced Study
Shenzhen University
Shenzhen, Guangdong, China

Zheng Li
Institute for Advanced Study
Shenzhen University
Shenzhen, Guangdong, China

Na Li
Key Laboratory of Physics and Technology for Advanced Batteries (Ministry of Education)
College of Physics, Jilin University
Changchun, P. R. China

Aijun Li
Key Laboratory of Physics and Technology for Advanced Batteries (Ministry of Education)
College of Physics, Jilin University
Changchun, P. R. China

Tao Lv
College of Physics and Optoelectronic Engineering
Shenzhen University
Shenzhen, China

Xiangbo Meng
Department of Mechanical Engineering
University of Arkansas
Fayetteville, Arkansas, USA

Md Masud Rana
Department of Electrical and Computer Engineering
Waterloo Institute for Nanotechnology, University of Waterloo
Waterloo, Ontario, Canada

Baoning Wang
Key Laboratory of Physics and Technology for Advanced Batteries (Ministry of Education)
College of Physics, Jilin University
Changchun, P. R. China

Xiao-Feng Wang
Key Laboratory of Physics and Technology for Advanced Batteries (Ministry of Education)
College of Physics, Jilin University
Changchun, P. R. China

Ziwei Wang
School of Chemistry and Chemical Engineering
Yangzhou University
Yangzhou, Jiangsu, China

Qin Xu
School of Chemistry and Chemical Engineering
Yangzhou University
Yangzhou, Jiangsu, China
State Key Laboratory of Analytical Chemistry for Life Science, School of Chemistry and Chemical Engineering
Nanjing University
Nanjing, China

Yuan Yao
Central China Normal University
Wuhan, Hubei, China

Yu-Jia Zeng
College of Physics and Optoelectronic Engineering
Shenzhen University
Shenzhen, China

Luhong Zhang
College of Physics and Optoelectronic Engineering
Shenzhen University
Shenzhen, China

Wenjie Zhao
Key Laboratory of Physics and Technology for Advanced Batteries (Ministry of Education)
College of Physics, Jilin University
Changchun, P. R. China

Wei-Wei Zhao
State Key Laboratory of Analytical Chemistry for Life Science, School of Chemistry and Chemical Engineering
Nanjing University
Nanjing, China

Kui Zhou
Institute for Advanced Study
Shenzhen University
Shenzhen, Guangdong, China

Ye Zhou
Institute for Advanced Study
Shenzhen University
Shenzhen, Guangdong, China

Yuan-Cheng Zhu
State Key Laboratory of Analytical Chemistry for Life Science, School of Chemistry and Chemical Engineering
Nanjing University
Nanjing, China

1 Introduction to Organic–Inorganic Heterojunction

Kui Zhou and Ye Zhou

Heterojunction is a structure formed by combining two different semiconductor materials together, which is an important material foundation for modern electronics and optoelectronics. Combining two or more materials with different original interfaces to form a heterojunction is critical to the design and manufacture of functional devices, so the research of heterojunctions has also become an important subject in the field of materials [1,2].

As early as 1951, A. I. Gubanov [3,4] performed a theoretical analysis of heterojunctions, but it was limited to the difficulty of heterojunction growth technology. In 1960, Anderson [5] made high-quality heterostructures junction for the first time and proposed a more detailed band diagram and theoretical model. In 1963, H. Kroemer and Z. I. Alferov independently proposed the principle of heterojunction lasers [6,7]. In 1969, H. Kroemer and Z. I. Alferov prepared heterojunction lasers that can continuously operate at room temperature [8–10]. This achievement has established the foundation for the development of modern optoelectronics, which was awarded the 2000 Nobel Prize in Physics.

In a heterojunction structure, due to the different electrical and optoelectrical parameters of the two semiconductor materials, such as the bandgap width, conductivity type, dielectric constant, refractive index, and absorption coefficient, etc., the behavior of electrons, the interaction of photons and electrons, and some other physical properties are different from those in a single semiconductor material. Therefore, organic–inorganic heterojunctions have attracted more interest and attention. The photoelectric characteristics of heterojunctions can generally be divided into two categories: one is the photocurrent or photovoltaic voltage generated by absorption of photons; the other is the emission of electrons due to current or electric field excitation. Many approaches can be used to create electron–hole pair in the heterojunction, mainly depending on the wavelength of the incident photons. There are usually two important absorption processes that affect the photoelectric properties of a heterojunction: one is the absorption of impurities or interface states to generate free electrons or holes; the other is the intrinsic absorption of electrons that transition from the valence band to the conduction band. Due to these processes, the free carriers generated at the interface or the transition region of the heterojunction cause the photogenerated current of the heterojunction. In addition to absorbing photons to generate free carriers, photovoltaic voltages can also be generated. The properties of the heterojunction are determined by the composed semiconductor material.

Semiconductor materials have been known since the discovery of electrical phenomena in the 18th century. At that time, the materials could be divided into three categories: conductors ($\rho \leq \sim 10^{-6}\ \Omega \cdot m$), insulators ($\rho \geq 10^{10}\ \Omega \cdot m$), and semiconductors in between. In 1879, the physicist Hall discovered the Hall effect while studying the conductive mechanism of metals. Since then, scientists have used it to study the conductive properties of semiconductor materials. Semiconductor materials are found to have two differently charged carriers that are fewer in number than metals but have higher mobility. Generally, semiconductors can be categorized as inorganic semiconductors and organic semiconductors. In 1910s, inorganic semiconductor materials were used to make cuprous oxide low-power rectifiers and selenium rectifiers. Although a lot of researches have been done on semiconductors, due to the lack of theory in nature, the trial research has made little progress. Until the early 1930s, due to the development of quantum mechanics and the development of the band concept, semiconductor materials provided a solid theoretical basis. The birth of the germanium transistor in 1948 led humans from the era of electron tubes to the semiconductor era. Into the 1960s, the development of silicon integrated circuits was successful, making a leap in the development of the semiconductor industry. While studying silicon-germanium materials, a lot of research was also awakened on other materials, and it was found that compounds formed by group III and group V elements and their multiple compounds are also semiconductors. In the 1970s, various epitaxial growth technologies were developed, and superlattice quantum well and strain layer composites were prepared.

Most inorganic semiconductor materials can be single crystals. Single crystals are formed by periodic repeating arrangements of closely spaced atoms. The distance between adjacent atoms is only a few tenths of a nanometer. Therefore, the electrons are different from the free electrons in the vacuum and the electrons in the isolated atom. The electrons in the crystal are in a so-called band state. The energy band is composed of many energy levels. A forbidden band is isolated between the valence band and the conduction band, and electrons are distributed on the energy levels in the energy band. The reason that solids can conduct electricity is the result of the directional movement of electrons under the action of an external electric field. From the perspective of energy band theory, it is the transition of electrons from one energy level to another. When an external electric field is applied, the electrons on the energy band occupied by the electrons can absorb energy from the external electric field and jump to the unoccupied energy level to form a current. This energy band is often called the conduction band. The full band that has been occupied by valence electrons is called the valence band, and the middle is the forbidden band. In semiconductors, both the holes in the valence band and the electrons in the conduction band participate in conduction, which is the biggest difference from metallic conductors.

Inorganic materials have the advantages of high light-dark conductivity, high carrier mobility, and long service life as optoelectronic materials, but they have narrow absorption bands, fewer families of materials, and high production costs. Although the appearance of the superlattice makes it possible to improve its optoelectronic performance, it also complicates the device preparation process, increases the device production cost, and makes it difficult to achieve industrial production.

Meanwhile, the research of semiconductor materials started with inorganic materials, but due to its own limitations, people soon thought of organic materials naturally, because there are many more types of organic materials than inorganic materials, and inorganic materials with semiconductor properties must be found from them. With the gradual deepening of human research on material science and engineering, people have synthesized a batch of organic materials with semiconductor characteristics and are trying to apply them to the field of traditional semiconductor devices. Compared with inorganic semiconductor materials, organic semiconductors can be processed by the solution method. Therefore, many non-traditional device processing technologies such as screen printing, inkjet printing, micro-contact printing, and stamps can be used for device fabrication, making it possible to fabricate large area semiconductor devices and lower the cost.

Organic semiconductors include molecular crystals, organic complexes, and polymers. Molecular crystal semiconductors have strong covalent bonds inside them, but due to the weak van der Waals forces, the molecules interact with each other and the distance between the molecules is large, which is not conducive to electronic exchange between molecules. Organic molecular crystal semiconductor bandgap $E_g = I_c - A_c$, where I_c is the ionization energy of the electrons in the crystal and A_c is electron affinity. Molecular crystals are easy to purify (purification methods include recrystallization, column chromatography, and meteorological transmission, etc.), and small molecules can simply obtain single crystal thin films, and this easily available ordering is very beneficial for charge transport. Organic complexes are a combination of a compound with a high electron affinity (electron acceptor) and a compound with a low ionization energy (electron donor), so it is also called charge transfer complex (CT complex) or donor–acceptor complex (DA complex). The charge transfer between the molecules of the material can greatly increase the conductivity of the material. The most common example of semiconducting polymer is a conjugated polymer. A conjugated polymer refers to a repeating unit of a polymer chain composed of an atomic combination polymer having π bond and sp2 hybridization. Some conjugated polymers have semiconductor and even metal properties. They have provided high-performance single-component devices (such as polymer thin-film transistors (TFTs)), and also multifunctional devices with other materials, such as bulk heterojunction (BHJ) solar cells and small molecule/polymer-based TFTs. Recent advances in polymer design, synthesis, and processing enabled remarkable progress in polymer-based device performance. As we know, organic semiconductors and inorganic semiconductors differ greatly in their properties. Generally, organic semiconductors are broadband materials with a bandgap of 2~3 eV, while the bandgap of inorganic semiconductors such as Ge, Si, and GaAs are only 0.66 eV, 1.12 eV, and 1.42 eV at room temperature. At present, most organic semiconductors in air condition exhibit hole transport properties, that is, hole accumulation occurs under negative gate bias; when the gate bias is positive, electron accumulation is rarely found. Very weak van der Waals forces in organic materials are intermolecular interaction forces, which makes organic materials have low dielectric constant and electron mobility. However, scientists can prepare a variety of materials with different optoelectronic properties by changing the positions and types of substituents in organic compounds, which enriches the types of semiconductor optoelectronic materials. At present, since organic semiconductor materials have adjustable energy bands and a wide variety, the great interest are low-cost solution-processed thin films on various robust or flexible substrates in large area scale, which makes them have broad application prospects. However, due to its poor photoelectric performance, material stability, and poor wear resistance, it seems difficult to really commercialize on a large scale.

Despite the demonstration of promising prototypes of inorganic or organic semiconductors, it still remains great challenge in their development and optimization for high-performance device. Particularly, it is becoming more and more difficult to integrate desired properties to individual materials to satisfy the increasing demands of multifunctionality for fundamental studies as well as device designs and optimization.

Considering the advantages and disadvantages of the two materials, the use of organic/inorganic heterostructures can take the advantages of the two parts which can be fully utilized to improve the photoelectric performance of the materials. With suitable synthesis technology, simplifying the production process may eventually lead to large-scale industrial production. Besides, some new electrical and optoelectrical properties are expected in pre-designed organic–inorganic heterojunctions due to interface effects including energy-level alignment, which is crucial for potential in optoelectrical application. Ideally, the relative position of the band edges is determined by only the Fermi levels E_F and the electron affinities χ. However, it has to be corrected in the real cases due to the following interface effects: (1) energy-level shift resulted from dipoles at the organic/inorganic interface by chemical reactions [11]; (2) band bending caused by pinning of the Fermi level at surface or interface states [12]; (3) the type and level of doping of the inorganic semiconductor that have a strong influence on the band alignment. The detail of the energy alignment will be discussed in Chapter 2.

In order to get good organic–inorganic heterojunction interface, the advanced surface deposition techniques are required. There are many growth methods for semiconductor heterojunctions. The methods commonly used in the early days are alloy method, sputtering method, vacuum evaporation method, solution growth method, and chemical vapor deposition method; but it is difficult to obtain high-quality heterojunctions. In recent years, three methods have been developed, which are relatively fast and easy to obtain good-quality heterojunctions. Molecular beam epitaxy (MBE) refers to epitaxial growth using atomic or molecular beams as a transport source under ultra-high vacuum conditions. During the growth, there is almost no collision between the molecules. The MBE method has the following characteristics: (1) The source and the substrate can be heated and controlled to make the growth at a low temperature; (2) The growth rate is low and easy to control, which is beneficial to the growth of multilayer heterojunctions; (3) During the growth process, the middle surface is in a vacuum, and the equipment can be used for in situ observation and analysis. However, this method requires high equipment and high production costs, and requires a large amount of experimental funds. Chemical vapor deposition is a method for crystal growth from the gas phase using chemical reactions. Among them, metal organic chemical vapor deposition (MOCVD) is developed by the use of metal organic compounds and trimethylaluminum as the source of the three group elements Ga and Al (TMGa and TMAl), and potassium cyanide as the source of the five group elements. An epitaxial layer is formed on the substrate by pyrolysis at 600–750°C. The reason why MOCVD is valued is that it has the following characteristics: (1) Each component participates in the growth and doping in the form of a gas, so the gas flow can be controlled to control the growth speed and the final control of the composition and thickness of the epitaxial layer; (2) The epitaxial layer is grown by thermal decomposition. It is a single-temperature region growth, easy to control with simple production equipment, and conducive to large-scale production; (3) The metal organic source is approximately proportional to the growth rate of the epitaxial layer, so the growth rate of the epitaxial layer can be adjusted by controlling its flow rate. The liquid phase epitaxy method is to place a single crystal substrate in a saturated or supersaturated solution to grow a single crystal layer on the single crystal substrate consistent with the substrate orientation. It is structurally the same as the original single crystal, but with a different material composition, forming a heterojunction. Its main advantages are as follows: (1) Liquid phase growth has a higher growth rate; (2) Growth in a saturated or supersaturated solution requires simple equipment; (3) High crystal purity, low epitaxial layer dislocation density, and good crystal integrity; (4) Wide choice of dopants for liquid phase growth; (5) Easy operation and straightforward industrial production.

Most importantly, atomic layer deposition (ALD) is a universal deposition method in inorganic semiconductor industry, which enables the control of thin film down to the atomic level through a gas-phase chemical reaction, since the ALD method can deposit various materials on almost any substrate. On the other hand, as its organic counterpart, molecular layer deposition (MLD), has opened up promising avenues for the fabrication of pure polymeric thin films or inorganic–organic hybrid thin films, which has many advantages such as lower deposition temperatures, tunable thermal stability, and improved mechanical properties [13–15]. However, the application of MLD is limited by the instability of the most organic precursors and the resulting polymeric layers. For this reason, the deposition temperatures are usually low and the corresponding temperature window may be either very narrow or nonexistent. The detail of MLD of organic–inorganic hybrid materials method will be discussed in Chapter 3.

As the molecular scale organic–inorganic heterojunction is obtained, the atomic-scale techniques, including scanning tunneling microscopy and scanning tunneling spectroscopy (STM and STS), are usually employed to characterize the atomic morphology and electronic properties of the heterojunction, respectively [16]. The STM results can tell us the structure information of the thin film such as structure geometry, crystal orientation, and even atomic defects in the film. The STS results can provide us the electrical information under different bias, for example, current-voltage and differential tunneling conductance spectra and STS mapping, etc. This information can assist to analyze the optoelectronic performance based on the organic–inorganic heterojunction. The detail of scanning tunneling microscope and spectroscope on organic–inorganic material heterojunction will be discussed in Chapter 4.

As high integration is required in the development of electronics and optoelectronics, the research of low-dimensional material synthesis and micro–nano processing technology is the mainstream trend and research focus. The low-dimensional materials include not only 0D nanodots and quantum dots, 1D nanowires and 2D materials but also low-dimensional organic semiconductors prepared by different processes such as pentacene, copper phthalate (CuPc), Per-3,4,9,10-tetrahydroacid dioxin (PTCDA) and dioctylbenzopyrenebenzobenzophenone (Cg-BTBT), fullerene (C_{60}), rubrene (Rubrene), etc. These low-dimensional materials come from a wide range of sources and have excellent electrical, optical, mechanical, and thermal properties. Moreover, in terms of forming heterojunction with organic semiconductors, the atomically flat surface of low-dimensional materials not only provides an ideal interface for high-efficiency charge separation, transfer, and transport but also act as a template for the epitaxial growth of organic semiconductors [1]. As a consequence, low-dimensional hybrid heterojunction generally exhibits enhanced electrical and optical performances compared to heterostructures consisting bulk phase of organic or inorganic materials. The 2D/organic heterojunctions has been reviewed recently [17]. Thus, the 0D nanodots and 1D nanowires heterojunctions will be discussed in Chapter 5 and Chapter 6, respxectively.

Furthermore, an individual molecule has been theoretically predicted as an active part in electronic or optoelectronic devices in 1974 [18]. The first conductance junction based on single molecule was fabricated experimentally in 1997 [19]. Molecular junctions are now extensively studied in both nanoscale devices and fundamental physical properties of molecule materials [20]. Moreover, advanced fabrication process combined with optical technologies has enabled optical experiments on current-carrying molecular junctions [21,22]. This progress connects molecular electronics with optical spectroscopy together, which has opened up an avenue of molecular optoelectronics [23]. Especially, the scanning tunneling microscope-induced light-emission technology can use the tunneling current of the STM as an atomic-scale source for induction of light emission from a single molecule. Thus, it has enabled the investigation of single-molecule properties at subnanometer spatial resolution. The detail of Electroluminescence of Organic Molecular Junction in Scanning Tunneling Microscope will be introduced in Chapter 7.

Based on the basic introduction above, we present recent advances of organic–inorganic semiconductor heterojunction in optoelectronic applications, for example, diodes, solar cells, light-emitting diodes, nanogenerators, resistive memories, photocatalyst, transistors, and wearable sensors in the following chapters of this book (Figure 1.1).

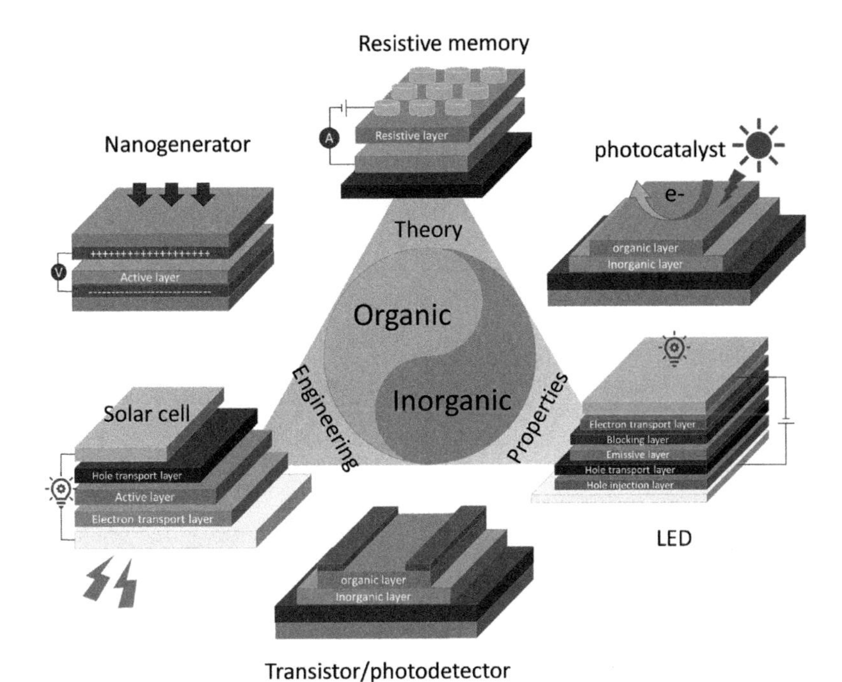

FIGURE 1.1 Organic–inorganic heterojunction and related electronic and optoelectronic applications.

REFERENCES

1. Y. L. Huang, Y. J. Zheng, Z. Song, D. Chi, A. T. S. Wee, S. Y. Quek, *Chemical Society Reviews* 2018, 47, 3241–3264.
2. Y. Liu, Y. Huang, X. Duan, *Nature* 2019, 567, 323–333.
3. A. Gubanov, *Zhurnal Eksperimental'noi i Teoreticheskoi Fiziki Zh. Eksper. Teor. Fiz.* 1951, 21, 721–730.
4. J. T. Calow, P. J. Deasley, S. J. T. Owen, P. W. Webb, *Journal of Materials Science* 1967, 2, 88–96.
5. R. L. Anderson, Experiments on Ge-GaAs Heterojunctions. In *Electronic Structure of Semiconductor Heterojunctions*, Margaritondo, G., Ed. Springer Netherlands: Dordrecht, 1988; pp 35–48.
6. H. Kroemer, *Proceedings of the IEEE* 1982, 70, 13–25.
7. Z. I. Alferov, *Reviews of Modern Physics* 2001, 73, 767-782.
8. Z. I. Alferov, V. Andreev, D. Garbuzov, Y. V. Zhilyaev, E. Morozov, E. Portnoi, V. Trofim, *Soviet Physics Semiconductors* 1971, 4, 1573–1575.
9. I. Hayashi, M. Panish, P. Foy, *IEEE Journal of Quantum Electronics* 1969, 5, 211–212.
10. H. Kressel, H. Nelson, *RCA Review* 1969, 30, 106–113.
11. H. Ishii, K. Sugiyama, E. Ito, K. Seki, *Advanced Materials* 1999, 11, 605–625.
12. S. M. Sze, K. K. Ng, *Physics of Semiconductor Devices*. John Wiley & Sons: Hoboken, NJ, 2006.
13. X. Li, A. Lushington, J. Liu, R. Li, X. Sun, *Chemical Communications* 2014, 50, 9757–9760.

14. X. Li, A. Lushington, Q. Sun, W. Xiao, J. Liu, B. Wang, Y. Ye, K. Nie, Y. Hu, Q. Xiao, R. Li, J. Guo, T.-K. Sham, X. Sun, *Nano Letters* 2016, 16, 3545–3549.
15. Y. Zhao, L. V. Goncharova, Q. Sun, X. Li, A. Lushington, B. Wang, R. Li, F. Dai, M. Cai, X. Sun, *Small Methods* 2018, 2, 1700417.
16. X. Liu, Z. Wei, I. Balla, A. J. Mannix, N. P. Guisinger, E. Luijten, M. C. Hersam, *Science Advances* 2017, 3, e1602356.
17. J. Sun, Y. Choi, Y. J. Choi, S. Kim, J.-H. Park, S. Lee, J. H. Cho, *Advanced Materials* 2019, 31, 1803831.
18. A. Aviram, M. A. Ratner, *Chemical Physics Letters* 1974, 29, 277–283.
19. M. A. Reed, C. Zhou, C. J. Muller, T. P. Burgin, J. M. Tour, *Science* 1997, 278, 252.
20. S. V. Aradhya, L. Venkataraman, *Nature Nanotechnology* 2013, 8, 399–410.
21. T. Shamai, Y. Selzer, *Chemical Society Reviews* 2011, 40, 2293–2305.
22. R. Zhang, Y. Zhang, Z. C. Dong, S. Jiang, C. Zhang, L. G. Chen, L. Zhang, Y. Liao, J. Aizpurua, Y. Luo, J. L. Yang, J. G. Hou, *Nature* 2013, 498, 82–86.
23. M. Galperin, *Chemical Society Reviews* 2017, 46, 4000–4019.

2 Energy-Level Alignment at Organic–Inorganic Heterojunctions

Sylke Blumstengel and Norbert Koch

CONTENTS

2.1 INTRODUCTION

The energy-level alignment between inorganic (ISC) and organic (OSC) semiconductors determines the direction and the efficiency of charge transfer and thus the functionality of the heterointerface in an optoelectronic device. For light-emitting applications, for example, a type-I interface (Figure 2.1) would be most suitable. In such a configuration, excitation energy could be transferred not only via excitonic coupling (dipole–dipole interaction) but also by simultaneous injection of electrons and holes into the lower bandgap material (Itskos et al. 2009; Bianchi et al. 2014; Schlesinger et al. 2015). Light-to-electrical energy conversion relies, on the other hand, on efficient exciton dissociation at heterojunctions; therefore, type-II interfaces with energy offsets larger than the exciton binding energy are required (Figure 2.1) (Oosterhout et al. 2011; Baeten et al. 2011; Eyer et al. 2017; Itskos et al. 2013). Another example is the use of the ISC as carrier injecting or extracting contact in organic electronic devices. Very widespread in applications is ZnO; however, its moderate work function (φ) in the range of about 3.6–4.5 eV often leads to a situation where the Fermi level (E_F), which is close to the conduction band (CB) in *n*-type ZnO, is located within the energy gap between the highest occupied molecular orbital

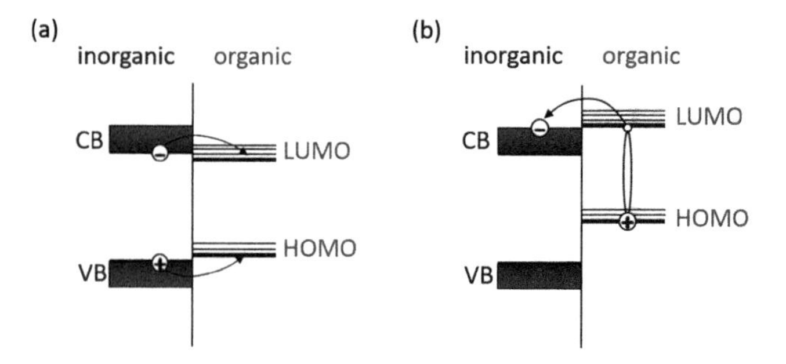

FIGURE 2.1 Schematic depiction of possible energy-level alignments between frontier molecular orbitals and semiconductor band edges at organic–inorganic heterojunctions: (a) type-I and (b) type-II. At type-I interfaces, electrons and holes are captured in the lower band-gap material while type-II interfaces facilitate electron–hole separation. VB: valance band, CB: conduction band.

(HOMO) level and the lowest unoccupied molecular orbital (LUMO) level of the OSC. This results in significant barriers for hole or electron injection (Bhosle et al. 2007; Kim et al. 2013; Bernède et al. 2008). These examples already illustrate the fundamental need of methods for tailoring the relative positions of the occupied and unoccupied energy levels at junctions of ISCs and OSCs to achieve the desired function and to design hybrid optoelectronic devices with indeed superior performance.

Due to the rich physics and chemistry involved in interface formation between OSC and ISC, for example, redistribution of charges, interface dipole formation, or chemical bonding, it is virtually impossible to predict the energy offset for a given material pair on the basis of tabulated values of ionization energy (IE) and electron affinity (EA). On the other hand, once the mechanisms responsible for level alignment are understood, this knowledge can be employed to engineer the level offsets at organic–inorganic semiconductor interfaces.

This chapter is organized as follows: To provide insight into the processes that govern the energy-level alignment at organic–inorganic semiconductor heterojunctions, a model junction (*p*-sexiphenyl and ZnO) is presented in Section 2.2 where, in particular, the influence of the termination of the inorganic semiconductor surface and the molecular film structure is discussed. With this basic understanding of the electronic structure at pristine ISC–OSC interfaces, we present methods to tune the interface energetics in order to control charge injection/transfer and achieve specific functions. Adjustment of the interface energetics is possible by the introduction of a suitable molecular interlayer between ISC and OSC. This concept was originally put forward in order to tune the Schottky barrier height of ISC- and OSC-metal junctions and later transferred to OSC-ITO junctions (Campbell et al. 1996; Vilan et al. 2000, 2010; de Boer et al. 2005; Koch et al. 2005a; Koh et al. 2006; Bröker et al. 2008; Alloway et al. 2009; Rangger et al. 2009; Niederhausen et al. 2011; Hotchkiss et al. 2012; Asyuda et al. 2020). We discuss two fundamental approaches: (i) Chemisorption of molecules with an electric dipole moment perpendicular to the interface on the inorganic semiconductor surface (Section 2.3) (Hotchkiss et al. 2011; Lange et al. 2014; Kedem et al. 2014; Timpel et al. 2014, 2015). The approach is exemplified for various ZnO surfaces functionalized with self-assembled monolayers (SAMs) with varying dipole moment. Changing the composition of the dipole layer allows a fine-tuning of the work function in a moderate energy range (≤1.5 eV); (ii) Introduction of a monolayer of strong molecular acceptors or donors (Schlesinger et al. 2013; Schlesinger et al. 2015, 2019a; Schultz et al. 2016). With this approach, a wider tuning of the work function of wide bandgap semiconductors (ZnO, GaN) is obtained spanning a range from 2.3 eV to 6.5 eV (Section 2.4). Substantial charge redistribution at the heterojunction is responsible for the strong modulation of φ. This also affects strongly the optical properties of the semiconductor surface, which in turn opens up an alternative way to study the electronic structure at organic–inorganic interfaces (Meisel et al. 2018). In Section 2.5, it is shown that the analysis of the evolution of the optical spectra of the ISC surface upon deposition of acceptor molecules provides an estimate of the magnitude of the electrostatic potential change. The effect of *p*-doping of an OSC on the energy-level alignment at the interface to the natively *n*-type ISC (ZnO) is discussed in Section 2.6 (Futscher et al. 2019), that is, a prototypical pn-junction. Finally, Section 2.7 demonstrates that the energy-level tuning via a molecular interlayer provides indeed the desired control over the charge transfer processes at the junction and its functionality (Schlesinger et al. 2015). In this example, a strong increase of the luminescence yield is achieved rendering such well-tailored heterojunctions suitable for light-emitting applications.

2.2 INTERFACE FORMATION BETWEEN ORGANIC AND INORGANIC SEMICONDUCTORS: 6P ON ZNO

The simplest model to predict how the energy levels of two semiconductors line up when brought into contact is based on an assumed vacuum level (E_0, where the electrostatic potential is set to zero) alignment across the interface. Accordingly, the electrostatic potential across the interface is constant and this implies that no charge density rearrangement occurs upon contact, that is, the limit of weak van der Waals interactions or physisorption. This model further neglects eventual doping of the semiconductors, as well as surface and interface states within the otherwise empty bandgaps. In that simple case, knowing the IE and EA of the two materials is sufficient to predict the offset between unoccupied (Δ_{unocc}) and occupied (Δ_{occ}) energy levels, as shown in Figure 2.2. While this situation is, in fact, often encountered at purely OSC heterojunctions, it is scarcely found when an ISC is involved. The main reason is that for ISCs there is a finite density of electrons spilling out into vacuum at the free surface, forming a net surface dipole that increases the sample work function, IE_{ISC}, and EA_{ISC}, by the same amount. This is in full analogy to the surface dipole of metals (Smoluchowski 1941), which is, however, larger in magnitude compared to ISCs. This electron density tailing out of the free surface becomes "pushed-back" by the adsorption of molecules, which effectively reduces the surface dipole and thus results in a reduction of electrostatic potential at the interface by Δ_{E0}. This mechanism has been first understood for metal–OSC interfaces (Koch et al. 2003), but it was also found for physisorptive ISC–OSC junctions (Greiner et al. 2012; Winkler et al. 2013; Schlesinger et al. 2019b). In literature, this phenomenon is often termed "interface dipole". However, the use of this term should be avoided when

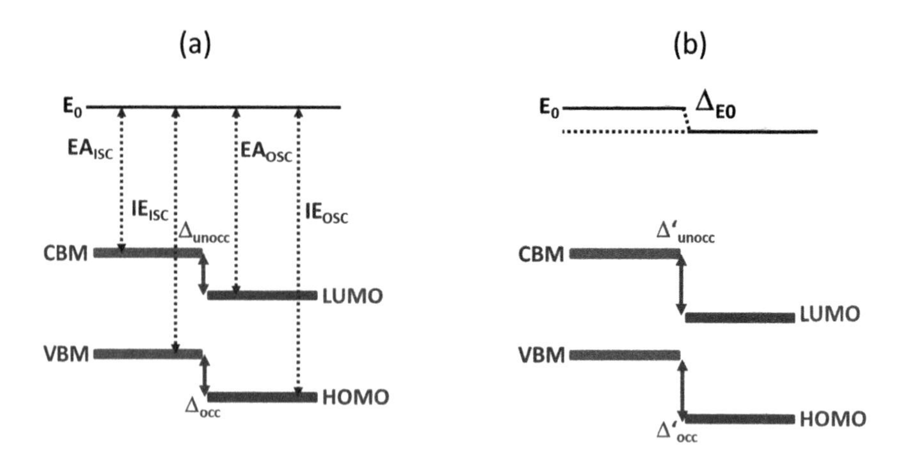

FIGURE 2.2 Schematic energy-level diagrams of an ISC--OSC heterojunction with (a) vacuum level alignment and (b) with an abrupt shift of the vacuum level (E_0) across the inter face by Δ_{E0}. The latter changes the energy offset between the unoccupied levels (Δ_{unocc}, Δ'_{unocc}) and occupied levels (Δ_{occ}, Δ'_{occ}) at the interface, compared to vacuum level alignment. VBM and CBM are the valence band maximum and conduction band minimum of the ISC, respectively.

describing the interfacial "push-back", because no dipole is formed upon contact of the materials, but instead an already existing one (the surface dipole) is modified, as experimentally observed as a change of sample φ. The consequence of this on the level alignment is a rigid shift of the HOMO and LUMO levels of the OSC compared to the ISC by Δ_{E0}, and thus different level offsets Δ'_{unocc} and Δ'_{occ}, as shown in Figure 2.2.

However, many actual ISC–OSC junctions can feature much more complexity. For instance, φ of the ISC surface depends on crystallographic orientation of the surface [just like φ of metals (Smoluchowski 1941)] and surface termination. Furthermore, IE_{OSC} and EA_{OSC} are not intrinsic material properties, but they depend on the relative orientation of the molecular assembly with respect to an interface (Duhm et al. 2008), with variations reaching up to 1 eV or more. The impact of both material-specific property variations on the interfacial energy levels is exemplified in the following for the physisorptive interfaces between *para*-sexiphenyl (6P) and three different ZnO surfaces (Schlesinger et al. 2019b).

The bare surface work function values are 3.6 eV for ZnO(0001), 3.9 eV for ZnO(101;$\bar{0}$), and 4.3 eV for ZnO(0001;$\bar{0}$), and all ZnO crystals are n-doped. The rod-like molecule 6P (chemical structure shown in the inset of Figure 2.3) exhibits a growth mode on all three ZnO surfaces that is characterized by a flat-lying (L) first layer and essentially upright standing (S) second and higher layers. This change in relative molecular orientation with respect to the substrate plane as function of layer is due to the fine balance of substrate-monolayer (dominant for the monolayer) and intermolecular interactions (dominant in multilayers). As noted above, the change in molecular orientation goes hand in hand with a change of the IE and EA of 6P by

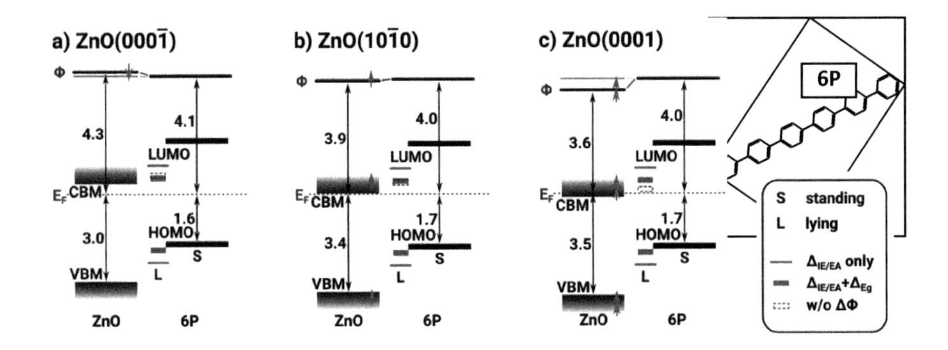

FIGURE 2.3 Schematic energy-level diagrams for 6P on (a) ZnO(0001;$\bar{0}$), (b) ZnO(101;$\bar{0}$), and (c) ZnO(0001). ZnO levels are for bare surfaces prior to adsorption. Adsorption-induced changes of ZnO levels (due to surface band bending changes) are indicated by arrows. The first 6P layer on ZnO is lying ("L") face-on, which leads to energy-level pinning on ZnO(101;$\bar{0}$) and ZnO(0001) raising φ to 4.0 eV. Subsequent layers of 6P are standing ("S") and vacuum aligned with the lying layer underneath. For the "L" layer the diagrams differentiate between the energy-level positions due to the different orientations alone ($\Delta_{IE/EA}$ – thin lines), and including the narrowing of the energy gap (boxes $\Delta_{IE/EA}$+Δ_{Eg} – thick lines); dashed boxes (w/o Δφ) indicate the LUMO states position before interfacial charge transfer due to E_F-pinning. Adapted from (Schlesinger et al. 2019b).

$\Delta_{IE/EA} \approx 0.8$ eV, which by itself would already have a notable effect on the interface energetics. In addition, the monolayer experiences two more effects: (i) the proximity to the ZnO (with a dielectric constant of ca. 8, compared to that of 6P of ca. 3) lowers the energy gap by ca. 0.3 eV due to increased dielectric screening compared to 6P molecules in the bulk; (ii) the inter-ring twist angle of 6P in the bulk is smaller for the adsorbed monolayer and that additionally lowers the energy gap by up to 0.3 eV (Koch et al. 2005b). In combination, the energy gap of the lying 6P monolayer on ZnO is reduced by up to $\Delta_{Eg} \approx 0.6$ eV compared to the bulk with an energy gap of ca. 3.5 eV (Hwang, Wan, and Kahn 2009).

Let us first consider the highest φ ZnO surface, that is, ZnO(0001;$\bar{0}$) with 4.3 eV. Deposition of 6P reduces φ by 0.2 eV and the Fermi level (E_F) of the n-doped ZnO is located well within the energy gap of 6P for both lying mono- and standing multilayer, as shown in Figure 2.3a. This situation thus corresponds to the scheme shown in Figure 2.2b, that is, an interface without charge transfer and only the "push-back" effect that lowers the electrostatic potential (in experiments observed as work function change). Turning toward the ZnO(101;$\bar{0}$) surface with an initial φ of 3.9 eV, we observe that 6P deposition slightly increases φ by 0.1 eV, and an even larger work function increase by 0.4 eV is found for ZnO(0001) with an initial φ of 3.6 eV. Since at least a small φ lowering is expected due to the push-back effect, the opposite change of the work function thus indicates that another mechanism must be at play for these two interfaces. The initial φ values of ZnO(101;$\bar{0}$) and ZnO(0001) are sufficiently low so that E_F would come to lie within the LUMO level manifold of the lying 6P layer due to the EA increase brought about by the gap narrowing effects of screening and reduced inter-ring twist angle (see preceding paragraph). To reach electronic equilibrium, electrons are thus transferred from the ZnO to 6P molecules. This results in an increase of φ, as seen in experiment, and concomitantly shifts the energy levels of all 6P molecules upward, as schematically shown in Figure 2.3. This interfacial charge transfer at a physisorptive interface, conventionally termed Fermi-level pinning (Yang et al. 2017; Mao et al. 2011), is driven by electronic equilibrium only so that no unoccupied levels come to lie below E_F.

The above examples demonstrate how the different simultaneously acting mechanisms of push-back, Fermi-level pinning, dielectric screening, molecular orientation change, and molecular conformation dependent energy gap narrowing are responsible for essentially identical level alignment between ZnO and multilayer 6P, despite varying starting conditions. It is rather difficult to access the information on the presence and properties of such molecularly thin interlayers, like the lying 6P layer here. If these details of the ISC–OSC junction go unnoticed, correlations between functional behavior of devices and assumed interfacial energy levels may lead to false conclusions for further development.

2.3 WORK FUNCTION TUNING OF ZNO VIA DIPOLE BEARING SELF-ASSEMBLED MONOLAYERS

The most straightforward way to tune the energy-level alignment between ISC and OSC is the introduction of a monolayer of dipole bearing molecules between the two materials. If these molecules are aligned, the resulting dipole layer just resembles a parallel plate capacitor and the potential change $\Delta\varphi$ perpendicular to the surface can be calculated by the Helmholtz equation

$$\Delta\varphi = \frac{e \cdot \sigma \cdot d}{\varepsilon_r \cdot \varepsilon_0} = \frac{e \cdot n \cdot \mu_\perp}{\varepsilon_r \cdot \varepsilon_0} \tag{2.1}$$

where e is the elementary charge, σ the area charge density, d the thickness of the dipole layer, ε_r and ε_0 are the relative and vacuum dielectric permittivity, respectively. n is the area density of the dipoles and $\mu_\perp$ the dipole moment component perpendicular to the interface. It comprises contributions of the molecular dipole moment as well as the bonding dipole.

Well-ordered dipole layers are achieved by the wet-chemical deposition of SAMs consisting of rod-like molecules bearing on one side an anchoring group capable of binding to the oxide surface and on the other side an either electron withdrawing or electron donating tail group. Both groups are connected by a molecular spacer, which, together with the anchoring group, determines the packing density and the structure of the monolayer.

The molecules discussed in the following possess as a molecular spacer a polarizable benzene ring and as anchor a phosphonic acid group. The phosphonate head group is known to have strong affinity to ZnO so that highly dense molecular films can be prepared (Ostapenko et al. 2016). A XPS analysis of the O 1s core level shows that the molecules adsorb on ZnO via a mixture of bidentate and tridentate binding configurations (Figure 2.4a) independent of the tail group (Timpel et al. 2015). The tuning of the work function of ZnO is exemplary demonstrated with the three phenyl

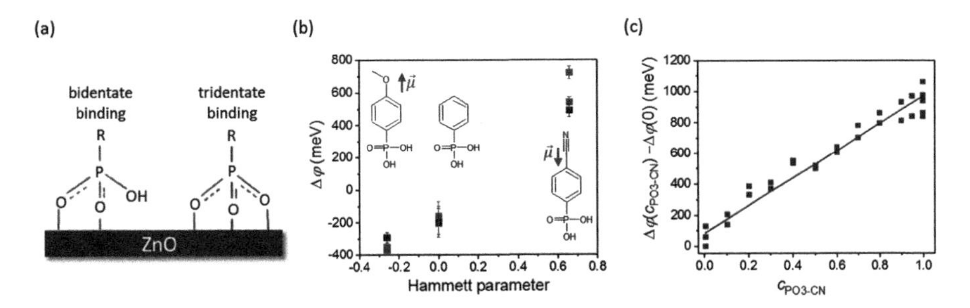

FIGURE 2.4 (a) Schematic illustration of the phosphonic acid binding modes on ZnO. (b) Work function modification Δφ as a function of the Hammett parameter of the molecules depicted in the inset. The direction of the molecular dipole moment is indicated as well. The data points refer to sol–gel ZnO (red), ALD ZnO (green), and MBE grown ZnO(000$\bar{1}$) (blue). (c) Work function tuning using SAMs of a mixture of PO_3-OCH_3 and PO_3-CN. The difference in the work function between the mixed SAM and the pure PO_3-OCH_3 SAM is plotted as a function of the partial concentration c_{PO3-CN} of PO_3-CN in the solution used to prepare the SAM. The total concentration of the two molecules is kept constant. (b) and (c) Adapted from (Kedem et al. 2014).

phosphonates: 4-methoxyphenyl phosphonic acid (PO3-OCH3), phenyl phosphonic acid (Phen-PO3), and 4-cyanophenyl phosphonic acid (PO3-CN), differing in the tail groups and thus in the molecular dipole moment (Kedem et al. 2014). The chemical structure of the molecules is shown in the inset of Figure 2.4b. To investigate if the ZnO surface termination, the morphology, and the intrinsic doping level affect the work function tuning, ZnO films are prepared by atomic layer deposition (ALD), sol–gel deposition, and molecular beam epitaxy (MBE). While ALD and sol–gel deposition results in polycrystalline films, MBE yields single crystalline O-terminated ZnO(000$\bar{1}$). Changes in the work function have been recorded by combined Kelvin Probe and UPS measurements. The results are summarized in Figure 2.4b where the work function changes $\Delta\varphi$ as a function of the Hammett substituent parameter of the molecules are depicted. This parameter describes the tendency of the functional group (OCH_3, H and CN in this case) to donate or withdraw electrons from the benzene ring and is thus a measure of the molecular dipole moment. As seen in Figure 2.4b, a linear correlation of the work function with the Hammett parameter is found for all ZnO morphologies, regardless of the totally achieved range, indicating that the effect of the molecules on the various types of ZnO films is essentially the same. The electron-donating methoxy tail group results in a positive dipole, pointing out from the surface, which decreases the work function, while the electron-withdrawing cyano tail group introduces a negative dipole, which increases the work function. The shift with the unsubstituted PO_3-Phen termination allows separation of the effect of the substituents alone from the effect of the dipole due to the bonding of the molecule to the surface, which changes the ZnO termination from Zn-OH to Zn-O-P, as well as any additional dipole caused by the unsubstituted molecule. A maximum work function span $\Delta\varphi$ of ~0.8…1.1 eV between the opposite dipoles is obtained. The small differences in the totally achieved $\Delta\varphi$ on the different ZnO surface terminations and morphologies can be related to a slight modification of dipole density and/or angle. To address the question, if the dipole layer is solely responsible for the work function modification or if a changed surface band bending within ZnO contributes, XPS measurements were performed (Kedem et al. 2014; Timpel et al. 2015). A shift in the Zn 2p or 3s peak positions due to assembly of the molecular layer would indicate a change in band bending. However, the Zn 2p and 3s peak positions stayed constant at the values of the bare ZnO (~1022 eV and ~140 eV, respectively). The major effect is thus extrinsic to ZnO and due to the molecular layer's dipole.

By mixing of PO_3-OCH_3 and PO_3-CN which possess opposite dipoles, the work function can be varied in a controllable manner according to the concentration ratio of the two molecules in the solution used to form the SAM as shown in Figure 2.4c. The work function is found to correlate linearly with the partial concentration of PO_3-CN in the solution. The advantage of such an approach is that only two types of molecules are required to span the entire modification range. The approach thus allows a continuous modification of the energy-level alignment at the heterointerface between ISC and OSC.

Introduction of dipolar SAMs presents a viable route to tune the work function of technologically relevant ZnO surface. The expansion of the functionality due to SAM modification has been demonstrated, for example, in planar hybrid photovoltaic diodes with the organic donor, N,N′-di(1-naphthyl)-N,N′-diphenyl-(1,1′-biphenyl)-4,4′-diamine (α-NPD) and ZnO as acceptor (Piersimoni et al. 2015). The open circuit voltage V_{OC} at organic–inorganic heterojunctions is correlated to the hybrid energy gap between CBM of the ISC and the HOMO of the OSC. By introduction of phenyl phosphonates with different tail groups as well as mixtures thereof, the hybrid energy gap at the α-NPD/ZnO heterojunction was varied by 300 meV. It was shown, that the V_{OC} of the devices changes by the same amount. Applying a forward bias to the devices, near-infrared electroluminescence is emitted stemming from the recombination of electrons in the ZnO CB with holes in HOMO of α-NPDs at the heterojunction. Similar to the V_{OC}, the peak position of the electroluminescence is also tunable by the introduction of the SAMs. Another example for the expansion of the functionality is the application of SAM-modified ZnO as injecting electrodes in optoelectronic devices (Lange et al. 2014). In that work, it is shown that ZnO can serve both as electron- and hole-injecting contact, and furthermore that the injection properties can be continuously altered from being strongly injection limited to Ohmic. Consequently, unipolar currents in P3HT and phenyl-C71-butyric acid methyl ester (PCBM)-based diodes could be tuned by several orders of magnitude just by controlling the ZnO work function with an appropriate SAM.

2.4 WORK FUNCTION TUNING WITH ELECTRON DONOR AND ACCEPTOR MOLECULES

Another approach to tune the work function of ISC surfaces, in order to modify the level alignment with a subsequently deposited OSC as discussed in Section 2.3, builds on a concept developed earlier to modify φ of metal electrodes for reducing charge injection barriers into OSCs. Molecules with strong electron donor and acceptor character chemisorb on metal surfaces involving pronounced charge transfer.

For instance, the strong molecular acceptor tetrafluoro-tetracyanoquinodimethane (F_4TCNQ) binds to the Ag(111) surface via orbital hybridization and bi-directional electron transfer involving numerous orbitals (Romaner et al. 2007). Overall, ca. one electron is transferred from the metal to F_4TCNQ, the positive counter-charge is located at the Ag surface. Consequently, each molecule-metal complex has a net dipole moment perpendicular to the interface, which increases φ according to the Helmholtz equation above. A decrease of the metal surface φ can be readily achieved by using a strong molecular donor (Bröker et al. 2008).

Such acceptor- or donor-induced interfacial charge transfer also occurs with many ISCs. For instance, Figure 2.5a reveals that huge φ increases can be achieved by depositing the acceptors 1,3,4,5,7,8-hexafluoro-tetracyanonaphthoquinodimethane (F_6TCNNQ) and 1,4,5,8,9,11-hexaazatriphenylenehexacarbonitrile (HATCN) onto ZnO and GaN. However, the mechanism leading to $\Delta\varphi$ is different from the metal case because the semiconducting nature of the material must be accounted for. First of all, no indications for pronounced chemisorption and orbital hybridization of such acceptors on semiconductor surfaces were found for the cases studied to date, including hydrogen-terminated Si surfaces (Wang et al. 2019), and thus integer charge transfer across the interface is considered (Schöttner et al. 2020) – in analogy to Fermi-level pinning discussed in Section 2.2. Next, it is generally observed that the overall $\Delta\varphi$ is in part due to molecule-induced (modified) surface band bending within the ISC, termed $\Delta\varphi_{BB}$ and exemplarily shown in Figure 2.5b for the same samples where $\Delta\varphi$ is shown in Figure 2.5a. At this point, we recall that ISCs are generally doped and possess a certain gap-state density of states (GDOS) at their surface, and

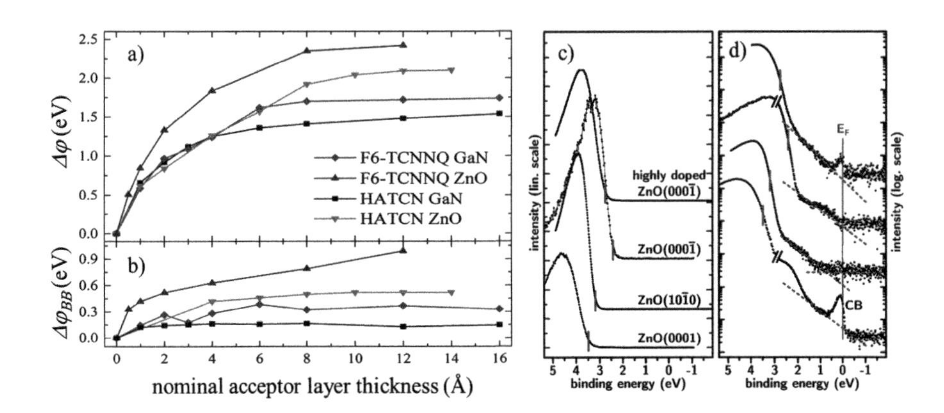

FIGURE 2.5 Work function change $\Delta\varphi$ (a) and change in surface band bending $\Delta\varphi_{BB}$ (b) for GaN(0001) and ZnO(0001) upon stepwise deposition of the acceptors HATCN and F_6TCNNQ. Valence photoemission spectra for different ZnO surfaces, plotted on (c) linear and (d) logarithmic intensity scale. The onset positions of the VBM are indicated with red markers. The breaks in (d) indicate concatenation of individually measured spectra. On logarithmic intensity scale a continuous (decaying toward E_F, set to zero) photoemission intensity in the forbidden energy gap region of ZnO (GDOS) is visible. For ZnO(101;$\bar{0}$) and the highly Ga-doped ZnO(0001;$\bar{0}$) face, the band tails fall off approximately exponentially (linear slope in logarithmic scale; indicated by dashed lines).

both can be the source of electrons (holes) that are transferred to the molecular acceptors (donors). In the following, the case of acceptors deposited on an n-doped ISC is discussed specifically, since this represents the most case studies to date, often with ZnO as ISC. The presence of a GDOS cannot be readily inferred from a photoemission valence spectrum plotted on a linear intensity scale (Figure 2.5c), as the high density of states (DOS) from the valence band dominates and the region between the VBM (indicated by red markers) and E_F appears empty. The same spectra plotted on logarithmic intensity scale (Figure 2.5d), however, allow directly observing that the region of the supposedly empty bandgap does feature a finite GDOS, whose energy and intensity distribution depends on details of the sample surface and may extend throughout the gap up to the CB. Here, for the highly doped ZnO(0001;$\bar{0}$) and Zn(0001) even part of the CB is filled with electrons.

Thus, in addition to electrons from ionized shallow donors from the bulk of the n-doped ISC, also electrons from this GDOS are transferred to the adsorbed molecular acceptors. Both, however, contribute differently to the overall work function change $\Delta\varphi$, as explained below.

The overall density of electrons transferred to the molecular acceptor layer is

$$\delta q = \frac{2eN_m}{e^{(\varphi_0 + \Delta\varphi - EA_m)/k_B T} + 1} \tag{2.2}$$

with e the elementary charge, N_m the area density of acceptors, φ_0 the bare ISC work function, EA_m the electron affinity of the acceptors, k_B the Boltzmann constant, and T the temperature. Thus, electron transfer stops as soon as no more LUMO levels of the acceptor layer are below E_F. The overall work function change has then two contributions, due to band bending in within the ISC ($\Delta\varphi_{BB}$) and an interface dipole between the charged acceptors and the charges in the ISC ($\Delta\varphi_{ID}$) (Schultz et al. 2016):

$$\Delta\varphi = \Delta\varphi_{BB} + \Delta\varphi_{ID} \tag{2.3}$$

with

$$\Delta\varphi_{BB} = \frac{\left(\delta q \dfrac{N_D}{N_D + N_{GDOS}}\right)^2}{2\varepsilon_0 \varepsilon N_D} \tag{2.4}$$

and

$$\Delta\varphi_{ID} = e\frac{\delta q}{\varepsilon_0} d_{eff} \tag{2.5}$$

where N_D is the donor density; N_{GDOS} is the density of surface gap states; ε_0 is the vacuum permittivity; ε is the dielectric constant of the inorganic semiconductor; and d_{eff} is an effective distance between the acceptor molecules and the surface of the inorganic semiconductor.

The conceptual partitioning of $\Delta\varphi$ into $\Delta\varphi_{BB}$ and $\Delta\varphi_{ID}$ is summarized in Figure 2.6. For practical ISCs one can state that the contribution of band bending to the overall work function becomes smaller for higher doping level (N_D) of the semiconductor,

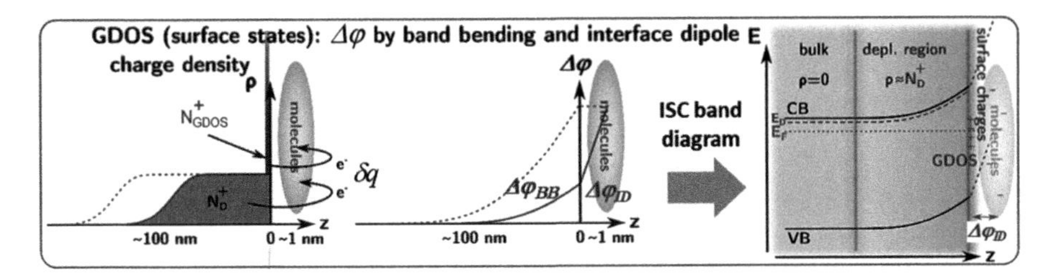

FIGURE 2.6 Schematic illustration of charge density (ρ) distribution, work function change (Δφ), and resulting energy-level diagrams for molecular acceptors adsorbed on an n-type ISC. Shallow donors in the bulk of the ISC with a binding energy ED below the CB.

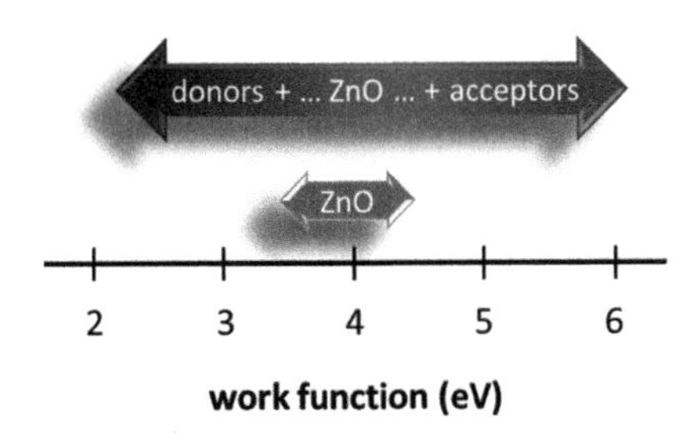

FIGURE 2.7 Representation of the achieved range of work function tuning by moelcular acceptors and donors for ZnO, compared to the work function range of the bare ISC.

and it is further reduced by an increasing gap-state density (N_{GDOS}). In the limit of a metal instead of a semiconductor, only $\Delta\varphi_{ID}$ is responsible for the work function change.

Considerations regarding molecular donors on ISCs to reduce the work function are analogous to the above. Overall, it is quite remarkable to which extent φ of ISCs can be tuned by molecular donors and acceptors, as shown in Figure 2.7 for ZnO. This, in turn, would allow for concomitant energy-level adjustment between an ISC and an OSC with a thin (monolayer range) interlayer of acceptors or donors, and tuning by over 1 eV in either direction has already been demonstrated (Schlesinger et al. 2013, 2015).

2.5 FINGERPRINT OF GROUND-STATE CHARGE TRANSFER IN THE OPTICAL SPECTRA OF ZNO-ACCEPTOR INTERFACES

As shown in the previous section, the huge work function increase induced at wide bandgap ISC surfaces by adsorption of acceptor molecules is due to charge redistribution. The change in the electrostatic potential is due to the formation of an interface dipole as well as a strong band bending in the ISC. In this section, it is shown that the electric field associated to the band bending (Figure 2.8a) massively changes the optical spectra of the ISC surface (Meisel et al. 2018). This intricate interplay between the electronic and the optical properties has to be, on the one hand, taken

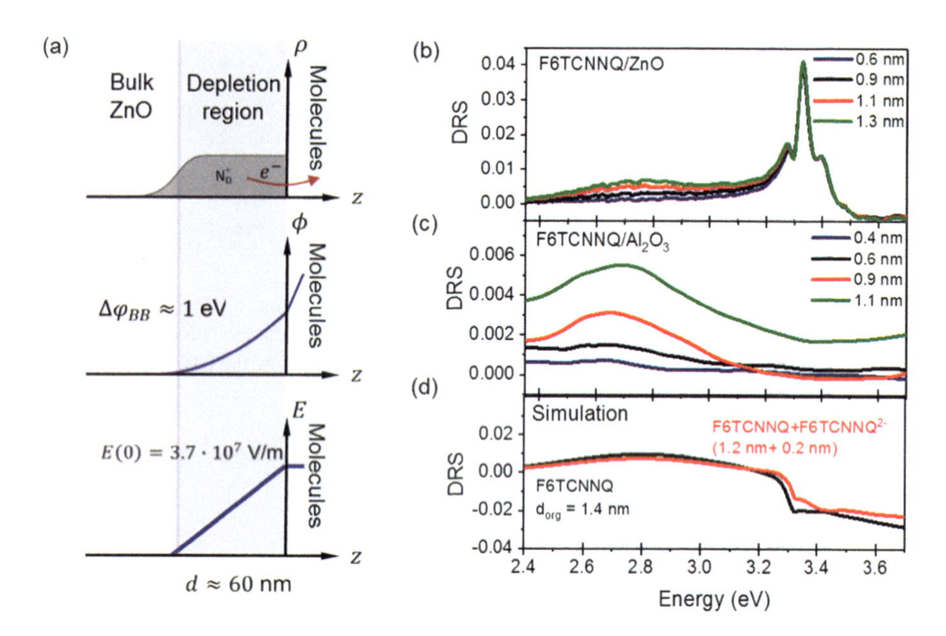

FIGURE 2.8 DR spectra of (a) the $F_6TCNNQ/ZnO(0001)$ and (b) the $F_6TCNNQ/Al_2O_3(0001)$ interface for increasing thickness of the organic layer. (c) Simulated DR spectra of a F_6TCNNQ/ZnO (black) and a $F_6TCNNQ/F_6TCNNQ^{2-}/ZnO$ (red) interface using the dielectric functions of the individual components. The simulations are performed using the transfer matrix method using the dielectric functions of the individual components. (b)–(d) Adapted from Meisel et al. (2018).

into account when designing heterojunctions for specific optoelectronic functions. On the other hand, if well understood, it can be employed to reveal the occurrence of charge redistribution at organic–inorganic heterojunctions and to study in situ and in real time the electronic structure evolution. We will show in the following examples that the analysis of the optical spectra can even yield an estimate of the magnitude of the electric field and consequently of the change in the electrostatic potential due to molecular adsorption. Optical spectroscopy thus provides a simple alternative to photoemission spectroscopy. The approach is presented for the example of the just discussed F_6TCNNQ-ZnO heterojunction.

The changes in the optical spectra are tracked by differential reflectance (DR) spectroscopy during the vacuum deposition of the acceptor molecules on the ZnO surface. The DR signal is defined as

$$\frac{\Delta R}{R} = \frac{R(d) - R(0)}{R(0)} \tag{2.6}$$

where $R(0)$ is the reflection spectrum of the pristine semiconductor surface recorded prior to the deposition of the molecules, while $R(d)$ refers to the reflection spectra of the surface covered by a molecular layer with the thickness d. Since the difference spectrum is measured in situ during the sublimation of the molecules, DR spectroscopy is highly sensitive and allows the detection of the signatures of even a fraction of a molecular monolayer (Proehl et al. 2005).

Figure 2.8b and c show the evolution of the DR spectra while F6TCNNQ layers grow on a ZnO(0001) surface and as reference on an inert Al_2O_3 substrate. The reference DR spectra feature a broad unstructured peak centered around 2.7 eV which corresponds to the absorption of the molecule. Assuming that the molecular layer thickness is much smaller than the wavelength of light $d \ll \lambda$, the DR spectra are approximately (McIntyre 1971)

$$\frac{\Delta R}{R} \approx -\frac{8\pi d}{\lambda} Im\left(\frac{1-\hat{\epsilon}_{\mathrm{org}}}{1-\hat{\epsilon}_{\mathrm{inorg}}}\right) \tag{2.7}$$

with $\hat{\epsilon}_{\mathrm{org}}$ and $\hat{\epsilon}_{\mathrm{inorg}}$ being the complex dielectric functions of the organic layer and of the inorganic substrate, respectively. It is apparent from the equation that in spectral regions where $\hat{\epsilon}_{\mathrm{inorg}}$ does not vary significantly (i.e., on the transparent Al_2O_3 in the whole considered spectral range), the DR spectra resemble the imaginary part $Im\,\hat{\epsilon}_{\mathrm{org}}$ of the organic adsorbate layer and thus its absorption spectrum. The broad feature is also visible in the DR spectra of the F_6TCNNQ/ZnO(0001) interface and can also here be assigned to the F_6TCNNQ absorption since ZnO is transparent in this spectral range. The slight blueshift is explained by a different packing of the molecules and a different dielectric environment. There are three additional sharp features in the spectral range between 3.2 and 3.4 eV which are absent in the reference spectra. They occur already at the lowest coverage and do not grow further in intensity upon deposition of organic material. These features cannot be reproduced by simulations of the DR spectra in the frame of a simple three-layer system consisting of a ZnO half space on one side, a F_6TCNNQ layer of thickness d with an oscillator at 2.7 eV accounting for the molecules absorption, and vacuum on the other side (Figure 2.8c). Even if a layer of charged F_6TCNNQ^{2-} molecules is introduced, which absorb in the spectral range of interest, that is between 3.2 and 3.4 eV, no agreement with the experimental spectra can be obtained (Figure 2.8d). These features must therefore be a sign for a modification of the dielectric function of ZnO as a consequence of the electronic interaction with F6TCNNQ. The photoemission spectroscopy experiments presented in Section 2.4 provide a straightforward explanation. Remember that the band bending at the ZnO surface changes by $\Delta\Phi_{\mathrm{BB}} \approx 1$ eV upon deposition of F_6TCNNQ (Schultz et al. 2016). The Schottky approximation yields a corresponding electric field $F(z = 0) \approx 3.7 \times 10^7$ V/m at the ZnO surface which drops linearly over the space charge region of a thickness $z \approx 55$ nm to zero (Figure 2.8a) assuming a donor density of $N_{\mathrm{D}} \approx 3 \times 10^{23}$ m^{-3} and a static dielectric constant of 8 which are reasonable values for the used ZnO. $F(z = 0)$ is thus comparable to the ionization field of the ZnO Wannier–Mott exciton which is $F_{\mathrm{I}} = 2.6 \times 10^7$ V/m. F_{I} corresponds to the potential drop of the effective Rydberg (exciton binding energy) over the exciton Bohr radius. The strong electric field present in the space charge region modifies the shape of the excitonic absorption edge of ZnO and causes the characteristic DR features between 3.2 and 3.4 eV. The quantitative analysis of the DR spectra provides an estimate of the magnitude of the field as briefly outlined in the following. Details on the calculations can be found elsewhere (Meisel et al. 2018).

Within the effective mass approximation, the ZnO Wannier–Mott excitons near the surface behave formally like hydrogen atoms in an electric field (Figure 2.9a). Calculated $Im\hat{\epsilon}_{ZnO}$ spectra as a function of the electric field strength are reported in Figure 2.9b. Qualitatively, the effect of the electric field on the absorption line shape can be understood as follows: At small field strengths $F \ll F_I$, the electric field leads to a widening of the Coulomb potential well causing a slight redshift of the excitonic transition energy (second order Stark effect). As the field strength increases but $F < F_I$, the rim of the Coulomb well further lowers so that the bound levels start to mix with continuum levels. This causes a blueshift and a broadening of the excitonic transition which is, finally, at $F \approx F_I$ completely smeared out. To simulate the DR spectra, the linearly decreasing electric field is approximated by step functions by subdividing the space charge region into thin layers of thickness z_i in which the electric field F_i is constant (Figure 2.9c). The transfer matrix method is applied to calculate the DR spectra of the multilayer system consisting of N ZnO layers with dielectric functions $\hat{\epsilon}_{ZnO}(F_i)$ and a F6TCNNQ layer described by the oscillator model using one oscillator whose parameters (resonance energy, line width, oscillator strength)

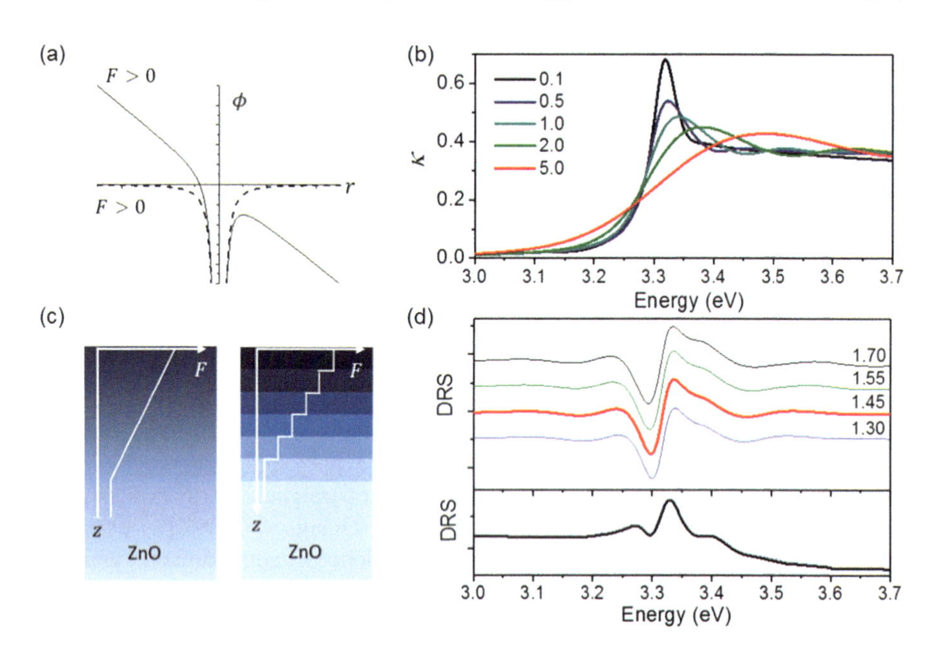

FIGURE 2.9 (a) Schematic depiction of the Coulomb potential of a Wannier–Mott exciton with and without applied electric field. (b) Calculated electric field dependence of the imaginary part of the complex refractive index κ of ZnO. (c) Schematic depiction of the electric field distribution in the near surface region of ZnO according to the Schottky approximation (left) and electric field distribution used in the calculation of the reflection spectra (right). (d) Simulated DR spectra of a F6TCNNQ/ZnO(0001) interface for different values of band bending and thus different ratios $F(z = 0)/F_I$ (upper panel). Experimental DR spectrum for a F6TCNNQ layer thickness d = 1.4 nm (lower panel). Adapted from Meisel et al. (2018).

were obtained by fitting the reference DR spectra. The stack is embedded between an infinitely thick ZnO layer with $\hat{\epsilon}_{ZnO}\left(F \rightarrow 0\right)$ on one side and vacuum on the other side.

The results of the simulations are depicted in Figure 2.9d in comparison to an experimental DR spectrum of F6TCNNQ/ZnO(0001). To be consistent, a slight downward band bending of ca. 0.2 eV (corresponding to $z \approx 25$ nm and $F(z = 0) \approx 1.65 \times 10^7$ V/m) at the pristine ZnO(0001) surface is accounted for in the calculation of $R(0)$. The reflectivity $R(d)$ of the F6TCNNQ/ZnO interface is calculated for different values of band bending and thus different $F(z = 0)$ and space charge layer widths. At a ratio $F(z = 0)/F_I = 1.4$ which corresponds within the Schottky depletion approximation to the change in band bending by $\Delta\Phi_{BB} \approx 1$ eV measured by PES, the comparison of the calculated DR spectra with an experimental spectrum shows a surprisingly good agreement despite the simple model used for the calculation of $\hat{\epsilon}_{ZnO}$ (Meisel et al. 2018). The principal features observed in the experimental spectrum are well visible in the simulated spectra. Their origin becomes clear considering the changes in the absorption line shape at the ZnO band edge (Figure 2.9b): In the presence of an electric field, the absorption below the band edge increases leading to a positive DR signal at 3.28 eV. The blueshift of the excitonic absorption peak produces the dispersive signal with a minimum at ca. 3.3 eV and a maximum at 3.33 eV. With a more comprehensive modeling of the electric field dependence of the optical functions of the semiconductor or a beforehand experimental determination of the same by electroreflectance measurements even better agreement could be achieved.

The example shows that in situ DR spectroscopy is a simple alternative method to reveal very sensitively charge redistribution at the heterointerface between ISC and acceptor molecules. Even minuscule electron transfer layer leads to a profound change in band bending at the semiconductor surface and this leaves a clear fingerprint in the optical spectra even at submonolayer coverage. A quantitative analysis of the spectra yields even an estimate of the magnitude of the fields.

2.6 ORGANIC–INORGANIC SEMICONDUCTOR PN-JUNCTION

Heterojunctions comprising ISCs only are often pn-junctions, with one ISC being n- and the other p-doped. Given the application relevance of pn-junctions, it is thus important to know whether inorganic–organic pn-junctions can be understood on the basis of established knowledge of their purely inorganic counterparts. Doping of organic semiconductors is readily achieved by mixing small molecules of strong electron donating or accepting character into the OSC matrix. Notably, such dopant molecules can be the same ones as used for tuning the work function of ISC surfaces, as discussed in Section 2.4. Therefore, it is important to avoid direct dopant molecule contact to the ISC, as the charge transfer and concomitant changes in the ISC surface band bending could potentially mask the effects occurring at the pn-junction. In the following example, where n-doped ZnO is combined with the p-doped organic hole-transport material N,N′-di(1-naphthyl)-N,N′-diphenyl-(1,1′-biphenyl)-4,4′-diamine (α-NPD), it was assured that direct contact of the dopant molecule 2,2′-(perfluoronaphthalene-2,6-diylidene)dimalononitrile (F_6TCNNQ) to ZnO does not dominate the interface energetics. A diagnostic observable for that is the unchanged density of surface hydroxyls of ZnO. As discussed in detail in Futscher et al. (2019), deposition of F_6TCNNQ alone onto the inorganic surface destabilizes surface hydroxyls due to the strong interfacial charge transfer, and their density is significantly reduced. For the pn-junction discussed here, the surface hydroxyl density remained constant upon deposition of the p-doped OSC onto the n-type ZnO.

As benchmark, the interface between n-type ZnO and the intrinsic OSC without dopant molecules is discussed. The bare ZnO(0001) surface exhibits downward band bending, and the CB actually is slightly below the Fermi level (E_F), as seen in Figure 2.10a. Deposition of intrinsic α-NPD leads to an abrupt reduction of the sample φ by about 0.4 eV. The energy levels within ZnO do not change, and those within the α-NPD layer are flat. These observations signify that this junction can be categorized as physisorptive with the push-back effect, in analogy to Figure 2.2b. The deposition of p-doped α-NPD changes the interface energy levels markedly (Figure 2.10b). The initial downward band bending near the ZnO surface changes to upward, with a total change of 0.4 eV, indicating a loss of electrons. The energy levels within the p-type organic layer are not flat but significant downward energy-level bending (ca. 0.85 eV) toward the interface occurs. As clearly observed from the energy-level diagram, the behavior of this inorganic–organic pn-junction fully reflects the expected trends for such a heterojunction, and can thus be described with established semiconductor models (Sze and Mattis 1970).

Since molecular doping of organic semiconductors is presently still investigated to unravel a complete understanding of the underlying mechanisms, it is noteworthy that the energy-level bending within the p-doped OSC layer can be used to extract

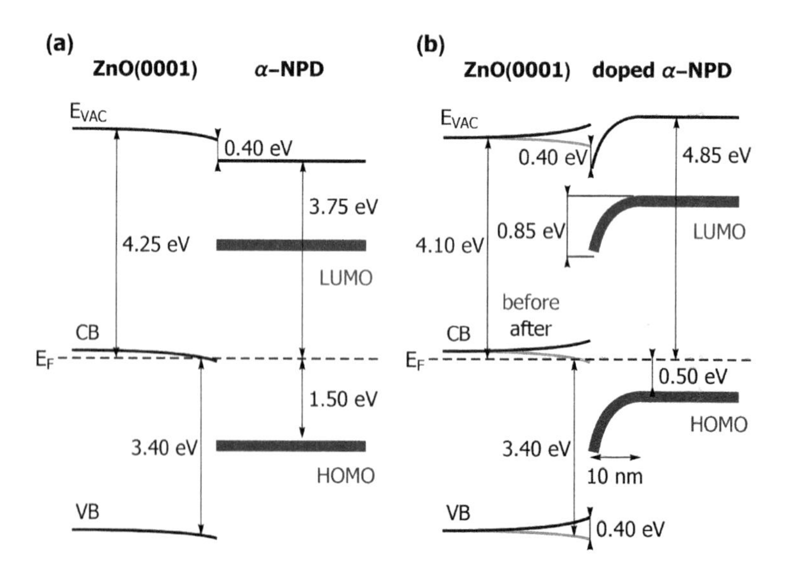

FIGURE 2.10 Energy-level diagrams at the interface between (a) intrinsic α-NPD and n-type ZnO(0001), and (b) α-NPD p-doped with 2 vol% F_6TCNNQ and n-type ZnO(0001). Gray-shaded energy levels indicate those of bare ZnO(0001) before OSC deposition. Reproduced with permission from Futscher et al. (2019).

valuable information about doping, such as an estimate of the doping efficiency of F_6TCNNQ for α-NPD, which is defined as the ratio of mobile charge carriers to the number of dopants present. Assuming that the dopant density in α-NPD is much higher than the dopant density in ZnO, the width of the accumulation layer w can be approximated by (Sze and Mattis 1970)

$$w \approx \sqrt{\frac{2\,\varepsilon\varepsilon_0}{e}\frac{V_b}{N_A}} \tag{2.8}$$

where V_b is the energy-level bending magnitude in α-NPD and N_A is the dopant density (here: 5×10^{19} cm^{-3}). When assuming the dielectric constant to be $\varepsilon = 3$, which is a value often used for organic semiconductors, using the measured energy-level bending of 0.85 eV and the accumulation layer width of ca. 10 nm, a doping efficiency of approximately 5% is derived, corresponding to a density of mobile holes of $(3 \pm 2) \times 10^{18}$ cm^{-3} within the p-doped α-NPD. This shows that only a small fraction of the dopants can introduce mobile holes that contribute to conductivity, which is one of the reasons that research on OSC doping is quite active at present.

2.7 ENERGY-LEVEL TUNED ORGANIC–INORGANIC HETEROJUNCTIONS FOR LIGHT-EMITTING APPLICATIONS

Section 2.2 focused on the increase of the work function of the ISC by assembly of molecular donors. There are, however, also applications where a substantial decrease is required. The EA of GaN and even more so of ZnO is much larger than that of common OSC leading to energy offsets between the CBM and the LUMO of about 1 eV. When such interfaces are used, for example in photovoltaic applications, the energy offset is much larger than that required to dissociate excitons squandering V_{OC}. In other applications, excited state charge transfer from the OSC's LUMO to the ISC's CB is altogether detrimental, for example for light-emitting applications as discussed in Section 2.1 (Blumstengel et al. 2008; Itskos et al. 2013; Bianchi et al. 2014). Realignment of the energy levels into a type-I interface via molecular interlayers is needed here. As shown in Section 2.3, inserting appropriate dipolar interlayers between inorganic and organic layers to shift the electrostatic potential is one way to tune the energy-level alignment. However, lowering the work function φ of ZnO using dipolar SAMs has been somewhat successful; the lowest values of φ achieved have been in the range of 3.5–4.3 eV (Ha et al. 2013; Kedem et al. 2014) which is insufficient to eliminate the large energy-level offsets with most OSC materials. An alternative way to lower φ of an ISC is the insertion of a monolayer of strong molecular donors engaging in charge transfer with the ISC and thus modifying the surface electrostatic potential, which then shifts the energy levels of a subsequently deposited organic semiconductor accordingly. In a n-type semiconductor, band bending cannot contribute in this configuration, thus achieving a substantial decrease of φ of wide bandgap semiconductors like ZnO and GaN which are natively n-type is much more challenging than increasing it by assembly of acceptor molecules. As donor molecule, $[RuCp^*mes]_2$ is selected since it has proven itself to be an efficient and air-stable n-dopant in organic electronics (Guo et al. 2012a, 2012b; Olthof et al.

2012; Giordano et al. 2015). In the following examples, it is demonstrated that the work function of ZnO can be decreased by about 2 eV by adsorption of this molecule. Due to this massive work function decrease, the energy-level alignment between ZnO and an OSC, a triply spiroannulated ladder-type quarterphenyl (L4P-sp3) in this case, could be changed from a type-II to a type-I. L4P-sp3 is a brightly emissive and photostable molecule (Kobin et al. 2017) which becomes, however, dark when in contact with ZnO. By tuning the energy-level alignment, interfacial exciton and charge transfer is regulated turning the heterojunction bright and rendering it very suitable for light-emitting applications.

Figure 2.11 summarizes PES data of [RuCp*mes]2-ZnO as well as L4P-sp3-[RuCp*mes]2-ZnO interfaces. Vacuum deposition of a ML of [RuCp*mes]2 on ZnO(0001) reduces φ by 1.5 eV to a value of 2.7 eV. The valence region of [RuCp*mes]2-ZnO does not feature detectable photoemission intensity in the ZnO bandgap. This implies that the [RuCp*mes]2 interlayer does not induce a large gap-state density that could act as exciton quencher at the interface and would therefore be detrimental for light-emitting applications. It has been shown that [RuCp*mes]2 and related dimers (Olthof et al. 2012) react with organic acceptors to form two acceptors anions and two monomeric cations, in this case, $[RuCp^*mes]^+$ (Guo et al. 2012b; Giordano et al. 2015). XPS in the C 1*s* and Ru 3*d* core level regions indicate the formation of $[RuCp^*mes]^+$ on ZnO as well (Schlesinger et al. 2015). On the other hand, a change of surface band bending upon deposition of [RuCp*mes]2 is not detected: The binding energies of the Zn 3*p* and O 1*s* core levels do not change after deposition of [RuCp*mes]2 (Schlesinger et al. 2015). Therefore, the origin of the φ

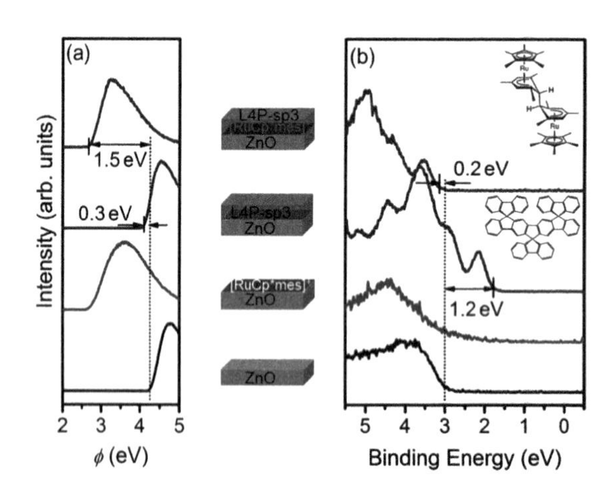

FIGURE 2.11 UPS spectra with and without a $[RuCp^*mes]_2$ interlayer. (a) φ derived from the SECO and (b) valence band region of UPS. The corresponding structures are shown in the center. The $[RuCp^*mes]_2$ coverage corresponds roughly to one ML, the L4P-sp3 layer is 5 nm thick. The structures of $[RuCp^*mes]_2$ (green) and the L4P-sp3 (blue) are shown as insets in (b). Energy-level diagrams of (c) ZnO/L4P-sp3 and (d) ZnO/L4P-sp3 with [RuCp*mes]2 interlayer. Energy values are referenced to the Fermi level and given in eV. The offsets Δ between the L4P-sp3 and ZnO energy levels are highlighted. Adapted from Schlesinger et al. (2015).

decrease is attributed solely to electron transfer from the interlayer to ZnO resulting in an interface dipole layer formed by the positively charged $[RuCp^*mes]^+$ layer and the negatively charged ZnO surface. This is fundamentally different from the situation encountered with a molecular acceptor interlayer, where a significant upward band bending in ZnO contributes to the φ increase (Section 2.4). The reason for that lies in the unintentional n-doping of ZnO. While molecular acceptors deplete the ZnO native donor levels, molecular donors fill the ZnO CB. The orders-of-magnitude larger density of states of the CB, as compared to that of the native donors, strongly pins the Fermi level as soon as the ZnO turns degenerate.

The change in the work function of ZnO upon adsorption of [RuCp*mes]2 results in a rigid shift of the L4P-sp3 levels with respect to those of the inorganic component (Figure 2.11a, b). For the untreated interface, the VBM of bare ZnO(0001) is at 3.0 eV below E_F implying slight upward band bending. Deposition of L4P-sp3 reduces φ by 0.3 eV (Figure 2.11a) as observed previously also for other molecules (Section 2.2). The L4P-sp3 HOMO binding energy onset is at 1.8 eV (Figure 2.11b). No new core level peaks arise in the O 1s and Zn 3p regions, thus evidencing physisorption of L4P-sp3 (Schlesinger et al. 2015). Furthermore, there are no L4P-sp3 thickness-dependent spectral changes (up to 5 nm nominal coverage) indicating the absence of band bending in the organic layer. The positions of the CBM of ZnO and the LUMO level of L4P-sp3 are estimated by adding the exciton binding energies to the optical bandgaps determined from the respective absorption spectra (3.25 eV for L4P-sp3). Under the assumption that the L4P-sp3 transport gap is only slightly wider (< 0.3 eV) than its optical gap, since the rigidified phenyls have a lower exciton binding energy compared to torsionally more flexible para-phenyls, the offset between the unoccupied frontier levels of the hybrid inorganic/organic system (HIOS) is estimated to be similar to that measured for the occupied levels. Hence, the HIOS energy-level alignment is of type-II with an offset between the respective filled/empty frontier levels of 1.2 eV (Figure 2.12a) favorable for charge separation yet unfavorable for light emission. On $[RuCp^*mes]^+$-covered ZnO(0001), L4P-sp3 deposition does not induce a φ change (Figure 2.11a). The onset of emission from the level associated with the HOMO of L4P-sp3 is 3.1 eV below E_F (Figure 2.11b). Consequently, the interlayer has aligned the filled/empty frontier levels of the HIOS, with an offset as little as 0.1 eV (Figure 2.12b). Notably, exposure of a sample consisting of 5 nm thick L4P-sp3 films on $[RuCp^*mes]^+$-ZnO(0001) to a nitrogen glove box atmosphere for 10 min results in change in the interface energy levels by less than 0.1 eV. Therefore, these optimized HIOS structures are remarkably robust with respect to handling outside of UHV conditions, which renders them of high practical relevance.

The goal of energy-level engineering is the improvement of the performance in hybrid optoelectronic devices. Depending on the application, the function of the heterojunction could be, for example, splitting of excitons and generation of a large V_{OC}, injection/blocking of charges or light emission. Each function requires a specific energy-level alignment between ISC and OSC. The present ZnO_L4P-sp3 material combination is suited for light-emitting applications where charge carriers are injected into the ISC and the light is emitted, following a Förster-type resonance energy transfer (FRET) step, from the OSC. Such excitation transfer scheme can provide some advantages over the traditional radiative transfer used for color

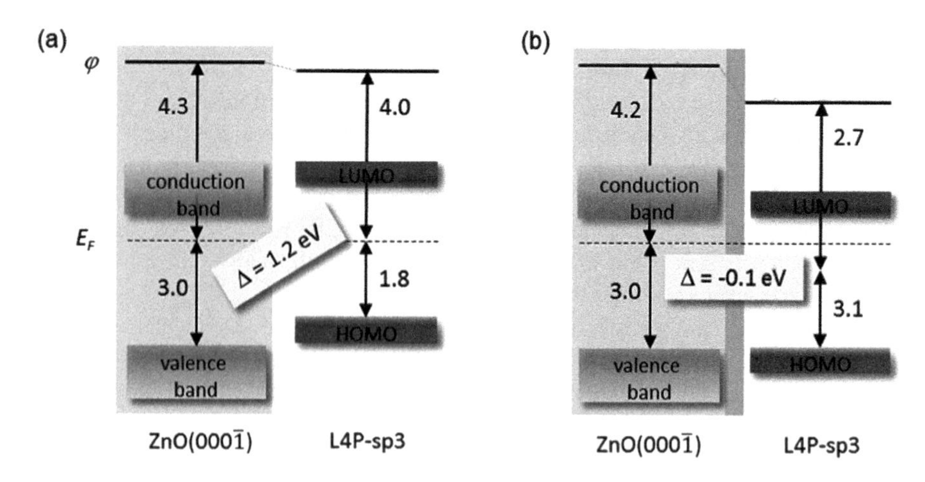

FIGURE 2.12 Energy-level diagrams of (a) ZnO/L4P-sp3 and (b) ZnO/L4P-sp3 with a [RuCp*mes]2 interlayer. Energy values are referenced to the Fermi-level E_F and given in eV. The offsets Δ between the L4P-sp3 and ZnO energy levels are highlighted. Adapted from (Schlesinger et al. 2015).

conversion in GaN-based light-emitting diodes in terms of efficiency, color rendering and response time (Ghataora et al. 2018). To achieve indeed superior performance engineering of the energy-level alignment proves essential as demonstrated in the example. The hybrid system is based on a ZnO/$Zn_{0.9}Mg_{0.1}O$ quantum well (QW) structure instead of a simple ZnO layer. The ZnO QW thickness is d_{QW}= 3.5 nm and the thickness of the top $Zn_{0.9}Mg_{0.1}O$ layer L_S= 2 nm. This layout guarantees that the excitons generated in the ISC are confined at a defined distance in close proximity to the interface with the OSC which is a prerequisite for efficient FRET. Using shadow masks during molecular deposition, three types of hybrid systems are prepared on the same ZnO/$Zn_{0.9}Mg_{0.1}O$ wafer (Figure 2.13a):

i) bare ZnO/$Zn_{0.9}Mg_{0.1}O$ QW (reference structure)
ii) 3 nm L4P-sp3 film on the QW structure (hybrid structure)
iii) ca. one ML $[RuCp^*mes]^+$ embedded between the QW structure and the 3 nm L4P-sp3 (optimized hybrid structure).

Efficient FRET alone requires solely spectral overlap between the PL spectrum of the QW donor and the absorption spectrum of the molecular acceptor. This condition is fulfilled in L4P-sp3-/ZnO/ZnMgO hybrid structure, as evident from Figure 2.13b, which compares the PL of structure (i) with the PL excitation (PLE) spectrum a reference 3 nm L4P-sp3 film deposited on sapphire. The PLE spectrum represents the absorption spectrum of the film. L4P-sp3 emits bright blue PL also shown Figure 2.13b. Consequently, excitons of the ZnO QW are efficiently converted into Frenkel excitons of L4P-sp3. The PL transients of the QW excitons depicted in Figure 2.13c provide an estimate for the FRET efficiency. The PL decay of the QW in absence of L4P-sp3 is in good approximation single-exponentially with a time constant of τ_{QW}= 205 ps.

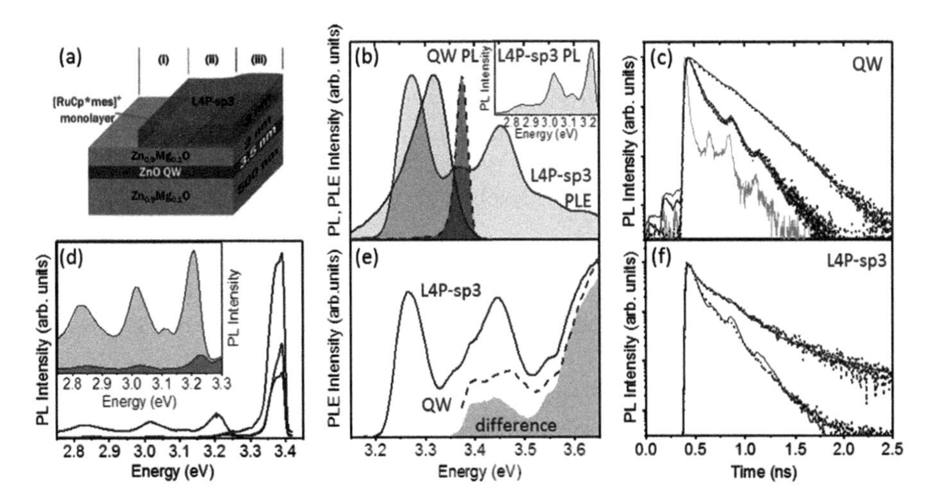

FIGURE 2.13 (a) Schematics and color code of the investigated structures: (i) reference structure (gray), (ii) hybrid structure (blue), and (iii) optimized hybrid structure (green). (b) PL and PLE of (i) and L4P-sp3 on sapphire (red). (c) Decay transients of QW excitons in (i), (ii), and (iii) as well as the instrument response function (light gray). The excitation energy is 3.46 eV. (d) PL spectra of (i), (ii), and (iii). (e) PLE of the optimized structure (iii) detecting at the $S_{1,\nu=1} \rightarrow S_{0,\nu=0}$ emission peak of L4P-sp3 and the low energy side of the QW emission. Yellow: Difference spectrum of L4P-sp3 in (iii) and on sapphire. (f). L4P-sp3 PL transients of (ii) and (iii). The excitation photon energy (3.3 eV) is below the ZnO absorption edge, excluding contributions from energy transfer. Solid curves in (c) and (d) are fits to the data by convoluting exponential transients (with up to two components) with the instrumental response. All experiments are performed at T = 5 K. Adapted from Schlesinger et al. (2015).

In the hybrid part (ii) covered with L4P-sp3, the QW decay time shortens to of τ_{HB}= 75 ps verifying the opening of an additional decay pathway for QW excitons due to FRET. Assuming that no parasitic losses on the QW side are introduced, a characteristic energy transfer time of τ_{FRET} = 115 ps and an efficiency of η_{FRET}= 0.65 are obtained. That means that approximately two out of three excitons generated in the ZnO QW are transferred to L4P-sp3. However, inspection of the PL spectrum shows that there is barely light produced by the organic layer (Figure 2.13d). The quantum yield of L4P-sp3 is sufficiently high in solid state and cannot serve as explanation. Rather, the unfavorable type-II energy-level alignment provides an effective pathway for the loss of excitation energy at the ZnMgO/L4P-sp3 interface. Once the excitons are transferred, they are rapidly quenched due to electron transfer from the L4P-sp3 LUMO to the ZnO CB. Hence, without optimization, such hybrid structure would be unsuited for light-emitting applications.The effect on the $[RuCp^*mes]^+$ interlayer on exciton dissociation and charge transfer is evaluated by analyzing PL transients of L4P-sp3 in structure (ii) and (iii). The excitation energy is set below the bandgap energy of ZnO excluding contributions from FRET. The short L4P-sp3 PL lifetime τ_H= 60 ps in the hybrid structure (ii) is recovered to a value of τ_H= 265 ps in the optimized hybrid structure (iii) (Figure 2.13e) signifying considerable suppression of exciton quenching. With an intrinsic decay time of L4P-sp3 of about τ_M= 500 ps obtained with a reference sample, the presented decay curves yield a characteristic time and efficiency of the charge separation process in the hybrid structure (ii) of τ_{CT}= 65 ps and η_{CT}= 0.9, respectively. That is, 9 out of 10 excitons generated (directly or indirectly) in L4P-sp3 are not converted into emitted light. In contrast, in the optimized hybrid structure (iii), the charge transfer efficiency is significantly reduced to η_{CT}= 0.45. Alignment of the frontier levels leads thus to a considerable reduction of interfacial exciton dissociation.

Finally, it needs to be assured that the $[RuCp^{*}mes]^{+}$ interlayer has no detrimental effect on the FRET efficiency. Substantial FRET from the QW to L4P-sp3 in structure (iii) is clearly demonstrated by PLE (Figure 2.13e). The spectrum taken at the $S_{1,\nu=1} \rightarrow S_{0,\nu=0}$ line of the molecular emission clearly shows the absorption features of the QW as well as that of the $Zn_{0.9}Mg_{0.1}O$ bandgap edge (> 3.55 eV), more obviously so when the difference spectrum with respect to L4P-sp3 on sapphire is constructed. The lifetime shortening of the QW excitons due to the L4P-sp3 overlayer is basically the same in structure (ii) and (iii) (Figure 2.13f) although the donor–acceptor spatial separation is widened by the ca. 0.3 nm thickness of the $[RuCp^{*}mes]^{+}$ interlayer. Thus, PLE and time-resolved PL data concordantly yield an efficiency of η_{CT}= 0.65 for the optimized structure (iii). Fully consistent with the increase in L4P-sp3 lifetime and the unchanged FRET, the molecular emission in the optimized hybrid structure (iii) increases by a factor of seven compared to structure (ii) (Figure 2.13d).

The yield of photons emitted by the L4P-sp3 layer per electron–hole pair generated in the QW (either by optical or electrical excitation) is $\eta = \eta_{FRET} \bullet \eta_{PL,\, L4p-sp3}$. The latter quantity is the emission yield of L4P-sp3 in the hybrid structure. As there is still residual exciton quenching at the ZnMgO interface, $\eta_{PL,\, L4p-sp3} \approx 0.55$ assuming that the intrinsic PL yield of L4P-sp3 approaches unity. Hence, the total luminescence yield of the hybrid structure (iii) is $\eta \approx 0.35$. At room temperature, the role of the $[RuCp^{*}mes]^{+}$ interlayer is even more crucial. Whereas L4P-sp3 emission from structure (ii) is no longer detectable, the signal remains bright in case of structure (iii). Despite this impressive improvement of radiative emission yield – particularly at room temperature – the present HIOS can further be optimized, since a considerable fraction of excitons is still not used for light emission. This might be traced back to interface states, which are too low in intensity to be directly revealed by photoemission. To further optimize such hybrid structures, work should focus on identifying and avoiding non-radiative side channels.

ACKNOWLEDGEMENTS

This work was funded by the Deutsche Forschungsgemeinschaft (DFG) - Projektnummer 182087777 - SFB 951.

REFERENCES

Alloway, Dana M, Amy L Graham, Xi Yang, Anoma Mudalige, Jr. Ramon Colorado, Vicki H Wysocki, Jeanne E Pemberton, T. Randall Lee, Ronald J Wysocki, and Neal R Armstrong. 2009. "Tuning the Effective Work Function of Gold and Silver Using ω-Functionalized Alkanethiols: Varying Surface Composition through Dilution and Choice of Terminal Groups." *The Journal of Physical Chemistry C* 113 (47): 20328. doi:10.1021/jp909494r.

Asyuda, Andika, Michael Gärtner, Xianglong Wan, Ines Burkhart, Torben Saßmannshausen, Andreas Terfort, and Michael Zharnikov. 2020. "Self-Assembled Monolayers with Embedded Dipole Moments for Work Function Engineering of Oxide Substrates." *The Journal of Physical Chemistry C* 124 (16): 8775–8785. doi:10.1021/acs.jpcc.0c00482.

Baeten, Linny, Bert Conings, Hans-Gerd Boyen, Jan D'Haen, An Hardy, Marc D'Olieslaeger, Jean V. Manca, and Marlies K. Van Bael. 2011. "Towards Efficient Hybrid Solar Cells Based on Fully Polymer Infiltrated ZnO Nanorod Arrays." *Advanced Materials* 23 (25): 2802. doi:10.1002/adma.201100414.

Bernède, J C, L Cattin, M Morsli, and Y Berredjem. 2008. "Ultra-Thin Metal Layer Passivation of the Transparent Conductive Anode in Organic Solar Cells." *Solar Energy Materials and Solar Cells* 92 (11): 1508–1515. doi:10.1016/j.solmat.2008.06.016.

Bhosle, V, J T Prater, Fan Yang, D Burk, S R Forrest, and J Narayan. 2007. "Gallium-Doped Zinc Oxide Films as Transparent Electrodes for Organic Solar Cell Applications." *Journal of Applied Physics* 102 (2). doi:10.1063/1.2750410.

Bianchi, F, S Sadofev, R Schlesinger, B Kobin, S Hecht, N Koch, F Henneberger, and S Blumstengel. 2014. "Cascade Energy Transfer versus Charge Separation in Ladder-Type Oligo(p-Phenylene)/ZnO Hybrid Structures for Light-Emitting Applications." *Applied Physics Letters* 105 (23): 233301. doi:10.1063/1.4903517.

Blumstengel, S, S Sadofev, C Xu, J. Puls, R L Johnson, H Glowatzki, N Koch, and F Henneberger. 2008. "Electronic Coupling in Organic-Inorganic Semiconductor Hybrid Structures with Type-II Energy Level Alignment." *Physical Review B - Condensed Matter and Materials Physics* 77 (8): 085323. doi:10.1103/PhysRevB.77.085323.

de Boer, B, A Hadipour, M M Mandoc, T van Woudenbergh, and P W M Blom. 2005. "Tuning of Metal Work Functions with Self-Assembled Monolayers." *Advanced Materials* 17 (5): 621. doi:10.1002/adma.200401216.

Bröker, B, R P Blum, J Frisch, A Vollmer, O T Hofmann, R Rieger, K Müllen, J P Rabe, E Zojer, and N Koch. 2008. "Gold Work Function Reduction by 2.2 EV with an Air-Stable Molecular Donor Layer." *Applied Physics Letters* 93 (24): 243303. doi:10.1063/1.3049616.

Campbell, I H, S Rubin, T A Zawodzinski, J D Kress, R L Martin, D L Smith, N N Barashkov, and J P Ferraris. 1996. "Controlling Schottky Energy Barriers in Organic Electronic Devices Using Self-Assembled Monolayers". *Physical Review B - Condensed Matter and Materials Physics* 54 (20): 14321.

Duhm, Steffen, Georg Heimel, Ingo Salzmann, Hendrik Glowatzki, Robert L Johnson, Antje Vollmer, Jürgen P Rabe, and Norbert Koch. 2008. "Orientation-Dependent Ionization Energies and Interface Dipoles in Ordered Molecular Assemblies." *Nature Materials* 7 (4): 326–332. doi:10.1038/nmat2119.

Eyer, Moritz, Johannes Frisch, Sergey Sadofev, Norbert Koch, Emil J W List-Kratochvil, and Sylke Blumstengel. 2017. "Role of Hybrid Charge Transfer States in the Charge Generation at ZnMgO/P3HT Heterojunctions." *Journal of Physical Chemistry C* 121 (40): 21955–21961. doi:10.1021/acs.jpcc.7b07293.

Futscher, Moritz H, Thorsten Schultz, Johannes Frisch, Maryline Ralaiarisoa, Ezzeldin Metwalli, Marco V Nardi, Peter Müller-Buschbaum, and Norbert Koch. 2019. "Electronic Properties of Hybrid Organic/Inorganic Semiconductor Pn-Junctions." *Journal of Physics Condensed Matter* 31 (6): 064002. doi:10.1088/1361-648X/aaf310.

Ghataora, Suneal, Richard M Smith, Modestos Athanasiou, and Tao Wang. 2018. "Electrically Injected Hybrid Organic/Inorganic III-Nitride White Light-Emitting Diodes with Nonradiative Förster Resonance Energy Transfer." *ACS Photonics* 5 (2): 642–647. doi:10.1021/acsphotonics.7b01291.

Giordano, Anthony J, Federico Pulvirenti, Talha M Khan, Canek Fuentes-Hernandez, Karttikay Moudgil, Jared H Delcamp, Bernard Kippelen, Stephen Barlow, and Seth R Marder. 2015. "Organometallic Dimers: Application to Work-Function Reduction of Conducting Oxides." *ACS Applied Materials & Interfaces* 7 (7): 4320. doi:10.1021/am5087648.

Greiner, Mark T, Michael G Helander, Wing Man Tang, Zhi Bin Wang, Jacky Qiu, and Zheng Hong Lu. 2012. "Universal Energy-Level Alignment of Molecules on Metal Oxides." *Nature Materials* 11 (1): 76–81. doi:10.1038/nmat3159.

Guo, Song, Sang Bok Kim, Swagat K Mohapatra, Yabing Qi, Tissa Sajoto, Antoine Kahn, Seth R Marder, and Stephen Barlow. 2012a. "N-Doping of Organic Electronic Materials Using Air-Stable Organometallics." *Advanced Materials* 24 (5): 699. doi:10.1002/adma.201103238.

Guo, Song, Swagat K Mohapatra, Alexander Romanov, Tatiana V Timofeeva, Kenneth I Hardcastle, Kada Yesudas, Chad Risko, Jean-Luc Brédas, Seth R Marder, and Stephen Barlow. 2012b. "N-Doping of Organic Electronic Materials Using Air-Stable Organometallics: A Mechanistic Study of Reduction by Dimeric Sandwich Compounds." *Chemistry—A European Journal* 18 (46): 14760. doi:10.1002/chem.201202591.

Ha, Ye Eun, Mi Young Jo, Juyun Park, Yong-Cheol Kang, Seong Il Yoo, and Joo Hyun Kim. 2013. "Inverted Type Polymer Solar Cells with Self-Assembled Mono Layer Treated ZnO." *The Journal of Physical Chemistry C* 117 (6): 2646. doi:10.1021/jp311148d.

Hotchkiss, Peter J, Simon C Jones, Sergio A Paniagua, Asha Sharma, Bernard Kippelen, Neal R Armstrong, and Seth R Marder. 2012. "The Modification of Indium Tin Oxide with Phosphonic Acids: Mechanism of Binding, Tuning of Surface Properties, and Potential for Use in Organic Electronic Applications." *Accounts of Chemical Research* 45 (3): 337. doi:10.1021/ar200119g.

Hotchkiss, Peter J, Michal Malicki, Anthony J Giordano, Neal R Armstrong, and Seth R Marder. 2011. "Characterization of Phosphonic Acid Binding to Zinc Oxide." *The Journal of Materials Chemistry* 21 (9): 3107. doi:10.1039/C0JM02829K.

Hwang, Jaehyung, Alan Wan, and Antoine Kahn. 2009. "Energetics of Metal-Organic Interfaces: New Experiments and Assessment of the Field." *Materials Science and Engineering R: Reports*. doi:10.1016/j.mser.2008.12.001.

Itskos, G, C R Belton, G Heliotis, I M Watson, M D Dawson, R Murray, and D D C Bradley. 2009. "White Light Emission via Cascade Förster Energy Transfer in (Ga, In)N Quantum Well/Polymer Blend Hybrid Structures." *Nanotechnology* 20 (27): 275207.

Itskos, G, X Xristodoulou, E Iliopoulos, S Ladas, S Kennou, M Neophytou, and S Choulis. 2013. "Electronic and Interface Properties of Polyfluorene Films on GaN for Hybrid Optoelectronic Applications." *Applied Physics Letters* 102 (6): 063303. doi:10.1063/1.4792211.

Kedem, Nir, Sylke Blumstengel, Fritz Henneberger, Hagai Cohen, Gary Hodes, and David Cahen. 2014. "Morphology-, Synthesis- and Doping-Independent Tuning of ZnO Work Function Using Phenylphosphonates." *Physical Chemistry Chemical Physics* 16 (18): 8310. doi:10.1039/C3CP55083D.

Kim, Yong Hyun, Jin Soo Kim, Won Mok Kim, Tae Yeon Seong, Jonghee Lee, Lars Müller-Meskamp, and Karl Leo. 2013. "Realizing the Potential of ZnO with Alternative Non-Metallic Co-Dopants as Electrode Materials for Small Molecule Optoelectronic Devices." *Advanced Functional Materials* 23 (29): 3645–3652. doi:10.1002/adfm.201202799.

Kobin, Björn, Jutta Schwarz, Beatrice Braun-Cula, Moritz Eyer, Anton Zykov, Stefan Kowarik, Sylke Blumstengel, and Stefan Hecht. 2017. "Spiro-Bridged Ladder-Type Oligo(Para-Phenylene)s: Fine Tuning Solid State Structure and Optical Properties." *Advanced Functional Materials* 27 (45): 1–10. doi:10.1002/adfm.201704077.

Koch, N, A Kahn, J Ghijsen, J J Pireaux, J Schwartz, R L Johnson, and A Elschner. 2003. "Conjugated Organic Molecules on Metal versus Polymer Electrodes: Demonstration of a Key Energy Level Alignment Mechanism." *Applied Physics Letters* 82 (1): 70–72. doi:10.1063/1.1532102.

Koch, N, S Duhm, J P Rabe, A Vollmer, and R L Johnson. 2005a. "Optimized Hole Injection with Strong Electron Acceptors at Organic-Metal Interfaces." *Physical Review Letters* 95 (23): 237601. doi:10.1103/PhysRevLett.95.237601.

Koch, Norbert, Georg Heimel, Jishan Wu, Egbert Zojer, Robert L Johnson, Jean Luc Brédas, Klaus Müllen, and Jürgen P Rabe. 2005b. "Influence of Molecular Conformation on Organic/Metal Interface Energetics." *Chemical Physics Letters* 413 (4–6): 390–395. doi:10.1016/j.cplett.2005.08.004.

Lange, Ilja, Sina Reiter, Michael Pätzel, Anton Zykov, Alexei Nefedov, Jana Hildebrandt, Stefan Hecht, et al. 2014. "Tuning the Work Function of Polar Zinc Oxide Surfaces Using Modified Phosphonic Acid Self-Assembled Monolayers." *Advanced Functional Materials* 24 (44): 7014. doi:10.1002/adfm.201401493.

Mao, Hong Ying, Fabio Bussolotti, Dong Chen Qi, Rui Wang, Satoshi Kera, Nobuo Ueno, Andrew Thye Shen Wee, and Wei Chen. 2011. "Mechanism of the Fermi Level Pinning at Organic Donor-Acceptor Heterojunction Interfaces." *Organic Electronics* 12 (3): 534–540. doi:10.1016/j.orgel.2011.01.003.

Mcintyre, D E Aspnes 1971. "Differential Reflection Spectroscopy of Very Thin Surface Films." *Surface Science* 24: 417–434.

Meisel, Tino, Mino Sparenberg, Marcel Gawek, Sergey Sadofev, Björn Kobin, Lutz Grubert, Stefan Hecht, Emil List-Kratochvil, and Sylke Blumstengel. 2018. "Fingerprint of Charge Redistribution in the Optical Spectra of Hybrid Inorganic/Organic Semiconductor Interfaces." *Journal of Physical Chemistry C* 122 (24): 12913–12919. doi:10.1021/acs.jpcc.8b03580.

Niederhausen, J, P Amsalem, J Frisch, A Wilke, A Vollmer, R Rieger, K Müllen, J P Rabe, and N Koch. 2011. "Tuning Hole-Injection Barriers at Organic/Metal Interfaces Exploiting the Orientation of a Molecular Acceptor Interlayer." *Physical Review B - Condensed Matter and Materials Physics* 84 (16): 165302. doi:10.1103/PhysRevB.84.165302.

Olthof, Selina, Shafigh Mehraeen, Swagat K. Mohapatra, Stephen Barlow, Veaceslav Coropceanu, Jean Luc Brédas, Seth R. Marder, and Antoine Kahn. 2012. "Ultralow Doping in Organic Semiconductors: Evidence of Trap Filling." *Physical Review Letters* 109 (17): 176601. doi:10.1103/PhysRevLett.109.176601.

Oosterhout, Stefan D, L Jan Anton Koster, Svetlana S van Bavel, Joachim Loos, Ole Stenzel, Ralf Thiedmann, Volker Schmidt, et al. 2011. "Controlling the Morphology and Efficiency of Hybrid ZnO:Polythiophene Solar Cells Via Side Chain Functionalization." *Advanced Functional Materials* 1 (1): 90. doi:10.1002/aenm.201000022.

Ostapenko, Alexandra, Tobias Klöffel, Jens Eußner, Klaus Harms, Stefanie Dehnen, Bernd Meyer, and Gregor Witte. 2016. "Etching of Crystalline ZnO Surfaces upon Phosphonic Acid Adsorption: Guidelines for the Realization of Well-Engineered Functional Self-Assembled Monolayers." *ACS Applied Materials and Interfaces* 8 (21): 13472–13483. doi:10.1021/acsami.6b02190.

Piersimoni, Fortunato, Raphael Schlesinger, Johannes Benduhn, Donato Spoltore, Sina Reiter, Ilja Lange, Norbert Koch, Koen Vandewal, and Dieter Neher. 2015. "Charge Transfer Absorption and Emission at ZnO/Organic Interfaces." *Journal of Physical Chemistry Letters* 6 (3): 500–504. doi:10.1021/jz502657z.

Proehl, Holger, Robert Nitsche, Thomas Dienel, Karl Leo, and Torsten Fritz. 2005. "In Situ Differential Reflectance Spectroscopy of Thin Crystalline Films of PTCDA on Different Substrates." *Physical Review B - Condensed Matter and Materials Physics* 71 (16): 1–14. doi:10.1103/PhysRevB.71.165207.

Rangger, Gerold M, Oliver T Hofmann, Lorenz Romaner, Georg Heimel, Benjamin Bröker, Ralf Peter Blum, Robert L Johnson, Norbert Koch, and Egbert Zojer. 2009. "F4TCNQ on Cu, Ag, and Au as Prototypical Example for a Strong Organic Acceptor on Coinage

Metals." *Physical Review B - Condensed Matter and Materials Physics* 79 (16): 165306. doi:10.1103/PhysRevB.79.165306.

Romaner, Lorenz, Georg Heimel, Jean Luc Brédas, Alexander Gerlach, Frank Schreiber, Robert L. Johnson, Jörg Zegenhagen, Steffen Duhm, Norbert Koch, and Egbert Zojer. 2007. "Impact of Bidirectional Charge Transfer and Molecular Distortions on the Electronic Structure of a Metal-Organic Interface." *Physical Review Letters* 99 (25): 256801. doi:10.1103/PhysRevLett.99.256801.

Schlesinger, R, F Bianchi, S Blumstengel, C Christodoulou, R Ovsyannikov, B Kobin, K Moudgil, et al. 2015. "Efficient Light Emission from Inorganic and Organic Semiconductor Hybrid Structures by Energy-Level Tuning." *Nature Communications* 6 (1): 6754. doi:10.1038/ncomms7754.

Schlesinger, Raphael, Fabio Bussolotti, Jinpeng Yang, Sergey Sadofev, Antje Vollmer, Sylke Blumstengel, Satoshi Kera, Nobuo Ueno, and Norbert Koch. 2019a. "Gap States Induce Soft Fermi Level Pinning upon Charge Transfer at ZnO/Molecular Acceptor Interfaces." *Physical Review Materials* 3 (7): 074601. doi:10.1103/PhysRevMaterials.3.074601.

Schlesinger, Raphael, Stefanie Winkler, Matthias Brandt, Sylke Blumstengel, Ruslan Ovsyannikov, Antje Vollmer, and Norbert Koch. 2019b. "Energy Level Alignment at Organic/Inorganic Semiconductor Heterojunctions: Fermi Level Pinning at the Molecular Interlayer with a Reduced Energy Gap." *Physical Chemistry Chemical Physics* 21 (27): 15072–15079. doi:10.1039/c9cp02763g.

Schlesinger, Raphael, Yong Xu, Oliver T. Hofmann, Stefanie Winkler, Johannes Frisch, Jens Niederhausen, Antje Vollmer, et al. 2013. "Controlling the Work Function of ZnO and the Energy-Level Alignment at the Interface to Organic Semiconductors with a Molecular Electron Acceptor." *Physical Review B - Condensed Matter and Materials Physics* 87 (15): 155311. doi:10.1103/PhysRevB.87.155311.

Schöttner, L, S Erker, R Schlesinger, Koch, A Nefedov, O T Hofmann, and C Wöll. 2020. "Doping-Induced Electron Transfer at Organic/Oxide Interfaces: Direct Evidence from Infrared Spectroscopy." *Journal of Physical Chemistry C* 124 (8): 4511–4516. doi:10.1021/acs.jpcc.9b08768.

Schultz, T, R Schlesinger, Niederhausen, F Henneberger, S Sadofev, S Blumstengel, A Vollmer, et al. 2016. "Tuning the Work Function of GaN with Organic Molecular Acceptors." *Physical Review B* 93 (12): 125309. doi:10.1103/PhysRevB.93.125309.

Koh Sharon E, Krystal D McDonald, David H Holt, and Charles S Dulcey, John A Chaney, and and Pehr E Pehrsson. 2006. "Phenylphosphonic Acid Functionalization of Indium Tin Oxide: Surface Chemistry and Work Functions." *Langmuir* 22 (14): 6249. doi:10.1021/la052379e.

Smoluchowski, R 1941. "Anisotropy of the Electronic Work Function of Metals." *Physical Review* 60 (9): 661–674. doi:10.1103/PhysRev.60.661.

Sze, Simon M, and Daniel C Mattis. 1970. "Physics of Semiconductor Devices." *Physics Today* 23 (6): 75–75. doi:10.1063/1.3022205.

Timpel, Melanie, Marco V Nardi, Stefan Krause, Giovanni Ligorio, Christos Christodoulou, Luca Pasquali, Angelo Giglia, et al. 2014. "Surface Modification of ZnO(0001)-Zn with Phosphonate-Based Self-Assembled Monolayers: Binding Modes, Orientation, and Work Function." *Chemistry of Materials* 26 (17): 5042–5050. doi:10.1021/cm502171m.

Timpel, Melanie, Marco V Nardi, Giovanni Ligorio, Berthold Wegner, Michael Pätzel, Björn Kobin, Stefan Hecht, and Norbert Koch. 2015. "Energy-Level Engineering at ZnO/Oligophenylene Interfaces with Phosphonate-Based Self-Assembled Monolayers." *ACS Applied Materials and Interfaces* 7 (22): 11900–11907. doi:10.1021/acsami.5b01669.

Vilan, A, A Shanzer, and D Cahen. 2000. "Molecular Control over Au/GaAs Diodes." *Nature* 404 (6774): 166.

Vilan, Ayelet, Omer Yaffe, Ariel Biller, Adi Salomon, Antoine Kahn, and David Cahen. 2010. "Molecules on Si: Electronics with Chemistry." *Advanced Materials* 22 (2): 140. doi:10.1002/adma.200901834.

Wang, Haiyuan, Sergey V Levchenko, Thorsten Schultz, Norbert Koch, Matthias Scheffler, and Mariana Rossi. 2019. "Modulation of the Work Function by the Atomic Structure of Strong Organic Electron Acceptors on H-Si(111)." *Advanced Electronic Materials* 5 (5): 1800891. doi:10.1002/aelm.201800891.

Winkler, Stefanie, Johannes Frisch, Raphael Schlesinger, Martin Oehzelt, Ralph Rieger, Joachim Räder, Jürgen P Rabe, Klaus Müllen, and Norbert Koch. 2013. "The Impact of Local Work Function Variations on Fermi Level Pinning of Organic Semiconductors." *Journal of Physical Chemistry C* 117 (43): 22285–22289. doi:10.1021/jp401919z.

Yang, Jin-Peng Peng, Lin-Tai Tai Shang, Fabio Bussolotti, Li-Wen Wen Cheng, Wen-Qing Qing Wang, Xiang-Hua Hua Zeng, Satoshi Kera, Yan-Qing Qing Li, Jian-Xin Xin Tang, and Nobuo Ueno. 2017. "Fermi-Level Pinning Appears upon Weak Electrode-Organic Contact without Gap States: A Universal Phenomenon." *Organic Electronics* 48 (September): 172–178. doi:10.1016/j.orgel.2017.06.005.

3 Molecular Layer Deposition of Organic–Inorganic Hybrid Materials

Xiangbo Meng

CONTENTS

3.1 INTRODUCTION

Nanoscience and nanotechnology are playing an ever-increasing importance in our society. Scientists and engineers have been striving to manipulate atoms and molecules precisely at will to develop ideal nanosized materials with exceptional properties.[1] To this end, various techniques have been devised for nanofabrication, such as mechanochemistry,[2,3] wet chemistry,[4] physical vapor deposition (PVD),[5] chemical vapor deposition (CVD),[6] atomic layer deposition (ALD),[7] and molecular layer deposition (MLD).[8] Among all the methods, recently MLD has been attracting more and more attention for growing pure polymeric and hybrid films.[8–10]

MLD was first coined in 1991 by Yoshimura and co-workers[11] exclusively for nanoscale films of organic materials, especially pure polymers and metal-based hybrid polymers.[9,12–15] This unique controllable technique was first demonstrated for synthesizing polyimides.[11] Subsequently, more polymeric films were developed via MLD, including polyazomethines,[16–19] polyureas,[20–24] polyamides,[25–28] poly (3,4-ethylenedioxythiophene)[29,30] polyimide-polyamides,[31] polythioureas,[32] polyethylene terephthalate,[33] and some others.[34,35] It was in 2008 when the first metal-based hybrid polymer was reported, which was an aluminum alkoxide (the so-called "alucone").[36] Thereafter, many more alucones[13,37–45] have been developed by MLD and also ignited research interests on other metalcones, resulting in mangancone,[45] zincones,[46–53] zircones,[54,55] titanicones,[56–59] hafnicones,[60] and vanadicone.[61] This greatly extended our capabilities in searching for advanced materials in a controllable mode. In this regard, some excellent review papers[8–10,12] have well documented MLD processes and their capabilities. Owing to its unlimited possibilities for new nanoscale polymeric films, MLD has exhibited great potentials for a large variety of applications, such as microelectronics,[62] catalysis,[63] energy conversion and storage,[64] organic magnets,[62] luminescent devices,[14] surface engineering,[10,65] and many others.[15] Recently, there has been an increasing interest in MLD polymeric and hybrid films for addressing issues in rechargeable batteries. In this context, organic–inorganic hybrid materials are particularly intriguing, ascribed to their desirable properties unattainable with the conventional materials.

This chapter focuses on introducing recent MLD research progresses on organic–inorganic hybrid materials, featuring their surface chemistry, growth characteristics, and film properties. Following this introductory section, we present some general basics commonly shared by MLD processes, including surface chemistry and growth characteristics, compared to those of ALD processes. The third part summarizes MLD processes of metalcones and the fourth part gives an account of other hybrid materials. In the last part, we conclude this chapter and give some outlook on future studies.

3.2 THE BASICS OF MLD

3.2.1 Surface Chemistry

MLD and ALD are two highly similar vapor-phase techniques for nanofabrication. They share the same operational principle to realize accurate controls over materials growth. They both commonly rely on alternative self-limiting surface reactions for materials growth. The former produces organic materials while the latter results in inorganic materials. They both grow materials accurately in a layer-by-layer mode. Figure 3.1a illustrates an ALD process for growing binary inorganic materials, while Figure 3.1b displays an MLD process for growing pure polymers using two homobifunctional precursors. In terms of surface reactions, for example, the model ALD process of Al_2O_3 using trimethylaluminum (TMA, $Al(CH_3)_3$) and H_2O can be described in Equations 3.1A and 3.1B as follows[66]:

$$|\text{–OH} + \text{Al}(\text{CH}_3)_3(\text{g}) \rightarrow |\text{–O} - \text{Al}(\text{CH}_3)_2 + \text{CH}_4(\text{g}) \tag{3.1A}$$

$$|\text{–O} - \text{Al}(\text{CH}_3)_2 + 2\text{H}_2\text{O}(\text{g}) \rightarrow |\text{–OAl}(\text{OH})_2 + 2\text{CH}_4(\text{g}) \tag{3.1B}$$

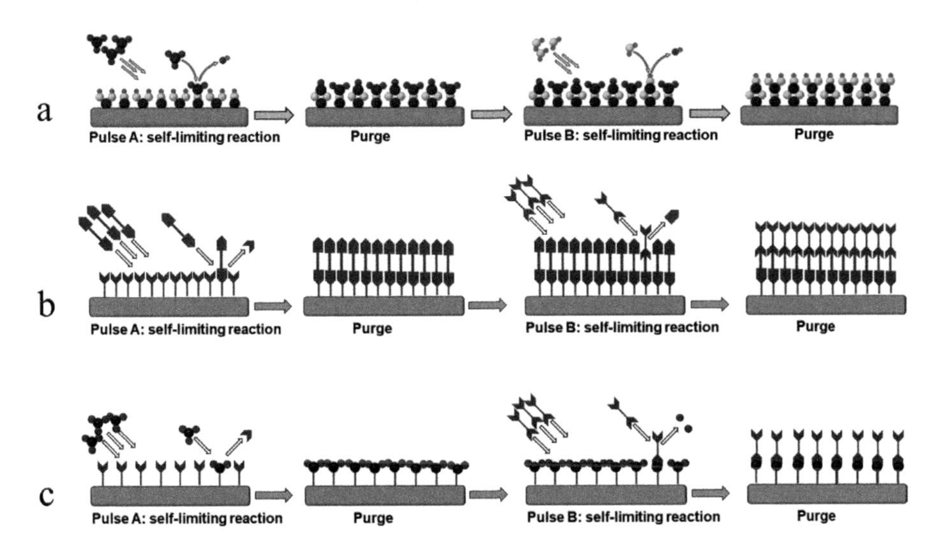

FIGURE 3.1 Illustrations of (a) ALD and MLD processes for growing (b) pure polymeric films and (c) organic–inorganic hybrid films.

where "|" indicates substrate surfaces while "(g)" signifies gas phases. The surface chemistry of the ALD Al_2O_3 is based on ligand exchanges between –OH and $-CH_3$ to arrange atoms accurately in a layer-by-layer mechanism. In addition to the ligand exchange mechanism as illustrated in Equations 3.1A and 3.1B, there are other mechanisms for ALD surface chemistry as well, such as dissociation and association.[67] The four steps in Figure 3.1a constitute one ALD cycle and they can repeat to build up films for desired thicknesses. The growth rate of ALD is described by growth per cycle (GPC), typically having a GPC of ~ 1 Å/cycle.[12] In the case of MLD of pure polymers (Figure 3.1b), one precursor first reacts with surface reactive groups via a corresponding linking chemistry to add a molecular layer on the substrate surface with new reactive sites.[62] Following a thorough purge, another precursor reacts with the new reactive sites with the production of another molecular layer and recovers the surface back to the initial reactive groups. Another full purge is performed to finish one MLD cycle. Through repeating the afore-discussed four steps, MLD can realize polymeric film growth accurately at the molecular level. The growth rate of MLD also is described by GPC. Using adipoyl chloride (AC) and 1,6-hexanediamine (HD) as precursors,[25,26] for example, an MLD process has been developed for growing nylon films linearly and the surface chemistry is described as follows:

$$|-NH_2 + ClCO(CH_2)_4COCl(g) \rightarrow |-NHCO(CH_2)_4COCl + HCl(g) \quad (3.2A)$$

$$\begin{aligned} &|-NHCO(CH_2)_4COCl + H_2N(CH_2)_6NH_2(g) \\ &\rightarrow |-NHCO(CH_2)_4CO-NH(CH_2)_6NH_2 + HCl(g) \end{aligned} \quad (3.2B)$$

The AC-HD MLD process could realize a GPC of 19 Å/cycle at 62 °C.[26] It is apparent that the molecular layers of $-CO(CH_2)_4CO-$ and $-NH(CH_2)_6NH-$ during the

MLD-nylon are much larger than the atomic layers of –Al– and –O– in the ALD-Al_2O_3. This underlies the higher GPC of the MLD nylon process of AC-HD. Additionally, many more pure polymeric materials via MLD recently have been summarized in literature.[8]

In addition to fabricating pure polymeric films as illustrated in Figure 3.1b, MLD also enables organic–inorganic hybrid materials by adopting an ALD precursor and an MLD precursor (Figure 3.1c), such as metal alkoxide materials (i.e., metalcones), in which diols can be used to couple with a metal precursor. Using TMA and ethylene glycol (EG, $HOCH_2CH_2OH$, a homobifunctional diol precursor), for instance, George's group first reported a metal-based hybrid polymer, an aluminum alkoxide (i.e., alucone) of $Al(OCH_2CH_2O)_2$ with the following surface chemistry[36]:

$$|{-OH} + Al(CH_3)_3(g) \rightarrow |{-O} - Al(CH_3)_2 + CH_4(g) \tag{3.3A}$$

$$|{-O} - Al(CH_3)_2 + 2HOCH_2CH_2OH(g) \rightarrow |{-OAl}(OCH_2CH_2OH)_2 + 2CH_4(g) \tag{3.3B}$$

Apparently, the molecular fragment of $-OCH_2CH_2O-$ attached in the MLD-alucone is far much larger than the atomic part of –O– in the ALD-Al_2O_3. This well explains that the resultant alucone has grown much faster than the ALD Al_2O_3, accounting for 4 Å/cycle at 85 °C for the MLD alkoxide[36] versus 1.3 Å/cycle for the ALD Al_2O_3 at 80 °C.[66] Through smartly selecting precursors for their functional groups and backbones, MLD enables different metalcones or hybrid materials with desired properties. Substituting EG with the aromatic hydroquinone (HQ, HOC_6H_4OH), for example, another alucone has been deposited and has its surface chemistry as follows[38]:

$$|{-OH} + Al(CH_3)_3(g) \rightarrow |{-O} - Al(CH_3)_2 + CH_4(g) \tag{3.4A}$$

$$|{-O} - Al(CH_3)_2 + 2HOC_6H_4OH(g) \rightarrow |{-OAl}(OC_6H_4OH)_2 + 2CH_4(g) \tag{3.4B}$$

This TMA-HQ MLD process exhibits a GPC of 4.1 Å/cycle at 150 °C.[38] Alucones with different backbones are expected to exhibit different properties. The aromatic backbone of HQ is expected to provide structural stability and contribute largely to the electrical properties of the resultant polymer films. To date, many more metalcones have been reported, including alucones,[13,36–45,53,65,68–84] zincones,[47–53,85,86] titanicones,[46–53,59,85,86] vanadicones,[61] zircones,[54,55] hafnicones,[60] mangancones,[45] metal quinolones,[87,88] and some other hybrid materials.[89–104]

To investigate the underlying mechanism of the many MLD processes, a suite of *in situ* techniques have been employed in previous studies. Fourier transform infrared spectroscopy (FTIR)[24,36,38,41–44,78] and quartz crystal microbalance (QCM)[36,38,39,42–44,54,56,93] are two widely utilized *in situ* instruments. They are very helpful to get insightful information on surface chemistry of MLD processes. FTIR spectra can clearly identify a molecular fingerprint after each surface reaction. QCM can definitively detect any molecular deposition in mass uptake and demonstrate a linear growth for any feasible MLD processes. On the other hand, quadrupole mass spectrometry (QMS)[69] is also very useful to detect any byproducts resulted from MLD surface reactions. All the data collected by the three *in situ* tools jointly help construct the underlying surface chemistry during a MLD process.

3.2.2 Growth Characteristics

Like ALD processes, MLD processes also are subject to three parameters, that is, precursor, temperature, and substrate. MLD precursors should be able to produce sufficient vapors easily. They should also be chemically stable at deposition temperatures and highly reactive to their coupled precursors. As discussed above using EG and HQ for alucones, MLD films are highly related to the precursors adopted, such as their structures, properties, and GPCs. On the other hand, substrates have impacts on MLD growth. In some cases, substrates should be pretreated for functionalization in order to initiate the growth of some polymeric films. On the contrary, sometimes substrates should be pretreated with a protective layer in order to resist any film growth.[105,106] Furthermore, deposition temperature often is critical for MLD film growth and it is worth noting that most of the MLD processes reported to date show a decreasing growth tendency with temperature. Ascribed to its unique growth mechanism, MLD produces uniform and conformal coatings over any shaped substrates.

3.3 MLD PROCESSES FOR ORGANIC–INORGANIC HYBRID METALCONES

Hybrid materials are very promising in a large variety of applications such as optics, electronics, mechanics, membranes, new energies, catalysis, and surface engineering. As recently summarized in our review article,[8] MLD has fabricated a variety of hybrid materials through adopting one typical ALD metal-containing precursor as the metal source and one MLD organic precursor. In addition, MLD processes can proceed with multiple precursors as well. The metal-containing ALD precursors have been collected in Figure 3.2 while the organic precursors are summarized in Figures 3.3 and 3.4 for growing hybrid materials. In this chapter, we focus on discussing MLD processes for growing metalcones. In terms of metal elements, there to date have been reported seven types of MLD metalcones, including alucones, titanicone, zincone, zircone, hafnicone, mangancone, and vanadicone.

3.3.1 Alucones

Alucones are polymeric aluminum alkoxide materials with carbon-containing backbones, that is, ···Al-O-R-O-Al···, rather than the Al-O-Al backbone associated with alumoxanes.[107] This type of polymer was first reported in solution-based methods by Schlenker in 1958.[107] In producing alucone using MLD, TMA is the predominant metal precursor while there are many choices for an organic precursor (Figures 3.5 and 3.6). These organic precursors can be divided into two classes: homobifunctional (i.e., EG, PPDA, TEA, LC, GL, HQ, BDO, and HDO) and heterobifunctional (i.e., EA and MA, GLY, and LAC) reactants. These two types of MLD precursors showed some distinct impacts on film growth characteristics.

3.3.1.1 Homobifunctional Organic Precursors

The first MAD alucone was reported through coupling TMA and EG (see Figure 3.5, AC-1).[36] The resultant alucone showed a temperature-dependent decreasing

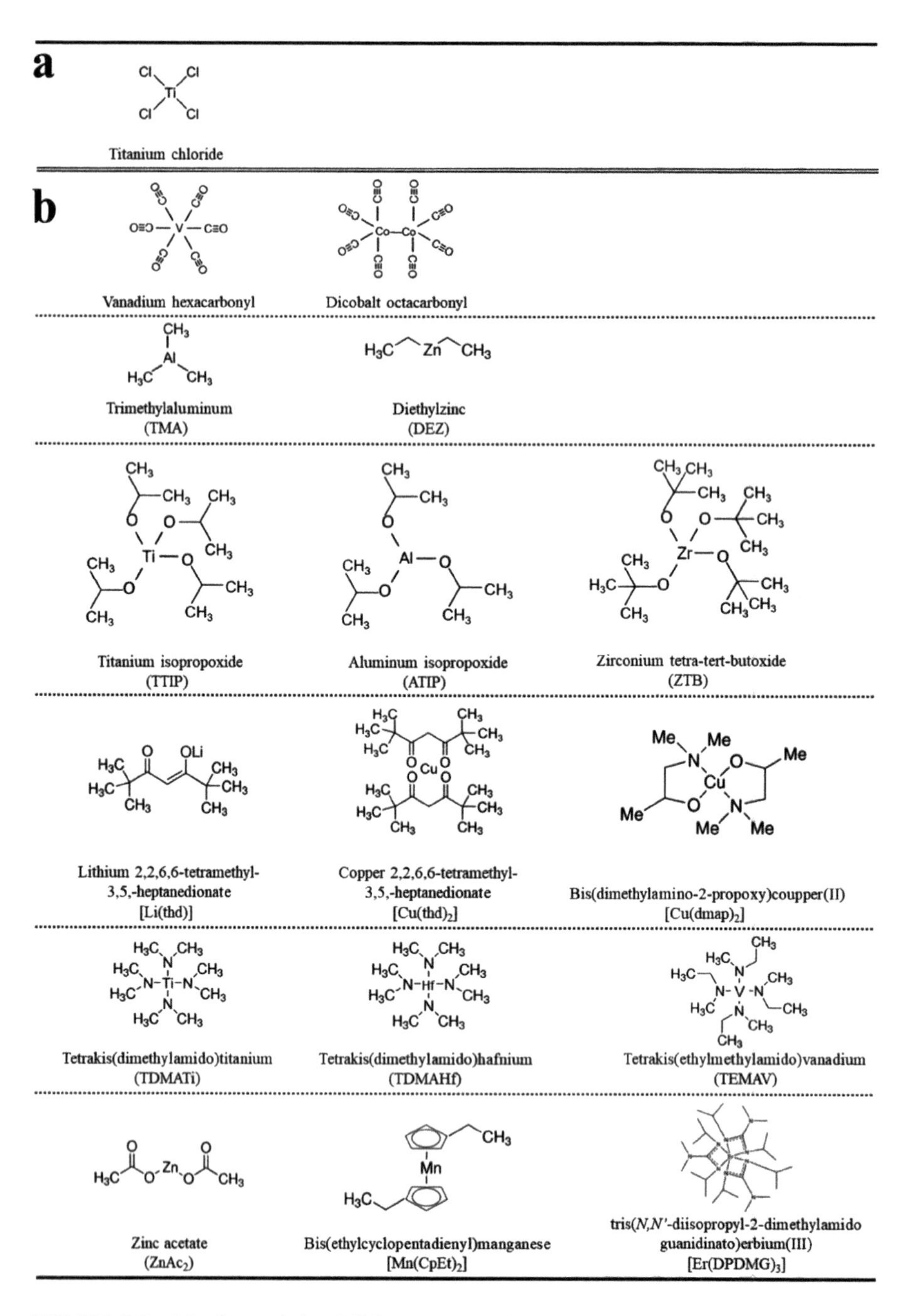

FIGURE 3.2 Metal-containing MLD precursors for organic–inorganic hybrid materials: (a) halides and (b) metal-organic compounds.

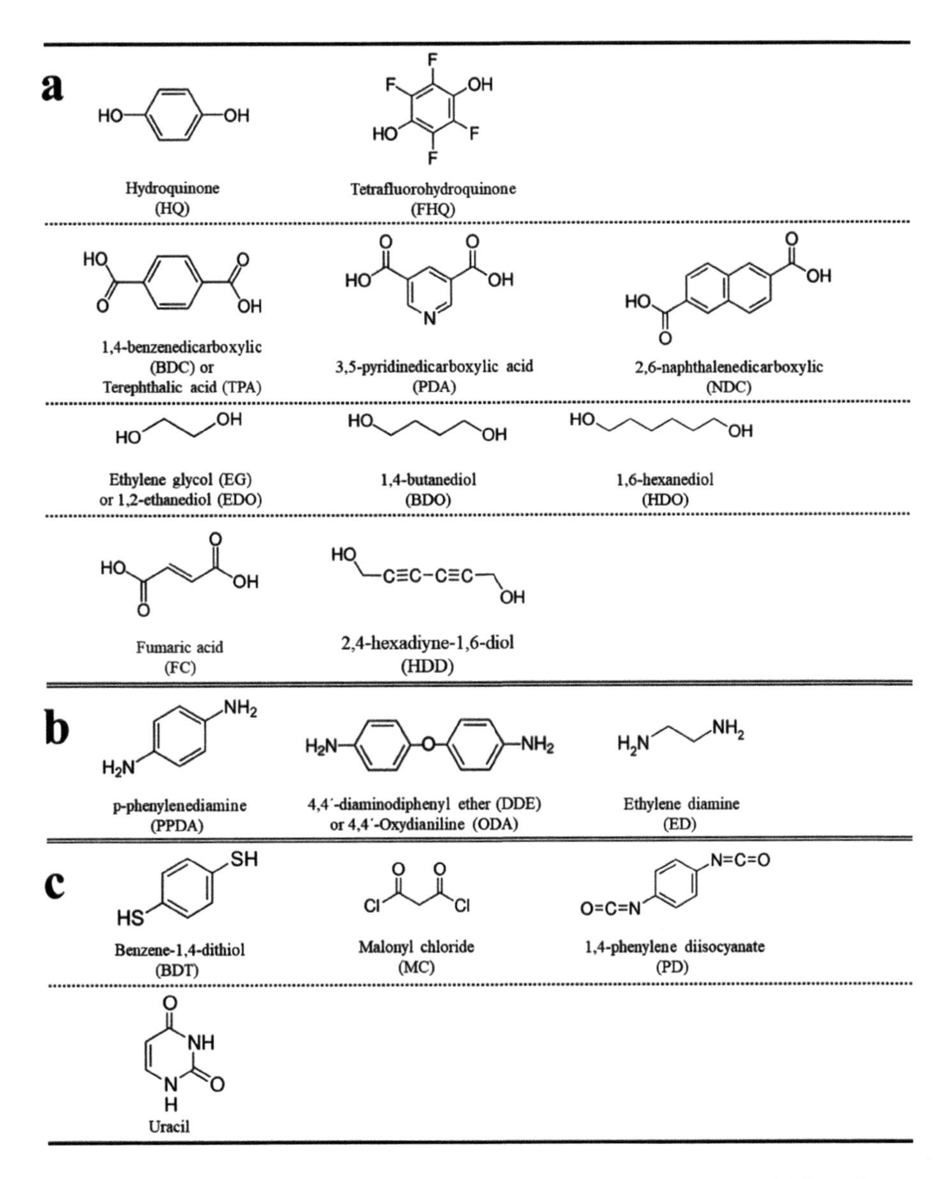

FIGURE 3.3 Homobifunctional MLD precursors for organic–inorganic hybrid films, featuring the reactive groups: (a) diol, (b) amine, (c) thiol, chloride, isocyanate, and uracil.

tendency in GPC from 4 Å/cycle at 85 °C to 0.4 Å/cycle at 175 °C. The surface reactions are shown in Equations of 3.3A and 3.3B. Using FTIR and QCM, Dameron et al.[36] also disclosed an alternative growth mechanism for this alucone MLD. At higher temperatures (e.g., 135 °C), there are no noticeable O–H stretching vibrations observed, suggesting the EG molecules might have reacted twice with –AlCH_3 species. Thus, it is believed that, during TMA-EG deposition, TMA molecules might have diffused into the alucone polymer film while reacted with –OH species. The EG

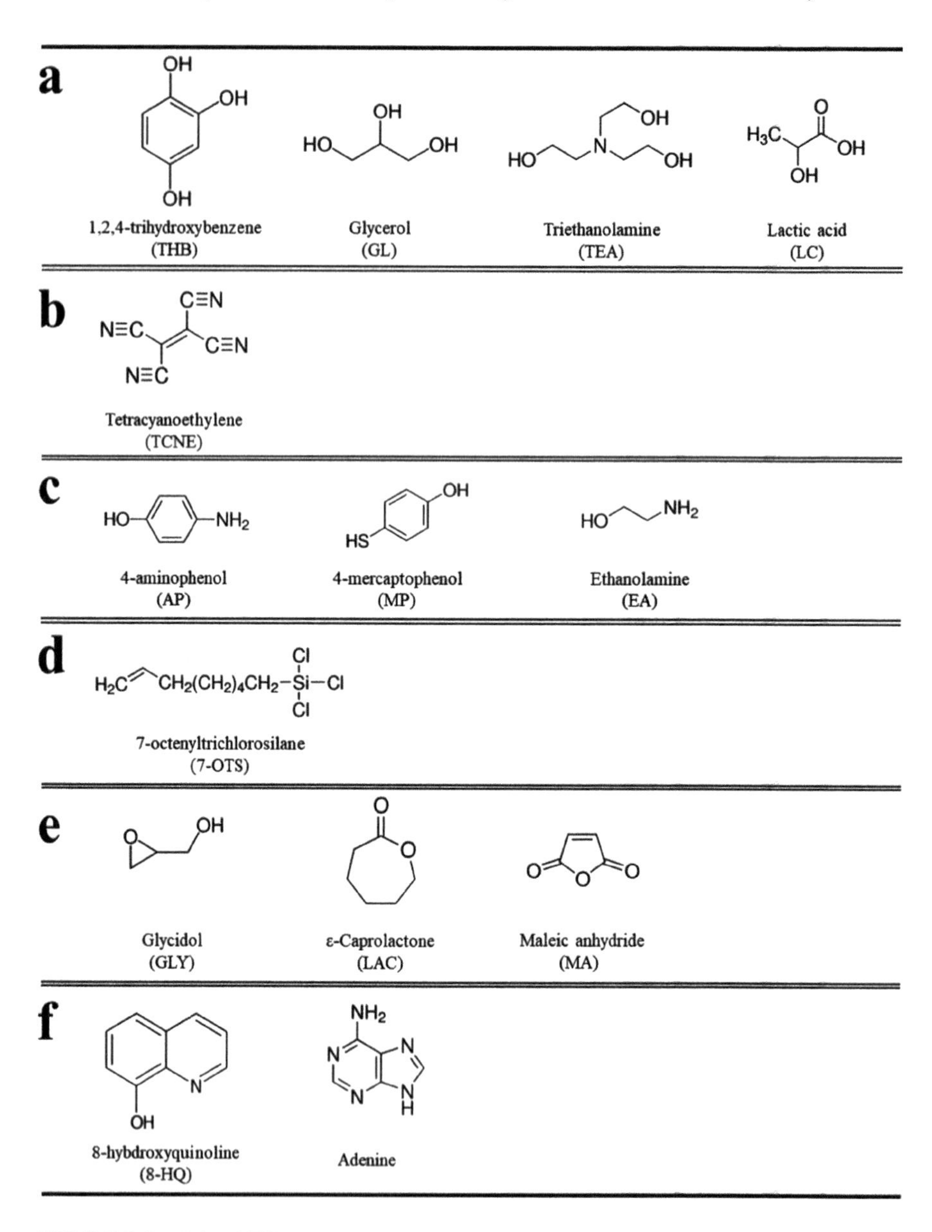

FIGURE 3.4 Other MLD precursors for organic–inorganic hybrid films, featuring the reactive groups: (a) three groups, (b) ethylene, (c-f) heterofunctional groups.

molecules might then react with $-AlCH_3$ species both on the alucone polymer surface and on TMA molecules in the alucone film. Dameron et al.[36] also revealed that the resultant MLD alucone films are not stable in air and their thickness decreases with time over the first 150 hours after their fabrication. At the meantime, the films' composition changes, due to either dehydration or dehydrogenation reactions.

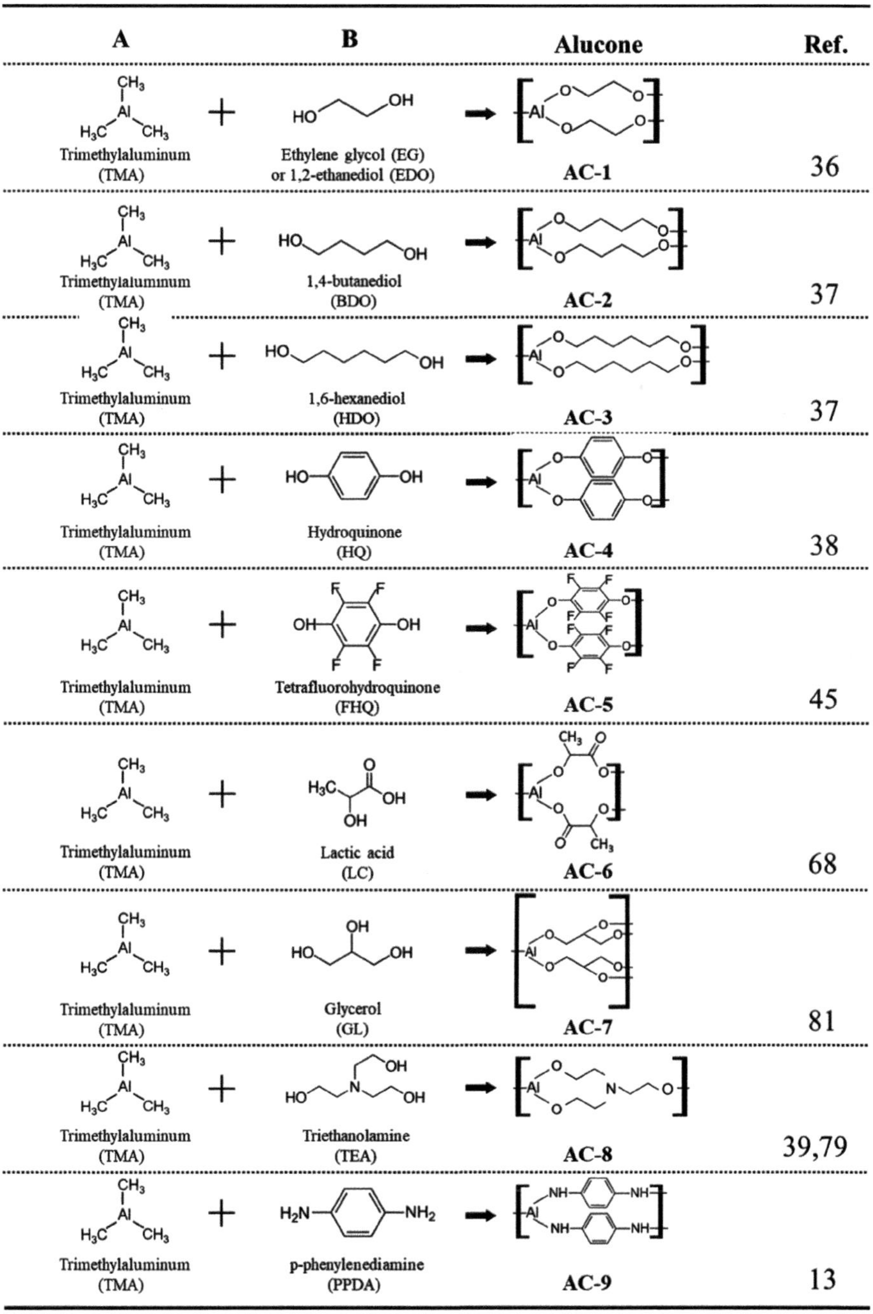

A		B		Alucone	Ref.
Trimethylaluminum (TMA)	+	Ethylene glycol (EG) or 1,2-ethanediol (EDO)	➡	AC-1	36
Trimethylaluminum (TMA)	+	1,4-butanediol (BDO)	➡	AC-2	37
Trimethylaluminum (TMA)	+	1,6-hexanediol (HDO)	➡	AC-3	37
Trimethylaluminum (TMA)	+	Hydroquinone (HQ)	➡	AC-4	38
Trimethylaluminum (TMA)	+	Tetrafluorohydroquinone (FHQ)	➡	AC-5	45
Trimethylaluminum (TMA)	+	Lactic acid (LC)	➡	AC-6	68
Trimethylaluminum (TMA)	+	Glycerol (GL)	➡	AC-7	81
Trimethylaluminum (TMA)	+	Triethanolamine (TEA)	➡	AC-8	39,79
Trimethylaluminum (TMA)	+	p-phenylenediamine (PPDA)	➡	AC-9	13

FIGURE 3.5 MLD processes for alucones by coupling TMA with homobifunctional precursors.

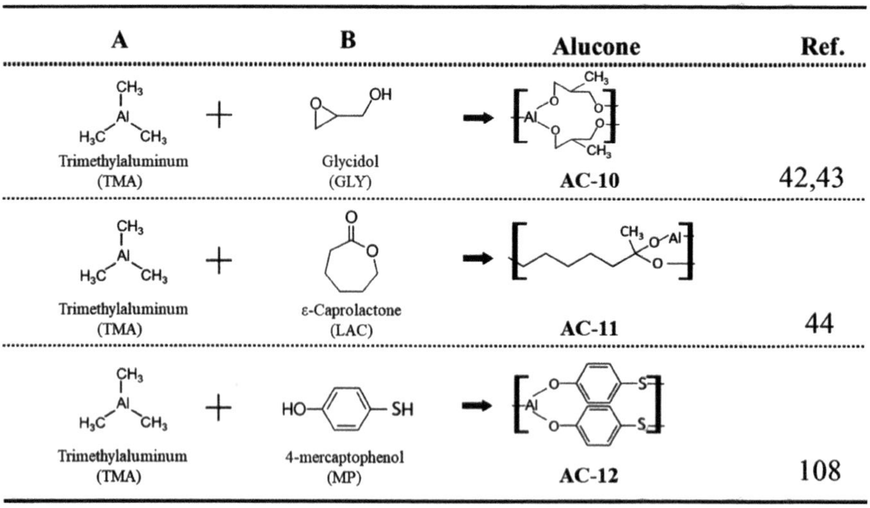

A		B		Alucone	Ref.
Trimethylaluminum (TMA)	+	Glycidol (GLY)	→	AC-10	42,43
Trimethylaluminum (TMA)	+	ε-Caprolactone (LAC)	→	AC-11	44
Trimethylaluminum (TMA)	+	4-mercaptophenol (MP)	→	AC-12	108

FIGURE 3.6 MLD processes for alucones by coupling TMA with heterofunctional precursors.

Similarly, BDO (see Figure 3.5, AC-2) and HDO (see Figure 3.5, AC-3) also were used to couple with TMA for growing alucones.[37] Park et al.[37] revealed that, due to the longer carbon chains of BDO and HDO than that of EG, they are prone to cause double reactions, that is, the two ends of EG both reacted with TMA species ($-Al(CH_3)_2$). Consequently, the resultant MLD films are more possible with holes formed by the double reactions.

Different from the flexible chain structure of EG, HQ is a homobifunctional diol but has a rigid structure. Choudhury and Sarkar[38] studied the MLD growth of TMA-HQ (see Figure 3.5, AC-4) using *in situ* FTIR and QCM and disclosed a linear growth in the range of 150–225 °C with a GPC of 4.1 Å/cycle at 150 °C and 3.5 Å/cycle at 225 °C. They also revealed that the TMA-HQ films are prone to degrade in air, but an ALD-Al_2O_3 capping layer can help improve their stability. With the same backbone as FQ, FHQ (Figure 3.5, AC-5) was also reported but the related MLD was not well studied.[45] In addition, PPDA has the same backbone as HQ but features two amines instead of hydroxyls. Zhou et al.[13] studied the TMA-PPDA growth (Figure 3.5, AC-9) and revealed that the MLD process of TMA-PPDA exhibited a GPC of 1.4 Å/cycle at 400 °C. In addition, the resultant TMA-PPDA films are air-sensitive and show a severe increase by 30% in thickness when exposed to air for 2 weeks.[13] Given the poor stability of the TMA-PPDA films in air, Zhou alloyed this alucone with ALD-Al_2O_3 and the resultant nanolaminates exhibited improved stability in air.[13] In addition, the 1:4 alucone/Al_2O_3 nanolaminates showed tunable electrical properties.[13]

TEA as a homobifunctional precursor was first preliminarily studied by Bahlawane et al.,[39] but was investigated with more efforts by Lemaire et al. (see Figure 3.5, AC-8).[79] It was revealed that the resultant amine-containing alucone showed a decreasing growth tendency from 6.7 Å/cycle at 150 °C to 0.8 Å/cycle at 195 °C.

In addition to the organic precursors discussed above, there are also some other homofunctional precursors exposed in literature, that is, LC (Figure 3.5, AC-6)[68] and GL (Figure 3.5, AC-7).[81]

3.3.1.2 Heterobifunctional Organic Precursors

Homobifunctional organic reactants (such as EG, BDO, and HDO) typically undergo a symmetric "double-end" surface reaction. This parasitic reaction process decreases the density of reactive sites available for the subsequent half-reaction and results in slow growth rates and poor material stability.[42] To this end, heterobifunctional reactants and ring-open reaction were suggested.[12] In this regard, Yoon et al.[41] first reported an ABC MLD process using TMA, EA, and MA as precursors, in which a heterobifunctional precursor EA and a ring-open reaction of MA were employed, as illustrated in Figure 3.7. The related surface chemistry could also be described as follows[41]:

$$|-OH + Al(CH_3)_3(g) \rightarrow |-O-Al(CH_3)_2 + CH_4(g) \tag{3.5A}$$

$$|-O-Al(CH_3)_2 + 2NH_2CH_2CH_2OH(g) \rightarrow |-OAl(OCH_2CH_2NH_2)_2 + 2CH_4(g) \tag{3.5B}$$

$$\begin{aligned} &|-OAl(OCH_2CH_2NH_2)_2 + C_4H_2O_3(g) \rightarrow |-OAl \\ &(OCH_2CH_2NH-COCHCHCOOH)_2(g) \end{aligned} \tag{3.5C}$$

One end of the heterobifunctional reactant might react preferentially to avoid the double reaction. Likewise, ring-open reactants might react and yield a new functional group upon ring-opening that did not react with the initial surface species. In the temperature range of 90–170 °C, Yoon et al.[41] monitored this three-step alucone film using *in situ* FTIR and found that the three sequential surface reactions displayed self-limiting growth, showing a decreasing growth trend with temperature from 24 Å/cycle at 90 °C to 4.0 Å/cycle at 170 °C. Using QCM in a subsequent study, the team revealed that there was a significant diffusion of TMA into the afore-deposited ABC films, contributing to an extraordinary mass gain.[82] It seemed that the pre-deposited ABC films have acted as a TMA reservoir during the TMA exposures. It was found that there were three regions for the three-step ABC film growth. The first region (~10 cycles on Al_2O_3 surface) was the formation region, in which the ABC film first reached a threshold thickness. In the second region, then, the ABC film growth showed an increase in mass gain per cycle (about another 20 cycles on Al_2O_3 surface). Once the ABC film was thicker than the TMA diffusion distance, the GPC enabled a constant steady state, that is, the third region (since the 30th cycle on Al_2O_3 surface). The diffused TMA molecules were believed to react with the following exposures of EA molecules, and the reaction was regarded as a CVD process. The CVD reaction most likely occurs close to the surface of the ABC film, for EA did not diffuse into the ABC film. In contrast, the third reaction of EA–MA was regarded only as a surface reaction, for MA was a much larger molecule and unable to diffuse into the ABC film. Seghete et al.[82] also disclosed that extended purging times were helpful to reduce the diffused TMA in the ABC film. Increasing growth temperatures higher than 130 °C is

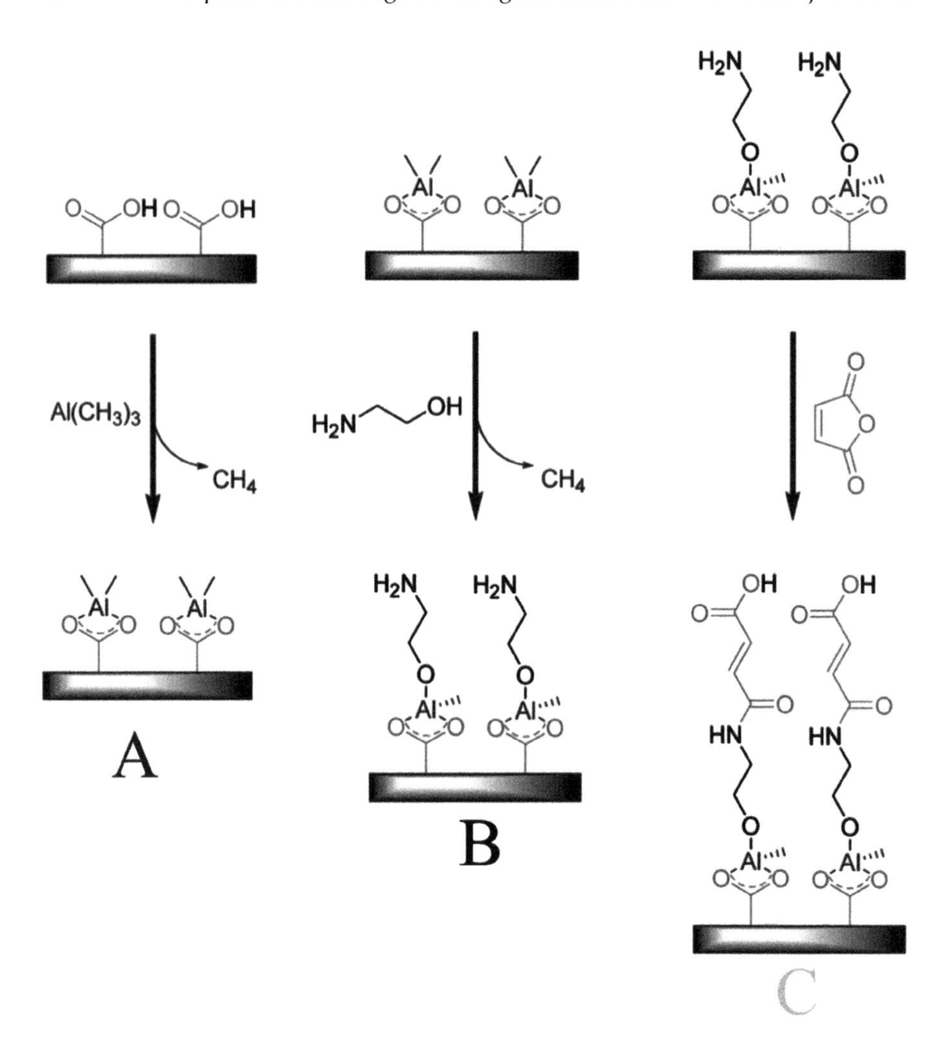

FIGURE 3.7 Schematic illustration of the three-step reaction sequence of the ABC alucone growth using (A) trimethylaluminum (TMA), (B) ethanolamine (EA), and (C) maleic anhydride (MA).[41] Reprinted with permission from Ref. 41. Copyright (2009) American Chemical Society.

another effective route to eliminate TMA diffusion into the ABC film. Furthermore, Seghete et al.[82] examined the ABC film's stability using XRR at 90 °C, disclosing that the ABC film has the largest change in the first 50–70 hours and then no change in thickness after 300 hours. They found that, after aging in air, the ABC film thickness decreased by 5.8%, in comparison to a decrease of 20% with the TMA-EG film.

In addition to the afore-discussed ABC MLD process, binary MLD processes have also been reported. Coupling GLY with TMA (see Figure 3.6, AC-10), for instance, two groups reported the same MLD process for an alucone independently almost at the same time.[42,43] In one of the two studies, Gong et al.[42] believed that, during the exposure of GLY, a ring-opening/transalkylation reaction occurred to produce (Al-O-CH_2-CH(CH_3)-CH_2-OH) or (Al-O-CH_2-CH(CH_2-CH_3)-OH) groups. The GLY could bond through Lewis acid/base interaction and by methyl elimination to form Al-O-C- bonds. The remaining hydroxyls then reacted with TMA during the next step. Despite its heterobifunctional structure, however, GLY was not successful in preventing double reactions from occurrence. Gong et al. [42] revealed that the TMA-GLY MLD decreased from 24 Å/cycle at 90 °C to 6 Å/cycle at 150 °C. They also disclosed that the resultant films of TMA-GLY were stable in air and had little change at an annealing temperature of 100 °C for 2 hours. However, the as-grown films lost hydroxyl groups at 200 °C and most of C-H species at 300 °C, associated with a film shrinkage during the annealing process. In another independent study, in comparison, Lee et al.[43] reported much lower growth rates of the TMA-GLY MLD, for example, a GPC of 1.3 Å/cycle at 125 °C. They noticed that the growth of TMA-GLY was very sensitive to the purging durations and shorter purges could dramatically increase GPCs. This may underlie the different GPCs reported by Lee et al.[43] and Gong et al.[42]. Interestingly, Lee et al.[43] proposed a different mechanism responsible for the TMA-GLY MLD. They believed that the hydroxyl group of GLY reacted with $AlCH_3$ surface species and produced alkoxy aluminum surface species (Figure 3.8). In addition, oxygen atom in the epoxy ring of GLY could coordinate to the Al atom via Lewis acid/base interactions (Figure 3.8b). This Lewis acid/base interaction weakened the C3-oxygen bond on the epoxy ring. In the subsequent exposure of TMA (Figure 3.8c), TMA could coordinate to the oxygen in the Al-O-C bond through another Lewis acid/base interaction. Methyl transferred from TMA to

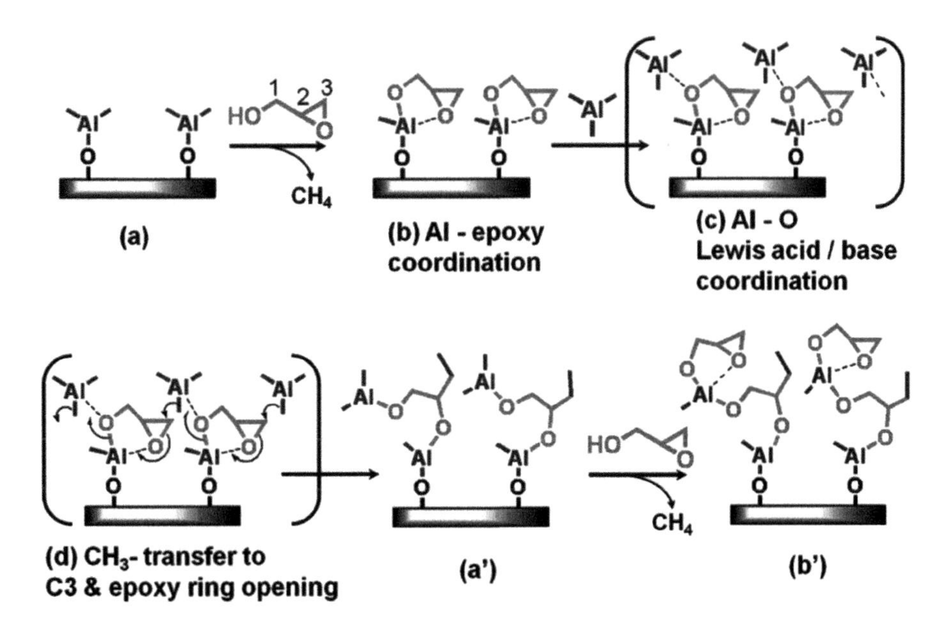

FIGURE 3.8 Proposed surface chemistry for the TMA-GLY reaction assuming a 1:1 TMA/GLY stoichiometry.[43] Reprinted with permission from Ref. 43. Copyright (2011) American Chemical Society.

position C3 on the epoxy ring (Figure 3.8d). This methyl transfer then formed a 1,2-butanediolate as displayed in Figure 3.8a′. Ring-opening after methyl transfer concurrently could form a new Al-O bond between the oxygen in the original epoxy ring and the neighboring Al atom. The proposed reaction sequence could then be repeated by the second GLY exposure, as illustrated in Figure 3.8b′. Lee et al.[43] also believed that Lewis acid/base interactions were important in the growth mechanism. There are two kinds of Lewis acid/base interactions: (1) one between the Al of the $AlCH_3$ surface species and the oxygen in the epoxy ring and (2) another between the Al of the TMA reactant and the oxygen in Al-O-C bond. Lee et al.[43] asserted that both the Lewis acid/base interactions were essential for the methyl transfer reaction from TMA to the epoxy ring. In particular, Lee et al.[43] conducted an additional experiment using DEZ and GLY to verify the importance of Lewis acid/base interactions. They revealed that there was no film growth with the DEZ-GLY, for the Zn atom in DEZ had much less Lewis acidity that is essential for efficient epoxy ring-opening. Lee et al.[43] further confirmed that the TMA-GLY films annealed at 300 or 500 °C for 24 hours had no remaining carbon, and had turned into porous Al_2O_3 films of 27% porosity, in terms of the density of ALD Al_2O_3 films.

Another heterofunctional precursor for MLD alucones is LAC (see Figure 3.6, AC-11) reported by Gong and Parsons.[44] They revealed that the growth of TMA-LAC exhibited a decreasing tendency, accounting for 0.75 Å/cycle at 60 °C and 0.08 Å/cycle at 120 °C. Gong and Parsons[44] explained the growth mechanism that, based on FTIR measurements, an Al-CH_3 Lewis acid site could catalyze the (C=O)-O-C ring-opening to form Al-O and C-CH_3 through methyl transfer from the aluminum and then the Al-O-C group was accessible for reaction during the next TMA exposure. Encouragingly, the resultant TMA-LAC films showed excellent stability in air over 30 days.

In a latest work, Baek et al. used MP to couple with TMA for growing an alucone (see Figure 3.6, AC-12) in the range of 100–200 °C.[108] MP is a heterofunctional organic precursor, featuring one -OH ligand and one -SH ligand on its two ends. The resultant alucone has a GPC of 2.0–2.5 Å/cycle in the temperature range. Very importantly, Baek et al. found that, different from other metalcones, this alucone are very stable in the atmospheric and humid air conditions. In addition, after annealing under vacuum at 300–750 °C, the annealed alucone films showed thermal polymerization and their carbon ring structures transformed into graphitic carbon flakes with improved electrical conductivity.

3.3.2 Titanicones

Titanicones are another type of popular metalcones investigated to date. There are two Ti sources used for growing titanicones, that is, $TiCl_4$ and TDMATi (see Figure 3.9). The organic precursors are EG, GL, TEA, and FA used in binary MLD processes (see Figure 3.9). In addition, there is a four-step ABCD MLD process for a titanicone using the sequence of $TiCl_4$-EA-MC-EA.[59]

Coupling $TiCl_4$ with EG (Figure 3.9, TC-1) and GL (Figure 3.9, TC-2), respectively, Abdulagatov et al. made a comparative study on the two MLD processes.[56] They revealed that the $TiCl_4$-EG MLD has a constant GPC of 4.5 Å/cycle in the range of 90–115 °C while the GPC decreases to 1.5 Å/cycle at 135 °C. In comparison, the

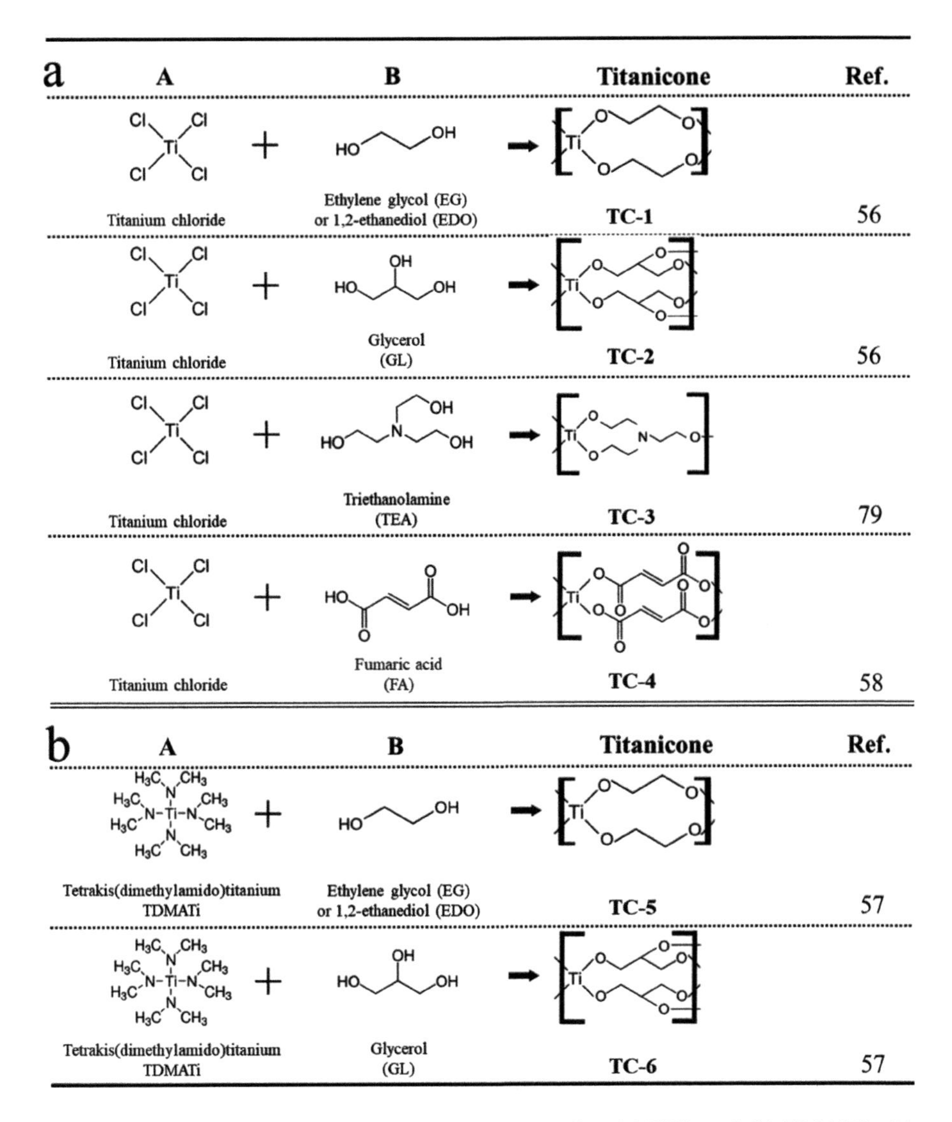

FIGURE 3.9 MLD processes for titanicones by coupling (a) $TiCl_4$ and (b) TDMATi with organic precursors.[8] Reprinted with permission from Ref. 8. Copyright (2017) The Royal Society of Chemistry.

$TiCl_4$-GL MLD needs higher temperatures of 130–210 °C and its GPC decreases from 2.8 Å/cycle at 130 °C to 2.1 Å/cycle at 210 °C. Working on 500-nm thick films with nanoindentation, Abdulagatov et al.[56] disclosed that the $TiCl_4$-EG film has an elastic modulus of ~8 GPa and a hardness of ~0.25 GPa while the $TiCl_4$-GL film has an elastic modulus of ~30 GPa and a hardness of ~2.62 GPa. It was further found that, compared to the $TiCl_4$-EG film, the $TiCl_4$-GL film has a much better thermal stability up to 250 °C, probably due to their higher network connectivity.

In another MLD process, Lemaire et al.[79] used TEA to couple with $TiCl_4$ (see Figure 3.9, TC-3) and compared the resultant titanicone with the alucone from the TMA-TEA MLD. They disclosed that, compared to the GPC of alucone that is 6.7 Å/cycle at 150 °C and 0.8 Å/cycle at 195 °C, the resultant titanicone enables a GPC of 5.2 Å/cycle at 150 °C and 2 Å/cycle at 195 °C. Lemaire et al.[79] also revealed that the TMA-TEA is more stable in ambient conditions, accounting for an increase by 5.1% after 24 hours and no more change over 500 hours versus a decrease of the $TiCl_4$-TEA films by 6.2% after 24 hours and 14.1% after 500 hours in air. In addition, Lemaire et al.[79] found that humidity has a significant impact on the $TiCl_4$-TEA films. The film thickness increases significantly from 207 to 346.9 nm with relative humidity from 10% to 65%. Particularly, the shrinking and welling behavior is reversible and repeatable unless the relative humidity exceeds ~65%, which leads to film degradation. In contrast, the TMA-TEA film thickness has much smaller change with humidity, having a change of ~3% with relative humidity from 10% to 65%. In another work,[58] Cao et al. reported $TiCl_4$ and FA as precursor for a hybrid Ti-polymer (see Figure 3.9, TC-4), showing a decreasing GPC of 1.1 Å/cycle at 180°C and 0.49 Å/cycle at 300 °C.

Alternatively, TDMATi has been used as a Ti precursor to couple with EG and GL, respectively. In a work, Van de Kerckhove et al.[57] studied the resultant MLD processes for growing titanicones. The researchers found that the TDMATi-EG MLD (Figure 3.9, TC-5) terminates after a few of cycles, due to severe double reactions of EG. In comparison, the TDMATi-GL MLD process (Figure 3.9, TC-6) can work efficiently and sustain self-limiting growth in the temperature range of 80–160 °C with a linearly decreasing growth tendency from 0.95 to 0.24 Å/cycle. Van de Kerckhove et al.[57] has attributed the decreasing GPCs to the aggravated desorption of TDMATi with temperature.

3.3.3 Zincones

Zincones are another type of metalcones resulting from zinc reactants and organic precursors. DEZ is dominantly used as the Zn source while there have six precursors reported to date, including EG, HQ, THB, HDD, GL, and AP (Figure 3.10a).

The first zincone MLD was reported by Yoon et al., using DEZ and EG as precursors (Figure 3.10a, ZC-1).[46] This MLD process showed a decreasing GPC from 4 Å/cycle at 90 °C to 0.25 Å/cycle at 170 °C. In surface chemistry, this process is similar to that of the TMA-EG and the DEZ-EG MLD processes. There is also some diffusion of DEZ into the zincone film. The diffused DEZ molecules can react with the subsequent EG exposure to form new zincone polymer chains. Furthermore, FTIR data revealed little O-H stretching vibration at 170 °C but evident O-H stretching vibration at 90 °C.[46] This suggested that most of the EG molecules reacted twice to $–ZnCH_2CH_3$ at 170 °C. Particularly, XPS analyses on zincone samples showed a decreasing Zn content with temperature, changing from 10.7% at 90 °C to 10.6, 9.9, 8.9% at 110, 130, and 150 °C, respectively. Furthermore, Yoon et al.[46] verified that the DEZ-EG films adsorb water after exposure to air and then become very stable for multiple weeks. The DEZ-EG MLD process has also been investigated by Peng et al.[47] They confirmed a decreasing GPC with temperature in the range of 100–170 °C but reported larger GPCs. Additionally, they also verified that the films are stable in dry air for

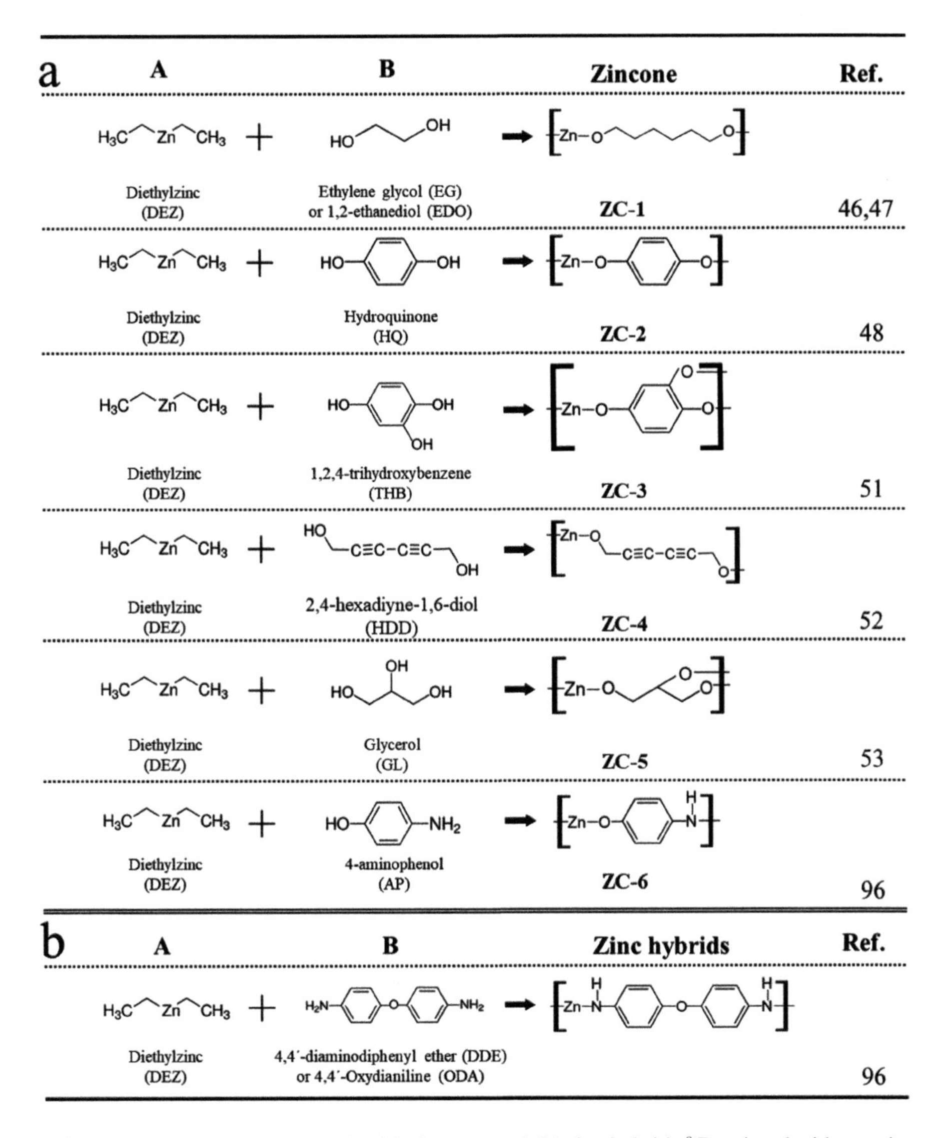

FIGURE 3.10 MLD processes for (a) zincones and (b) zinc hybrids.[8] Reprinted with permission from Ref. 8. Copyright (2017) The Royal Society of Chemistry.

3 days. Peng et al.[47] also observed that the DEZ-EG films are prone to adsorb water under ambient conditions for 1 hour but have no more changes up to 12 hours. This indicates that the reaction between the DEZ-EG films and moisture from ambient environment proceeded quickly, resulting in hydrolyzed DEZ-EG films. A further annealing on the hydrolyzed DEZ-EG films at 100 °C for 2 hours produced a structure similar to the ALD ZnO.

Another MLD process of zincones is DEZ-HQ. Due to the conjugated structure of HQ, the resultant DEZ-HQ was expected to produce electrically conductive polymers.[48] In addition, the rigid structure of HQ was expected to reduce the possibility of double reactions. Yoon et al. studied the DEZ-HQ MLD (see Figure 3.10a, ZC-2) in the temperature range of 130–170 °C and revealed a linear growth with a GPC of 1.6 Å/cycle at 150 °C.[48] Subsequently, Yoon et al. explored conductive hybrid materials by combining ALD-ZnO with the MLD zincone of DEZ-HQ for various nanolaminates (alloys).[49,50] They disclosed that, in comparison to the conductivity of 14 S/cm of ALD ZnO, the ZnO:zincone alloys of 1:1 and 2:2 cycle ratios yielded conductivity of 116 and 170 S/cm, respectively. Thus, these studies paved a venue to tune electrical conductivity of hybrid materials through combining ALD with MLD rationally. Other researchers also confirmed that the alloys of ALD-ZnO/MLD-zincone are tunable in electrical properties.[85,86]

Other zincones include DEZ-THB (Figure 3.10a, ZC-3),[51] DEZ-HDD (Figure 3.10a, ZC-4),[52] DEZ-GL (Figure 3.10a, ZC-5),[53] DEZ-AP (Figure 3.10a, ZC-6).[96] The DEZ-THB could realize a GPC of 2.6 Å/cycle in the range of 100–160 °C and the DEZ-HDD process exhibited a GPC of 5.2 Å/cycle in the range of 100–150 °C. Particularly, the DEZ-HDD films showed excellent stability in air up to 400 °C. There also has a zinc-based hybrid material reported using DEZ-ODA (Figure 3.10b).[96]

3.3.4 Other Metalcones

Additionally, some other metalcones have also been reported, including zircones, hafnicones, mangancones, and vanadicones (Figure 3.11). One MLD process of zircons was fulfilled using ZTB and EG as precursors (Figure 3.11a),[54,55] exhibiting a decreasing GPC from 1.6 Å/cycle at 105 °C to 0.3 Å/cycle at 195 °C. Using TDMAHf as the hafnium (Hf) precursor to couple with EG, Lee et al.[60] succeeded a hafnicone MLD (see Figure 3.11b), having a decreasing GPC from 1.2 Å/cycle at 105 °C to 0.4 Å/cycle at 205 °C. In addition, a MLD process of mangancones has been reported using $Mn(CpEt)_2$ and EG as precursors (see Figure 3.11c).[45] Furthermore, two processes have been reported in a study[61] for vanadicones, using either EG or GL to couple with TEMAV (Figure 3.11d). In the study, it was found that the TEMAV-EG couple was not successful for growing vanadicone, possibly due to double reactions of EG. In contrast, the TEMAV-GL couple (see Figure 3.11d) enabled a reliable process for vanadicones, showing a linearly decreasing growth tendency from ~1.2 Å/cycle at 80 °C to ~0.5 Å/cycle at 180 °C.

More recently, $Cu(dmap)_2$ was used to grow Cu-based metalcones (Cu-cones) with either HQ or BDC (or TPA) (see Figure 3.11e) via MLD.[109] The Cu-HQ metalcone (Figure 3.11e, CC-1) was conducted in the range of 100–120 °C with a GPC of 1.0–1.5 Å/cycle while the Cu-BDC metalcone (Figure 3.11e, CC-2) was performed at 160 °C with a GPC of 2.6 Å/cycle. Furthermore, $Cu(dmap)_2$ also has been used to couple with three other organic precursors for growing other hybrid materials via MLD in the same study. The three other organic precursors are PPDA, ODA, and BDT (see Figure 3.3). The Cu-ODA was deposited in the range of 140–220 °C with a constant GPC of 2.4 Å/cycle and the Cu-PPDA hybrid was investigated in the range of 70–200 °C with a constant GPC of 1.7 Å/cycle. The Cu-BDT was studied in the range

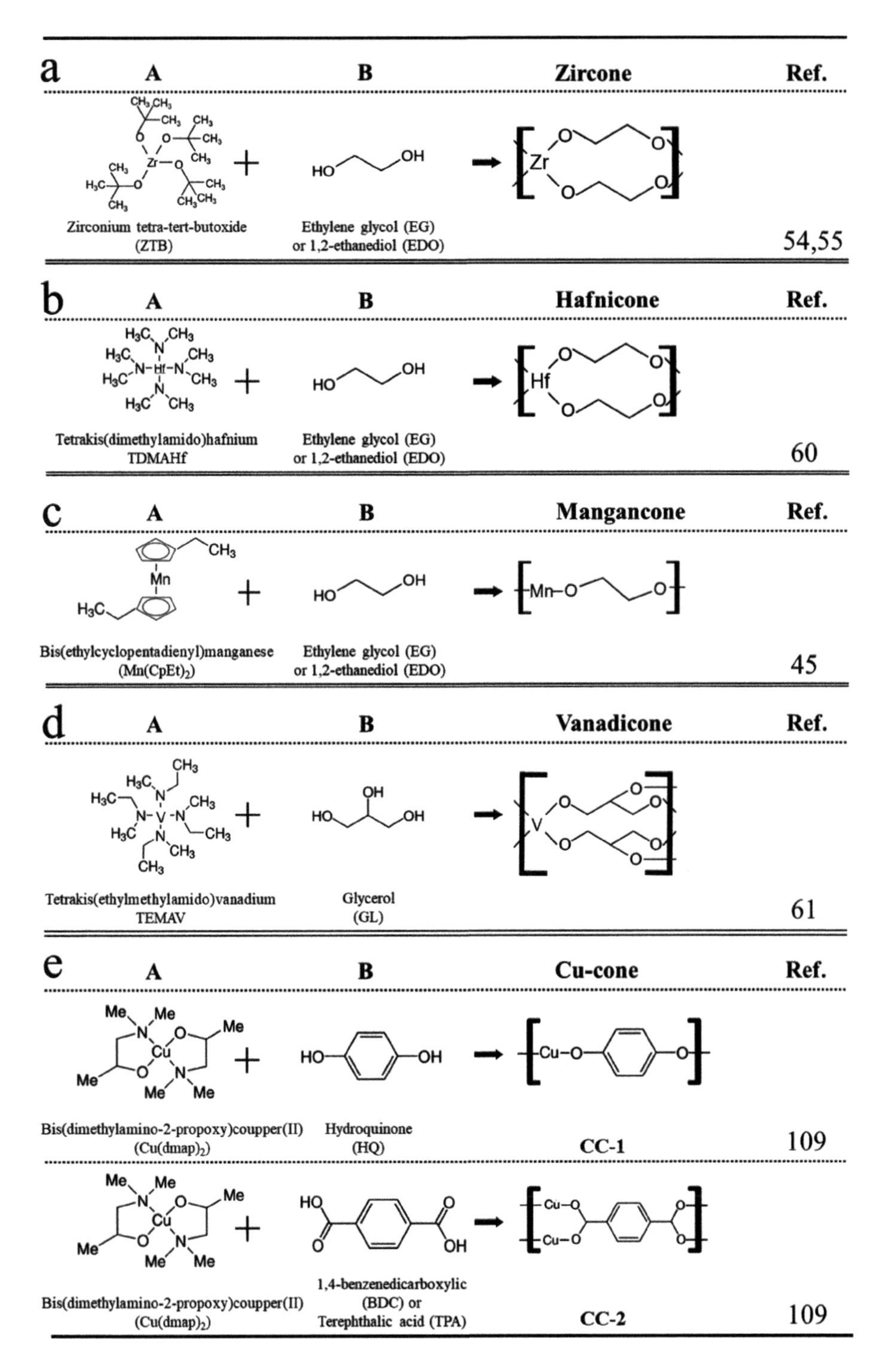

FIGURE 3.11 MLD processes for (a) zircones, (b) hafnicones, (c) mangancones, (d) vanadicones, and (e) Cu-cones.

of 80–140 °C, showing a decreasing GPC from 1.9 Å/cycle at 80 °C to 0.8 Å/cycle at 140 °C. All these Cu-based hybrid materials are amorphous or nanocrystalline.

3.4 OTHER HYBRID MATERIALS

Besides metalcones discussed above, some other types of hybrid materials have also been reported, as summarized in Figures 3.12 and 3.13 using one metal precursor and one organic precursor. In addition, there have been some MLD processes using multiple precursors for growing hybrid materials, which provide alternative solutions for tunable fabrication of novel polymeric films, as summarized in Figure 3.14.

3.4.1 Luminescent Hybrid Materials

Metal quinolones (Mq_x) are very promising hybrid luminescent materials. Nilsen et al.[87,110] first developed MLD processes of Mq_x. Using 8-HQ to couple with TMA, DEZ, or $TiCl_4$, Nilsen et al.[87] synthesized Alq_3, Znq_2, or Tiq_4, respectively (see Figure 3.12a). The three metal quinolones exhibit decreasing GPCs with temperature, 4–7 Å/cycle at 80 °C but no growth at 200 °C. The Alq_3 was examined for stability at 85 °C and it was found that the Alq_3 is stable to water at 85 °C.[87] In a later study, Räupke et al. demonstrated the luminescent performance of MLD Alq_3 films.[88]

Subsequently, another team studied luminescent hybrid materials using three metal precursors (i.e., Na(thd), $Ba(thd)_2$, and $La(thd)_3$) and two organic precursors (uracil and adenine).[111,112] For the metal precursors, thd is 2,2,6,6-tetramethyl-3,5-heptadionate. Uracil is one of the four nucleobases (NBs) in the nucleic acid of ribonucleic acid (RNA) while adenine is one of the four NBs in the nucleic acid of deoxyribonucleic acid (DNA). Among them, the $Ba(thd)_2$-uracil pair has been well studied and exhibited a nearly constant GPC of ~2.8 Å/cycle in the range of 260–320 °C. They found that the metal precursors have evident influence on the crystallinity of the resultant luminescent hybrid metal-NB materials. Specifically, the La-NBs are amorphous, the Ba-NBs are at least partially crystalline, and the Na-NBs are well crystalline. These metal-NB thin films were tested for their luminescent properties. Particularly, both the Na-NB and Ba-NB films show intense photoluminescence in blue and green wavelengths. This work has significant implications for MLD-deposited hybrid materials in luminescent applications.

Another luminescent work was recently conducted on erbium-based hybrid materials. Using $Er(DPDMG)_3$ (Figure 3.2) with PDA (Figure 3.3) as precursors, Mai et al. developed an MLD process for growing Er-di-3,5-pyridinecarboxylate [$Er_2(3,5\text{-}PDC)_3$] films.[113] In the range of 250–265 °C, this MLD process exhibits a constant GPC of ~6.4 Å/cycle. The resultant $Er_2(3,5\text{-}PDC)_3$ films were confirmed for their promising UV absorption properties and a photoluminescence at 1535 nm for a 325-nm excitation.

3.4.2 Metal-Organic Frameworks (MOFs)

Metal-organic frameworks (MOFs), also known as porous coordination polymers (PCPs), are crystalline coordination polymers built from inorganic nodes (bridging ligands) and organic linkers.[114,115] One of the most remarkable features of MOFs is

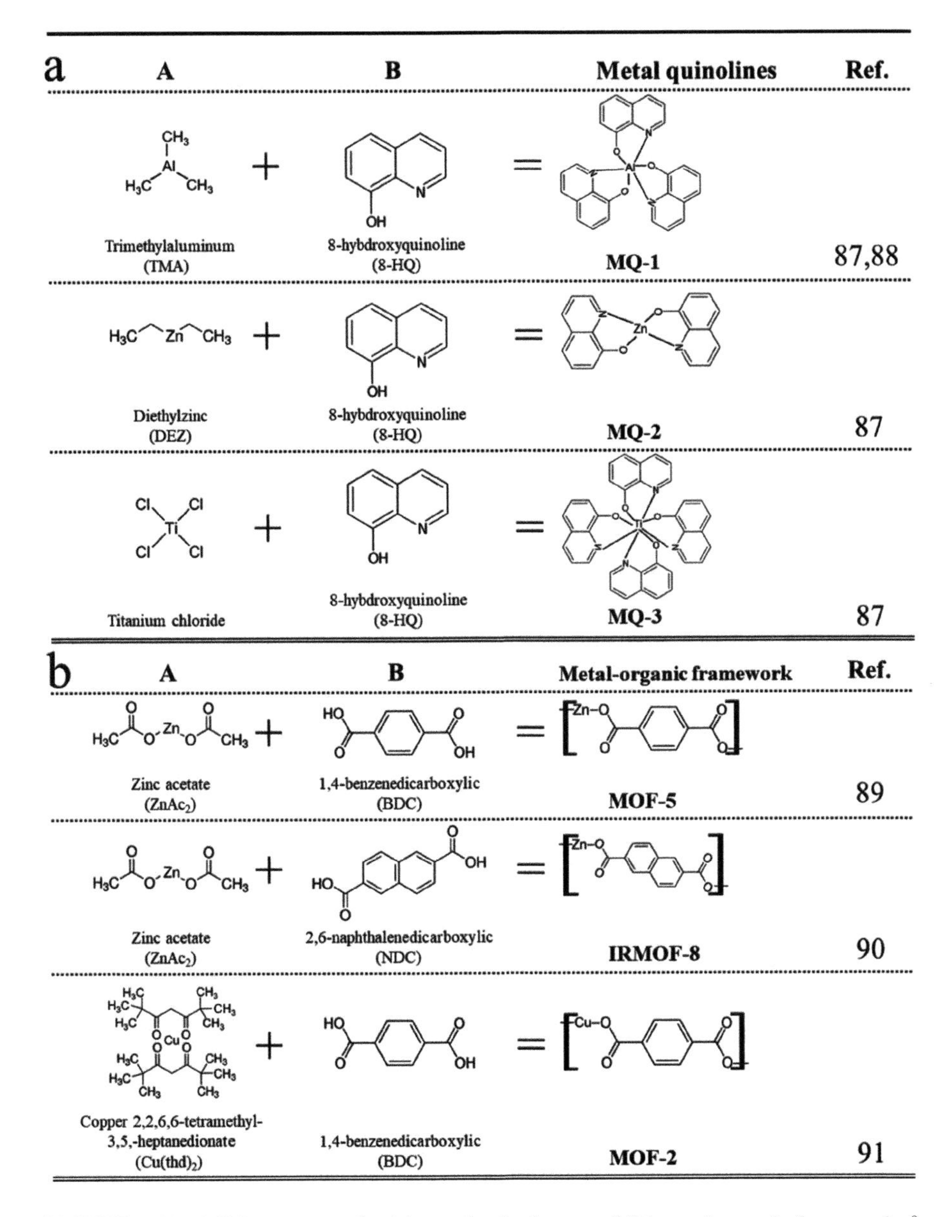

FIGURE 3.12 MLD processes for (a) metal quinolones and (b) metal-organic frameworks.[8] Reprinted with permission from Ref. 8. Copyright (2017) The Royal Society of Chemistry.

their extremely high porosity, consequently featuring their high surface areas and high pore volumes in uniformly tailorable pores. In addition, MOF's mechanical and chemical properties are also tunable by carefully selecting the linker molecule or introducing guest molecules into the pores. These unique characteristics make MOFs very promising in many applications, such as gas storage and separation,[116–119]

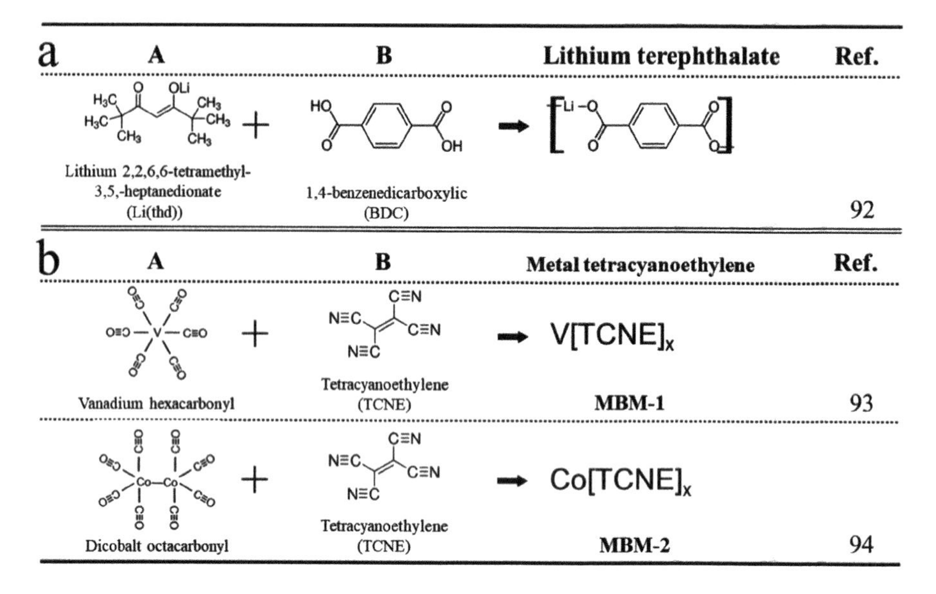

FIGURE 3.13 MLD processes for (a) lithium terephthalate and (b) metal tetracyanoethylene.[8] Reprinted with permission from Ref. 8. Copyright (2017) The Royal Society of Chemistry.

luminescence,[120] drug delivery,[121,122] energy conversion and storage,[123,124] and catalysis.[114,125,126]

Salmi et al.[89] initiated the first study of MLD for growing MOFs (see Figure 3.12b, MOF-5). In the work, the researchers adopted $ZnAc_2$ and BDC as precursors for MOF-5, an isoreticular MOF (also known as IRMOF-1) having a cubic framework structure and consisting of ZnO_4 clusters connected with rigid benzene dicarboxylate linkers. The resultant films exhibited a decreasing tendency in GPC from 6.5 Å/cycle at 225 °C to 4.5 Å/cycle at 325 °C. There is no film growth at 350 °C. Unfortunately, in the whole deposition temperature range, no crystalline phases have been grown. Surprisingly, however, the as-deposited amorphous films could crystalize into an unidentified structure at room temperature under a relative humidity of 60%. This moisture-induced structure was confirmed with no porosity but could be recrystallized into the MOF-5 phase in an autoclave with *N,N*-dimethylformamide (DMF) at 150 °C. In a subsequent study, Salmi et al. [90] used NDC to couple with $ZnAc_2$ for growing IRMOF-8 (Figure 3.12b, IRMOF-8). This MLD of $ZnAc_2$-NDC also exhibited a linear decreasing GPC from 4.9 Å/cycle at 260 °C to 4.0 Å/cycle at 300 °C. Again, the as-deposited films were amorphous but could be crystallized into an unknown structure at room temperature under a relative humidity of 70%. The unknown crystallized structure was further recrystallized in an autoclave with DMF at 150 °C and turned into IRMOF-8. The crystallization in humid air was confirmed being critical, for the as-deposited films could not be directly turned into IRMOF-8 only via the second crystallization. Furthermore, Salmi et al.[90] verified the films via the recrystallization are porous and they successfully loaded Pd into the porous films uniformly.

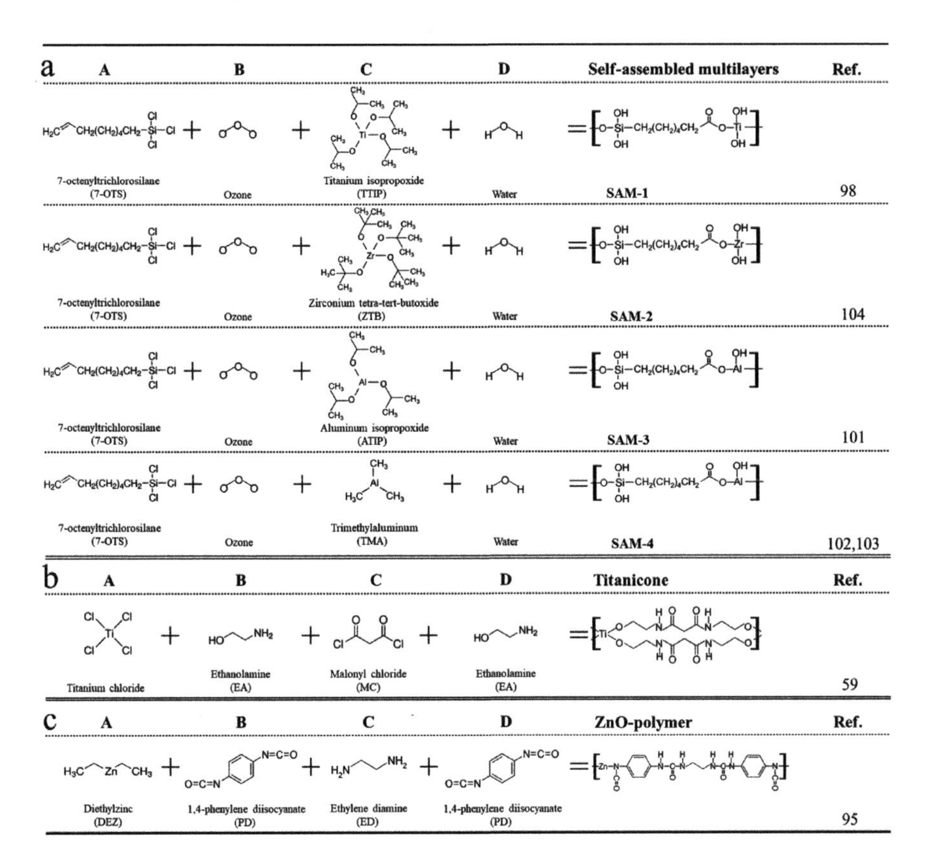

FIGURE 3.14 Four-step ABCD MLD processes for organic–inorganic hybrid materials of (a) self-assembled monolayers, (b) titanicones, and (c) ZnO-polymers.[8] Reprinted with permission from Ref. 8. Copyright (2017) The Royal Society of Chemistry.

A more recent work of MLD MOFs was reported by Ahvenniemi and Karppinen[91] for copper(II)terephthalate (i.e., Cu-TPA, also MOF-2) using $Cu(thd)_2$ and BDC as precursors (see Figure 3.12b, MOF-2). Very encouragingly, this MLD process of $Cn(thd)_2$-BDC can directly grow porous crystalline Cu-TPA films in the range of 180–195 °C. However, a higher temperature of 260 °C led to amorphous films. In the whole temperature range of 180–260 °C, the GPC decreased with temperature from 3 to 0.2 Å/cycle.

3.4.3 Energy-Storage Materials

Using Li(thd) and TPA as precursors, Nisula and Karppinen[92] developed a lithium-containing crystalline hybrid polymer, LiTP, lithium terephthalate (LTP) (see Figure 3.13a). They revealed that the MLD process of LTP sustained a self-limiting growth during the temperature of 200–280 °C with a decreasing tendency from ~4.0 to 1.0 Å/cycle. Electrochemical testing confirmed that these LTP hybrid films are promising anodes in lithium-ion microbatteries. Subsequently, this team further synthesized hybrid materials of metal-TP, using $Ca(thd)_2$, Na(thd), K(thd), $Mg(thd)_2$, $Sr(thd)_2$, $Ba(thd)_2$, and $La(thd)_3$ to couple with TPA as precursors, respectively.[127,128]

3.4.4 Organic Magnets

Organic molecule-based magnets (MBMs) are a relatively new class of magnetic materials and first emerged in 1985.[129,130] Compared to conventional metallurgic and ceramic magnets, the main benefits of MBMs are usually associated with their lightweight, mechanical flexibility, tunable color or transparency, low-temperature processing, solubility, and compatibility with polymers and other classes of molecular materials.[129] The M[TCNE] (M = 3d metal; TCNE = tetracyanoethylene) complexes represent one of the most interesting classes of MBMs, possessing numerous compositions and structures with varying dimensionalities of magnetic coupling from one-dimensional (1D) inorganic polymer chains and two-dimensional (2D) layers to three-dimensional (3D) networks and amorphous solids.[129,130] Kao et al. reported an MLD attempt to synthesize an MBM, $V[TCNE]_x$ using tetracyanoethylene (TCNE) and vanadium hexacarbonyl ($V(CO)_6$) at room temperature (Figure 3.13b, MBM-1).[93] $V[TCNE]_x$ is a promising MBM for the future organic/molecular electronic, magnetic, and spintronic applications. They described the MLD process of $V(CO)_6$-TCNE as follows:

$$|-\mathrm{TCNE}+x\mathrm{V}(\mathrm{CO})_6(g)\rightarrow|-\mathrm{TCNE}-\left(\mathrm{V}(\mathrm{CO})_y\right)_x+x(6-y)\mathrm{CO}(g) \quad (3.6\mathrm{A})$$

$$|-\mathrm{TCNE}-\left(\mathrm{V}(\mathrm{CO})_y\right)_x+z\mathrm{TCNE}(g)\rightarrow|-\mathrm{TCNE}-\mathrm{V}_x-\mathrm{TCNE}_z+xy\mathrm{CO}(g) \quad (3.6\mathrm{B})$$

Kao et al. revealed a linear growth with a GPC of 9.8 Å/cycle. In a subsequent effort, Kao et al. further synthesized two more MBMs using $Co_2(CO)_8$ as the Co source, that is, $Co[TCNE]_x$ (see Figure 3.13b, MBM-2) and $Co_xV_y[TCNE]_z$ at room temperature.[94]

3.4.5 Complex MLD Processes

Not limited to two precursors (one metal precursor and one organic precursor, as shown in Figure 3.1c), there are some MLD processes combining multiple precursors for hybrid materials. In this regard, Sung's group first developed four-step MLD processes for growing self-assembled monolayers (SAMs), in which organic fragments are connected by metal oxides, including TO_x-SAMs (Figure 3.14a, SAM-1),[98] ZrO_x-SAMs (Figure 3.14a, SAM-2),[104] AlO_x-SAMs (Figure 3.14a, SAM-3[101] and SAM-4[102,103]). All the MO_x-SAMs reported exhibited good mechanical flexibility and stability, excellent insulating properties, and relatively high dielectric capacitances with a high dielectric strength, showing promising applications as nanohybrid dielectrics.[99,100,103]

In addition to SAMs, Knez's group reported an MLD process consisting of a four-step procedure of $TiCl_4$-EA-MC-EA to produce a titanicone (Figure 3.14b).[59] Using this coupling combination, Chen et al.[59] demonstrated the high flexibility of MLD processes for a wide range of hybrid materials. At 100 °C, this four-step MLD process receives a growth rate of 6 Å/cycle. The resultant Ti-hybrid polymer films were later used for producing nitrogen-doped porous TiO_2 nanotubes for improved photocatalytic activity. Additionally, Qin's group reported another four-step MLD procedure of DEZ-PD-EDA-PD for a Zn-based hybrid polymer (Figure 3.14c)[95] with a GPC of 5 Å/cycle at 100 °C.

3.4.6 Organic–Inorganic Hybrid Nanolaminates by MLD and ALD

Besides the strategies discussed above, there has been another strategy combining ALD and MLD for organic–inorganic hybrid nanolaminates. Compared to MLD, ALD is a relatively well-developed technique with a large variety of inorganic materials reported to date. Thus, there is an unlimited research space for developing function-oriented nanolaminates through combining ALD and MLD. For instance, combining the MLD process of TiO_x-SAMs with the ALD process of TTIP-H_2O, Lee et al. demonstrated that various TiO_x-SAMs/TiO_2 nanolaminate films (see Figure 3.15) can be accurately realized in a highly tunable fabrication mode.[98] The thickness of SAMs and TiO_2 nanolayers in each sample could be controlled by adjusting the number of MLD and ALD cycles. The GPCs of the MLD SAMs and the ALD TiO_2 GPC are ~11 and 0.6 Å/cycle, respectively. Similarly, the nanolaminates of AlO_x-SAMs/Al_2O_3, AlO_x-SAMs/ZnO, AlO_x-SAMs/Al_2O_3/AlO_x-SAMs/ZnO were also reported.[102]

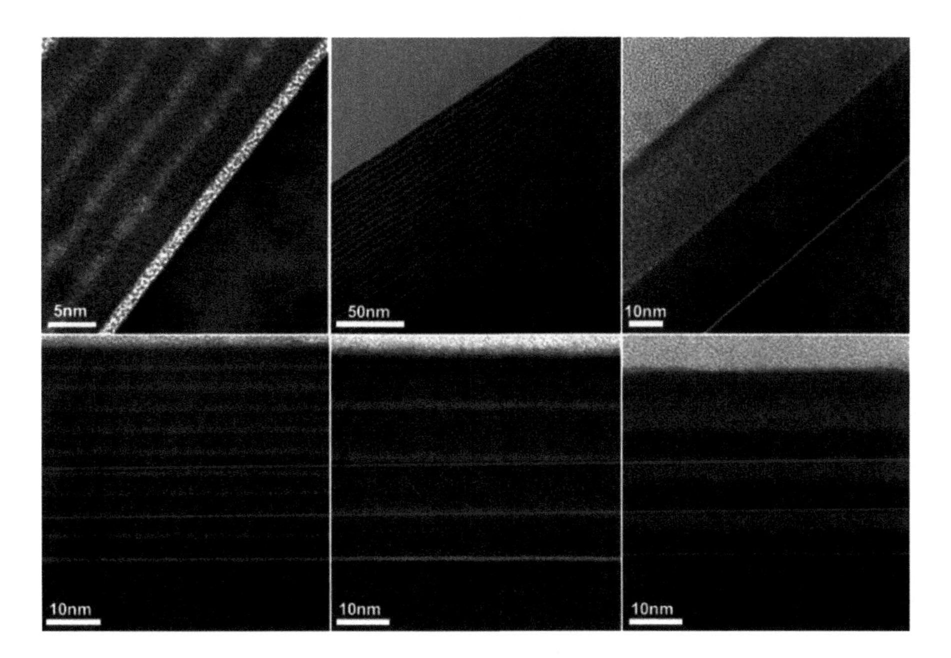

FIGURE 3.15 TEM images of self-assembled organic multilayer/TiO_2 nanolaminate films fabricated using MLD and ALD (white is the SAMs layer and black is the TiO_2 layer).[98] Reprinted with permission from Ref. 98. Copyright (2007) American Chemical Society.

In addition, combining ALD-ZnO (using DEZ and water) and MLD processes of zinc-based polymeric films (DEZ-HQ, DEZ-AP, and DEZ-ODA), a series of nanolaminates of ZnO/zinc-polymeric film have been developed.[96] Along with the increasing needs in new materials, it is expected that more and more hybrid nanolaminates will be reported with exceptional properties.

3.5 CONCLUSIONS

This work focuses on summarizing the recent progresses of MLD in developing inorganic–organic hybrid materials. MLD exhibits highly flexible capabilities in synthesizing novel hybrid materials with exceptional properties. Firstly, MLD can develop hybrid materials through combining one ALD metal precursor with one MLD organic precursor. To date, most of the hybrid materials have been produced by this strategy. Secondly, MLD can synthesize hybrid materials through combining multiple precursors. In this regard, just a few cases have been reported. Thirdly, MLD can also combine with ALD for an unlimited amount of nanolaminates with controllable thicknesses for inorganic and organic layers. All these have determined that MLD is a very attractive technique for novel materials with desirable properties. These capabilities of MLD will enable providing new solutions to many urgent challenges in many applications.

Although all these exciting progresses have been made with MLD so far, it should be noted that many more efforts are urgently needed for expanding MLD. On one hand, current MLD processes and resultant hybrid films are still unsatisfactory, mainly due to their precursors. The ideal precursors should meet high volatility, high stability, and high reactivity. In addition, the precursors should be able to avoid double-end reactions. On the other hand, the stability of organic–inorganic hybrid materials by MLD to date remains an issue. Most of MLD hybrid materials are sensitive to air, humidity, and/or heat. The instability of MLD-deposited hybrid materials has limited their applications. In comparison, ALD has gained a much wider recognition in many areas, such as new energies,[131–134] catalysis,[135,136] and new materials.[137,138] The last but not the least, MLD to date is restricted to a limited range of applications and has not received wide implementation. With the development of ideal MLD precursors and the improvement of resultant organic materials, we expect that the applications of MLD, in addition to those discussed in this review, will be greatly expanded into the areas of semiconductor,[139] sensor,[140] optics,[141] biology,[142] medical[143,144] and smart materials.[145]

ACKNOWLEDGEMENTS

Funding for this research was provided by the Center for Advanced Surface Engineering, under the National Science Foundation grant no. OIA-1457888 and the Arkansas EPSCoR Program, ASSET III.

REFERENCES

1. Mansoori, G. A.; Soelaiman, T. F., Nanotechnology—an introduction for the standards community. *J. ASTM Int.* 2005, 2 (6), 1–22.
2. James, S. L.; Adams, C. J.; Bolm, C.; Braga, D.; Collier, P.; Friščić, T.; Grepioni, F.; Harris, K. D. M.; Hyett, G.; Jones, W.; Krebs, A.; Mack, J.; Maini, L.; Orpen, A. G.;

Parkin, I. P.; Shearouse, W. C.; Steed, J. W.; Waddell, D. C., Mechanochemistry: Opportunities for new and cleaner synthesis. *Chem. Soc. Rev.* 2012, 41 (1), 413–447.
3. Takacs, L., The historical development of mechanochemistry. *Chem. Soc. Rev.* 2013, 42 (18), 7649–7659.
4. Dahl, J. A.; Maddux, B. L. S.; Hutchison, J. E., Toward Greener Nanosynthesis. *Chem. Rev.* 2007, 107 (6), 2228–2269.
5. Singh, J.; Wolfe, D. E., Review Nano and macro-structured component fabrication by electron beam-physical vapor deposition (EB-PVD). *J. Mater. Sci.* 2005, 40 (1), 1–26.
6. Cai, Z.; Liu, B.; Zou, X.; Cheng, H.-M., Chemical vapor deposition growth and applications of two-dimensional materials and their heterostructures. *Chem. Rev.* 2018, 118 (13), 6091–6133.
7. George, S. M., Atomic layer deposition: An overview. *Chem. Rev.* 2010, 110 (1), 111–131.
8. Meng, X., An overview of molecular layer deposition for organic and organic-inorganic hybrid materials: Mechanisms, growth characteristics, and promising applications. *J. Mater. Chem. A* 2017, 5 (35), 18326–18378.
9. George, S. M.; Lee, B. H.; Yoon, B.; Abdulagatov, A. I.; Hall, R. A., Metalcones: Hybrid organic–inorganic films fabricated using atomic and molecular layer deposition techniques. *J. Nanosci. Nanotechnol.* 2011, 11 (9), 7948–7955.
10. Lee, B. H.; Yoon, B.; Abdulagatov, A. I.; Hall, R. A.; George, S. M., Growth and properties of hybrid organic-inorganic metalcone films using molecular layer deposition techniques. *Adv. Funct. Mater.* 2013, 23 (5), 532–546.
11. Yoshimura, T.; Tatsuura, S.; Sotoyama, W., Polymer films formed with monolayer growth steps by molecular layer deposition. *Appl. Phys. Lett.* 1991, 59 (4), 482–484.
12. George, S. M.; Yoon, B.; Dameron, A. A., Surface chemistry for molecular layer deposition of organic and hybrid organic–inorganic polymers. *Acc. Chem. Res.* 2009, 42 (4), 498–508.
13. Zhou, W.; Leem, J.; Park, I.; Li, Y.; Jin, Z.; Min, Y.-S., Charge trapping behavior in organic-inorganic alloy films grown by molecular layer deposition from trimethylaluminum, p-phenylenediamine and water. *J. Mater. Chem.* 2012, 22 (45), 23935–23943.
14. Sundberg, P.; Karppinen, M., Organic and inorganic–organic thin film structures by molecular layer deposition: A review. *Beilstein J. Nanotechnol.* 2014, 5, 1104–1136.
15. Gregorczyk, K.; Knez, M., Hybrid nanomaterials through molecular and atomic layer deposition: Top down, bottom up, and in-between approaches to new materials. *Prog. Mater Sci.* 2016, 75, 1–37.
16. Yoshimura, T.; Tatsuura, S.; Sotoyama, W.; Matsuura, A.; Hayano, T., Quantum wire and dot formation by chemical vapor deposition and molecular layer deposition of one-dimensional conjugated polymer. *Appl. Phys. Lett.* 1992, 60 (3), 268–270.
17. Yoshimura, T.; Oshima, A.; Kim, D.-I.; Morita, Y., Quantum dot formation in polymer wires by three-molecule molecular layer deposition (MLD) and applications to electro-optic/photovoltaic devices. *ECS Trans.* 2009, 25 (4), 15–25.
18. Yoshimura, T.; Ebihara, R.; Oshima, A., Polymer wires with quantum dots grown by molecular layer deposition of three source molecules for sensitized photovoltaics. *J. Vac. Sci. Technol. A* 2011, 29 (5), 051510.
19. Yoshimura, T.; Ishii, S., Effect of quantum dot length on the degree of electron localization in polymer wires grown by molecular layer deposition. *J. Vac. Sci. Technol. A* 2013, 31 (3), 031501.
20. Kim, A.; Filler, M. A.; Kim, S.; Bent, S. F., Layer-by-layer growth on Ge(100) via spontaneous urea coupling reactions. *J. Am. Chem. Soc.* 2005, 127 (16), 6123–6132.
21. Loscutoff, P. W.; Zhou, H.; Clendenning, S. B.; Bent, S. F., Formation of organic nanoscale laminates and blends by molecular layer deposition. *ACS Nano* 2010, 4 (1), 331–341.
22. Zhou, H.; Bent, S. F., Molecular layer deposition of functional thin films for advanced lithographic patterning. *ACS Appl. Mater. Interfaces* 2011, 3 (2), 505–511.

23. Zhou, H.; Toney, M. F.; Bent, S. F., Cross-linked ultrathin polyurea films via molecular layer deposition. *Macromolecules* 2013, 46 (14), 5638–5643.
24. Park, Y.-S.; Choi, S.-E.; Kim, H.; Lee, J. S., Fine-tunable absorption of uniformly aligned polyurea thin films for optical filters using sequentially self-limited molecular layer deposition. *ACS Appl. Mater. Interfaces* 2016, 8 (18), 11788–11795.
25. Shao, H.-I.; Umemoto, S.; Kikutani, T.; Okui, N., Layer-by-layer polycondensation of nylon 66 by alternating vapour deposition polymerization. *Polymer* 1997, 38 (2), 459–462.
26. Du, Y.; George, S. M., Molecular layer deposition of nylon 66 films examined using in situ FTIR spectroscopy. *J. Phys. Chem. C* 2007, 111 (24), 8509–8517.
27. Adamczyk, N. M.; Dameron, A. A.; George, S. M., Molecular layer deposition of poly(p-phenylene terephthalamide) films using terephthaloyl chloride and p-phenylenediamine. *Langmuir* 2008, 24 (5), 2081–2089.
28. Peng, Q.; Efimenko, K.; Genzer, J.; Parsons, G. N., Oligomer orientation in vapor-molecular-layer-deposited alkyl-aromatic polyamide films. *Langmuir* 2012, 28 (28), 10464–10470.
29. Atanasov, S. E.; Losego, M. D.; Gong, B.; Sachet, E.; Maria, J.-P.; Williams, P. S.; Parsons, G. N., Highly conductive and conformal poly(3,4-ethylenedioxythiophene) (PEDOT) thin films via oxidative molecular layer deposition. *Chem. Mater.* 2014, 26 (11), 3471–3478.
30. Kim, D. H.; Atanasov, S. E.; Lemaire, P.; Lee, K.; Parsons, G. N., Platinum-free cathode for dye-sensitized solar cells using poly(3,4-ethylenedioxythiophene) (PEDOT) formed via oxidative molecular layer deposition. *ACS Appl. Mater. Interfaces* 2015, 7 (7), 3866–3870.
31. Miyamae, T.; Tsukagoshi, K.; Matsuoka, O.; Yamamoto, S.; Nozoye, H., Preparation of polyimide-polyamide random copolymer thin film by sequential vapor deposition polymerization. *Jpn. J. Appl. Phys.* 2002, 41 (2R), 746.
32. Loscutoff, P. W.; Lee, H.-B.-R.; Bent, S. F., Deposition of ultrathin polythiourea films by molecular layer deposition. *Chem. Mater.* 2010, 22 (19), 5563–5569.
33. Ivanova, T. V.; Maydannik, P. S.; Cameron, D. C., Molecular layer deposition of polyethylene terephthalate thin films. *J. Vac. Sci. Technol. A* 2012, 30 (1), 01A121.
34. Li, Y.-h.; Wang, D.; Buriak, J. M., Molecular layer deposition of thiol–ene multilayers on semiconductor surfaces. *Langmuir* 2010, 26 (2), 1232–1238.
35. Vasudevan, S. A.; Xu, Y.; Karwal, S.; van Ostaay, H. G. M. E.; Meesters, G. M. H.; Talebi, M.; Sudholter, E. J. R.; Ruud van Ommen, J., Controlled release from protein particles encapsulated by molecular layer deposition. *Chem. Commun.* 2015, 51 (63), 12540–12543.
36. Dameron, A. A.; Seghete, D.; Burton, B. B.; Davidson, S. D.; Cavanagh, A. S.; Bertrand, J. A.; George, S. M., Molecular layer deposition of alucone polymer films using trimethylaluminum and ethylene glycol. *Chem. Mater.* 2008, 20 (10), 3315–3326.
37. Park, Y.-S.; Kim, H.; Cho, B.; Lee, C.; Choi, S.-E.; Sung, M. M.; Lee, J. S., Intramolecular and Intermolecular Interactions in Hybrid Organic–Inorganic Alucone Films Grown by Molecular Layer Deposition. *ACS Appl. Mater. Interfaces* 2016, 8 (27), 17489–17498.
38. Choudhury, D.; Sarkar, S. K.; Mahuli, N., Molecular layer deposition of alucone films using trimethylaluminum and hydroquinone. *J. Vac. Sci. Technol. A* 2015, 33 (1), 01A115.
39. Bahlawane, N.; Arl, D.; Thomann, J.-S.; Adjeroud, N.; Menguelti, K.; Lenoble, D., Molecular layer deposition of amine-containing alucone thin films. *Surf. Coat. Technol.* 2013, 230, 101–105.
40. Miller, D. C.; Foster, R. R.; Jen, S.-H.; Bertrand, J. A.; Seghete, D.; Yoon, B.; Lee, Y.-C.; George, S. M.; Dunn, M. L., Thermomechanical properties of aluminum alkoxide (alucone) films created using molecular layer deposition. *Acta Mater.* 2009, 57 (17), 5083–5092.

41. Yoon, B.; Seghete, D.; Cavanagh, A. S.; George, S. M., Molecular layer deposition of hybrid organic–inorganic alucone polymer films using a three-step ABC reaction sequence. *Chem. Mater.* 2009, 21 (22), 5365–5374.
42. Gong, B.; Peng, Q.; Parsons, G. N., Conformal organic–inorganic hybrid network polymer thin films by molecular layer deposition using trimethylaluminum and glycidol. *J. Phys. Chem. B* 2011, 115 (19), 5930–5938.
43. Lee, Y.; Yoon, B.; Cavanagh, A. S.; George, S. M., Molecular layer deposition of aluminum alkoxide polymer films using trimethylaluminum and glycidol. *Langmuir* 2011, 27 (24), 15155–15164.
44. Gong, B.; Parsons, G. N., Caprolactone ring-opening molecular layer deposition of organic-aluminum oxide polymer films. *ECS J. Solid State Sci. Technol.* 2012, 1 (4), P210-P215.
45. Abdulagatov, A. I.; Terauds, K. E.; Travis, J. J.; Cavanagh, A. S.; Raj, R.; George, S. M., Pyrolysis of titanicone molecular layer deposition films as precursors for conducting TiO_2/carbon composite films. *J. Phys. Chem. C* 2013, 117 (34), 17442–17450.
46. Yoon, B.; O'Patchen, J. L.; Seghete, D.; Cavanagh, A. S.; George, S. M., Molecular layer deposition of hybrid organic-inorganic polymer films using diethylzinc and ethylene glycol. *Chem. Vap. Deposition* 2009, 15 (4–6), 112–121.
47. Peng, Q.; Gong, B.; VanGundy, R. M.; Parsons, G. N., "Zincone" zinc oxide–organic hybrid polymer thin films formed by molecular layer deposition. *Chem. Mater.* 2009, 21 (5), 820–830.
48. Yoon, B.; Lee, Y.; Derk, A.; Musgrave, C.; George, S., Molecular layer deposition of conductive hybrid organic-inorganic thin films using diethylzinc and hydroquinone. *ECS Trans.* 2011, 33 (27), 191–195.
49. Yoon, B.; Lee, B. H.; George, S. M., Molecular layer deposition of flexible, transparent and conductive hybrid organic-inorganic thin films. *ECS Trans.* 2011, 41 (2), 271–277.
50. Yoon, B.; Lee, B. H.; George, S. M., Highly conductive and transparent hybrid organic–inorganic zincone thin films using atomic and molecular layer deposition. *J. Phys. Chem. C* 2012, 116 (46), 24784–24791.
51. Han, K. S.; Sung, M. M., Molecular layer deposition of organic–inorganic hybrid films using diethylzinc and trihydroxybenzene. *J. Nanosci. Nanotechnol.* 2014, 14 (8), 6137–6142.
52. Cho, S.; Han, G.; Kim, K.; Sung, M. M., High-performance two-dimensional polydiacetylene with a hybrid inorganic–organic structure. *Angew. Chem. Int. Ed.* 2011, 50 (12), 2742–2746.
53. Brown, J. J.; Hall, R. A.; Kladitis, P. E.; George, S. M.; Bright, V. M., Molecular layer deposition on carbon nanotubes. *ACS Nano* 2013, 7 (9), 7812–7823.
54. Lee, B. H.; Anderson, V. R.; George, S. M., Molecular layer deposition of zircone and ZrO_2/zircone alloy films: Growth and properties. *Chem. Vap. Deposition* 2013, 19 (4–6), 204–212.
55. Hall, R. A.; George, S. M.; Kim, Y.; Hwang, W.; Samberg, M. E.; Monteiro-Riviere, N. A.; Narayan, R. J., Growth of zircone on nanoporous alumina using molecular layer deposition. *JOM* 2014, 66 (4), 649–653.
56. Abdulagatov, A. I.; Hall, R. A.; Sutherland, J. L.; Lee, B. H.; Cavanagh, A. S.; George, S. M., Molecular layer deposition of titanicone films using $TiCl_4$ and ethylene glycol or glycerol: Growth and properties. *Chem. Mater.* 2012, 24 (15), 2854–2863.
57. Van de Kerckhove, K.; Mattelaer, F.; Deduytsche, D.; Vereecken, P. M.; Dendooven, J.; Detavernier, C., Molecular layer deposition of "titanicone", a titanium-based hybrid material, as an electrode for lithium-ion batteries. *Dalton Trans.* 2016, 45 (3), 1176–1184.
58. Cao, Y.-Q.; Zhu, L.; Li, X.; Cao, Z.-Y.; Wu, D.; Li, A.-D., Growth characteristics of Ti-based fumaric acid hybrid thin films by molecular layer deposition. *Dalton Trans.* 2015, 44 (33), 14782–14792.

59. Chen, C.; Li, P.; Wang, G.; Yu, Y.; Duan, F.; Chen, C.; Song, W.; Qin, Y.; Knez, M., Nanoporous nitrogen-doped titanium dioxide with excellent photocatalytic activity under visible light irradiation produced by molecular layer deposition. *Angew. Chem. Int. Ed.* 2013, 52 (35), 9196–9200.
60. Lee, B. H.; Anderson, V. R.; George, S. M., Growth and properties of hafnicone and HfO_2/hafnicone nanolaminate and alloy films using molecular layer deposition techniques. *ACS Appl. Mater. Interfaces* 2014, 6 (19), 16880–16887.
61. Van de Kerckhove, K.; Mattelaer, F.; Dendooven, J.; Detavernier, C., Molecular layer deposition of "vanadicone", a vanadium-based hybrid material, as an electrode for lithium-ion batteries. *Dalton Trans.* 2017, 46, 4542–4553.
62. Zhou, H.; Bent, S. F., Fabrication of organic interfacial layers by molecular layer deposition: Present status and future opportunities. *J. Vac. Sci. Technol. A* 2013, 31 (4), 040801.
63. Sarkar, D.; Ishchuk, S.; Taffa, D. H.; Kaynan, N.; Berke, B. A.; Bendikov, T.; Yerushalmi, R., Oxygen-deficient titania with adjustable band positions and defects; molecular layer deposition of hybrid organic–inorganic thin films as precursors for enhanced photocatalysis. *J. Phys. Chem. C* 2016, 120 (7), 3853–3862.
64. Ban, C.; George, S. M., Molecular layer deposition for surface modification of lithium-ion battery electrodes. *Adv. Mater. Interfaces* 2016, 3, 1600762.
65. Gong, B.; Spagnola, J. C.; Parsons, G. N., Hydrophilic mechanical buffer layers and stable hydrophilic finishes on polydimethylsiloxane using combined sequential vapor infiltration and atomic/molecular layer deposition. *J. Vac. Sci. Technol. A* 2012, 30 (1), 01A156.
66. Groner, M. D.; Fabreguette, F. H.; Elam, J. W.; George, S. M., Low-temperature Al_2O_3 atomic layer deposition. *Chem. Mater.* 2004, 16 (4), 639–645.
67. Puurunen, R. L., Surface chemistry of atomic layer deposition: A case study for the trimethylaluminum/water process. *J. Appl. Phys.* 2005, 97 (12), 121301.
68. Vähä-Nissi, M.; Sievänen, J.; Salo, E.; Heikkilä, P.; Kenttä, E.; Johansson, L.-S.; Koskinen, J. T.; Harlin, A., Atomic and molecular layer deposition for surface modification. *J. Solid State Chem.* 2014, 214, 7–11.
69. Liang, X.; King, D. M.; Li, P.; George, S. M.; Weimer, A. W., Nanocoating hybrid polymer films on large quantities of cohesive nanoparticles by molecular layer deposition. *AICHE J.* 2009, 55 (4), 1030–1039.
70. Seghete, D.; Davidson, B. D.; Hall, R. A.; Chang, Y. J.; Bright, V. M.; George, S. M., Sacrificial layers for air gaps in NEMS using alucone molecular layer deposition. *Sens. Actuators A Phys.* 2009, 155 (1), 8–15.
71. Liang, X.; Yu, M.; Li, J.; Jiang, Y.-B.; Weimer, A. W., Ultra-thin microporous-mesoporous metal oxide films prepared by molecular layer deposition (MLD). *Chem. Commun.* 2009, (46), 7140–7142.
72. Qin, Y.; Yang, Y.; Scholz, R.; Pippel, E.; Lu, X.; Knez, M., Unexpected oxidation behavior of Cu nanoparticles embedded in porous alumina films produced by molecular layer deposition. *Nano Lett.* 2011, 11 (6), 2503–2509.
73. Lee, B. H.; Yoon, B.; Anderson, V. R.; George, S. M., Alucone Alloys with Tunable Properties Using Alucone Molecular Layer Deposition and Al_2O_3 Atomic Layer Deposition. *J. Phys. Chem. C* 2012, 116 (5), 3250–3257.
74. Jen, S.-H.; Lee, B. H.; George, S. M.; McLean, R. S.; Carcia, P. F., Critical tensile strain and water vapor transmission rate for nanolaminate films grown using Al2O3 atomic layer deposition and alucone molecular layer deposition. *Appl. Phys. Lett.* 2012, 101 (23), 234103.
75. Loebl, A. J.; Oldham, C. J.; Devine, C. K.; Gong, B.; Atanasov, S. E.; Parsons, G. N.; Fedkiw, P. S., Solid Electrolyte Interphase on Lithium-Ion Carbon Nanofiber Electrodes by Atomic and Molecular Layer Deposition. *J. Electrochem. Soc.* 2013, 160 (11), A1971-A1978.

76. Li, X.; Lushington, A.; Liu, J.; Li, R.; Sun, X., Superior stable sulfur cathodes of Li-S batteries enabled by molecular layer deposition. *Chem. Commun.* 2014, 50 (68), 9757–9760.
77. Lushington, A.; Liu, J.; Bannis, M. N.; Xiao, B.; Lawes, S.; Li, R.; Sun, X., A novel approach in controlling the conductivity of thin films using molecular layer deposition. *Appl. Surf. Sci.* 2015, 357, Part B, 1319–1324.
78. DuMont, J. W.; George, S. M., Pyrolysis of alucone molecular layer deposition films studied using in situ transmission fourier transform infrared spectroscopy. *J. Phys. Chem. C* 2015, 119 (26), 14603–14612.
79. Lemaire, P. C.; Oldham, C. J.; Parsons, G. N., Rapid visible color change and physical swelling during water exposure in triethanolamine-metalcone films formed by molecular layer deposition. *J. Vac. Sci. Technol. A* 2016, 34 (1), 01A134.
80. Oldham, C. J.; Gong, B.; Spagnola, J. C.; Jur, J. S.; Senecal, K. J.; Godfrey, T. A.; Parsons, G. N., Encapsulation and chemical resistance of electrospun nylon nanofibers coated using integrated atomic and molecular layer deposition. *J. Electrochem. Soc.* 2011, 158 (9), D549-D556.
81. Piper, D. M.; Travis, J. J.; Young, M.; Son, S.-B.; Kim, S. C.; Oh, K. H.; George, S. M.; Ban, C.; Lee, S.-H., Reversible high-capacity Si nanocomposite anodes for lithium-ion batteries enabled by molecular layer deposition. *Adv. Mater.* 2014, 26 (10), 1596–1601.
82. Seghete, D.; Hall, R. A.; Yoon, B.; George, S. M., Importance of trimethylaluminum diffusion in three-step ABC molecular layer deposition using trimethylaluminum, ethanolamine, and maleic anhydride. *Langmuir* 2010, 26 (24), 19045–19051.
83. Liang, X.; Evanko, B. W.; Izar, A.; King, D. M.; Jiang, Y.-B.; Weimer, A. W., Ultrathin highly porous alumina films prepared by alucone ABC molecular layer deposition (MLD). *Microporous Mesoporous Mater.* 2013, 168, 178–182.
84. Gould, T. D.; Izar, A.; Weimer, A. W.; Falconer, J. L.; Medlin, J. W., Stabilizing Ni catalysts by molecular layer deposition for harsh, dry reforming conditions. *ACS Catal.* 2014, 4 (8), 2714–2717.
85. Huang, J.; Lucero, A. T.; Cheng, L.; Hwang, H. J.; Ha, M.-W.; Kim, J., Hydroquinone-ZnO nano-laminate deposited by molecular-atomic layer deposition. *Appl. Phys. Lett.* 2015, 106 (12), 123101.
86. Huang, J.; Zhang, H.; Lucero, A.; Cheng, L.; Kc, S.; Wang, J.; Hsu, J.; Cho, K.; Kim, J., Organic-inorganic hybrid semiconductor thin films deposited using molecular-atomic layer deposition (MALD). *J. Mater. Chem. C* 2016, 4 (12), 2382–2389.
87. Nilsen, O.; Haug, K. R.; Finstad, T.; Fjellvåg, H., Molecular hybrid structures by atomic layer deposition – Deposition of Alq_3, Znq_2 and Tiq_4 (q=8-hydroxyquinoline). *Chem. Vap. Deposition* 2013, 19 (4–6), 174–179.
88. Räupke, A.; Albrecht, F.; Maibach, J.; Behrendt, A.; Polywka, A.; Heiderhoff, R.; Helzel, J.; Rabe, T.; Johannes, H.-H.; Kowalsky, W.; Mankel, E.; Mayer, T.; Görrn, P.; Riedl, T., Conformal and highly luminescent monolayers of Alq_3 prepared by gas-phase molecular layer deposition. *ACS Appl. Mater. Interfaces* 2014, 6 (2), 1193–1199.
89. Salmi, L. D.; Heikkilä, M. J.; Puukilainen, E.; Sajavaara, T.; Grosso, D.; Ritala, M., Studies on atomic layer deposition of MOF-5 thin films. *Microporous Mesoporous Mater.* 2013, 182, 147–154.
90. Salmi, L. D.; Heikkilä, M. J.; Vehkamäki, M.; Puukilainen, E.; Ritala, M.; Sajavaara, T., Studies on atomic layer deposition of IRMOF-8 thin films. *J. Vac. Sci. Technol. A* 2015, 33 (1), 01A121.
91. Ahvenniemi, E.; Karppinen, M., Atomic/molecular layer deposition: A direct gas-phase route to crystalline metal-organic framework thin films. *Chem. Commun.* 2016, 52 (6), 1139–1142.

92. Nisula, M.; Karppinen, M., Atomic/molecular layer deposition of lithium terephthalate thin films as high rate capability Li-ion battery anodes. *Nano Lett.* 2016, 16 (2), 1276–1281.
93. Kao, C.-Y.; Yoo, J.-W.; Min, Y.; Epstein, A. J., Molecular layer deposition of an organic-based magnetic semiconducting laminate. *ACS Appl. Mater. Interfaces* 2012, 4 (1), 137–141.
94. Kao, C.-Y.; Li, B.; Lu, Y.; Yoo, J.-W.; Epstein, A. J., Thin films of organic-based magnetic materials of vanadium and cobalt tetracyanoethylene by molecular layer deposition. *J. Mater. Chem. C* 2014, 2 (30), 6171–6176.
95. Zhang, B.; Chen, Y.; Li, J.; Pippel, E.; Yang, H.; Gao, Z.; Qin, Y., High efficiency Cu-ZnO hydrogenation catalyst: The tailoring of Cu-ZnO interface sites by molecular layer deposition. *ACS Catal.* 2015, 5 (9), 5567–5573.
96. Tynell, T.; Yamauchi, H.; Karppinen, M., Hybrid inorganic–organic superlattice structures with atomic layer deposition/molecular layer deposition. *J. Vac. Sci. Technol. A* 2014, 32 (1), 01A105.
97. Tynell, T.; Karppinen, M., ZnO: Hydroquinone superlattice structures fabricated by atomic/molecular layer deposition. *Thin Solid Films* 2014, 551, 23–26.
98. Lee, B. H.; Ryu, M. K.; Choi, S.-Y.; Lee, K.-H.; Im, S.; Sung, M. M., Rapid vapor-phase fabrication of organic–inorganic hybrid superlattices with monolayer precision. *J. Am. Chem. Soc.* 2007, 129 (51), 16034–16041.
99. Cha, S. H.; Oh, M. S.; Lee, K. H.; Im, S.; Lee, B. H.; Sung, M. M., Electrically stable low voltage ZnO transistors with organic/inorganic nanohybrid dielectrics. *Appl. Phys. Lett.* 2008, 92 (2), 023506.
100. Lee, K. H.; Choi, J.-M.; Im, S.; Lee, B. H.; Im, K. K.; Sung, M. M.; Lee, S., Low-voltage-driven pentacene thin-film transistor with an organic-inorganic nanohybrid dielectric. *Appl. Phys. Lett.* 2007, 91 (12), 123502.
101. Lee, B. H.; Lee, K. H.; Im, S.; Sung, M. M., Vapor-phase molecular layer deposition of self-assembled multilayers for organic thin-film transistor. *J. Nanosci. Nanotechnol.* 2009, 9 (12), 6962–6967.
102. Huang, J.; Lee, M.; Lucero, A.; Kim, J., Organic-inorganic hybrid nano-laminates fabricated by ozone-assisted molecular-atomic layer deposition. *Chem. Vap. Deposition* 2013, 19 (4–6), 142–148.
103. Huang, J.; Lee, M.; Lucero, A. T.; Cheng, L.; Ha, M.-W.; Kim, J., 7-Octenyltrichrolosilane/trimethyaluminum hybrid dielectrics fabricated by molecular-atomic layer deposition on ZnO thin film transistors. *Jpn. J. Appl. Phys.* 2016, 55 (6S1), 06GK04.
104. Lee, B. H.; Im, K. K.; Lee, K. H.; Im, S.; Sung, M. M., Molecular layer deposition of ZrO_2-based organic–inorganic nanohybrid thin films for organic thin film transistors. *Thin Solid Films* 2009, 517 (14), 4056–4060.
105. Hashemi, F. S. M.; Prasittichai, C.; Bent, S. F., A new resist for area selective atomic and molecular layer deposition on metal–dielectric patterns. *J. Phys. Chem. C* 2014, 118 (20), 10957–10962.
106. Prasittichai, C.; Pickrahn, K. L.; Minaye Hashemi, F. S.; Bergsman, D. S.; Bent, S. F., Improving area-selective molecular layer deposition by selective SAM removal. *ACS Appl. Mater. Interfaces* 2014, 6 (20), 17831–17836.
107. McMahon, C. N.; Alemany, L.; Callender, R. L.; Bott, S. G.; Barron, A. R., Reaction of $Al(^tBu)_3$ with ethylene glycol: Intermediates to aluminum alkoxide (alucone) preceramic polymers. *Chem. Mater.* 1999, 11 (11), 3181–3188.
108. Baek, G.; Lee, S.; Lee, J.-H.; Park, J.-S., Air-stable alucone thin films deposited by molecular layer deposition using a 4-mercaptophenol organic reactant. *J. Vac. Sci. Technol. A* 2020, 38 (2), 022411.
109. Hagen, D. J.; Mai, L.; Devi, A.; Sainio, J.; Karppinen, M., Atomic/molecular layer deposition of Cu–organic thin films. *Dalton Trans.* 2018, 47 (44), 15791–15800.

110. Nilsen, O.; Klepper, K.; Nielsen, H.; Fjellvaåg, H., Deposition of organic-inorganic hybrid materials by atomic layer deposition. *ECS Trans.* 2008, 16 (4), 3–14.
111. Giedraityte, Z.; Sainio, J.; Hagen, D.; Karppinen, M., Luminescent metal-nucleobase network thin films by atomic/molecular layer deposition. *J. Phys. Chem. C* 2017, 121 (32), 17538–17545.
112. Giedraityte, Z.; Lopez-Acevedo, O.; Espinosa Leal, L. A.; Pale, V.; Sainio, J.; Tripathi, T. S.; Karppinen, M., Three-dimensional uracil network with sodium as a linker. *J. Phys. Chem. C* 2016, 120 (46), 26342–26349.
113. Mai, L.; Giedraityte, Z.; Schmidt, M.; Rogalla, D.; Scholz, S.; Wieck, A. D.; Devi, A.; Karppinen, M., Atomic/molecular layer deposition of hybrid inorganic–organic thin films from erbium guanidinate precursor. *J. Mater. Sci.* 2017, 52 (11), 6216–6224.
114. Isaeva, V. I.; Kustov, L. M., The application of metal-organic frameworks in catalysis (Review). *Pet. Chem.* 2010, 50 (3), 167–180.
115. Liu, X.-W.; Sun, T.-J.; Hu, J.-L.; Wang, S.-D., Composites of metal-organic frameworks and carbon-based materials: Preparations, functionalities and applications. *J. Mater. Chem. A* 2016, 4 (10), 3584–3616.
116. Murray, L. J.; Dinca, M.; Long, J. R., Hydrogen storage in metal-organic frameworks. *Chem. Soc. Rev.* 2009, 38 (5), 1294–1314.
117. Suh, M. P.; Park, H. J.; Prasad, T. K.; Lim, D.-W., Hydrogen storage in metal–organic frameworks. *Chem. Rev.* 2012, 112 (2), 782–835.
118. Khan, N. A.; Hasan, Z.; Jhung, S. H., Adsorptive removal of hazardous materials using metal-organic frameworks (MOFs): A review. *J. Hazard. Mater.* 2013, 244–245, 444–456.
119. Sabouni, R.; Kazemian, H.; Rohani, S., Carbon dioxide capturing technologies: A review focusing on metal organic framework materials (MOFs). *Environ. Sci. Pollut. Res.* 2014, 21 (8), 5427–5449.
120. Allendorf, M. D.; Bauer, C. A.; Bhakta, R. K.; Houk, R. J. T., Luminescent metal-organic frameworks. *Chem. Soc. Rev.* 2009, 38 (5), 1330–1352.
121. Horcajada, P.; Chalati, T.; Serre, C.; Gillet, B.; Sebrie, C.; Baati, T.; Eubank, J. F.; Heurtaux, D.; Clayette, P.; Kreuz, C.; Chang, J.-S.; Hwang, Y. K.; Marsaud, V.; Bories, P.-N.; Cynober, L.; Gil, S.; Ferey, G.; Couvreur, P.; Gref, R., Porous metal-organic-framework nanoscale carriers as a potential platform for drug delivery and imaging. *Nat Mater* 2010, 9 (2), 172–178.
122. Vallet-Regí, M.; Balas, F.; Arcos, D., Mesoporous materials for drug delivery. *Angew. Chem. Int. Ed.* 2007, 46 (40), 7548–7558.
123. Wang, L.; Han, Y.; Feng, X.; Zhou, J.; Qi, P.; Wang, B., Metal–organic frameworks for energy storage: Batteries and supercapacitors. *Coord. Chem. Rev.* 2016, 307, Part 2, 361–381.
124. Xia, W.; Mahmood, A.; Zou, R.; Xu, Q., Metal-organic frameworks and their derived nanostructures for electrochemical energy storage and conversion. *Energy Environ. Sci.* 2015, 8 (7), 1837–1866.
125. Mahmood, A.; Guo, W.; Tabassum, H.; Zou, R., Metal-organic framework-based nanomaterials for electrocatalysis. *Adv. Energy Mater.* 2016, 6 (17), 1600423.
126. Gkaniatsou, E.; Sicard, C.; Ricoux, R.; Mahy, J.-P.; Steunou, N.; Serre, C., Metal-organic frameworks: A novel host platform for enzymatic catalysis and detection. *Mater. Horiz.* 2017, 4, 55–63.
127. Ahvenniemi, E.; Karppinen, M., In Situ atomic/molecular layer-by-layer deposition of inorganic–organic coordination network thin films from gaseous precursors. *Chem. Mater.* 2016, 28 (17), 6260–6265.
128. Penttinen, J.; Nisula, M.; Karppinen, M., Atomic/molecular layer deposition of s-block metal carboxylate coordination network thin films. *Chem. –Eur. J.* 2017, 23 (72), 18225–18231.

129. Olson, C. S.; Gangopadhyay, S.; Hoang, K.; Alema, F.; Kilina, S.; Pokhodnya, K., Magnetic exchange in MnII[TCNE] (TCNE = tetracyanoethylene) molecule-based magnets with two- and three-dimensional magnetic networks. *J. Phys. Chem. C* 2015, 119 (44), 25036–25046.
130. Miller, J. S., Magnetically ordered molecule-based materials. *Chem. Soc. Rev.* 2011, 40 (6), 3266–3296.
131. Meng, X.; Wang, X.; Geng, D.; Ozgit-Akgun, C.; Schneider, N.; Elam, J. W., Atomic layer deposition for nanomaterial synthesis and functionalization in energy technology. *Mater. Horiz.* 2017, 4 (2), 133–154.
132. Meng, X.; Yang, X. Q.; Sun, X. L., Emerging applications of atomic layer deposition for lithium-ion battery studies. *Adv. Mater.* 2012, 24 (27), 3589–3615.
133. Sun, Q.; Lau, K. C.; Geng, D.; Meng, X., Atomic and molecular layer deposition for superior lithium-sulfur batteries: Strategies, performance, and mechanisms. *Batteries Supercaps* 2018, 1 (2), 41–68.
134. Meng, X., Atomic-scale surface modifications and novel electrode designs for high-performance sodium-ion batteries via atomic layer deposition. *J. Mater. Chem. A* 2017, 5, 10127–10149.
135. O'Neill, B. J.; Jackson, D. H. K.; Lee, J.; Canlas, C.; Stair, P. C.; Marshall, C. L.; Elam, J. W.; Kuech, T. F.; Dumesic, J. A.; Huber, G. W., Catalyst design with atomic layer deposition. *ACS Catal.* 2015, 5 (3), 1804–1825.
136. Rimoldi, M.; Bernales, V.; Borycz, J.; Vjunov, A.; Gallington, L. C.; Platero-Prats, A. E.; Kim, I. S.; Fulton, J. L.; Martinson, A. B. F.; Lercher, J. A.; Chapman, K. W.; Cramer, C. J.; Gagliardi, L.; Hupp, J. T.; Farha, O. K., Atomic layer deposition in a metal–organic framework: Synthesis, characterization, and performance of a solid acid. *Chem. Mater.* 2017, 29 (3), 1058–1068.
137. Cai, J.; Han, X.; Wang, X.; Meng, X., Atomic layer deposition of two-dimensional layered materials: Processes, growth mechanisms, and characteristics. *Matter* 2020, 2, 587–630.
138. Zheng, X.; Lee, H.; Weisgraber, T. H.; Shusteff, M.; DeOtte, J.; Duoss, E. B.; Kuntz, J. D.; Biener, M. M.; Ge, Q.; Jackson, J. A.; Kucheyev, S. O.; Fang, N. X.; Spadaccini, C. M., Ultralight, ultrastiff mechanical metamaterials. *Science* 2014, 344 (6190), 1373–1377.
139. Lim, J. A.; Liu, F.; Ferdous, S.; Muthukumar, M.; Briseno, A. L., Polymer semiconductor crystals. *Mater. Today* 2010, 13 (5), 14–24.
140. Adhikari, B.; Majumdar, S., Polymers in sensor applications. *Prog. Polym. Sci.* 2004, 29 (7), 699–766.
141. Kara, P., Polymer optical fiber sensors—a review. *Smart Mater. Struct.* 2011, 20 (1), 013002.
142. Bajpai, A. K.; Bajpai, J.; Saini, R.; Gupta, R., Responsive polymers in biology and technology. *Polym. Rev.* 2011, 51 (1), 53–97.
143. Jagur-Grodzinski, J., Polymers for tissue engineering, medical devices, and regenerative medicine. Concise general review of recent studies. *Polym. Adv. Technol.* 2006, 17 (6), 395–418.
144. Maitz, M. F., Applications of synthetic polymers in clinical medicine. *Biosurf. Biotribol.* 2015, 1 (3), 161–176.
145. Jochum, F. D.; Theato, P., Temperature- and light-responsive smart polymer materials. *Chem. Soc. Rev.* 2013, 42 (17), 7468–7483.

4 Scanning Tunneling Microscope and Spectroscope on Organic–Inorganic Material Heterojunction

Sadaf Bashir Khan and Shern Long Lee

CONTENTS

4.1 INTRODUCTION

The operating mechanism of photovoltaic devices (PVs) comprises photogeneration of excitons, corresponding dissociation, and transference of holes and electrons to the opposite electrodes [1]. The effectiveness of these stages takes place in a sequence and can be influenced by active materials, exciton dissociation, which needs a junction in the form of a metal/organic or an organic/organic interface. The narrative of PVs symbolizes a junction between a stratum of donor and a stratum of acceptor molecules [2] such as a quantum dots (QDs) [QDs in a conjugated polymer matrix to form a hybrid bulk heterojunction (BHJ)] or BHJ [between electron-donor and electron-acceptor molecules/polymers] [3,4]. Generally, heterojunction is defined as a contact of two materials having different electrical characteristics. These materials include polymers, elements, tiny molecules, or composites. Commonly, in a heterojunction, a donor is electron-rich than the acceptor. When the donor is excited via external stimuli such as light or electric pulse, it generates a negative charge in the acceptor, whereas the donor turns into a positive charge. The positive charges transfer via hole hopping and the negative charges via electron hopping, which takes place during charge transfer between molecules.

Consequently, charges transport in the organic strata, which force charge to exchange in the electrodes. In hybrid BHJs, QDs, nanorods [range of II–VI, IV–VI], ternary or even quaternary semiconductors (SCs) have been integrated into appropriate polymer matrices [5–8]. Among nanostructures having different configurations, nanorods have an additional benefit as they offer one-dimensional (1D) conduction passageways for carriers in comparison to hopping transference via QDs. Among all junctions, exciton dissociation was confirmed via appropriate material selection, to generate a staggering bandgap or type-II band alignment between energy levels of the polymer-nanostructure at the interface [9].

Regarding this, scanning tunneling microscopy (STM) has made noteworthy contributions in comprehending the subprocesses underlying in various practical applications, specifically in photovoltaics. These procedures permit confined exploration of the materials' electrical or optical characteristics. The spatially resolved measurements of surface photovoltage and photocurrent have been predominantly beneficial in comprehending the charge generation and separation in optoelectronic devices. In thin-film inorganic solar cells, charge separation does not take place at a heterojunction as expected, but instead, it occurs at a homojunction buried ~50 nm within the absorbing layer [studied via Kelvin probe force microscopy (KPFM)] [10,11]. In organic photovoltaics, photocurrent has contributed to the understanding of the interplay concerning processing situations, blend morphology, configuration, or device enactment. This functional imaging technique discriminates STM from complementary structural characterization methods. The other techniques include SPM or atomic force microscopy (AFM). The AFM subdivided into different kinds, including photoconductive AFM, nearfield scanning optical microscopy, KPFM and time-resolved electric force microscopy.

The band edges of low-dimensional nanostructures have been considered as chief influencing factors in designing an optoelectronic device. Different methods have been followed to monitor band edges such as doping, alloying or quantum confinement effect via lessening the length of active materials along with one to three directions [12–15]. The junctions generated between two semiconductors are in the form of type-I, [13,16] type-II/staggered [17–19], type-III or type-II broken gap [20–22]. The band-offset is characteristically thought-provoking system as far as the interface energy levels are concerned. Commonly, the band-offsets are engendered in core-shell nanoparticles, nanorods, nanowires, or in pn-junctions form in a nanorod [23–28]. Henceforth, the interface offsets besides band edges are essential aspects of nanostructure engineering. STM so far remained an essential method in identifying the band edges of lower-dimensional coordination systems [29–33]. The capability to evaluate the nanostructures band edges has made the procedure distinctive in different aspects.

The nanostructures extended from QDs toward intricate core-shells, hybrid core-shell arrangements in which the organic molecules generate a shell layer on inorganic nanocrystals [32,34]. In scientific exploration focused on practical applications, STM has been considered to observe the composite materials interface, that is, bulk-heterojunction solar cells [13,35,36]. STM is a beneficial surface analytical technique that probes the local density of states (LDOS) having a high spatial and energetic resolution. In STM, a sharp metallic tip mechanically cut/chemically etched is brought into vicinity (~0.3–1.0 nm) to a conducting or semiconducting surface. Generally, the tip is made of platinum–iridium (Pt/Ir) or tungsten (W) due to their chemical steadiness and mechanical resilience. An electrical bias is applied between the tip and the conducting substrate having ultrathin films. The electrons tunnel through the tip-sample gap, while skimming through the thin-film surface. The STM electronics display this current through a feedback loop to maintain a tip-sample distance. The applied voltage polarity governs the direction of tunneling current. In the two bias directions, the tip might tunnel electrons to vacant states or accept electrons from filled states of the SC. Thus, the interconnection between the tip and the base electrode completes the circuit. A tunneling development happens when the magnitude of bias permits an energetic configuration between the tip work function and energy levels of the SC. At low temperature and low sample bias, the density of states (DOS) can be considered to be proportional to the tunneling conductance. The voltage dependence of tunneling current hence provides the location of the energy levels.

In the differential conductance (dI/dV) spectrum, CB and VB edges appear as first peaks in the two bias directions. A dI/dV spectrum allows the derivation of the energy levels of the SC at the undergoing specified point of measurement. As it is a restricted localized mode of measurement, one may record many such spectra on various points of the SC monolayer. The acquired energies are plotted as histograms, which gives information regarding the location of CB and VB edges of the SC. One may deduce the DOS in the structure from the histograms distribution extending within the bandgap. As well, dI/dV versus voltage plots helps in energy-level mapping over the desired area. The nanoscale domains can be identified, and their corresponding local electronic characteristics can be probed simultaneously through dI/dV imaging. STM also helps in establishing a localized measurement model that applies to a single domain. In dI/dV images, the size of domains can be as small as a subnanometer due to tip shape. Hence, the material morphology under examination may depend on the substrate, information on the domains, substrate materials, and the externally applied parameters. The main objective of the present chapter is to deliver and demonstrate the materials science community with the competences, considerations, capabilities, and limitations linked with STM explorations focusing on optoelectronic organic–inorganic semiconductor heterojunction.

4.2 BAND MAPPING; ACROSS A PN-JUNCTION IN A NANOROD

Earlier Amlan J. Pal and his colleagues investigate the band edges across generate-junction in a nanorod [37]. A single junction was established between Cu_2S (p-type) and CdS (n-type) via controlled cationic exchange progression of CdS nanorods.

They illustrate nanorods comprising single materials and the distinct junction in a nanorod using an ultrahigh vacuum (UHV) scanning tunneling microscope (UHV-STM) at room temperature determining the conduction band (CB) and valence band (VB) edges at various points across the junction. They explore the band diagram of nanorod junctions to explore the salient structures of a diode, that is, p and n sections, depletion region, and band bending. The two distinctly designed nanostructures generate a junction grown via colloidal synthesis routes [38,39]. Between two distinct materials, the organic ligands act as stabilizers that perturb the interface or the junction itself [40]. The junction bands in a nanorod were mapped and visualize the electric field across the depletion region of the diode formed in a single nanorod through STM. The optical absorption spectra of CdS and Cu_2S nanorods, and $CdS|Cu_2S$ junctions is presented in Figure 4.1a [i]. Two different reaction times

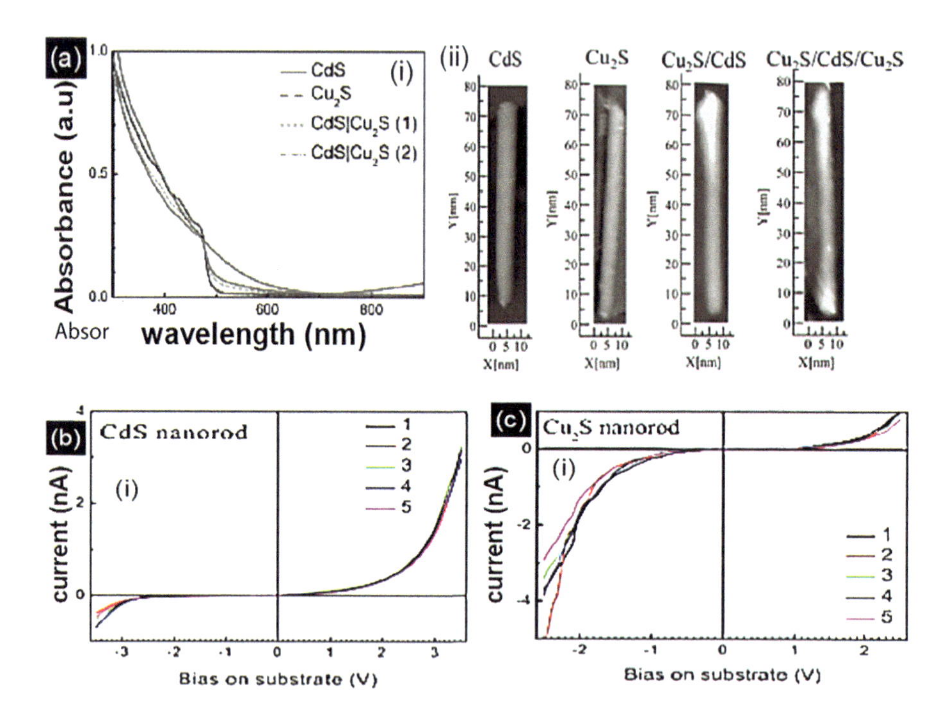

FIGURE 4.1 (a) Optical absorption spectra of CdS and Cu_2S nanorods and $CdS|Cu_2S$ junctions, [i] In the nanorod junctions, (1) $CdS|Cu_2S$ (7 Min reaction time) and (2) $CdS|Cu_2S$ (10 Min) [for the cationic exchange process which controls the length of Cu_2S section in nanorod junctions], [ii] STM topography of CdS, Cu_2S nanorods, $Cu_2S|CdS$, and [iii] $Cu_2S|CdS|Cu$ S nanorod junctions. Set-points for the approach of the tip were 0.4 nA at 2.0 V, which were used during STS measurements. (b) [i] STM topography of a CdS nanorod, [ii] displaying the points at which tunneling current versus voltage characteristics was noted, and [iii] DOS spectra of the I–V characteristics at the spots on the nanorod. The broken lines specify the conduction and VB edge's location. (c) [i] STM topography of a CdS nanorod, [ii] displaying the points at which tunneling current versus voltage characteristics was noted, and [iii] DOS spectra of the I–V characteristics at the spots on the nanorod. The broken lines specify the conduction and VB edges location.

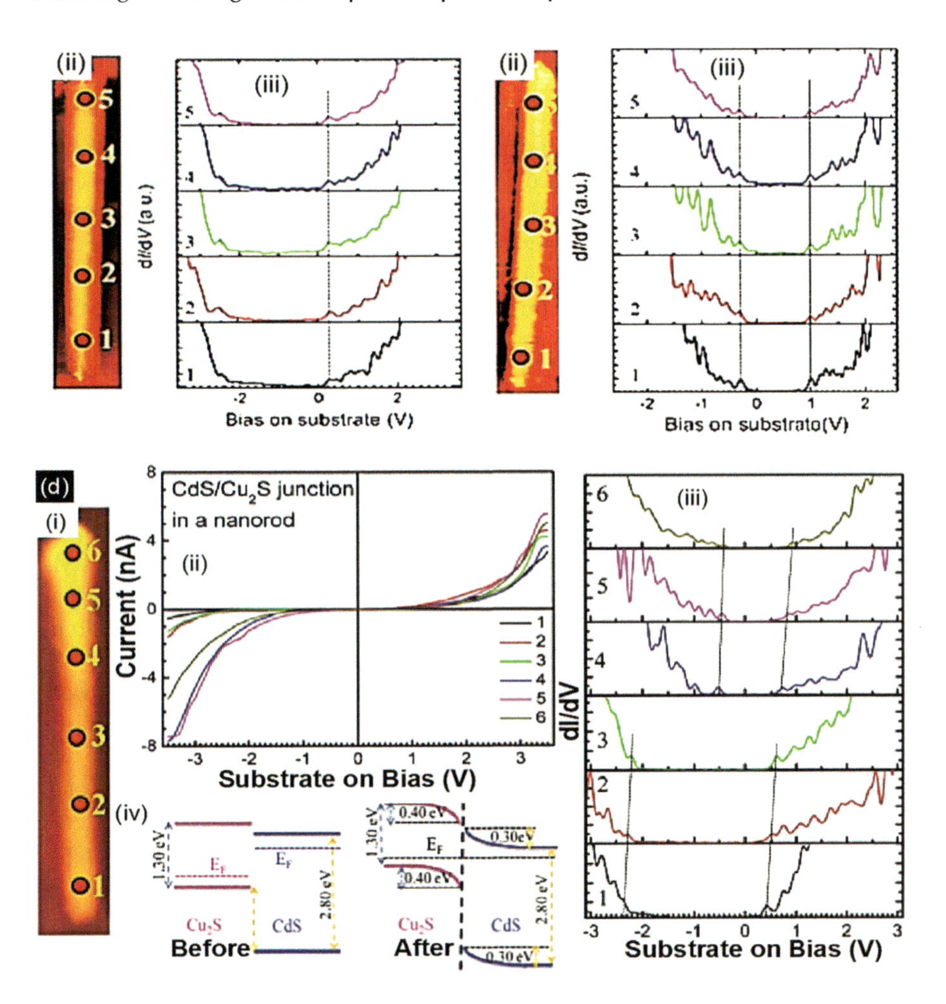

FIGURE 4.1 (CONTINUED) (d) [i] STM topography of a CdS|Cu_2S nanorod junction showing the spots at which [ii] tunneling current versus voltage characteristics was recorded, [iii] DOS spectra of the I–V characteristics at the spots on the nanorod. The broken lines represent CB and VB edges; [iv] Schematic band diagrams of Cu2S and CdS nanorods before and after the formation of a pn-junction. (Reproduced with permission [37]).

(7 Min, 10 Min) were selected for the cationic exchange process in nanorod junctions, which impacts and control the length of Cu_2S in the CdS|Cu_2S junction. The optical absorption spectra demonstrate that toward short wavelength region, CdS shows absorption, however one of Cu_2S shows absorption to the near-IR region. The absorbance in the 475–650 nm wavelengths appears due to Cu_2S; due to the localized surface plasmon resonance arising because of copper deficiencies [38,41,42]. The CdS|Cu_2S junction's spectra were the sum of individual ones demonstrating the absorbance in the 475–650 nm regions upsurge with the growth in the length of Cu_2S in CdS|Cu_2S nanorod junctions.

4.2.1 Nanorods and Junctions Characterization: Tunneling Current and Density of States

Figure 4.1a [ii] shows the STM topography of CdS, Cu_2S nanorods and CdS|Cu_2S junctions. The junctions encompassed Cu_2S|CdS|Cu_2S nanorods, Cu_2S present at ends of CdS. The features of CdS and Cu_2S were homogenous and uniform. The STM imaging was carried out in a constant-current mode. The Cu_2S appears brighter than the CdS due to the higher conductivity of Cu_2S, which enhances tip-to-nanorod distance in comparison to CdS. The variance in brightness and broadness leads to the classification of materials in a nanorod. The CdS and Cu_2S nanorods diameter or their sections in a junction were nearly about 5 nm [43–47]. The probing of STM tip across the nanorods at different points helps in locating the conduction and VB edges. Therefore, the tunneling current versus applied voltage (I–V) was calculated via the STM tip. The tunneling current was noted after locating the STM tip overhead; the sample deactivating the scanning and feedback controls. Since the present structure looks like a double barrier tunnel junction (DBTJ) connecting the tip nanorod and nanorod substrate tunnel barriers [45,48]. The measurements were performed at various tip nanorod distances via different set-points. To evaluate the band edges, the DOS were estimated too. The bias was applied concerning the tip. That is why at the positive voltage region, the electrons can be introduced from the tip to the nanorod, symbolize the position of CBs.

The I–V characteristics and STM topography at five different positions were recorded on CdS and Cu_2S nanorods, as shown in Figure 4.1b and c [i, ii, iii]. In I–V characteristics, the asymmetry arises due to variation in the work function of the two electrodes. In the case of CdS nanorods, higher current in the positive voltage region is observed (Figure 4.1b) due to electrons insertion via tip into the CB of the n-type material, in comparison to the current in the negative voltage region. Correspondingly, for the p-type Cu_2S, at the negative voltage region, the current remains higher due to facile hole injection to (i.e., electron extraction from the VB) (Figure 4.1c). The I–V characteristics across the nanorod nearly remain the same. It did not vary from one point to another, because single compositional material was being characterized along the nanorod. The DOS spectra of the nanorods demonstrated the location of conduction and VB edges in CdS and Cu_2S. The positive and negative voltage at which peaks appear in DOS spectra signifies the conduction and VB's location. The experimental result demonstrates that in CdS, the CB is positioned closer to the Fermi energy (fixed at 0 V). Likewise, in Cu_2S, the VB is nearer to its Fermi energy. The DOS band, therefore, agrees with the n- and p-type characteristics of CdS and Cu_2S nanorods. The position of band edges and DOS spectra intensity remain consistent throughout across nanorods. According to reported literature, the bandgaps of CdS and Cu_2S are nearly 2.8 and 1.3 eV, respectively. The DOS depicts that Cu_2S bandgap is lower than that of CdS. Besides this, the materials form a type-II band alignment when a junction formation takes place between CdS|Cu_2S (single junction in a nanorod).

The STM topography of CdS|Cu_2S nanorod junctions is shown in Figure 4.1d [i]. The recorded I–V characteristics and corresponding DOS spectra (for determination of conduction and VB edges) on a single junction nanorod at various points having a smaller interval of distance are demonstrated in Figure 4.1d [ii, iii]. It is observed that the nanorod terminations resemble the individual material. On an individual point on the nanorod, at least 50 I–V characteristics were recorded to determine the conduction and VB edges from the DOS spectra. Figure 4.1d [iv] schematically represents the conduction and VB edges of a single pn-junction in a nanorod. The band edges at the terminals resemble the discrete nanorods, that is, CdS and Cu_2S (Figure 4.1b and c). STM helps in mapping a single pn-junction signifying the conduction and VBs formed in a single nanorod having a length of about 70 nm. At the interface, band edges show the band bending along with the depletion region of a PN-junction diode. The outcomes allow one to envision the band locations along with the interface (between CdS and Cu_2S sections in a pn-junction). The energy levels of the pn-junction at the interface coordinated well, like a conventional diode junction having a band bending of 0.3–0.4 eV. The experimental results hence demonstrate that the STM is a powerful technique in mapping across a pn-junction in a single nanorod the band edges via recording tunneling current at different points on the nanorods at room temperature. One can easily locate the DOS spectra, conduction and VB edges across the Cu_2S and the CdS nanorods useful for heterojunctions optoelectronic devices.

Accordingly, the mechanism of controlling and regulating the electronic, optical and magnetic characteristics of semiconductors (SCs) is correspondingly accomplished predominantly by the amalgamation of impurity atoms, that is, dopants, which modify the semiconductor's electronic configuration and properties via creating surplus electrons (n-type) or holes (p-type). The systematic and accurate doping of SCs is the foundation of engendering photovoltaics, microelectronics, optoelectronic devices, sensors, or laser applications. Even though innovative, progressive approaches for the precise doping of fabricated SC assemblies have developed recently. Noteworthy progress has been made in this field, yet the SC doping at a nanoscale level still faces significant challenges in the semiconductor engineering [49–51]. Attaining highly oriented, organized, and ordered dopant distribution and in nanostructures is extremely important to enhance SCs optoelectrical performance. Recently, a pn-junction comprising the composite material was formed via introducing nanorods in a polymer matrix to form hybrid bulk heterojunction (BHJ) solar cells [52]. The pn-junctions were established in n-type CdS nanorods via an organized cationic exchange method. The p-type (Cu_2S) generates at one end of CdS due to the selective reactivity of crystalline planes of the nanorods. In the nanorod, an epitaxial connection between Cu_2S and CdS SC creates a depletion region forming a pn-junction, which is a classic illustration of type-II band alignment. The junction parted charge carriers during illumination via a drift of minority carriers across the depletion region. Thus, hybrid BHJs centered on Cu_2S|CdS pn-junction nanorods in a polymer matrix represented as efficient solar cells as compared to similar BHJ devices with nanorods of individual materials (i.e., is Cu_2S or CdS). The variation in the length of nanorods helps in controlling charge separation and carrier transport to enhance solar cell efficiency. The optimum proficiency of 3.7% at 1 sun intensity in BHJs was observed, comprising Cu_2S|CdS (40:60, length-wise) pn-junctions in a P3HT matrix [52]. The efficiency is considerably developed in comparison to BHJ devices with nanorods of n-type CdS/p-type Cu_2S. Besides this, optimized dopants (bismuth) insertion in lead sulfide (PbS) QDs establishes a BHJ solar cell having higher efficiency studied via STS generating a type-II BAND alignment [53]. Similar investigations on substrate solution were carried out via STM to explore the mechanism and examine electrical mapping or thermal stress, such as, analysis of the lateral and vertical phase separation of the photoactive layer in organic solar cells comprising the solution-based polymer (donor) and fullerene (acceptor) in BHJ [54]. Amalgamation of an imidazolium-substituted polythiophene interlayer increased BJH efficiency significantly due to additional incorporated electric field promoting charge transfer [55]. In short, STM is appropriate to analyze semiconductors, metals, or biological samples competently [56].

4.2.2 Parallel PN-Junctions across Nanowires via One-Step Ex Situ Doping

Earlier, a single-step technique was employed for the transformation of undoped silicon NWs (SiNWs) into nanoscale building blocks introducing parallel pn-junctions having homogenous and consistent dopant distributions [57]. The specified

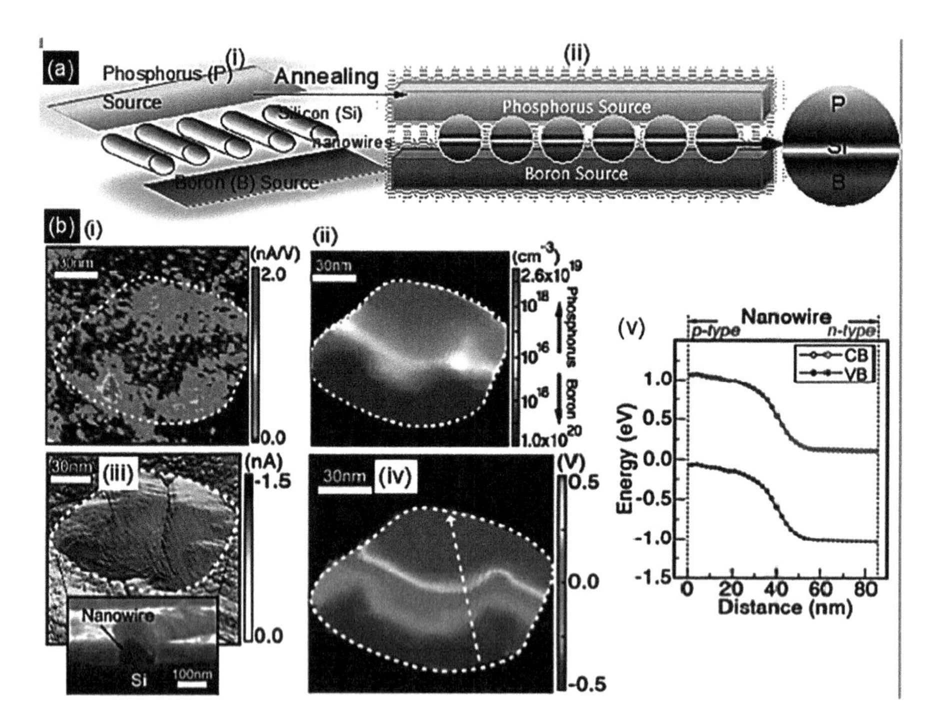

FIGURE 4.2 (a) Schematic of [i] parallel pn-junction configuration formation across oriented NWs. [ii] Intrinsic Si NWs are transferred to a pretreated substrate with a monolayer containing boron functionalities and then covered by a second pretreated substrate with a phosphorus-containing annealed monolayer resulting in the controlled decomposition of the monolayers and dopant diffusion onto the NWs. (b) [i] STM/S spectroscopic dI/dV image recorded at −3.5 V sample bias, [ii] dopant concentration distribution across the NW junction, [iii] typical cross-sectional STM image of the NW obtained at a sample bias of 3.0 V. Inset: SEM secondary electron image displaying the geometrical structure of the NW from the cross-sectional view, [iv] electrostatic potential distribution mapping, obtained by analyzing the STM/S data that demonstrates the formation of a junction in the NW, [v] CB edge and VB edge switched from p- to n-type across the NW, corresponding to the B- and P-doped poles, respectively, along the dashed arrow trajectory in (iv). The white dashed oval curve outlines the NW boundary. (Reproduced with permission [57])

alignment of the resultant asymmetric configuration takes place comparative to the macroscopic framework. A parallel pn-junction formation takes place via a one-step process that is applied to SiNWs (Figure 4.2a [i & ii]). The doping method depends on a monolayer contact doping (MLCD) process, which develops monolayers as a source for dopant atoms. The MLCD method was earlier established to distribute extremely well-ordered development of doping elements on Si. The dopant concentration varies from several orders of magnitude from the modified interface into the bulk. The analogous pn-junction was measured via STM to acquire the electronic spatial characteristics of the junction generated across the NWs to reveal the association between the nanoscale topography and local electronic configuration via the cross-sectional profile of the NW at 100 K. The topographic STM imaging was accompanied at a continuous current of 0.4 nA having sample voltage of −0.3 V. The spectra were attained by operating STM in the current imaging tunneling spectroscopy (CITS) mode, various tunneling current descriptions were acquired at different sample bias voltages, Vs (4.0 to −4.0 V). The STM images and cross-sectional analyses of a SiNW are presented in Figure 4.2b [i,ii,iii,iv].

Figure 4.2a shows the formation of parallel pn-junctions across the SiNW by the one-step doping process. The Si-doped with B and P engenders a p- and n-type electronic characteristics, as demonstrated by the shift in the CB edge relative to the Fermi level (Figure 4.2b [v]). The two-dimensional (2D) cross-sectional spatial reliance of the DOS, at −3.5 V, for the parallel pn-junction configuration doped Si NW is presented in Figure 4.2b [i]. The STM/S spectroscopic data deliver dopant distributions quantitative mapping, electrostatic potential, and local band structures within the doped NWs (Figure 4.2b [ii, iv, v]). At the P-doped pole, the dopant surface concentration was 2.6×10^{19} cm^3, whereas that at the B-doped pole was 1×10^{20} cm^3. The doping concentration at the NW was 3 orders of reduced magnitude, demonstrating a compensated area approximately 10^{17} cm^{-3} n- and p-type dopant concentrations at NW core regions. Generally, the one-step NW doping procedure generated parallel n^--i-p^+ junctions through the NWs. Specifically, the n-type dopant dispersal displayed a diffusion profile with a leading dopant concentration at the NW interface interaction region, having a monotonic decline in dopant from the NW exterior, including core, which shows consistency. The post-growth and doping method proposes to tailor surface chemistry and control over dopant dispersal generating asymmetric configuration alignment that is important for employing nanoscale building blocks useful in electronic applications.

4.3 INTERFACIAL BAND MAPPING ACROSS VERTICALLY PHASED SEPARATED POLYMER/FULLERENE HYBRID SOLAR CELLS

Above and beyond this, STM also helps in investigating the BHJ polymer solar cells. Usually, polymer solar cells gathered significant consideration in the fabrication of low-cost and mechanically flexible PVs as they support solution processing and patterning on flexible supports. The intensively explored and investigated materials for BHJ polymer solar cells comprise poly(3-hexylthiophene) (P3HT) and fullerene derivative phenyl-C61-butyric acid methyl ester (PCBM) blends, having power conversion of 4–5% [58–60]. In recent times high photovoltaic efficiencies (7–8%) have been acquired via integrating s semiconducting polymers having small bandgap [6]. In semiconducting polymer electron acceptors generally, amalgam with polymers to form BHJs with a nanoscale interpenetrating donor/acceptor linkage due to short exciton diffusion length (<20 nm) [61–63]. In a conventional BHJ polymer solar cell, electron-hole pair excitons characterize the governing photogenerated species. These excitons offer energetically favorable passageway, that is, from polymer (donor) to an electron-accepting species (acceptor). The device performance is directly influenced via adjusted phase-separated donor–acceptor configuration of the BHJ, which confirms sufficient dissociations of photogenerated excitons and consistent pathways for transferring charge carriers toward the electrodes [64,65]. The standard device structure generally comprises polymer/fullerene intermingling and is sandwiched between PEDOT: PSS-coated indium tin oxide (ITO) acts as anode and a low work function metal as a cathode [66,67]. At the donor/acceptor interfaces, the photoactive layer/electrode interfaces and interfacial energy band assemblies impact the efficiency of photoinduced charge separation, the transport or collection, and the resulting device performance. Previously, numerous high-resolution techniques have been used to analyze the nanoscale morphology of BHJ solar cells, such as SPM [66,68,69], transmission electron microscopy (TEM) [70,71], three-dimensional (3D) electron tomography demonstrating the morphological constraints of both lateral and vertical statistics. SPM explores the quantitative association between nanoscale morphologies and the local electronic properties [70,72]. Earlier, BHJ morphology had been explored only through the thin-film upper surface. The photogenerated charge carriers usually transport toward the opposite electrodes across the film thickness relatively parallel along a thin film surface. This issue requires a considerate association between the cross-sectional morphology and the local electronic structures in the vertical direction of BHJ materials. STM thus plays an influential role in spatially providing the DOS statistics about the heterojunction organic solar cells. A detailed analysis has been conducted on optimized P3HT: PCBM BHJ solar cell and the interfacial band-mapping images were investigated via cross-sectional scanning tunneling microscope (XSTM) under an UHV chamber. The evaluations envision the vertically phase-separated BHJ configuration of a P3HT: PCBM hybrid solar cell at a subnanometer resolution via STM having base pressure of 5×10^{-11} Torr. STM facilitates the direct interfacial band alignments at the donor–acceptor boundaries. The interfaces between the P3H T: PCBM amalgams and PEDOT: PSS layer having atomic scale spatial resolution. The investigation discloses the interaction between the nanomorphology of vertical phase-segregations and localized interfacial electronic configurations of the polymer/fullerene BHJ, which helps in boosting up charge generation, transportation, and collection of polymer solar cell devices. A Si(100) wafer was designated as the supportive substrate since it displays cleaved surface [73–75]. Initially, a 30 nm thick layer of PEDOT:PSS (Baytron P 4083) was spun-cast onto the Si(100) substrate to acquire a high-quality cross-sectional polymer/fullerene hybrid samples. The samples were moved into a nitrogen-purged glove box for succeeding deposition after baking the PEDOT: PSS films (120 °C for 1 h). The photoactive layer having thickness of 100 nm was then set down on top of the PEDOT: PSS layer via spin coating [1.0:0.8 weight ratio blend of P3HT: PCBM dissolved in chlorobenzene]. The prepared sample was thermally annealed at 150 °C for

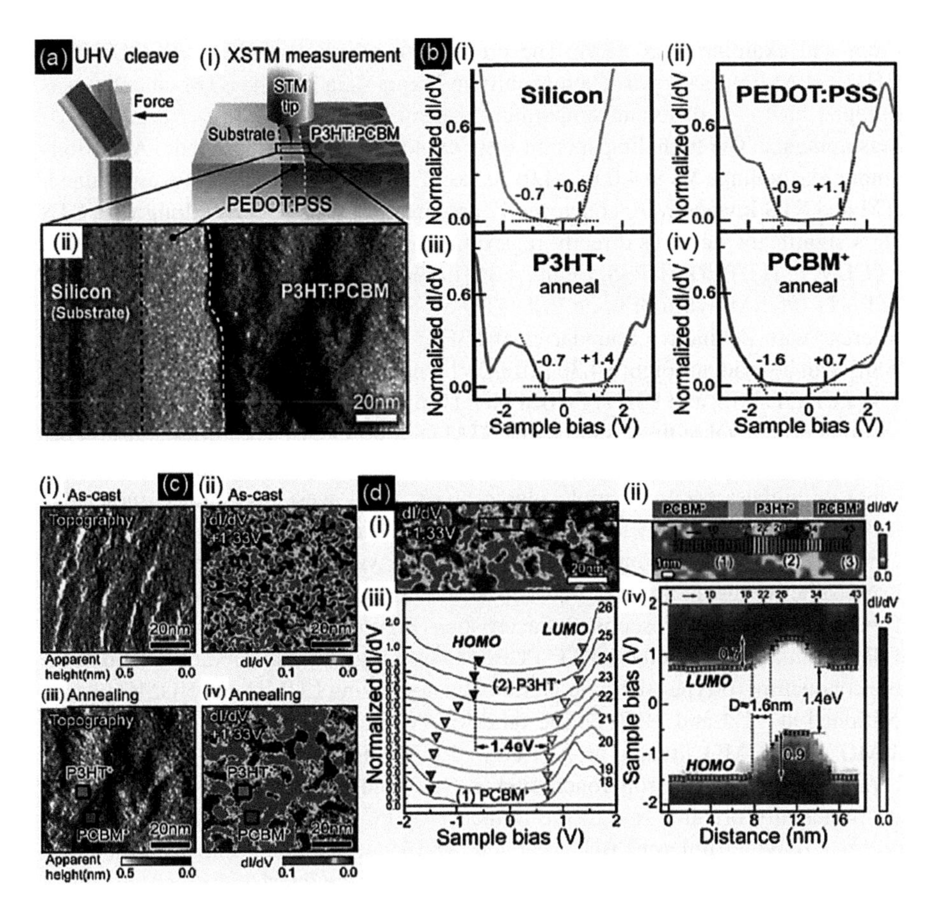

FIGURE 4.3 (a)[i] The graphic of the cleaving sample in XSTM measurements; [ii] cross-sectional STM topography image of the Si/PEDOT:PSS/P3HT:PCBM film. The image was taken at Vs of +4 V and tunneling current of 150 pA. (b) Specifically normalized dI/dV curves of [i] silicon (Si), [ii] PEDOT:PSS, (iii) P3HT-rich (P3HT+), and (iv) PCBM-rich (PCBM+) region of the thermally annealed sample (current onsets in filled and empty states were indicated each by dark ticks). (c) Cross-sectional STM images of [i] as-cast, [iii] the thermally annealed P3HT:PCBM sample. Normalized dI/dV images probed at +1.33 V sample bias of [ii] the as-cast and [iv] the thermally annealed P3HT: PCBM sample. The regions colored by green and red in the dI/dV images are electronically identified and represented as the portions of P3HT+ and PCBM+. (d) [i] Normalized dI/dV images of the P3HT: PCBM active layer with the thermal annealing treatment, [ii] magnified sliced image across the PCBM+(1)/(P3HT+)(2)/PCBM+(3) heterojunction, [iii] Local density of states (LDOS) measurements from PCBM+ to P3HT+ across the interfacial region are indicated by red, green, and yellow curves, [iv] Atomic-scale evolution of band alignment across the P3HT+/PCBM+ heterointerface. Scientific position numbering is made in panel b, and the corresponding numbers by the electronic curves in panel c, and above the band structure in panel d are also indicated (Reproduced with permission [76]).

5 min and examined via STM. The cross-sectional slice of the Si/PEDOT:PSS/P3HT:PCBM film is shown schematically in Figure 4.3a [i]. The STM chamber was equipped having a different temperature system, which helps in performing STS measurements. The tunneling spectra were explored via the CITS mode. At various sample bias voltage Vs (+4.0 to −4.0) series of tunnel current images was attained. STM and STS images were concurrently acquired at ~100 K. STM along with STS offers significant statistics directly regarding local electronic structure at interfaces of PCBM:P3HT/PEDOT:PSS. Figure 4.3a [ii] represents the typical XSTM image of the P3HT: PCBM hybrid film on PEDOT: PSS/Si (100) [Vs (+4.0 V)]. Three different areas with distinctive boundaries at P3HT: PCBM/PEDOT: PSS and PEDOT: PSS/Si can be evident. Figure 4.3b [i,ii,iii,iv] represents the normalized dI/dV curves of Si, PEDOT: PSS, and P3HT: PCBM blend film. Zero sample bias demonstrates the location Fermi level of the system in the STM measurements. The current onsets corresponding to valence and conduction bands (VB/CB) [the occupied/unoccupied states] or highest occupied molecular orbitals and lowest unoccupied molecular orbitals (HOMO/LUMO)) limits are extracted and indicated by tick marks Figure 4.3b [76]. The particular onset energy calculated was expected to be ±0.10 eV. A clear distinctiveness at the edge between Si and PEDOT: PSS can be notable because of their tunneling spectroscopy characteristics (Figure 4.3b). Besides this, two sorts of dI/dV curves seem in the P3HT: PCBM hybrid region. One curvature validates a hole conducting (p-type) semiconductor behavior having LUMO and HOMO phases positioned at +1.4 and −0.7 V. The other dI/dV spectrum possesses onset bias of LUMO and HOMO situated at +0.7 and −1.6 V (Figure 4.3b [iv]), displaying the characteristics of an electron-conducting (n-type) semiconductor. This kind of curvature principally oriented due to the influence of PCBM. Discussing the tunneling spectra of polymer/fullerene BHJ solar cell, XSTM measurements having a subnanometer resolution device make it promising to explore the interfacial electronic characteristics of the P3HT (donors) and PCBM (acceptors) including the interfaces between the P3HT:PCBM amalgam blends and the PEDOT: PSS stratum. Figure 4.3c [i, ii,iii,iv] represents the XSTM topography images of the P3HT: PCBM hybrid films before and after the annealing treatment. The thermally annealed sample displays a clear phase-separated morphology in comparison to as-cast sample having identifiable P3HT and PCBM dominant sections in a thin film. The mapping of the corresponding tunneling conductance was noted concurrently with topographical images via attaining the differential tunneling current (dI/dV) as a function of the sample bias (Figure 4.3c [ii],[iv]). The PCBM dI/dV spectrum demonstrates the emblematic characteristics of n-type (electron-conducting) semiconductor. The areas having increased tunneling current verified at +1.33 V sample bias. It is below the characteristic current onset of P3HT predominantly linked with the contribution of PCBM. The red color (Figure 4.3c [ii], [iv]) signifies the PCBM-rich (PCBM+) regions, while the green characterizes the P3HT-rich (P3HT+) region. In the as-cast P3HT: PCBM hybrid, the PCBM molecules are uniformly scattered within the P3HT matrix having smaller domain sizes (2–4 nm). The thermal treatment generates PCBM aggregates that give rise to bigger clusters having domain width of 10–20 nm. These domains, along the vertical direction, transform into an interpenetrated linkage within the P3HT matrix. It helps in generating the passageways prerequisite for charge transport. A larger current density and power conversion efficiency of (~4.2%)

is observed in the annealed samples in comparison to the as-cast sample (~2.4%). Furthermore, the morphological investigation attained from the XSTM measurement displays excellent agreement with the observation of the cross-sectional TEM images [77,78]. The XSTM result demonstrates that at subnanometer, STM is a distinctive tool to analyze the vertical nanoscale morphology and locally corresponding electronic properties crossways the layer width. It is observed that extent of fullerene aggregation results in fluctuation in electronic behaviors influencing the corresponding energy stages in the P3HT: PCBM samples. Figure 4.3d displays the interfacial electronic band mapping via spatial spectroscopic through the P3HT and PCBM heterojunctions. Figure 4.3d [i] displays the normalized dI/dV images of the P3HT: PCBM active stratum after thermal annealing treatment. A transversely small image of PCBM+/P3HT+/PCBM+ heterojunction is enlarged in Figure 4.3d [ii]. The zones number as (1), (3) symbolize the PCBM+ domains, while the zone number (2) displays the P3HT+ domain. To investigate the subnanometer electronic properties, the colored solid bars of Figure 4.4b indicate the scanning profile positions with a spatial separation of 0.4 nm across the regions numbered from (1) to (3). The solid colored sections in Figure 4.3d [ii] explore the subnanometer electronic characteristics. It designates the scanning profile having a spatial separation of 0.4 nm through the regions [numbered from (1) to (3)]. Figure 4.4c displays the characteristic tunneling spectra attained across the P3HT: PCBM heterojunction using differential tunneling current (dI/dV) as a function of the sample bias. The estimated positions of HOMO and LUMO (filled and empty states) resulting from the offsets of the tunneling current were extracted and showed (solid and dashed triangle marks) in Figure 4.3d [iii]. Based on the electronic characteristics of these spatial spectroscopic measurements, Figure 4.3d [iv] demonstrates the XSTM mapping image of DOS and the band configuration through the heterointerfaces of PCBM+/P3HT+/PCBM+. The P3HT and PCBM, HOMO and LUMO stage represents a typical type-II heterojunction. The offset in LUMO that lies between P3HT and PCBM is roughly 0.7 eV. It is higher than the binding energy of excitons in polymer (~0.2–0.5 eV) [78–80]. It proposes that charge separation might occur due to interface when electron transference from P3HT to PCBM takes place. However, HOMO offset level that lies between P3HT and PCBM is 0.9 eV. Generally, the energy variance between the HOMO and LUMO levels of P3HT and PCBM is associated with open-circuit voltage (V_{oc}) of a donor–acceptor BHJ solar cell (i.e., 1.4 eV) [79–81]. At P3HT/PCBM heterojunction, the band energy diagram at atomic-scale resolution based on XSTM measurement shows a steady trend [82–84]. Prominently, the atomic-scale progression of the local electronic structure across the P3HT/PCBM interface makes it promising to envisage the discrete band bending and electronic configurations at the interface. The P3HT and PCBM domain has an expected width of 1.6 nm. The characteristic diffusion length of excitons in a conjugated polymer is 10 nm[9]. The mean domain width of P3HT is 10–20 nm (Figure 4.3d [iv]), which proposes that excitons reach the interfaces during its lifetime. Hence, the large potential gradient emerging at the P3HT/PCBM boundary leads toward charge separation of photogenerated excitons and its transformation into free electrons or holes results in ultrafast electron transfer at the polymer/fullerene heterojunction interface [85]. At the interface between the PCBM+ and P3HT+ domains, the normalized dI/dV curvature has the position of the LUMO near to the Fermi level, signifying an n-type electronic characteristic. At the interface, the electronic configurations and DOS are nearer to those at the PCBM+ region than at the P3HT+ region. It implies that electron transfer is efficient and more enhanced than hole transport at the interface [85,86]. Preceding research has also suggested that significant charge transfer via interfacial bridge states is observed from the excited state of P3HT to the ground state of fullerene atoms when the photoactive layer is exposed to illumination [87,88]. Thus, the distinctiveness of STM measurements makes it applicable to explore the interfacial electronic structures between

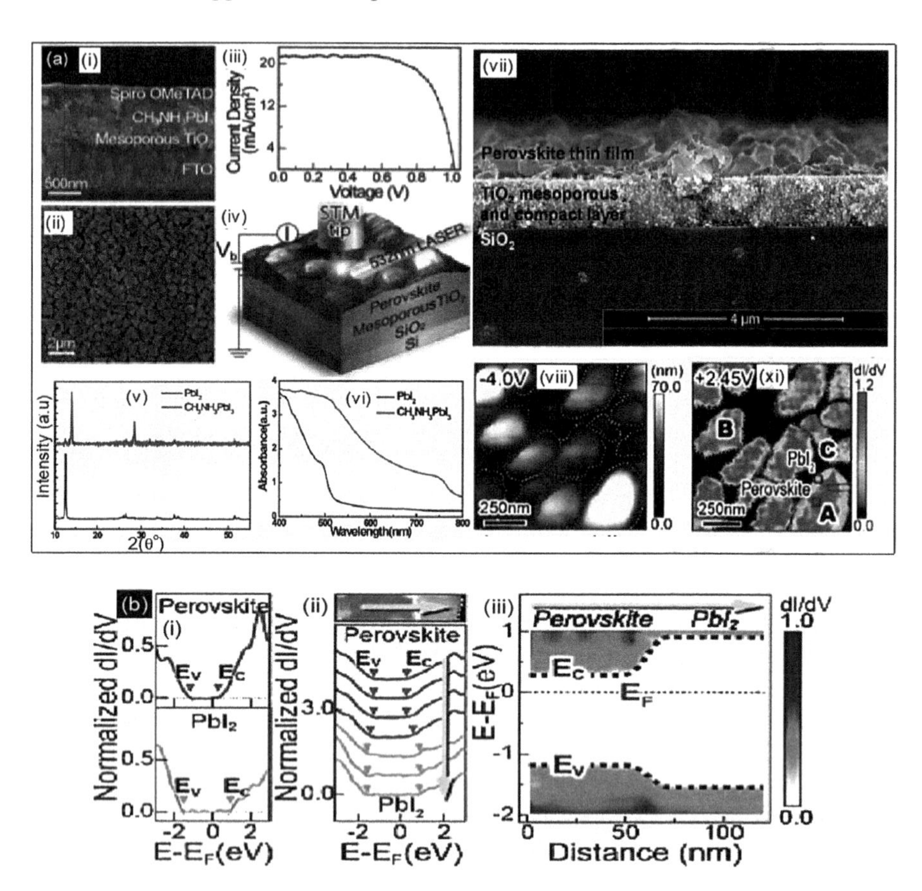

FIGURE 4.4 (a) [i,ii] SEM top [before deposition of spiro-OMeTAD] and cross-sectional image of a $CH_3NH_3PbI_3$ perovskite solar cell; [iii] current–voltage curves for $CH_3NH_3PbI_3$ perovskite solar cell; [iv] schematic description of the LM-STM measurement; [v] XRD characterization of the perovskite thin film after thermal annealing; [vi] absorption spectra of $CH_3NH_3PbI_3$ and PbI_2; [vii] cross-sectional scanning electron microscope image of $CH_3NH_3PbI_3$perovskite solar cell with the architecture consisting of Si substrate/SiO_2/ $CH_3NH_3PbI_3$; [viii] STM morphology image; and [xi] mapping image of the normalized dI/dV spectra of the perovskite grains. (b) [i] Normalized dI/dV curves (lower and upper panels) obtained at the outer part of the grain (blue square in grain A) and in the grain interior (red square in grain A); [ii] point-to-point electronic dI/dV curves acquired across the perovskite/ PbI_2 heterojunction in grain A; [iii] mapping image of the band alignment across the perovskite/ PbI_2 heterointerface in grain A by extracting the locations of the VB and CB edges; and

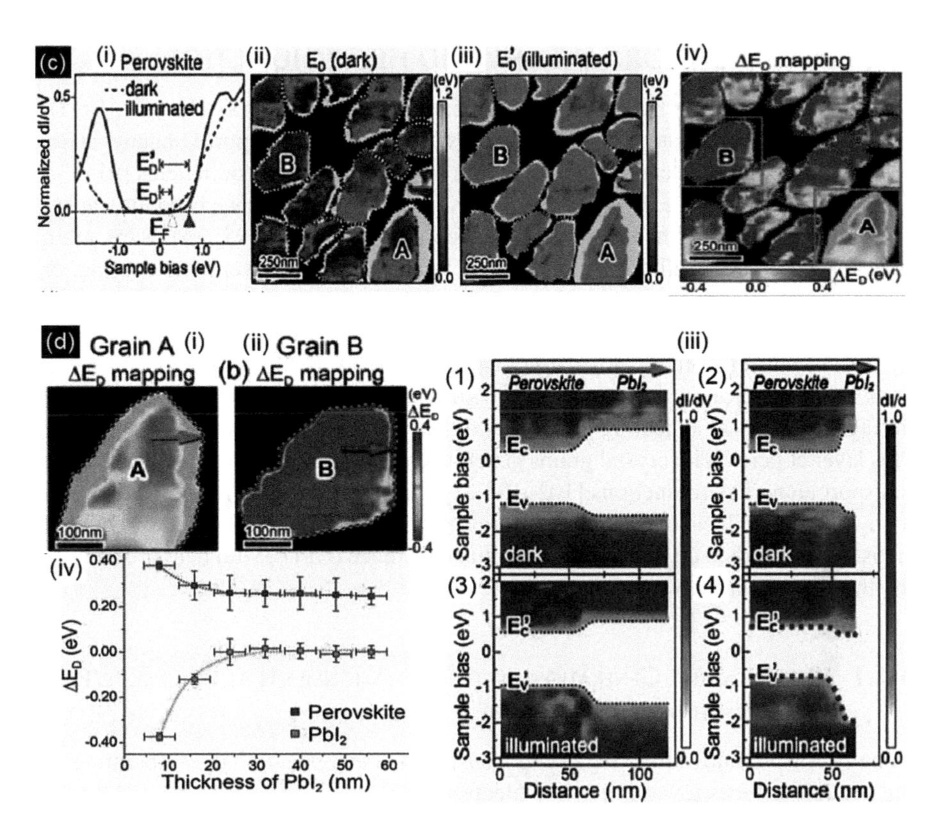

FIGURE 4.4 (CONTINUED) (c) [i] dI/dV curves of perovskites in the dark (dashed curve) and during light illumination (solid curve) conditions. Imaging of the spatial distribution of the ED and ED′ values derived for the perovskite grains [ii] in the dark and [iii] under illumination, [iv] A mapping image of the spatial distributions of ΔED values (ΔED = ED′ − ED) observed for the perovskite grains under light illumination. (d) Perovskite grains [i] grain A and [ii] grain B; [iii] PbI_2 thickness influence in the dark and under light illumination; (1) and (3) correspond to the band alignments across the heterojunctions (green arrows) of grain A consisting of a thicker PbI_2 (~45 nm); (2) and (4) correspond to the band alignments across the heterojunctions (green arrows) of grain B consisting of a thinner (~8 nm) PbI_2 layer; [iv] Thickness dependence of PbI_2 passivation layers on the ΔED energy shifts (acquired from statistical analyses of 12 perovskite crystal grains during light on/off) (Reproduced with permission [101]).

PEDOT: PSS and P3HT: PCBM composite layers. STM helps in determining the direct visualization of interfacial band-mapping descriptions of an optimized P3HT/PCBM BHJ solar cell through the film thickness in an UHV compartment. The exclusive benefit of XSTM is to illustrate BHJ solar cells and explore the quantitative relation between vertical morphologies corresponding to local interfacial electronic characteristics at an atomic-scale resolution sandwiched between the two opposite electrodes. Hence, this methodology has potentials to characterize and optimize nanoscale phase-separated organic hybrid photovoltaic amalgams for local carrier production, transportation, and collection.

4.4 ORGANIC–INORGANIC HYBRID HETEROJUNCTION

The organic, inorganic hybrid perovskite materials [$CH_3NH_3PbX_3$ (X = Cl, Br, I)] have also shown enhanced power conversion efficiencies (approximately 20%) [89,90] due to efficient carrier mobility, high optical absorption coefficients [91–93], and stable electron–hole diffusion lengths. Numerous methods have been directed to assemble perovskite films via monitoring the morphologies [94,95], exposure, grain dimensions, and crystallinity extent, via thermal treatment, solvent engineering, or integration of minor extents of additives [96–98]. Besides this, during light exposure, the characteristic intrinsic polycrystallinity of perovskite films predominately influences transport efficiency and charge generation [99,100]. Chen and his group fellows testified perovskite crystal grains with the PbI_2 layer grown via precipitation during post-thermal annealing [101]. The device performance enhances because the PbI_2 layer at perovskite crystal grains generates passivation, which influences carrier transport along heterojunctions [102–104]. Earlier AFM and KPFM have been used to investigate the local electronic structures of perovskite PVs. Both methods deliver statistics established on the contact potential difference (CPD), with help in concluding the corresponding work functions [105–107].

4.4.1 Photocarrier Generations and Band Alignments at Perovskite/PbI_2 Heterointerfaces

Comprehensive knowledge regarding transport behavior and charge separation at the heterointerfaces toward opposite electrodes via enlightening the mechanism of interfacial band alignments of VBs and CBs at perovskite crystals is a requisite to enhance the efficiency, strength, and stability of perovskite-based PVs. STM, along with scanning tunneling spectroscopy (STS), has the exclusive benefit of determining the local electronic density of states (LDOS) at the atomic scale (spatial resolution at a nanometer scale), interfacial band alignments, and charge generation behaviors. A light modulated (LM) STM method has been employed to explore the correlation at heterointerfaces between interfacial electronic structures and compositional distributions of $CH_3NH_3PbI_3$ perovskite crystals and PbI_2 thin film (passivation layer). Figure 4.4a [i,ii] demonstrates a top-view and cross-sectional scanning electron microscope (SEM) image of a $CH_3NH_3PbI_3$ perovskite solar cell consisting of FTO/TiO_2/$CH_3NH_3PbI_3$/spiro-OMeTAD and made up via sequential deposition process (two-step process). A top-view SEM image [before deposition of a hole transporting layer (HTL) of spiro-OMeTAD] exposes a granular morphology encompassing many isolated $CH_3NH_3PbI_3$ grains (having dimensions of 200–600 nm), analogous to morphological opinions in preceding information. The mesoscopic $CH_3NH_3PbI_3$ perovskite solar cell demonstrated promising device performance having V_{oc} of 1.02 V, a current density (J_{sc}) of 21.2 mA/cm^2, PCE of 15.06%, and fill factor (FF) of 69.6% (Figure 4.4a [iii]). LM-STM [under UHV] were used (laser diode, $\lambda = 532$ nm, power intensity = 3 mW/cm^2) to execute experiments of photoinduced carrier generation and band alignments at heterointerfaces between the $CH_3NH_3PbI_3$ perovskite crystals and the PbI_2 passivation layers at nanometer spatial resolution (Figure 4.4a [iv]). The X-ray diffraction (XRD) analyses were carried out after thermal annealing (Figure 4.4a [v]). Strong diffraction peaks were detected at 14.2° and 28.5°, from (110) and (220) planes of the tetragonal ($CH_3NH_3PbI_3$) perovskite lattice. A lower diffraction peak at 12° that corresponds to the (001) plane of the PbI_2 lattice demonstrates the existence of PbI_2 in the $CH_3NH_3PbI_3$ perovskite thin film. A distinctive difference is observed at the absorption spectra of $CH_3NH_3PbI_3$ and PbI_2 (Figure 4.4a [vi]). The light illumination at $\lambda = 532$ nm was used for excitation to assure photocarrier generation (electrons and holes) that primarily takes place in $CH_3NH_3PbI_3$ region rather in the PbI_2 region. For the LM-STM measurement, the FTO substrate was substituted with a 550 μm silicon substrate; because the FTO electrode for perovskite solar cells is characteristically deposited on a 2 mm glass substrate, which is too dense to be compatible in STM sample holder. A 300 nm compact SiO_2 stratum was thermally grown up via wet oxygen on topmost layer of Si substrate to preclude the photocurrent resulting from Si substrate contributing to the tunneling current (Figure 4.4a [vii]). The device configuration comprises TiO_2 dense stratum, mesoporous TiO_2 scaffold, and $CH_3NH_3PbI_3$ photoactive coating sequentially deposited. Figure 4.4a [viii] shows a topographic STM image of the perovskite film measured in the dark. The perovskite crystal grains are bordered via the dashed lines. In the perovskite film, numeral grains having different dimensions (i.e., 200–600 nm) were perceived, which is analogous to SEM top-view image (Figure 4.4a [ii]). The normalized differential conductances (dI/dV) via STS were carried out to determine the compositional distribution within individual grain at room temperature to analyze DOS at perovskite grains. Figure 4.4a [xi] displays the mapping image of the normalized dI/dV spectra of the perovskite grains noted at the sample bias of +2.45 V. At the perovskite grains, the electronic image (Figure 4.4a [xi]) demonstrates two distinctive local electronic behaviors. Generally, stronger dI/dV signals appear in grain interiors (indicated via red color), while lower dI/dV signals appear at the exterior of grains (displayed via blue color). In the STS measurement, an undetectable electronic signal seems to appear due to the height variance among grains, which is noticed as the signal from the background, displayed as black (Figure 4.4a [xi]). Figure 4.4b [i] demonstrates two representatives normalized dI/dV curves (as a function of the sample bias) at the grain interior and the exterior part of grain A (Figure 4.4a [xi]), highlighted with red and blue squares. In the STM measurement, position of the "0" sample bias represents the Fermi level (EF) of the system.

Figure 4.4a [xi] demonstrates grains A, B, and C. Grain A comprises thicker (30–50 nm) PbI_2 layer, grain B consists of a thinner (5–8 nm) PbI_2 layer, and grain C comprises the mix phases of $CH_3NH_3PbI_3$ and PbI_2 at the corresponding perovskite crystal grains. The STM mapping image of the normalized dI/dV spectra acquired via STS measurement is shown in Figure 4.4a [xi]. These were consistent with the preceding assumption consequences that thin PbI_2 passivation films may be predominantly precipitated nearby the perovskite crystal grains after thermal annealing. The VB and CB edges corresponding to occupied and unoccupied states during current onsets were extracted as indicated via triangular marks (Figure 4.4b [i]). The grain interior showed more conductive feature in comparison to the exterior part of the grain demonstrated via characteristic dI/dV curve. The energy positions of the VB and CB edges were at −1.2 and +0.3 V, signifying an expected bandgap of 1.5 eV, as shown in Figure 4.4b [ii]. These curves were primarily distributed in the interior of the perovskite grains, primarily resulting due to perovskite $CH_3NH_3PbI_3$ according to the earlier reported values of the bandgap size. The representative dI/dV curves were perceived at the exterior of grain A, which displays low conductivity. The onset biases of the VB and CB edges were situated at −1.5 and +0.9 V, displaying a bandgap of 2.4 eV (Figure 4.4b [i]). This kind of curve can be ascribed due to the presence of PbI_2 thin stratum. At the perovskite crystal grains, the dissemination of the respective constituents can be distinguished from the corresponding normalized dI/dV curves due to the distinctive electronic characteristics between perovskite $CH_3NH_3PbI_3$ and PbI_2. Figure 4.4b [i] demonstrates the progression of tunneling spectra as a function of the sample bias attained across the perovskite/PbI_2 heterojunction at grain A (denoted by the rectangular box (Figure 4.4a [xi]) to analyze the band configuration between the perovskite crystals and the PbI_2 of separable perovskite grains. According to the electronic physiognomies of the spatially reliant spectroscopic curves (Figure 4.4b [ii]), the locations of the VB and CB edges are designated by triangular mark yielding a mapping image of the band configuration through the perovskite/PbI_2 heterointerface in grain A as shown in Figure 4.4b [iii]. The band energy of the VB and CB edges for perovskite/PbI_2 were nearly 0.3 and 0.6 eV, clarifying a particular type-I heterojunction at the perovskite/PbI_2 edge.

Besides this, $CH_3NH_3PbI_3$ in the interior of grain and PbI_2 at the peripheral part of the grain displayed intrinsic n-type electronic behavior. The Fermi levels are close to the corresponding CB edges, showing consistency with preceding reported results [88,108]. Generally, the conventional methods define the ionization potentials or electron affinities of discrete constituents to acquire energy illustrations of a heterojunction arrangement [108]. Thus, STM helps indirectly envisaging spatially resolved band alignment and electronic configurations across the $CH_3NH_3PbI_3/PbI_2$ heterojunctions.

4.4.2 Photocarrier Generations of Perovskites during Illumination

Generally, in semiconductors, the energy differences (EDs) between band (CB and VB) edges and Fermi levels can be used to evaluate the carrier concentrations and carrier type. The LM-STM method is used to conduct real-space interpretations of the photoinduced charge transfer and band bending at the perovskite crystals during its exposure to illumination. The ED shift between the energetic positions of the CB edge (EC) and Fermi level (EF) is mathematically expressed as ED = EC − EF. This expression is used to characterize photoinduced charge transference behaviors in perovskite grains during light exposure. The expressions ED and ED′, as shown in Figure 4.4c [i], indicate the separation between the CB boundaries (EC or EC′) and the Fermi level (EF) before and after light exposure. The position of the CB boundary (EC) [of $CH_3NH_3PbI_3$ part (denoted by the open red triangular symbol)] shifts toward a higher energy level [(EC′, indicated by the solid red triangular symbol)] during light illumination, as a result, increasing the values of ED up to ED′. The development of the LDOS acquired via the dI/dV curve (at a negative sample bias) after light illumination was also perceived. This improvement was primarily resulted due to amplified tunneling current from the VB to the tip, demonstrating an enlarged carrier density of photoinduced holes in the $CH_3NH_3PbI_3$ region of the perovskite grains. Figure 4.4c [ii, iii] displays the distributions of the ED and ED′, which were acquired via the individual normalized dI/dV spectra for the perovskite grains during dark and light illumination. During light exposure, photoexcited carriers were engendered and separated in the perovskite crystals, ensuing the difference between ED and ED′. Therefore, the transferals of ΔED values, defined as $\Delta ED = ED' - ED$, directly interrelated with the difference of photogenerated carrier concentrations or carrier types, which creates under light illumination. Positive ΔED values are associated with upsurges in hole concentrations. As the ED between the CB edge and the Fermi level enhances. However, negative ΔED values are interconnected with a rise in electron concentrations as the CB edge is nearby to the Fermi level after light exposure.

Figure 4.4c [iv] displays ΔED mapping image of the spatial distributions of perovskite grains under light exposure. In the grain interiors, the ΔED values were maximally positive (colored red-orange). The exterior parts of these grains were either negative (colored blue) or nearly zero (colored gray). Notably, the ΔED values mapping image at these perovskite grains displays a connection to the compositional distributions of $CH_3NH_3PbI_3$ and PbI_2, demonstrating that the shift of the ΔED values and the corresponding charge transfer behaviors be influenced by the chemical constituents of perovskite grains. Generally, $CH_3NH_3PbI_3$ perovskite material performs as an ambipolar SC having long electron–hole diffusion lengths. Even though the intrinsic $CH_3NH_3PbI_3$ demonstrates a Fermi level close to the CB edge [108,109]. The experimental result demonstrates that photogenerated holes were efficiently extracted in comparison to electrons in $CH_3NH_3PbI_3$-based perovskite solar cells [22] The positive ΔED values were perceived in the grain interiors ($CH_3NH_3PbI_3$ regions) chiefly accredited to the increased hole concentrations in these regions during under light exposure. These carriers can be effectively collected by introducing HTL (spiro-OMeTAD), contributing to hole current [110,111]. Besides this, Figure 4.4c [iv] exposes that the magnitudes of positive ΔED values and the resultant hole concentrations can also be influenced by the thickness of the PbI_2 passivation layers of the corresponding grains. For example, grain B (Figure 4.4c [iii]) encompassing a thin PbI_2 layer shows higher positive ΔED values specifying a significant increment in the hole concentrations in grain interiors during light illumination. Beside this, negative ΔED values were perceived at the adjoining interface of grain B. It is due to increase in electron concentration, signifying improvement in charge separation occurs at the heterointerface of grain B. Thus, the ΔED values remain consistent at the exterior of grain A (comprising thicker PbI_2 layer), specifying insignificant efficacy in charge separation at the heterointerface. In the central region ($CH_3NH_3PbI_3$) of grain A, positive ΔED [+0.25 eV] values were observed which were lower than those detected in the internal regions of grain B [+0.40 eV]. Furthermore, the average ΔED value at the peripheral region (PbI_2 part) of the grain A was approximately −0.03 eV. The ΔED value at the peripheral region of grain B was around −0.34 eV, specifying efficient charge separation at the heterointerface. Hence, these experimental results suggest that the photoinduced charge generation is strongly influenced by the passivation layers at the perovskite crystal grains.

4.4.3 Band Alignments of Perovskites during Illumination

The perovskite/PbI_2 heterojunctions of grains A and B were observed to explore the impact of the PbI_2 layer on the photocarriers' generation, separation at the heterointerface, and the interfacial band alignments across perovskite/PbI_2 heterojunctions as shown in Figure 4.4d [i – grain A, ii – grain B]). Figure 4.4d [iii(1–4)] demonstrates the interfacial band alignments at grains A and B during dark and light exposure resulting from the typical curves of the point-to-point STS spectroscopic statistics. During illumination, in grain B, a significant downward band bending was detected in the PbI_2 region (Figure 4.4d [iii (2, 4)]). The energy shifts in the PbI_2 region of grain A (Figure 4.4d [iii(1,3)] remain nearly consistent. It clearly shows that the photogenerated electrons were efficaciously transported from the central region ($CH_3NH_3PbI_3$) part to the peripheral (PbI_2) region in grain B comprising a thinner PbI_2 layer. In grain A, the thicker PbI_2 layer hinders the photoinduced charge transfer across the perovskite/PbI_2 interface and also upsurges the possibility of carrier recombination. Given that, the upsurge in the hole concentration in the central region of grain A was significantly inferior to that in grain B during light illumination. The VB and CB boundaries shift toward high energy levels approximately 0.25 Ev (ΔED = +0.25 eV) in grain A, and 0.40 eV in grain B (ΔED = +0.40 eV) in the perovskite region. Though the variation of a thickness (PbI_2 passivation layers) at the outer regions of perovskite crystal grains is yet not exclusively known. One possibility is the occurrence of nonuniform decomposition of $CH_3NH_3PbI_3$ during thermal treatment, which takes place at a moderate temperature via releasing $CH_3NH_3PbI_3$ and precipitating PbI_2 phase or defective sites occurrence [112–114]. Thus, LM-STM measurements are a useful technique to acquire the spatially resolved chemical compositions and interfacial band alignments simultaneously to link the photocarrier generation behaviors at the heterointerfaces of perovskite crystals [115–117].

4.4.4 PbI_2 Layer Thickness Dependence of ΔED

Figure 4.4d [iv] demonstrates the thickness of the PbI_2 passivation layers' dependence on the ΔED energy shifts attained via statistical investigations of 12 perovskite crystal grains. A noticeable photoinduced charge separation and transfer efficiencies at the heterointerfaces between the $CH_3NH_3PbI_3$ and PbI_2 passivation layers can be experiential. When the thickness of PbI_2 layer is <20 nm, positive ΔED values corresponding to increased hole concentrations at $CH_3NH_3PbI_3$ regions (grain centers) and negative ΔED values corresponding to increased electron concentrations at the PbI_2 region (grains exterior) can be perceived concurrently. When PbI_2 passivation layer >20 nm, the ΔED values in the PbI_2 regions were negligible, and the ΔED values at the $CH_3NH_3PbI_3$ regions (grain interiors) were unaffected. It signifies minor charge separation occurrence at the $CH_3NH_3PbI_3$/PbI_2 heterointerfaces of perovskite grains. Hence, an ideal strategy of PbI_2 passivation layers at perovskite crystals can enable charge separation and prevent back recombination, therefore improving the device performance of perovskite solar cells. STM offers and benefits in understanding efficient charge separation and transport of photogenerated carriers at the heterointerfaces of perovskite crystal grains. The electrons can be transported through thin PbI_2 passivation layers to the electron transporting layer (TiO_2) and holes in the $CH_3NH_3PbI_3$ region can be collected via the introduction of HTL (spiro-OMeTAD) resulting low recombination loss.

Besides this, synchronized STM and synchrotron X-ray measurements were also executed in a gas environment [118]. A technologically advanced apparatus was progressed appropriately for executing the local X-ray absorption measurements via transforming a previously established high-pressure STM and X-ray scattering device [119]. Numerous methods have been made to accomplish the prerequisite sensitivity to measure the small X-ray-induced alteration of the tunneling current that transmits the local X-ray absorption signal [120,121]. STM/STS and synchrotron X-ray diffraction were also studied on clustered C_3N_4 nanoparticles (nanoflakes) [122]. The morphological features and electronic characteristics were investigated. The experiential nanoflakes bandgap exposes the occurrence of diverse phases dominating the graphitic C_3N_4 abundant phase. Moreover, STM-electroluminescence (STMEL) is noticed in C_3N_4 nanoflakes set down on a gold substrate. The tunneling current generates photons, three times more energetic than the tunneling electrons. A nonlinear optical phenomenon or a localized state were the two possible representations to clarify the perceived photon emission energy [122]. STM also helps in assessing the defects, twists, distortions, and deformations in materials at the nanoscale. These defects can be modified to enhance device performance. Earlier defects were found via STM in low-temperature grown GaAs crystals [123]. These crystals exhibit a photoinduced alteration to a metastable state at 90 K. The excitation spectrums were close to the photoquenching spectra of EL2 centers in bulk GaAs crystals according to simulation and experimental results. Few defects were not affected via light illumination, demonstrating that the EL2 family origin is due to environmental influence [123]. Thus, STM is an influential device that permits a direct observation at atomic arrangements of solid surfaces, including the point defects even underneath the crystal surface. Feenstra and his colleagues explore and image the As-related defects in LT-GaAs samples [124]. Similar studies were performed by various research groups determining the satellite peaks via STM in GaAs (110) [125], low-temperature-grown gallium arsenide (LTG: GaAs) possessing mid gap states [126,127] or estimating donor band states positioned at near E(v)+0.5 eV [124]. Further investigation reveals that the surface antisite defects are electrically inactive, having a confined defect state, which contributes to rise a distinct feature in STM images [128]. A phosphorus dopant near clean Si (100) - (2 × 1) surfaces displaying the charge-induced band bending interlinked with the dopants shows enhancement in both filled and empty states [129]. Such as the observation of substitutional phosphorus in the top layer and interstitial phosphorus on clean Si (100)-(2x1) via STM [130] or substitutional geometry and strain impact in stratums of phosphorus on Si(111) [131]. The exploration of bulk dopants, defects, and buried interfaces atom-by-atom in real space has developed via a cross-sectional STM method. According to a fundamental viewpoint, the presence of dopants, defects occurrence, and interfaces in semiconductors possess much importance as they are essential constituents of semiconductor devices as perceived from theoretical analysis via simulation and experimental evidences.

4.5 OUTLOOK AND UPCOMING CHALLENGES

STM has been verified to be an exceptional apparatus for the determination of DOS of SCs via band mapping across junctions. The heterojunctions established on organic or inorganic semiconductors in the form of organic/organic, organic/inorganic, or inorganic/inorganic interfaces can be easily studied through STM. The band illustration computes via this technique will infer the separation of carriers and pathways for electrons and holes to the opposite electrodes. In a heterojunction, the materials morphology/topography can be explored through STS/STM. The dI/dV imaging method, at the nanoscale, can be identified easily because of the energy-level mapping. In heterojunction, the domain identification and the corresponding correlating morphology and dynamic characteristics can be attained. It is due to continual passageways for electrons and holes, which decreases the recombination loss yielding efficient transport routes. The energy band diagrams and domains of donor and acceptor materials illustration help in attaining novel insights toward device performance enhancement.

In photovoltaics, the application of STM is a moderately new area of research. The bands' illustrations analyzed from STM are according to the associated Fermi energy. The materials, CB and VB energies, have allowed drawing the exact band diagram. The cross-sectional STS–STM arrangement helps in developing domains along with the device depth. In this framework, the band diagram of numerous heterojunctions for different optoelectronic devices, such as solar cell applications, could be drawn precisely. The acquired type-II band alignment for PVs grounded on forthcoming materials can be drawn formerly before implementing the device practically. Thus, the functionality of devices, at a commercial level and industrial scale, beginning from batteries and fuel cells toward the high-tech devices, includes various processes comprising electrocatalysts, corrosion, electroactive polymers, molecular machines, or biological systems. The majority of them are underpinned via an intricate network of spatially separated electrochemical progressions and supramolecular assemblies linked via covalent, hydrogen or van der Walls forces. The local probing and exploration of these molecular behaviors on the scale of individual structural components and inhomogeneity, from grains to imperfections at the atomic and molecular level, is mandatory to attain advancement in this field.

Earlier, researchers have established numerous methods and systems for the macroscopic characterization of these interconnected dependent networks. Though, these procedures were usually restricted to macroscopically materials responses, the contributions of distinct components are averaged out. The mechanisms, that is, local reaction, nucleation rates, lateral transport of ionic species between reaction sites, remain unfamiliar and unnoticed. Besides this, at the nanoscale, these mechanisms are generally unrealistic due to detection confinement and sensitivity. STM method is beneficial for observation, developing topographic or functional imaging modalities. In STM, additional prospects are unlocked via the manifestation of multiple (noninvasive) approaches sensitive to different parameters. Though, nanoscale confinement also facilitates new imaging styles when the probe turns out to be a dynamic, active part of the procedure, both persuading and identifying responses locally.

The related mechanisms often significantly vary from the macroscopic correspondents ever since the local electroneutrality, surface responses, polarizability, etc. become important. STM technique can offer an understanding of the local substrate solvent functionalities at atomic scales specifically in optoelectronic devices for future technical advancement in BHJ devices.

REFERENCES

1. C.W. Tang, Two-layer organic photovoltaic cell, *Applied Physics Letters*, 48 (1986) 183–185.
2. N.S. Sariciftci, D. Braun, C. Zhang, V.I. Srdanov, A.J. Heeger, G. Stucky, F. Wudl, Semiconducting polymer-buckminsterfullerene heterojunctions: Diodes, photodiodes, and photovoltaic cells, *Applied Physics Letters*, 62 (1993) 585–587.
3. M.C. Scharber, N.S. Sariciftci, Efficiency of bulk-heterojunction organic solar cells, *Progress in Polymer Science*, 38 (2013) 1929–1940.
4. A.J. Heeger, 25th anniversary article: Bulk heterojunction solar cells: Understanding the mechanism of operation, *Advanced Materials*, 26 (2014) 10–28.
5. M.J. Greaney, S. Das, D.H. Webber, S.E. Bradforth, R.L. Brutchey, Improving open circuit potential in hybrid P3HT:CdSe bulk heterojunction solar cells via colloidal tert-butylthiol ligand exchange, *ACS Nano*, 6 (2012) 4222–4230.
6. S. Ren, L.Y. Chang, S.K. Lim, J. Zhao, M. Smith, N. Zhao, V. Buloviä, M. Bawendi, S. Gradecak, Inorganic-organic hybrid solar cell: Bridging quantum dots to conjugated polymer nanowires, *Nano Letters*, 11 (2011) 3998–4002.
7. M. Kus, F. Ozel, S. Buyukcelebi, A. Aljabour, A. Erdogan, M. Ersoz, N.S. Sariciftci, Colloidal $CuZnSnSe_{4-x}S_x$ nanocrystals for hybrid solar cells, *Optical Materials*, 39 (2015) 103–109.
8. J.Y. Choi, I. Kim, G.E. Jabbour, Enhanced surface passivation of colloidal CdSe nanocrystals for improved efficiency of nanocrystal/polymer hybrid solar cells, *IEEE Journal of Photovoltaics*, 6 (2016) 1–7.
9. C. Yong, W. Gang, D. Pan, Colloidal synthesis and optical properties of metastable wurtzite I3-III-IV-VI5 (Cu3InSnS5) nanocrystals, *Crystengcomm*, 15 (2013) 10459–10463.
10. Y. Gu, W. Jie, L. Li, Y. Xu, Y. Yang, J. Ren, G. Zha, T. Wang, L. Xu, Y. He, Te inclusion-induced electrical field perturbation in CdZnTe single crystals revealed by Kelvin probe force microscopy, *Micron*, 88 (2016) 48–53.
11. X.N. Zhang, H.M. Hu, Investigation of interfaces of ionic liquid via Kelvin probe force microscopy at room temperature, *Acta Physica Sinica*, 32 (2016) 1722–1726.
12. A.D. Yoffe, Low-dimensional systems: Quantum size effects and electronic properties of semiconductor microcrystallites (zero-dimensional systems) and some quasi-two-dimensional systems, *Advances in Physics*, 42 (2002) 173–262.
13. A. Majdabadi, M.R. Gaeeni, M.S. Ghamsari, M.H. Majles-Ara, Investigation of stability and nonlinear optical properties CdSe colloidal nanocrystals, *Journal of Laser Applications*, 27 (2015) 022010.
14. A.D. Yoffe, Low-dimensional systems: Quantum size effects and electronic properties of semiconductor microcrystallites (zero-dimensional systems) and some quasi-two-dimensional systems, *Advances in Physics*, 51 (2002) 799–890.
15. C. Xie, Q.M. Wu, R.N. Li, G.C. Gu, X. Zhang, N. Li, R. Berndt, J.R. KröGer, Z.Y. Shen, S.M. Hou, Isolated supramolecules on surfaces studied with scanning tunneling microscopy, *Chinese Chemical Letters*, 27 (2016) 807–812.
16. A. Ashrafi, Band offsets at ZnO/SiC heterojunction: Heterointerface in band alignment, *Surface Science*, 604 (2010) L63–L66.

17. Y. Tak, H. Kim, D. Lee, K. Yong, Type-II CdS nanoparticle-ZnO nanowire heterostructure arrays fabricated by a solution process: Enhanced photocatalytic activity, *Chemical Communications*, 8 (2008) 4585.
18. N.J. Borys, M.J. Walter, J. Huang, D.V. Talapin, J.M. Lupton, The role of particle morphology in interfacial energy transfer in CdSe/CdS heterostructure nanocrystals, *Science*, 330 (2010) 1371–1374.
19. T. Teranishi, D. Inui, T. Yoshinaga, M. Saruyama, M. Kanehara, M. Sakamoto, A. Furube, Crystal structure-selective formation and carrier dynamics of type-II CdS–Cu31S16 heterodimers, *Journal of Materials Chemistry C*, 1 (2013) 3391–3394.
20. A.N. Baranov, N. Bertru, Y. Cuminal, G. Boissier, C. Alibert, A. Joullié, Observation of room-temperature laser emission from type III InAs/GaSb multiple quantum well structures, *Applied Physics Letters*, 71 (1998) 735–737.
21. B.K. Sharma, N. Khare, S. Ahmad, A ZnO/PEDOT:PSS based inorganic/organic hetrojunction, *Solid State Communications*, 149 (2009) 771–774.
22. A.A. Toropov, I.V. Sedova, O.G. Lyublinskaya, S.V. Sorokin, A.A. Sitnikova, S.V. Ivanov, J.P. Bergman, B. Monemar, F. Donatini, S.D. Le, Coexistence of type-I and type-II band lineups in Cd(Te,Se)/ZnSe quantum-dot structures, *Applied Physics Letters*, 89 (2006) 551.
23. E.J. Tyrrell, S. Tomić, Effect of correlation and dielectric confinement on $1S_{1/2}{}^{(e)}nS_{3/2}{}^{(h)}$ excitons in CdTe/CdSe and CdSe/CdTe type-II quantum dots, *Journal of Physical Chemistry C*, 119 (2015) 12720–12730.
24. S. Christodoulou, G. Vaccaro, V. Pinchetti, F.D. Donato, J.Q. Grim, A. Casu, A. Genovese, G. Vicidomini, A. Diaspro, S. Brovelli, Synthesis of highly luminescent wurtzite CdSe/CdS giant-shell nanocrystals using a fast continuous injection route, *Journal of Materials Chemistry C*, 2 (2014) 3439–3447.
25. A.M. Saad, M.M. Bakr, I.M. Azzouz, M.T.H.A. Kana, Effect of temperature and pumping power on the photoluminescence properties of type-II CdTe/CdSe core-shell QDs, *Applied Surface Science*, 257 (2011) 8634–8639.
26. S. Dey, A. Pal, Tuning of band-edges in type-I core-shell nanocrystals through band-offset engineering: Selective quantum confinement effect, *RSC Advances*, 3 (2013) 13225–13231.
27. S. Dey, K. Mohanta, A.J. Pal, Co-occurrence of conductance switching and magnetization: Tuning of electrical bistability of Fe_3O_4 quantum dots by magnetic field, *Chemical Physics Letters*, 492 (2010) 281–284.
28. K. Papatryfonos, G. Rodary, C. David, F. Lelarge, A. Ramdane, J.C. Girard, One-dimensional nature of InAs/InP quantum dashes revealed by scanning tunneling spectroscopy, *Nano Letters*, 15 (2015) 4488.
29. L. Wang, Q. Chen, G.B. Pan, L.J. Wan, S. Zhang, X. Zhan, B.H. Northrop, P.J. Stang, Nanopatterning of donor/acceptor hybrid supramolecular architectures on highly oriented pyrolytic graphite: A scanning tunneling microscopy study, *Journal of the American Chemical Society*, 130 (2008) 13433–13441.
30. W. Ling, Q. Chen, G.B. Pan, L.J. Wan, S. Zhang, X. Zhan, B.H. Northrop, P.J. Stang, Nanopatterning of donor/acceptor hybrid supramolecular architectures on highly oriented pyrolytic graphite: A scanning tunneling microscopy study, *Journal of the American Chemical Society*, 130 (2008) 13433–13441.
31. S.S. Li, B.H. Northrop, Q.H. Yuan, L.J. Wan, P.J. Strang, ChemInform abstract: Surface confined metallosupramolecular architectures: Formation and scanning tunneling microscopy characterization, *Cheminform*, 40 (2010) 249–259.
32. T.Y. Chien, L.F. Kourkoutis, J. Chakhalian, B. Gray, M. Kareev, N.P. Guisinger, D.A. Muller, J.W. Freeland, Visualizing short-range charge transfer at the interfaces between ferromagnetic and superconducting oxides, *Nature Communications*, 4 (2013) 2336.

33. H. Dai, S. Wang, I. Hisaki, S. Nakagawa, N. Ikenaka, K. Deng, X. Xiao, Q. Zeng, On-surface self-assembly of a C3-symmetric π-conjugated molecule family studied by STM: Two-dimensional nanoporous frameworks, *Chemistry – An Asian Journal*, 12 (2017) 2558–2564.
34. T.Y. Chien, J. Chakhalian, J.W. Freeland, N.P. Guisinger, Cross-sectional scanning tunneling microscopy applied to complex oxide interfaces, *Advanced Functional Materials*, 23 (2013) 2565–2575.
35. D.B. Dougherty, P. Maksymovych, J.T.Y. Jr, Direct STM evidence for Cu-benzoate surface complexes on Cu(1 1 0), *Surface Science*, 600 (2006) 4484–4491.
36. S. de Feyter, A. Miura, H. Uji-I, P. Jonkheijm, A.P.H.J. Schenning, E.W. Meijer, Z. Chen, F. Würthner, F. Schuurmans, J. Van Esch, Supramolecular chemistry at the liquid/solid interface a scanning tunneling microscopy approach, *Solid State Phenomena*, 121–123 (2007) 369–372.
37. A. Bera, S. Dey, A.J. Pal, Band mapping across a pn-junction in a nanorod by scanning tunneling microscopy, *Nano Letters*, 14 (2014) 2000–2005.
38. B. Sadtler, D.O. Demchenko, H. Zheng, S.M. Hughes, M.G. Merkle, U. Dahmen, L.W. Wang, A.P. Alivisatos, Selective facet reactivity during cation exchange in cadmium sulfide nanorods, *Journal of the American Chemical Society*, 131 (2009) 5285–5293.
39. D.O. Demchenko, R.D. Robinson, B. Sadtler, C.K. Erdonmez, A.P. Alivisatos, L.W. Wang, Formation mechanism and properties of CdS-Ag_2S nanorod superlattices, *ACS Nano*, 2 (2008) 627–636.
40. B. Freitag, G. Knippels, S. Kujawa, M.V.D. Stam, D. Hubert, P.C. Tiemeijer, C. Kisielowski, P. Denes, A. Minor, U. Dahmen, First performance measurements and application results of a new high brightness Schottky field emitter for HR-S/TEM at 80–300kV acceleration voltage, *Microscopy Society of America*, 14 (2008) 1370–1371.
41. A.E. Saunders, A. Ghezelbash, P. Sood, B.A. Korgel, Synthesis of high aspect ratio quantum-size CdS nanorods and their surface-dependent photoluminescence, *Langmuir the Acs Journal of Surfaces & Colloids*, 24 (2008) 9043.
42. K.M. Ryan, S. Singh, P. Liu, A. Singh, Assembly of binary, ternary and quaternary compound semiconductor nanorods: From local to device scale ordering influenced by surface charge, *Crystengcomm*, 16 (2014) 9446–9454.
43. K. Vinokurov, Y. Bekenstein, V. Gutkin, I. Popov, O. Millo, U. Banin, Rhodium growth on Cu2S nanocrystals yielding hybrid nanoscale inorganic cages and their synergistic properties, *Crystengcomm*, 16 (2014) 9506–9512.
44. U. Banin, Y. Benshahar, K. Vinokurov, Hybrid semiconductor? Metal nanoparticles: From architecture to function, *Chemistry of Materials*, 26 (2013) 97–110.
45. U. Banin, O. Millo, Tunneling and optical spectroscopy of semiconductor nanocrystals, *Annual Review of Physical Chemistry*, 54 (2003) 465–492.
46. D. Katz, T. Wizansky, O. Millo, E. Rothenberg, T. Mokari, U. Banin, Size dependent tunneling and optical spectroscopy of CdSe quantum rods, *Physical Review Letters*, 89 (2002) 086801.
47. O. Millo, D. Katz, Y.W. Cao, U. Banin, Tunneling and optical spectroscopy of InAs and InAs/ZnSe core/shell nanocrystalline quantum dots, *Physica Status Solidi B*, 224 (2010) 271–276.
48. A. Antanovich, A. Prudnikau, V. Gurin, M. Artemyev, Cd/Hg cationic substitution in magic-sized CdSe clusters: Optical characterization and theoretical studies, *Chemical Physics*, 455 (2015) 32–40.
49. T.J. Kempa, B. Tian, D.R. Kim, J. Hu, X. Zheng, C.M. Lieber, Single and tandem axial p-i-n nanowire photovoltaic devices, *Nano Letters*, 8 (2008) 3456.
50. S. Yu, B. Witzigmann, A high efficiency dual-junction solar cell implemented as a nanowire array, *Optics Express*, 21 (2013) 167.

51. J.C. Ho, Y. Roie, Z.A. Jacobson, F. Zhiyong, R.L. Alley, J. Ali, Controlled nanoscale doping of semiconductors via molecular monolayers, *Nature Materials*, 7 (2008) 62.
52. U. Dasgupta, A. Bera, A.J. Pal, pn-Junction nanorods in a polymer matrix: A paradigm shift from conventional hybrid bulk-heterojunction solar cells, *Solar Energy Materials & Solar Cells*, 143 (2015) 319–325.
53. S.K. Saha, B. Abhijit, A.J. Pal, Improvement in PbS-based hybrid bulk-heterojunction solar cells through band alignment via bismuth doping in the nanocrystals, *ACS Applied Materials & Interfaces*, 7 (2015) 8886–8893.
54. R.C. Chintala, J. Tait, P. Eyben, E. Voroshazi, S. Surana, C. Fleischmann, T. Conard, W. Vandervorst, Insights into the nanoscale lateral and vertical phase separation in organic bulk heterojunctions via scanning probe microscopy, *Nanoscale*, 8 (2016) 3629–3637.
55. J. Drijkoningen, J. Kesters, T. Vangerven, E. Bourgeois, L. Lutsen, D. Vanderzande, W. Maes, J. D'Haen, J. Manca, Investigating the role of efficiency enhancing interlayers for bulk heterojunction solar cells by scanning probe microscopy, *Organic Electronics*, 15 (2014) 1282–1289.
56. A. Cricenti, R. Generosi, Air operating atomic force-scanning tunneling microscope suitable to study semiconductors, metals, and biological samples, *Review of Scientific Instruments*, 66 (1995) 2843–2847.
57. O. Hazut, B.-C. Huang, A. Pantzer, I. Amit, Y. Rosenwaks, A. Kohn, C.-S. Chang, Y.-P. Chiu, R. Yerushalmi, Parallel p–n junctions across nanowires by one-step ex situ doping, *ACS Nano*, 8 (2014) 8357–8362.
58. O. Hazut, R. Yerushalmi, Direct Dopant patterning by remote monolayer doping enabled by monolayer fragmentation study, *Langmuir*, 33 (2017) 5371–5377.
59. F. Patolsky, G. Zheng, O. Hayden, M. Lakadamyali, X. Zhuang, C.M. Lieber, Electrical detection of single viruses, *Proceedings of the National Academy of Sciences of the United States of America*, 828 (2004) 14017–14022.
60. F. Patolsky, G. Zheng, C.M. Lieber, Nanowire sensors for medicine and the life sciences, *Nanomedicine*, 1 (2006) 51–65.
61. P. Taheri, H.M. Fahad, M. Tosun, M. Hettick, D. Kiriya, K. Chen, A. Javey, Nanoscale junction formation by gas-phase monolayer doping, *ACS Applied Materials & Interfaces*, 9 (2017) 20648–20655.
62. Z. Sun, O. Hazut, B.-C. Huang, Y.-P. Chiu, C.-S. Chang, R. Yerushalmi, L.J. Lauhon, D.N. Seidman, Dopant diffusion and activation in silicon nanowires fabricated by ex situ doping: A correlative study via atom-probe tomography and scanning tunneling spectroscopy, *Nano Letters*, 16 (2016) 4490–4500.
63. C. Celle, C.L. Mouchet, E. RouvièRe, J.P. Simonato, Controlled in situ n-doping of silicon nanowires during VLS growth and their characterization by scanning spreading resistance microscopy, *Journal of Physical Chemistry C*, 114 (2010) 760–765.
64. E. Koren, J.K. Hyun, U. Givan, E.R. Hemesath, L.J. Lauhon, Y. Rosenwaks, Obtaining uniform dopant distributions in VLS-grown Si nanowires, *Nano Letters*, 11 (2011) 183.
65. U. Givan, J.K. Hyun, E. Koren, J.S. Hammond, D.F. Paul, L.J. Lauhon, Y. Rosenwaks, Direct measurement of inhomogeneous longitudinal dopant distribution in SiNWs using nano-probe scanning auger microscopy, *MRS Proceedings*, 1349 (2011) mrss11-1349-dd1308-1305.
66. W. Yang, A. Broski, J. Wu, Q. Fan, L. Wen, Characteristics of transparent, PEDOT:PSS coated Indium-Tin-Oxide (ITO) microelectrodes, *IEEE Transactions on Nanotechnology*, 185–186 (2017) 1–1. https://doi.org/10.1016/j.synthmet.2013.10.005
67. E. Andreoli, K.S. Liao, A. Haldar, N.J. Alley, S.A. Curran, PPy:PSS as alternative to PEDOT:PSS in organic photovoltaics, *Synthetic Metals*, 185–186 (2013) 71–78.
68. N. Misra, C.P. Grigoropoulos, D.P. Stumbo, J.N. Miller, Laser activation of dopants for nanowire devices on glass and plastic, *Applied Physics Letters*, 93 (2008) 5213.

69. O. Hazut, A. Agarwala, T. Subramani, S. Waichman, R. Yerushalmi, Monolayer contact doping of silicon surfaces and nanowires using organophosphorus compounds, *The Journal of Visualized Experiments*, 82 (2013) 50770.
70. H. Lin, J. Lagoute, V. Repain, C. Chacon, Y. Girard, J.S. Lauret, R. Arenal, F. Ducastelle, S. Rousset, A. Loiseau, Coupled study by TEM/EELS and STM/STS of electronic properties of C- and CN_x-nanotubes, *Comptes rendus - Physique*, 12 (2011) 909–920.
71. N.A. Feoktistov, V.G. Golubev, J.L. Hutchison, D.A. Kurdyukov, A.B. Pevtsov, R. Schwarz, J.S.L.M. Sorokin, TEM and HREM study of silicon and platinum nanoscale ensembles in 3D dielectric opal matrix, *MRS Proceedings*, 609 (2000) A24.24.
72. F. Cardon, W.P. Gomes, W. Dekeyser, *Photovoltaic and Photoelectrochemical Solar Energy Conversion*, Springer, Boston, MA, 1981. Editors Affiliation: Laboratory for Crystallography and the Study of Solids and Laboratory for Physical ChemistryState University of GentGentBelgium Online ISBN 978-1-4615-9233-4. https://doi.org/10.1007/978-1-4615-9233-4
73. Z. Gan, D.E. Perea, J. Yoo, S. Tom Picraux, D.J. Smith, M.R. Mccartney, Mapping electrostatic profiles across axial p-n junctions in Si nanowires using off-axis electron holography, *Applied Physics Letters*, 103 (2013) 625.
74. Z. Gan, D.J. Smith, M.R. Mccartney, D.E.P.S.T. Picraux, Mapping the electrostatic profile across axial p-n junctions in Si nanowires using off-axis electron holography, *Microscopy & Microanalysis*, 18 (2012) 1826–1827.
75. Z. Gan, D.E. Perea, J. Yoo, H. Yang, R.J. Colby, J.E. Barker, G. Meng, S.X. Mao, C. Wang, S.T. Picraux, Characterization of electrical properties in axial Si-Ge nanowire heterojunctions using off-axis electron holography and atom-probe tomography, *Journal of Applied Physics*, 120 (2016) 617.
76. R.M. Feenstra, Tunneling spectroscopy of the (110) surface of direct-gap III-V semiconductors, *Physical Review B - Condensed Matter and Materials Physics*, 50 (1994) 4561–4570.
77. A.A. Herzing, L.J. Richter, I.M. Anderson, 3D nanoscale characterization of thin-film organic photovoltaic device structures via spectroscopic contrast in the TEM 1, *The Journal of Physical Chemistry C*, 114 (2010) 17501–17508.
78. L. Drummy, R. Davis, D. Moore, M. Durstock, R. Vaia, J. Hsu, Molecular-scale and nanoscale morphology of P3HT:PCBM bulk heterojunctions: Energy-filtered TEM and low-dose HREM, in: *Chem. Mater*, 23(3)2011, 907–912.
79. J.A. Mcleod, A.L. Pitman, E.Z. Kurmaev, L.D. Finkelstein, I.S. Zhidkov, A. Savva, A. Moewes, Linking the HOMO-LUMO gap to torsional disorder in P3HT/PCBM blends, *Journal of Chemical Physics*, 143 (2015) 593.
80. G. Garcia-Belmonte, P.P. Boix, J. Bisquert, M. Sessolo, H.J. Bolink, Simultaneous determination of carrier lifetime and electron density-of-states in P3HT:PCBM organic solar cells under illumination by impedance spectroscopy, *Solar Energy Materials & Solar Cells*, 94 (2010) 366–375.
81. Y. Shen, L. Scudiero, M.C. Gupta, Temperature dependence of HOMO-LUMO levels and open circuit voltage for P3HT:PCBM organic solar cells, *MRS Proceedings*, 1360 (2011) 51–59.
82. O.P. Dimitriev, D.A. Blank, C. Ganser, C. Teichert, Effect of the polymer chain arrangement on exciton and polaron dynamics in P3HT and P3HT:PCBM films, *The Journal of Physical Chemistry C*, 122 (2018) 17096–17109.
83. C.J. Brabec, A. Cravino, D. Meissner, N.S. Sariciftci, T. Fromherz, M.T. Rispens, L. Sanchez, J.C. Hummelen, Origin of the open circuit voltage of plastic solar cells, *Advanced Functional Materials*, 11 (2010) 374–380.
84. M.C. Shih, B.C. Huang, C.C. Lin, S.S. Li, H.A. Chen, Y.P. Chiu, C.W. Chen, Atomic-scale interfacial band mapping across vertically phased-separated polymer/fullerene hybrid solar cells, *Nano Letters*, 13 (2013) 2387–2392.

85. C.J. Brabec, G. Zerza, G. Cerullo, S.D. Silvestri, S. Luzzati, J.C. Hummelen, S. Sariciftci, Tracing photoinduced electron transfer process in conjugated polymer/fullerene bulk heterojunctions in real time, *Chemical Physics Letters*, 340 (2001) 232–236.
86. H. Xu, T. Xiao, J. Li, J. Mai, X. Lu, Z. Ni, In situ probing of the charge transport process at the polymer/fullerene heterojunction interface, *Journal of Physical Chemistry C*, 119 (2015) 25598–25605.
87. L.J.A. Koster, V.D. Mihailetchi, P.W.M. Blom, Bimolecular recombination in polymer/fullerene bulk heterojunction solar cells, *Applied Physics Letters*, 88 (2006) 85.
88. Y.K. And, J.C. Grossman, Insights on interfacial charge transfer across P3HT/fullerene photovoltaic heterojunction from Ab initio calculations, *Nano Letters*, 7 (2007) 1967–1972.
89. M.A. Green, K. Emery, Y. Hishikawa, W. Warta, E.D. Dunlop, Solar cell efficiency tables (Version 45), *Progress in Photovoltaics Research & Applications*, 23 (2015) 1–9.
90. S. Heo, H.I. Lee, T. Song, J.B. Park, D.S. Ko, J.G. Chung, K.H. Kim, S.H. Kim, D.J. Yun, Y.N. Ham, Direct band gap measurement of Cu(In,Ga)(Se,S)2 thin films using high-resolution reflection electron energy loss spectroscopy, *Applied Physics Letters*, 106 (2015) 145–150.
91. C. Wehrenfennig, M. Liu, H.J. Snaith, M.B. Johnston, L.M. Herz, Homogeneous emission line broadening in the organo lead halide perovskite $CH_3NH_3PbI_{3-x}Cl_x$, *Journal of Physical Chemistry Letters*, 5 (2014) 1300–1306.
92. C. Wehrenfennig, M. Liu, H.J. Snaith, M.B. Johnston, L.M. Herz, Charge carrier recombination channels in the low-temperature phase of organic-inorganic lead halide perovskite thin films, *APL Materials*, 2 (2014) 591–647.
93. C. Wehrenfennig, G.E. Eperon, M.B. Johnston, H.J. Snaith, L.M. Herz, High charge carrier mobilities and lifetimes in organolead trihalide perovskites, *Advanced Materials*, 26 (2014) 1584–1589.
94. H. Cho, C. Wolf, J.S. Kim, H.J. Yun, J.S. Bae, H. Kim, J.M. Heo, S. Ahn, T.W. Lee, High-efficiency solution-processed inorganic metal halide perovskite light-emitting diodes, *Advanced Materials*, 29 (2017) 1700579.
95. H. Tsai, W. Nie, J.C. Blancon, C.C. Stoumpos, R. Asadpour, B. Harutyunyan, A.J. Neukirch, R. Verduzco, J.J. Crochet, S. Tretiak, High-efficiency two-dimensional Ruddlesden–Popper perovskite solar cells, *Nature*, 536 (2016) 312–316.
96. A.J. Neukirch, W. Nie, J.C. Blancon, K. Appavoo, H. Tsai, M.Y. Sfeir, C. Katan, L. Pedesseau, J. Even, J.J. Crochet, Polaron stabilization by cooperative lattice distortion and cation rotations in hybrid perovskite materials, *Nano Letters*, 16 (2016) 3809–3816.
97. J.C. Blancon, W. Nie, A.J. Neukirch, G. Gupta, S. Tretiak, L. Cognet, A.D. Mohite, J.J. Crochet, The effects of electronic impurities and electron–hole recombination dynamics on large-grain organic–inorganic perovskite photovoltaic efficiencies, *Advanced Functional Materials*, 26 (2016) 4283–4292.
98. A. Buin, P. Pietsch, J. Xu, O. Voznyy, A.H. Ip, R. Comin, E.H. Sargent, Materials processing routes to trap-free halide perovskites, *Nano Letters*, 14 (2014) 6281–6286.
99. M. Song, J.H. Park, S.K. Chang, D.H. Kim, Y.C. Kang, S.H. Jin, W.Y. Jin, J.W. Kang, Highly flexible and transparent conducting silver nanowire/ZnO composite film for organic solar cells, *Nano Research*, 7 (2014) 1370–1379.
100. Y.H. Kim, C. Wolf, H. Cho, S.H. Jeong, T.W. Lee, Highly efficient, simplified, solution-processed thermally activated delayed-fluorescence organic light-emitting diodes, *Advanced Materials*, 28 (2016) 734–741.
101. M.C. Shih, S.S. Li, C.H. Hsieh, Y.C. Wang, H.D. Yang, Y.P. Chiu, C.S. Chang, C.W. Chen, Spatially resolved imaging on photocarrier generations and band alignments at perovskite/PbI2 heterointerfaces of perovskite solar cells by light-modulated scanning tunneling microscopy, *Nano Letters*, 17 (2017) 1154.

102. N.A. Davydova, V.A. Bibik, *Optical Phonons in Layer PbI2 Crystals with Stacking Faults*, (1995). https://doi.org/10.1117/12.226218
103. S. Mabrouk, A. Dubey, W. Zhang, N. Adhikari, B. Bahrami, M.N. Hasan, S. Yang, Q. Qiao, Increased efficiency for perovskite photovoltaics via doping the PbI_2 Layer, *Journal of Physical Chemistry C*, 120 (2016) 24577–24582.
104. C. Ying, C. Shi, N. Wu, J. Zhang, M. Wang, A two-layer structured PbI_2 thin film for efficient planar perovskite solar cells, *Nanoscale*, 7 (2015) 12092–12095.
105. E. Edri, S. Kirmayer, A. Henning, S. Mukhopadhyay, K. Gartsman, Y. Rosenwaks, G. Hodes, D. Cahen, Why lead methylammonium tri-iodide perovskite-based solar cells require a mesoporous electron transporting scaffold (but not necessarily a hole conductor), *Nano Letters*, 14 (2014) 1000–1004.
106. A.R. Yusoff, M.K. Nazeeruddin, Organohalide lead perovskites for photovoltaic applications, *Journal of Physical Chemistry Letters*, 7 (2014) 2448–2463.
107. D. Liu, J. Yang, T.L. Kelly, Compact layer free perovskite solar cells with 13.5% efficiency, *Journal of the American Chemical Society*, 136 (2014) 17116–17122.
108. P. Schulz, Interface energetics in organo-metal halide perovskite-based photovoltaic cells, *Energy & Environmental Science*, 7 (2014) 1377–1381.
109. A. Gupta, F.J. Owens, K.V. Rao, Z. Iqbal, J.M.O. Guille, R. Ahuja, High-temperature ferromagnetism in Cu-doped GaP by SQUID magnetometry and ferromagnetic resonance measurements, *Physical Review B*, 74 (2006) 224449.
110. H. Zhou, C. Qi, L. Gang, L. Song, T. Song, H.S. Duan, Z. Hong, J. You, Y. Liu, Y. Yang, [Report] Interface engineering of highly efficient perovskite solar cells, *American Association for the Advancement of Science*, 345 (2014), 542–546.
111. Q. Chen, H. Zhou, T.B. Song, S. Luo, Z. Hong, H.S. Duan, L. Dou, Y. Liu, Y. Yang, Controllable self-induced passivation of hybrid lead iodide perovskites toward high performance solar cells, *Nano Letters*, 14 (2014) 4158.
112. M. Shirayama, M. Kato, T. Miyadera, T. Sugita, T. Fujiseki, S. Hara, H. Kadowaki, D. Murata, M. Chikamatsu, H. Fujiwara, Degradation mechanism of $CH_3NH_3PbI_3$ perovskite materials upon exposure to humid air, *Journal of Applied Physics*, 119 (2016) 10–356.
113. T. Supasai, N. Rujisamphan, K. Ullrich, A. Chemseddine, Formation of a passivating $CH_3NH_3PbI_3/PbI_2$ interface during moderate heating of $CH_3NH_3PbI_3$ layers, *Applied Physics Letters*, 103 (2013) 1739.
114. S.R. Raga, M.C. Jung, M.V. Lee, M.R. Leyden, Y. Kato, Y. Qi, Influence of air annealing on high efficiency planar structure perovskite solar cells, *Chemistry of Materials*, 27 (2015) 150203172552005.
115. K. Fukuda, M. Nishizawa, T. Tada, L. Bolotov, K. Suzuki, S. Sato, H. Arimoto, T. Kanayama, *Simulation of light-illuminated STM measurements*, in: *International Conference on Simulation of Semiconductor Processes & Devices*, Yokohama, Japan, September 9–11, 2014.
116. F. Endres, N. Borisenko, S.Z. El Abedin, R. Hayes, R. Atkin, The interface ionic liquid(s)/electrode(s): In situ STM and AFM measurements, *Faraday Discussions*, 154 (2011) 221–233.
117. T. Matsushima, T. Okuda, T. Eguchi, M. Ono, A. Harasawa, T. Wakita, A. Kataoka, M. Hamada, A. Kamoshida, Y. Hasegawa, Development and trial measurement of synchrotron-radiation-light-illuminated scanning tunneling microscope, *Review of Scientific Instruments*, 75 (2004) 2149–2153.
118. R.V. Mom, W.G. Onderwaater, M.J. Rost, M. Jankowski, S. Wenzel, L. Jacobse, P.F.A. Alkemade, V. Vandalon, M.A. van Spronsen, M. van Weeren, B. Crama, P. van der Tuijn, R. Felici, W.M.M. Kessels, F. Carlà, J.W.M. Frenken, I.M.N. Groot, Simultaneous scanning tunneling microscopy and synchrotron X-ray measurements in a gas environment, *Ultramicroscopy*, 182 (2017) 233–242.

119. W.G. Onderwaater, P.C.V.D. Tuijn, R.V. Mom, M.A.V. Spronsen, S.B. Roobol, A. Saedi, J. Drnec, H. Isern, F. Carla, T. Dufrane, Combined scanning probe microscopy and x-ray scattering instrument for in situ catalysis investigations, *Review of Scientific Instruments*, 87 (2016) 113705.
120. N. Shirato, M. Cummings, H. Kersell, Y. Li, B. Stripe, D. Rosenmann, S.W. Hla, V. Rose, Elemental fingerprinting of materials with sensitivity at the atomic limit, *Nano Letters*, 14 (2014) 6499.
121. A. Dilullo, N. Shirato, M. Cummings, H. Kersell, S.W. Hla, V. Rose, *Direct elemental and magnetic contrast of magnetic thin films and nanoparticles measured by synchrotron X-ray scanning tunneling microscopy and spectroscopy*, in: *APS March Meeting 2015, Bulletin of the American Physical Society*, San Antono, TX, 60, March 2–6, 2015.
122. E.P. Andrade, B.B.A. Costa, C.R. Chaves, A.M. de Paula, L.A. Cury, A. Malachias, G.A.M. Safar, STM-electroluminescence from clustered C3N4 nanodomains synthesized via green chemistry process, *Ultrasonics Sonochemistry*, 40 (2018) 742–747.
123. K. Maeda, A. Hida, Y. Iguchi, Y. Mera, T. Fujiwara, STM nanospectroscopic studies of individual As-antisite defects in GaAs, *Materials Science in Semiconductor Processing*, 6 (2003) 253–256.
124. R.M. Feenstra, J.M. Woodall, G.D. Pettit, Observation of bulk defects by scanning tunneling microscopy and spectroscopy: Arsenic antisite defects in GaAs, *Physical Review Letters*, 71 (1993) 1176.
125. R.B. Capaz, K. Cho, J.D. Joannopoulos, Signatures of bulk and surface arsenic antisite defects in GaAs(110), *Physical Review Letters*, 75 (1995) 1811–1814.
126. R.M. Feenstra, J.M. Woodall, G.D. Pettit, Scanning tunneling microscopy and spectroscopy of arsenic antisite defects in GaAs, in: *Materials Science Forum*, 143–147 (1993), 1311–1318. https://doi.org/10.4028/www.scientific.net/msf.143-147.1311.
127. N.P. Chen, D.B. Janes, Distribution model of arsenic antisite defects in LTG:GaAs, *Journal of Physics & Chemistry of Solids*, 69 (2008) 325–329.
128. P. Ebert, P. Quadbeck, K. Urban, B. Henninger, K. Horn, G. Schwarz, J. Neugebauer, M. Scheffler, Identification of surface anion antisite defects in (110) surfaces of III–V semiconductors, *Applied Physics Letters*, 79 (2001) 2877–2879.
129. G.W. Brown, H. Grube, M.E. Hawley, Observation of buried phosphorus dopants near clean Si(100)-(2x1) with scanning tunneling microscopy, *Physical Review B*, 70 (2004) 121301.
130. G.W. Brown, B.P. Uberuaga, H. Grube, M.E. Hawley, S.R. Schofield, N.J. Curson, M.Y. Simmons, R.G. Clark, Observation of substitutional and interstitial phosphorus on clean Si (100) – (2 × 1) with scanning tunneling microscopy, *Physical Review B*, 72 (2005) 195323.
131. L. Vitali, M.G. Ramsey, F.P. Netzer, Substitutional geometry and strain effects in overlayers of phosphorus on Si(111), *Physical Review B*, 57 (1998) 15376–15384.

5 Organic–Inorganic Semiconducting Nanomaterial Heterojunctions

Jie Guan, Ziwei Wang, Yuan-Cheng Zhu, Wei-Wei Zhao, and Qin Xu

CONTENTS

5.1 OVERVIEW

Semiconductors have gained much attention in their preparation and application because of their important roles in several technologies such as catalysts [1], light-emitting diodes [2], waveguides [3], nanoscale electronic devices [4], laser technology [5], solar cells [6], chemical sensors, and biosensors [7]. When the sizes of semiconductors are reduced to the nanoscale, the properties of physical and chemical conductivity and optical characteristics (absorption coefficient and refractive index) change dramatically, resulting in unique natures because of the quantum effects. Due to these characteristics, semiconductive nanomaterials can interact with other materials (e.g., polymers), which are becoming functional materials in the fields of photocatalysis, light capture, sensors, and imaging. Nanotechnology will effectively promote the further development of semiconductor nanomaterials.[8].

Generally, people divide semiconductors into inorganic semiconductors and organic semiconductors according to the composition of materials. According to the width of the bandgap, inorganic semiconductors are divided into wide and narrow bandgap semiconductors [9]. Wide bandgap semiconductors mainly include diamond, SiC and IIB sulfides, oxides, and selenides. These kinds of materials have large bandgap width, high electron drift saturation speed, high thermal conductivity, low dielectric constant, and other characteristics and are suitable for manufacturing high-power, high-temperature, and high-density integrated electronic devices. They can also make use of their large forbidden bandwidth to produce light-detection devices and light-emitting devices. Narrow bandgap semiconductor materials are mainly group IV–VI compounds [10]. Narrow bandgap materials are sensitive to the external environment due to their small forbidden bandwidths and are suitable for manufacturing sensitive devices and detect pieces. However, its narrow absorption band, fewer optional materials, and high production costs make the device preparation process complicated, the device production cost high, and it is difficult to achieve industrial production. All in all, although inorganic semiconductors have low bandgap (metal conductive) properties and good thermo mechanical stability, a number of issues including harsh processing conditions, limited light absorption, low quantum yield, and so on have hindered the development of inorganic semiconductors.

Various methods have been developed to solve these problems to date. One of the best strategies is to construct heterojunction. In 1960, Anderson [11] made high-quality heterojunctions for the first time, laying the foundation for the development of heterojunctions. Unlike inorganic semiconductors, organic semiconductors typically have a bandgap of 2–3 ev and have low dielectric constant and electron mobility [12]. In addition, the semiconductor characteristics of organic molecules with high extinction coefficient and light resistance are impressive. They can not only absorb a wide range of visible light and provide practical functions like photosensitizers but also offer good process conditions to prepare high-quality semiconductor nanodot heterojunctions. The heterojunction structure could overcome the poor photoelectric properties, material instability, and weak wear resistance of organic and inorganic semiconductors.

In general, the restrictions on lattice matching limit the choice of specific inorganic semiconductors, which can be avoided via the combination of inorganic semiconductors without large strains. However, an attractive feature of the heterojunction structure is that there are no such restrictions for the organic–inorganic semiconductor material combinations [13]. As long as the organic components are changed slightly, the optical and electronic properties of the semiconductor will effect drastically, which is a great advantage of the organic–inorganic heterojunction. In addition, different combinations of organic and inorganic semiconductors can make ideal properties or applications different. In the past 20 years, many research groups have been working on the organic–inorganic nanodots heterojunctions. The organic materials that have been studied include P3HT, PEDOT, PTCDA, PANI, and so on, while the inorganic semiconductor materials include ZnO, CdSe, CdS, and TiO_2, Si et al. Figure 5.1 shows the chemical structure of the heterojunction organic polymers that appear in this chapter. The preparation of organic–inorganic semiconductor hybrid heterojunctions is generally divided into two cases. One is to directly synthesize

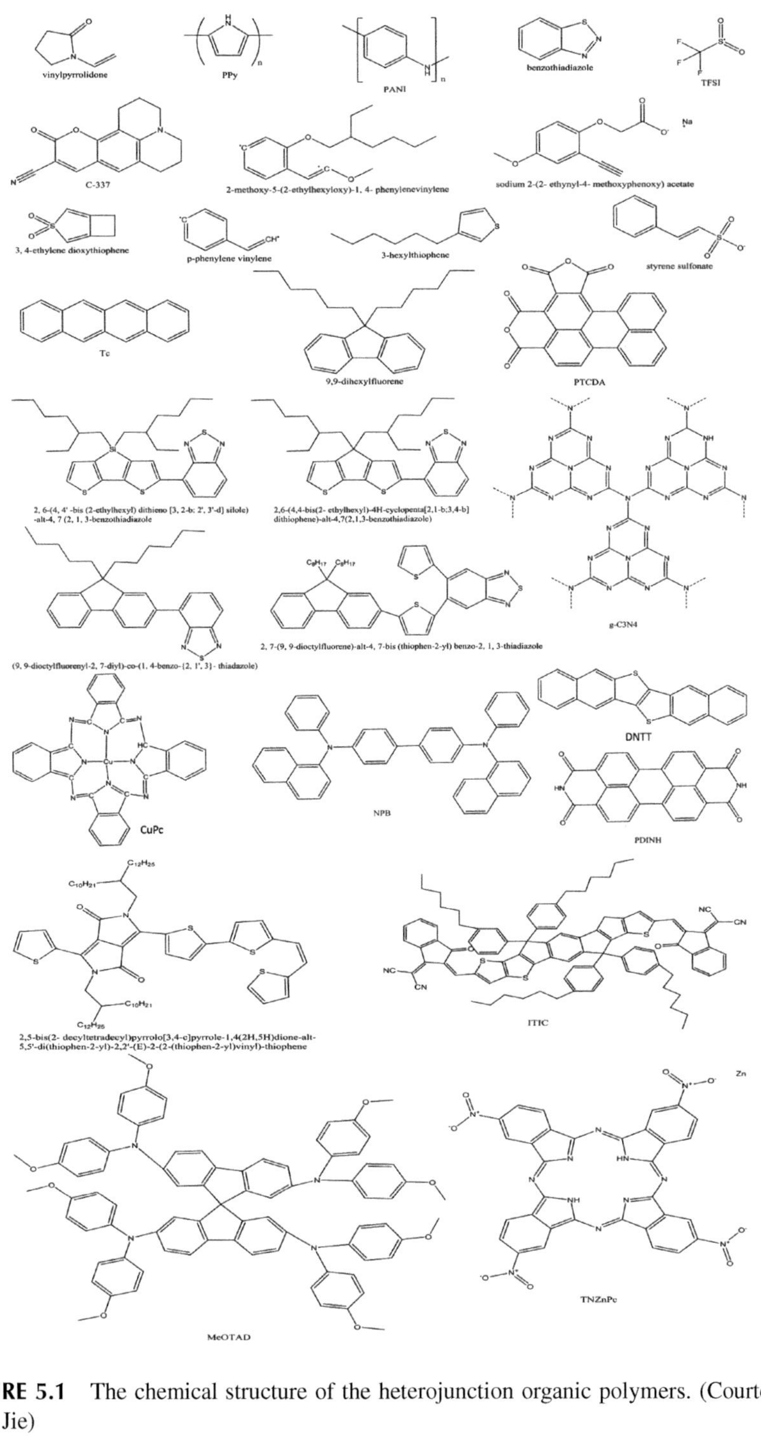

FIGURE 5.1 The chemical structure of the heterojunction organic polymers. (Courtesy of Guan Jie)

organic–inorganic nanoparticle semiconductor composites by methods such as self-assembly [14], blending [15], and solvothermal methods [16], etc. The other is to use two-step method: The first step is to synthesize inorganic semiconductor nanoparticles, and the second step is to deposit [17] or spin-coat [18] the organic semiconductor nanoparticles on the inorganic semiconductor nanoparticles to form a heterojunction structure. In the following sections, a brief summary of the work performed by different research groups related to the organic–inorganic semiconducting nanodots heterojunctions will be given with emphasis on the inorganic and organic materials used, device performances, and their potential applications.

5.2 HETEROJUNCTION OF CD-BASED INORGANIC SEMICONDUCTOR

Group II–VI semiconductor nanomaterials mainly refer to compounds composed of group IIB metal elements Cd, Hg and group VI elements S, Se, Te, such as CdS, Cd Te, Cd Se, etc. The inorganic semiconductors that make up the Cd heterojunction are mainly CdSe and CdS. This type of compound semiconductor has a wide variation in forbidden bandwidth and has the advantages of direct transition energy band structure. Therefore, it has broad applications in solid-state light emitting, laser, infrared detector, solar cells, piezoelectric, and other devices.

Mao, L., et al. [19] use inorganic CdS QDs combined with organic polymer TPP-doped PFBT nanoparticles to construct molecularly imprinted polymer photoelectrochemical sensors for α-Solanine. Specifically, polymer nanoparticles and inorganic quantum dots are used as electron donors and acceptors, respectively, to form organic–inorganic nanoparticle p-n heterojunctions, which enhances the signal response of the sensor. And combined with molecular imprinting technology, the sensor overcomes the selective lack of photochemistry. Thus, the detection limit of α-Solanine by molecularly imprinted sensors based on organic–inorganic heterojunctions is 6.5 pg mL^{-1}, and the linear range is 0.01 to 1000 ng mL^{-1}.

The improvement of the organic/inorganic hybrid open-circuit potential of a solar cell can be achieved by reasonably introducing small but strongly bound electron-donor ligand, which can increase the LUMO energy level of the nanocrystalline acceptor phase and increase the energy transfer of the polymer HOMO. Greaney, M.J., et al. [20] blend P3HT and CdSe nanocrystals treated with tBT to prepare organic/inorganic hybrid solar cells, which always exhibits the highest Voc at 0.80 V. In addition, the P3HT: CdSe-tBT device power conversion efficiency reaches 1.9%.

Modular manufacturing can provide a multifunctional platform for the preparation of nanocrystal/polymer hybrid solar cells, which is easy to apply to different semiconductor nanocrystals and other conductive polymers. Lim, J., et al. [21] use the method to fabricate nanocrystal/polymer hybrid solar cells by completing the surface modification of nanocrystals and the implantation of semiconducting polymer P3HT based on the assembly of breakwater-like CdSe tetrapod nanocrystal networks, whose conversion efficiency is 2.24% and Voc is 0.63 V.

Constructing a highly efficient bulk heterojunction is essential for the hybrid organic/inorganic solar cells. Xu, W., et al. [22] prepare a new mixed structure using

P3HT nanowires and CdSe nanocrystals for holes and electrons, respectively. P3HT nanowires enhance long-wave absorption and carrier transport in the heterojunction active layer. Therefore, the structural effect of the inorganic–organic heterojunction makes the PCE of the hybrid solar cell reach 1.7%, which is increased by 42% compared to the traditional P3HT molecular solar cell.

A simplified ligand sphere model is proposed to explain the improvement of power conversion efficiency. Zhou, Y., et al. [23] use conjugated polymer P3HT and non-ligand exchanged CdSe QDs to prepare heterojunction hybrid solar cells. The key factor to improve equipment performance is simple and fast acid-assisted cleaning procedure for CdSe QDs. Under AM1.5G light, the PCE of the solar cell has been improved by about 2%, which is currently the highest value based on CdSe QDs photovoltaic devices.

Khan et al. [24] synthesize a well-crystallized wurtzite CdSe nanocrystals at 270 °C by a pyrolytic synthesis, whose diameter is about 6 nm and bandgap is 1.85 eV. Then, the synthesized CdSe nanocrystals and P3HT were blended in a chloroform solvent to fabricate CdSe heterojunction solar cells, whose maximum power conversion efficiency reaches 1.41%.

Tan, F., et al. [25] synthesize organic and inorganic hybrid heterojunction solar cells using core-shell CdSe/PbS nanocrystals as electron acceptors and P3HT as electron donors. Compared with pure CdSe nanocrystals, core-shell nanocrystals have better performance of splitting photogenerated excitons in P3HT. Compared with 1.15% of P3HT: CdSe, the photovoltaic performance of P3HT: CdSe/PbS hybrid solar cells is significantly improved to 2.02%, which proves the promise of a new type of nanomaterial as electron acceptors and transmission medium for efficient hybrid solar cells.

Lai, L.-H., et al. [26] prepare for the first time an organic–inorganic hybrid photocathode composed of CdSe and P3HT. The EDT-treated CdSe: P3HT (10: 1 (w/w)) hybrid heterojunction shows efficient hydrogen generation and water reduction ability. Compared with the reversible hydrogen electrode, EQE is 15% and the photocurrent at 0 V is -1.24 mA/cm^2. Under the light of AM1.5G, the unprecedented Voc is 0.85V in mild electrolytes.

Saha, S.K., et al. [27] combine n-type CdSe nanomaterials with p-type CuPc materials to prepare organic–inorganic heterojunction for solar cells, whose short-circuit current is 2 mA/cm^2, open-circuit voltage is 0.46 V, fill factor is 0.34, and power conversion efficiency is 0.32%.

Wang, Q., et al. [28] first utilize semiconductor organic PDs and inorganic QDs for construction of organic–inorganic heterojunctions for PEC bioanalysis. As shown in Figure 5.2, they use CdS QDs, CdTe QDs, and TPP-doped PFBT PDs to form different types of organic–inorganic nanodots heterojunctions. Under light, all heterojunctions exhibit different PEC behaviors, of which CdTe QDs/PDs heterojunction shows significant photocurrent enhancement. Therefore, they use CdTe QDs/PDs and L-cysteine as electrodes and target molecules, respectively. The corresponding linear equation $y = 1.28 \times 10^{-7} - 1.27 \times 10^{-7}$ is obtained by linear fitting, proving the proposed system applicability to PEC bioanalysis.

Zhang, X.H., et al. [29] use modified conjugated BE flake and CdS nanorod to fabricate a new type organic/inorganic material for photocatalytic water cracking to

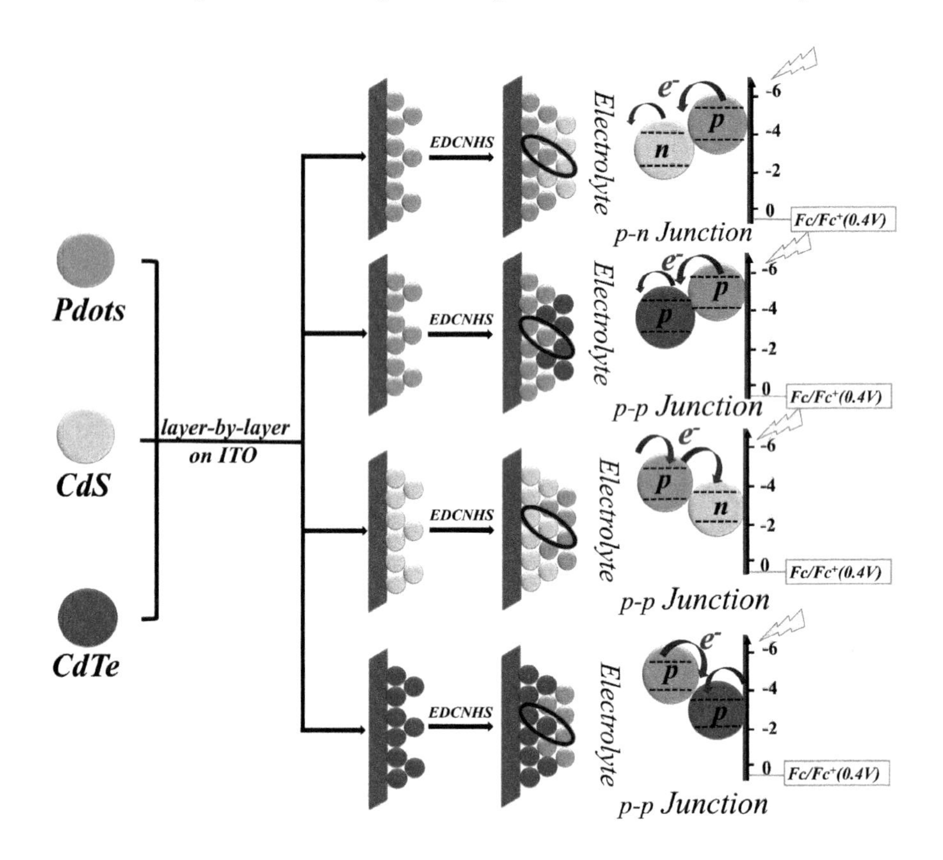

FIGURE 5.2 Different types of organic–inorganic nanodots heterojunctions. (Source: Analytical Chemistry, 2018. **90**(6): p. 3759–3765, with permission)

H_2. Pt was deposited on this heterojunction through an in situ photodeposition technique to form the internal Z-scheme. The Z-scheme BE-CdS heterojunction light-absorption region is from 400 to 700 nm and possesses a rapid photoelectric separation rate. The photocatalytic performances show that the H_2 production yield of BE-CdS photocatalyst is 8.3 times and 23.3 times higher than that of pure CdS and BE catalysts, respectively.

Ren, S., et al. [30] use chemical grafting and ligand exchange processes to prepare organic–inorganic heterojunction hybrid solar cells with P3HT NWs and CdS QDs. Due to the electronic interaction between the heterojunction donor and acceptor elements, both the short-circuit current and the open-circuit voltage of the solar cell are improved. In addition, compared with conventional CdS/P3HT heterojunction, the maximum power conversion efficiency of the solar cell prepared by this method is improved by 4.1% under AM 1.5 sunlight.

Yuan, K., et al [31] prepare LC/CdS hybrid nanocomposites by thermally decomposing cadmium xanthate precursor in LC small-molecule polymer (e.g., PBTTT, TP) in situ, which allows CdS nanocrystals to grow directly. The results show that thermotropic LC can be used as a tissue template for nanoscale organic/inorganic composite nanocomposites. What's more, devices based on PBTTT/CdS nanocomposites achieved an optimal PCE of 1.23%.

Lin, H.W., et al. [32] use a simple template-assisted electrochemical co-deposition method to construct CdS/PPV heterojunction-based photodetectors. Compared with a single CdS element, the photoelectric response of the uniformly mixed CdS/PPV NWs array is significantly enhanced due to the good energy level matching between PPV and CdS, and the switching ratio is increased by 17 times. Moreover, CdS/PPV NWs array has obvious spectral selectivity, especially under 545 nm green light, which has broad application prospects.

5.3 HETEROJUNCTION NANODOTS OF ZN-BASED INORGANIC SEMICONDUCTORS

Group VI Zn semiconductors with wide bandgap are attractive materials due to their low production costs, nontoxicity, and so on. ZnO is the most typical semiconductor, which has fast carrier mobility (~205–1000 $cm^2\ V^{-1}\ s^{-1}$) and significant electron affinity (4.2 eV). In addition, ZnO is easier to crystallize and is used in large-scale synthesis. However, the wide direct bandgap (3.4 eV) of ZnO will limit the response of most of the spectrum, and in the photocatalytic reaction, ZnO is corroded and deactivated due to the low degree of charge carrier separation and acid/alkaline medium, which affect its wide application. Heterojunction structure provides a route to overcome some of the shortcomings.

The microwave-assisted hydrothermal synthesis can prepare inexpensive nanomaterials in a short time, which is a low-cost method. Sharma, R., et al. [33] fabricate ITO/PEDOT: PSS/PFO-DBT: ZnO/Al with ZnO nanomaterials synthesized by microwave-assisted hydrothermal method and polymer PFO-DBT as acceptors and donors, respectively. The devices based on PFO-DBT/ZnO composite nanomaterial exhibit significant open-circuit voltage, power conversion efficiency, and short-circuit current density, which have potential application prospects for non-fullerene-based polymer solar cells. The best PCE can reach 1.41%.

Pei, J., et al. [34] use a triphenylamine derivative P3HT to modify ZnO to form a photosensitive layer, which is sandwiched between the FTO substrate and a hole transport material (PEDOT: PSS) to prepare a low-cost organic–inorganic heterojunction solar cell. Due to the interaction characteristics of organic and inorganic components on the same platform, the conversion efficiency of solar cells based on ZnO/P3HT has been significantly improved, and the increase is about 1.41%.

Ouyang, B.S., et al. [35] prepare a self-powered UV photodetector based on p-P3HT/n-ZnO NWs array heterojunction, which exhibits low cost, fast recovery and response speed, large photoconductive gain, etc and has great application prospects in light imaging. The specific detectivity and responsivity of the device is 3.7×10^7 Jones and 125 $\mu A\ W^{-1}$, the photoconductive gain of 4.2×10^{-4}. In addition, under the temperature difference of −15.4 °C, the response of the photodetector can be improved by 21.9% because of the thermal photoelectric effect.

Interface engineering design through a self-assembled single layer is an effective method to adjust the work function and electronic properties of the surface, which improve the charge injection efficiency and device performance of microelectronic applications. Pandey, R.K., et al. [36] modify the surface of the ZnO film with self-assembled monolayer before forming a hybrid ZnO/P3HT heterojunction diode. The study of charge transfer characteristics and device parameters demonstrates that comparing unmodified ZnO/P3HT diodes, before the formation of the heterojunction, the modification of ZnO can improve the electronic parameters and the charge injection efficiency of the device. For example, the rectification ratio at ± 3V is increased by ~ 10 times, and the forward current density at 3 V is increased by ~ 32 times. This research has laid a foundation for further improving the performance and parameters of electronic devices.

Tai, H.L., et al. [37] fabricate a P3HT-ZnO nanoparticle-based mixed film room temperature formaldehyde gas sensor via a spraying process. the P3HT/ZnO heterojunction structure as the active layer of a thin-film transistor not only improves the sensing performance of HCHO but also makes the sensor have a higher response value and better reversibility (more than two times that of general sensors), attributed to three-dimensional porous morphology of the P3HT-ZnO hybrid film and the accumulation of organic–inorganic heterojunction structure.

Tonpe, D.A., et al. [38] fabricate PANI/ZnO nanorods-based hybrid solar cell device in two steps. They first synthesize ZnO films on the ITO substrate by soft chemical synthesis, and then anneal them at 300 °C. Next, they use a doctor-blade method to chemisorb PANI on ZnO film's surface to form a heterojunction structure. Due to faster interface charge transfer and the effective charge separation in the initial ZnO films, the ZnO films do not need to be annealed, and the light conversion efficiency of the prepared PANI/ZnO films-based device is higher, which is 1.8 times that of the treated film.

Xie, Y.R., et al. [39] Fabricated a p-n heterojunction diode using organic spiro-MeOTAD and inorganic ZnO NWs array. Under zero bias illumination, the maximum photosensitivity was observed to be 1.32 mA/W. The photocurrent time response of the diode is fast, consistent, and repeatable. The rise and decay time constants are 0.12 and 0.06 s, respectively, under the blue light irradiance (410 nm, 75 mW/cm^2). These results indicate that ZnO/spiro-MeOTAD heterojunction can be candidate materials for efficient self-powered blue light detectors.

Li, X.Y., et al. [40] use an easy-to-operate hydrothermal method to grow a vertically aligned n-type ZnO nanorod array on ITO conductive glass and spin-coat p-type PFH on its surface to prepare a PFH/ZnO nanorod heterojunction photodetector, which exhibits a self-powered characteristic. Under the light of 405 nm and the bias voltage of 0V, the current of the device decreases as the light is turned on, which is a negative current. When the dark environment of the light is turned off, the current can be restored to 0 A.

Song, J., et al. [41] synthesize ZnO nanoparticles using a simple low-temperature wet-chemical method and spin-coat p-type organic material PFH on ZnO nanoparticles to prepare organic/inorganic p-n heterojunction. Under illumination, the exciton can be quickly and efficiently transferred between interfaces. A fast and stable photoelectric response was observed and reverse bias can improve the generated photocurrent and increase the ratio of photocurrent to dark current.

Mokhtari, H., et al. [42] prepare inorganic n-type ZnO and organic PVP layer by a sol–gel spinning-coating method at 2000 rpm to construct an Au/PVP/ZnO/Si/Al semiconductor heterojunction diode. They study the current-voltage characteristics of semiconductor heterojunction diodes under dark conditions, whose ideal factor and barrier height are 4.4 and 0.57 eV, respectively.

Difficulties in design and manufacture processes of omnidirectionally stretchable photodetectors limit stretchability in multiaxial directions. Here, Trung, T.Q., et al. [43] directly grow ZnO nanorods on the PEDOT: PSS transport layer to prepare an organic–inorganic p-n heterojunction photodetector, which includes a three-dimensional micropattern stretchable substrate. The detector is highly responsive to ultraviolet light and has good stability under multiaxial and uniaxial strains up to 30%, where it can perform 15,000 tensile cycles.

Luka, G., et al. [44] prepare organic–inorganic heterojunctions using typical p-type small molecule Tc and n-type semiconductors ZnO to investigate the influence of the electrical properties of inorganic layer ZnO on the electrical behavior of ZnO-Tc hybrid heterojunctions.. Under different deposition conditions, when the carrier concentration in the n-type ZnO film gradually increases from 10^{15} cm^{-3} to 10^{19} cm^{-3}, the reverse current in the heterojunction structure decreases, which improves its asymmetry.

Padma, N., et al. [45] synthesize n-type ZnO/p-type CuPc nanocomposites by solution method and study the electrical properties of ZnO/CuPc nanocomposite films. Different methods are used to estimate series resistance, interface barrier height, and the ideal factor. It is found that those of nanocomposite film diodes are much lower than the double-layer structure. This proves that the heterojunction structure is beneficial to the efficient operation of optoelectronic devices.

In addition to ZnO, other inorganic Zn-containing semiconductors have also been studied. For example, Chaudhary, D.K., et al. [46] synthesize stable $CuZn_2AlS_4$ nanocrystals with an easy-to-use hydrothermal method, which are firstly used as electron acceptor and photoactive material to apply for organic–inorganic hybrid solar cells. They use MEH-PPV and P3HT as electron-donor materials, optimize the mixing ratio of inorganic and organic substances, and study the photovoltaic performance of $CuZn_2AlS_4$-based devices in classic and inverted device structures. The efficiency of devices without $CuZn_2AlS_4$ nanocrystals is 1.72%, while the efficiency of devices with $CuZn_2AlS_4$ nanocrystals is 3.25%.

Chen, G.-H., et al. [47] add cubic ZnSe NWs with polymer-like winding structures to the MEH-PPV matrix to form an inorganic–organic p-n heterojunction and apply them to light-emitting diodes. The entangled nanostructure of ZnSe and the conductive polymer hybridize to form an interpenetrating network, which provides a continuous path for the carrier. Compared with the single-layer MEH-PPV light-emitting diode, the electroluminescence intensity of the heterojunction diode device is significantly increased by more than 330%, and the driving voltage is reduced by about 26%.

5.4 HETEROJUNCTION OF TI-BASED INORGANIC SEMICONDUCTORS

The inorganic semiconductor constituting the Ti-based heterojunction is mainly TiO_2. TiO_2 nano-semiconductor material has good ultraviolet light selectivity due to its wide forbidden bandwidth. At the same time, it also has the advantages of stable chemical properties, strong weather resistance, high electron injection rate, and conductive properties that other semiconductor materials do not have, and thus plays an increasingly important role in optoelectronics. However, TiO_2 can only absorb UV light because of its wide bandgap. Therefore, the construction of the organic–inorganic hybrid system can not only improve the utilization of sunlight but also improve the separation efficiency of holes (h^+) and photoelectrons (e^-).

Li, X., et al. [48] prepare PDINH/TiO_2 organic–inorganic heterojunction photocatalysts based on PDINH and TiO_2 P25. Compared with pure PDINH and TiO_2 photocatalysts, the photocatalytic hydrogen production activity and photocurrent intensity are both higher. In addition, under 365 nm irradiation, the apparent quantum efficiency of H_2 produced by PDINH/TiO_2 above 0.5% is as high as 70.69%.

Xiao, G., et al. [49] use a facile and green one-step calcination procedure to synthesize a highly efficient TiO_2 @ g-C_3N_4 hybrid photocatalyst. The heterostructure is formed between the two semiconductors through a strong interfacial action. Compared with pure g- C_3N_4 and TiO_2 photocatalysts, the heterostructure photocatalysts have higher photocatalytic activity for the degradation of methyl orange. Under visible light, the removal of several typical highly toxic environmental pollutants (methyl orange, rhodamine B, phenol,) also shows excellent performance, the conversion rate is over 90%, and it can be reused for five cycles with preserving 86% of its activity.

The TNZnPc/TiO_2 organic–inorganic heterostructures are promising candidate materials for wastewater treatment. Zhang, Q., et al. [50] successfully prepare TNZnPc/TiO_2 organic–inorganic heterostructures using a simple combination method of electrospinning technique and solvothermal processing. The prepared heterostructures exhibit excellent photocatalytic efficiency for degradation of RB solution, reaching about 89% in 100 minutes.

Semiconductor in situ fabrication technology can be applied in the fields of solar cells, photocatalytic CO_2, and so on. Hou, H.J., et al. [51] synthesize conjugated microporous organic compounds BBT and TiO_2 by in situ polymerization to prepare efficient BBT/TiO_2 heterojunction photocatalysts. The photoactivities of BBT/TiO_2 photocatalysts are 18.0 and 20.4 times that of pure BBT for photocatalytic hydrogen production and degradation of ciprofloxacin under optimal conditions.

Qin, Y., et al. [52] introduce GQDs as an intermediate buffer layer to the hybrid heterojunction to form a cascaded energy level structure. On the one hand, the ternary blending structure enhances the ability to capture high-energy photons. On the other hand, due to the existence of GQDs, photo-excited carriers can flow directly to the circuit. Compared with the traditional TiO_2/P3HT device, the PCE of the ternary blend system increased by 3.16%. Although this work is not yet complete, these advantages are sufficient to support it as a promising candidate buffer material for photovoltaic applications.

Szkoda, M., et al. [53] propose a new method for preparing organic–inorganic heterojunctions of TiO_2 nanotubes and PEDOT. They electrochemically synthesized PEDOT nested with different Mehcf and deposited them on titanium dioxide nanotubes. The experiments show that the photocurrent of the TiO_2 nanotube/PEDOT heterojunction is stable, and its photocurrent intensity is more than 5 times that of pure TiO_2.

Zhang, D.Z., et al. [54] develop a heterojunction photoconductive UV detector based on NPB-coated TiO_2 nanowire arrays. They first fill the TiO_2 NWs arrays gaps with a static infusion method, while using dynamic solution cleaning to remove the unwelcomed top NPB layer. Under ultraviolet light, the accumulation and dissociation of photogenerated carriers between the p-n heterojunctions can eliminate the loss effect, thereby improving the carrier density and photoconductivity. Compared with devices with an NPB top layer, devices without NPB top layer achieve an enhanced light current to the dark current ratio of 1.67×10^4.

Sharma, S., et al. [55] fabricate F-Cl-TiO_2/CSA-PANI p-n heterostructures by spin coating on ITO-coated glass substrate at room temperature, which is sensitive to ultraviolet light and has a considerable light-dark current contrast at a voltage of −1 V. In addition, they also exhibit large response rate of 24.9A/W, special ideal factor of 1.9, light guide gain of 84.59 and reasonable rise and fall time of 0.6 and 1.6 s, respectively. Therefore, the F-Cl-TiO_2/CSA-PANI heterostructure c is promising for light-detection applications.

Subramanian, E., et al. [56] prepare PANI-TiO_2/SnO_2 organic–inorganic heterojunction material films by dispersing metal oxides (TiO_2/SnO_2) and PANI in ethanol and coating them with oxidants in situ on printed circuit boards. In previous work, PANI/SnO_2 was in the form of pellet, when the analyte concentration is 1650 ppm, the sensitivity to benzene is 16.93% and to toluene is 19.53%. However, the current results show that the sensitivity of the PANI-TiO_2/SnO_2 film formed on the printed circuit board is 130% at a benzene content of 140 ppm and 300% at a toluene content of 70 ppm.

In addition to TiO_2, other inorganic Ti-containing inorganic nanomaterials have also been used for the construction of organic–inorganic heterojunction material. Jarka, P., et al. [57] prepare p-n heterojunction films by physical vapor deposition using PTCDA and TIPOc material as the electron acceptor and donor, respectively. The I_{SC} and V_{OC} values of the generated organic photovoltaic cells are the highest, 0.570 V and 220 nA, respectively, and the proportions of the elements in each layer are close to equal (60% of PTDCA and 40% of TIPOc).

Rakibuddin, M., et al. [58] synthesize organic–inorganic hybrid photocatalysts based on g-C_3N_4/N-doped $LaTiO_3$ by a sol--gel polymerization composite method. They wrap N-doped $LaTiO_3$ nanoparticles with g-C_3N_4 nanoflakes to form a heterojunction structure. Compared with pure g-C_3N_4 and n-doped $LaTiO_3$ photocatalysts, the g-C_3N_4/N-doped $LaTiO_3$ heterojunctions not only can absorb visible light in a wider wavelength region but also have stronger photocatalytic and photocurrent activities.

Yazdani, D., et al. [59] prepare g-C_3N_4/NiTi-LDH heterojunction films by facile hydrothermal synthesis, which are used to prepare a fixed-bed photoreactor on coated glass for photocatalytic degradation of methyl orange. Compared with pure g- C_3N_4 and NiTi-LDH films, the g-C_3N_4/NiTi-LDH films have higher photocatalytic activity, which can achieve the maximum methyl orange degradation efficiency (100%) in 240 minutes.

Kumar, A., et al. [60] synthesize organic–inorganic heterojunction g-C_3N_4/$CaTiO_3$ photocatalyst on the surface of g-C_3N_4 nanosheets by simply depositing $CaTiO_3$ nanoflakes. The highest photocatalytic activity is g-C_3N_4/$CaTiO_3$ heterojunction prepared by mixing g-C_3N_4 and $CaTiO_3$ at a ratio of 1:1, which is used to degrade BPA that is degraded by 47% in sunlight after 120 minutes. Therefore, this study can provide insights into the development of two-dimensional heterojunction photocatalysts.

5.5 HETEROJUNCTION OF SI-BASED INORGANIC SEMICONDUCTORS

Si is an indispensable material for modern complementary metal oxide semiconductor processes, and III-V semiconductors are widely used in optoelectronics, ultra-high-speed microelectronics, and ultra-high-frequency microwaves. For a long time, scientists have tried to epitaxial high-quality III-V semiconductors on Si substrates. However, due to the lattice mismatch, the grown III-V semiconductors are of poor quality. Organic semiconductors can hopefully solve this problem.

Nawar, A.M., et al. [61] use organic material C-337 and inorganic material n-Si (111) to prepare organic–inorganic hybrid solar cells. The results have proven that the Au/C-337/n-Si/Sb-Au heterojunction solar cells show good photovoltaic performance at 90 m W/cm^2 light intensity; and that the fill factor is 0.59 I_{SC} is 5.42 mA, V_{OC} is 0.55 V, and PCE IS 4.39%.

El-Nahass, M.M., et al. [62] fabricate n-type SO/p-type Si heterojunction solar cells based on SO thin films deposited on P-Si substrates using conventional thermal evaporation. The device responds well to illumination. The I_{SC}, V_{OC}, PCE, and fill factor of solar cells are 56.6 mA/cm^2, 0.6V, 1.2%, and 0.32%, respectively. In addition, the Au/SO/Si/Al heterojunction can respond well to light, which can be used as a candidate material for photovoltaic devices for visible light detection.

Mahato, S., et al. [63] fabricate PEDOT: PSS/Si hybrid heterojunction solar cells with p-type conducting polymer PEDOT: PSS and n-type crystalline Si, whose I_{SC} is 29.7 mA/cm^2, V_{OC} is 0.56 V, PCE is 8.5%, and fill factor is 49.7%.

By physically cutting and controlling the appropriate substrate size to enhance charge trapping, the performance of organic–inorganic hybrid solar cells is greatly improved. Zhang, J., et al. [64] prepare organic–inorganic solar cells with n-type silicon (100) and PEDOT: PSS based on the above principles, under AM 1.5G illumination at 100 mW/cm^2, the photoelectronic conversion capacity of which reaches 12.1%.

Harraz, F.A., et al. [65] use electrochemical technology to permeate conductive polymer PPy into PSi matrix under constant current conditions to form a heterojunction structure, which significantly improved the photoluminescence spectrum emission and conductivity. The impedance characteristics of the prepared Psi- and PPy-filled PSi are measured under dark conditions of 0.0 V. The signal amplitude for measurement frequency range of 100 MHz to 100 kHz is 5 mV. In addition, under visible light irradiation, the luminous intensity of PPy electropolymerized PSi is 1.5 times that of unmodified PSi.

5.6 HETEROJUNCTION OF PEROVSKITE-BASED INORGANIC SEMICONDUCTORS

Recently, organic–inorganic hybrid perovskites have attracted widespread attention from researchers. Especially alkylammonium lead halides, $(RNH_3) PbX_3$ (R = alkyl, X = Cl, Br, I) exhibit excellent properties as hole and electronic conductors as well as photosensitizers due to their particular hybrid organic–inorganic perovskite structure.

Luo, J., et al. [66] prepare $MAPbI_3$-TiO_2 heterojunction nanoarrays, which can efficiently and stably evolve hydrogen in solution. Electrons are injected into TiO_2 from $MAPbI_3$, and then transported to the counter electrode along the one-dimensional TiO_2 nanochannel to form hydrogen. The photocurrent density $MAPbI_3$-TiO_2 PEC is 1.75 mA cm^{-2}, and the rate of hydrogen production reaches 33.3 μmol cm^{-2} h^{-1} under AM 1.5G illumination.

Zhou, H.W., et al. [67] have successfully based low-temperature carbon electrodes to prepare M-TiO_2/$CH_3NH_3PbI_3$/C heterojunction solar cells, which can not only reduce costs but also simplify the manufacturing process of heterojunction solar cells. The M-TiO_2/$CH_3NH_3PbI_3$/C heterojunction solar cells have good stability, and that unencapsulated can be left for at least 2000 h in dark environment and the PCE reaches 9%.

Zhang, W.N., et al. [68] synthesize the heterojunction $MASnI_3$/TiO_2 composites successfully via a wet-chemical method. To our knowledge, it could be the first report on the application of $MASnI_3$ in the field of photocatalysis. For the degradation of rhodamine B, the $MASnI_3$/TiO_2 materials have better photocatalytic properties under light illumination than pure $MASnI_3$ and pure TiO_2. The percentage of residual rhodamine B decreases to about 3% after 40 minutes of irradiation, which illustrates an approximately completed degradation of RhB on the $MASnI_3$/TiO_2 (1:9) samples.

Zhang, X.X., et al. [69] prepare a-$MAMnI_3$ thin films via the spin coating method, and then fill a-$MAMnI_3$ into the mesoporous TiO_2 layer to form mesoscopic a-$MAMnI_3$/TiO_2 heterojunction. The photovoltage reaches 300 mV under AM 1.5 G simulated illumination. Compared with UV-responsive MA_2MnCl_4 materials, the a-$MAMnI_3$-based devices present obvious response to visible light (530 nm green) and ultraviolet light.

Jeong, H., et al. [70] base their study on $MAPbI_3$ perovskite nanomaterials and low-bandgap polymer PCPDTBT-PCBM to construct organic–inorganic ternary hybrid heterojunctions. They continuously spin-coated $MAPbI_3$ nanodots on ITO/PEDOT: PSS substrates. Compared with traditional bulk heterojunction equipments, the improved photocurrent and fill factor result in a 28% increase in efficiency.

Rahaq, Y., et al. [71] provide a simple and feasible way to prepare reproducible perovskite solar cells. They combine $MAPbI_3$ films by two-step deposition method with the polymer to fabricate $MAPbI_3$/PCBM heterojunction. The perovskite films prepared by adding MAI solution with 1.0 wt% to react for one minute is better, solar cells based on which exhibit an average performance that the PCE is generally 13.63%, up to 15.01%.

Luo, L.B., et al. [72] construct heterojunctions based on Cs-doped $FAPbI_3$ perovskite and organic DNTT for photoelectric detection, the response rate and specific detection rate of which are 778 AW^{-1} and 1.04×10^{13} Jones, respectively. In addition, the Cs-doped $FAPbI_3$/DNTT photodetector has good stability and broadband response, whose range can be from deep ultraviolet to near-infrared light.

Chen, Y., et al. [73] successfully synthesize $MASnI_3$/SnO_2 nanocomposite sensor to detect NO_2 in industries such as petroleum, petrochemicals, automobile exhaust, chemical plants, etc. by calcining crystalline SnO_2 directly on the surface of $MASnI_3$.

In the absence of UV light stimulation, the response of sensor based on $MASnI_3$/SnO_2 to NO_2 of a concentration of 5 ppm is 17.2 ± 0.5, which increases to 240.6 ± 5.1 in the presence of UV light stimulation. In addition, under UV light irradiation, the fitted curve of $MASnI_3/SnO_2$ nanocomposite sensor to NO_2 in the concentration range of 0.5–25 ppm is consistent with Langmuir's law of adsorption and the linear range is 0.5–10 ppm.

Chen, H.-Y., et al. [74] use chemical vapor deposition to prepare high-quality p-type semiconductor materials, Se microwires. Based on this, they fabricate p-n heterojunction photodetectors combining the perovskite material $CH_3NH_3PbCl_3$. Under 5V bias, the peak response of the p-n heterojunction photodetector at 600 nm is 850% higher than that of the single Se microtube, which is 23.8 mA/W.

5.7 HETEROJUNCTION OF AG-BASED INORGANIC SEMICONDUCTORS

Inspired by the development and autocatalytic process of the silver halide latent image, the application of silver and silver compounds as photocatalytic materials in the photocatalytic process is expected to amplify the photoquantum efficiency of the photocatalyst and improve the photocatalytic reaction efficiency [75]. Therefore, Ag series semiconductors have also attracted much attention.

The g-C_3N_4/Ag_3PO_4 nanocomposite photocatalyst is very promising for energy applications and purification of industrial wastewater. Nagajyothi, P.C., et al. [76] synthesize organic–inorganic g-C_3N_4/Ag_3PO_4 hybrid heterojunction photocatalyst by simple one-step hydrothermal, which is applied to hydrogen production and degradation of MB pollutants. Compared with the hydrogen production rate and the degradation rate of MB of g-C_3N_4/Ag_3PO_4, nanocomposites are significantly improved, that are 44.5 and 1.14 times of Ag_3PO_4 photocatalyst.

Heterogeneous hybridization opens a new way for visible light to efficiently drive photocatalysts to repair environment. Matheswaran, P., et al. [77] synthesize environmentally friendly g-C_3N_4/AgCl heterojunction, the photocatalytic degradation of MB and rhodamine B by which is 10 times and 51 times that of g-C_3N_4 photocatalyst, respectively. It not only enhances its catalytic activity but also shows considerable stability during degradation, and is expected to become a candidate for dye degradation.

Mondal, P., et al. [78] fabricate PANI/Ag_2MoO_4 organic–inorganic heterojunction nanocatalyst via in situ deposition of Ag_2MoO_4 on PANI polymer, that Ag_2MoO_4 hybridizes with PANI to form a p-n heterojunction, achieving efficient photocarrier separation and migration. The PANI/Ag_2MoO_4 heterojunctions have strong photocatalytic activity. When the Congo red dye is degraded under ultraviolet light, the photocatalytic degradation capacity of pure Ag_2MoO_4 is very poor with only 2.91%, while as the percentage of PANI increases, the degradation rate of nanocomposites also increases, and when the percentage is 20%, the adsorption efficiency is up to 25%.

Ag/TiO_2@PPy is a promising material for photodegradable organic pollutants. Kumar, R, et al. [79] assemble a self-stabilized p-type Ag-Ag_2O semiconductor on rutile TiO_2@PPy substrate to synthesize an Ag/TiO_2@PPy heterojunction. The Ag/TiO_2@

PPy heterojunction has strong photocatalytic activity against MB dyes in both ultraviolet and visible light. Low concentrations of MB are degraded by almost 100% under visible light and UV, as the MB concentration increases from 4 to 20 mg/L, the degradation rate under UV light decreases from 99.62% to 83.9%, and the efficiency decreases from 99.37% to 18.85% under visible light.

5.8 HETEROJUNCTION OF BI-BASED INORGANIC SEMICONDUCTORS

In recent years, Bi series semiconductors have attracted great interest from researchers due to their unique electronic structure, excellent visible light-absorption ability, and high organic degradation ability. The Bi-based semiconductors are mainly: Bi_2WO_6, BiOX (X = Cl, Br, I), and so on.

Zhu, L.-B., et al. [80] fabricate PPy@ Bi_2WO_6 organic–inorganic heterojunction for CK-MB PEC analysis. As the concentration of CK-MB increases from 0.5 to 2000 ng mL^{-1}, the photocurrent of the PPy@ Bi_2WO_6 heterojunction sensor gradually decreases, that the linear regression equation is $I\ (\mu A) = -0.4406 \log c\ (ng\ mL^{-1}) + 1.759$ (S/N = 3), the correlation coefficient and detection limits are 0.995 and 0.16 ng mL^{-1}, respectively.

Chang, C., et al. [81] synthesize organic–inorganic three-dimensional g-C_3N_4/BiOI heterojunction photocatalysts via solvothermal method, which has excellent catalytic activity under visible light and can be used for the degradation of BPA in water. Compared with pure g-C_3N_4 and BiOI, the photoelectrochemical and photocatalytic properties of the heterojunction are improved. When the g-C_3N_4 ratio is 10%, the photocatalytic performance is the best that k_{obs} and the intensity of photocurrent are 1.6 times and1.5 times than that of pure BiOI, 3.4 times and 2.0 times than that of g-C_3N_4, respectively, under simulated visible light conditions.

Ma, S., et al. [82] use Li-TFSI-doped BiI_3 and light-absorbing polymer PTB7-Th to prepare organic–inorganic BiI3/polymer heterojunction solar cells. The light-absorption capacity of Li-TFSI-doped BiI_3/PTB7-Th heterojunction is enhanced, improving the EQE of the device, which is extended from 650 to 750 nm. In addition, optimizing the polymer content, the efficiency of the solar cell is 1.03% and I_{SC} of the device can be significantly increased from 1.3 to 3.7 mA cm^{-2}, which is up to 7.8 mA cm^{-2}.

5.9 HETEROJUNCTION OF PB-BASED INORGANIC SEMICONDUCTORS

Group IV–VI compounds are a kind of semiconductor material with a narrow bandgap of cubic structure. Especially PbS has superior electrical and optical properties, which has broad application prospects in some high-tech fields such as photovoltaic conversion materials, solar cells, photovoltaic devices, sensing, and detection [83].

Wurst, K.M., et al. [84] prepare a heterojunction hybrid optical switch based on polymer PAE-1 and PbS nanocrystals. The infrared excitation of PbS nanocrystals can transfer the charge in PAE and cause the continuous formation of polarons, so that complex operations such as processing two optical signals simultaneously in a switch can be achieved. The light response of PbS-PAE-1 to 500–800 nm weak light stimulation is easily amplified by the trigger signal of 1100–1400 nm, which is very meaningful for optical communication.

Nam, M., et al.[85] prepare efficient hybrid bulk heterojunction solar cells based on PSBTBT and PbS QDs, which have high photovoltaic efficiency. Since PbS QDs can absorb near-infrared light and PSBTBT is a low-bandgap polymer, the hybrid PSBTBT/PbS nanocomposite devices respond to a wide spectrum from ultraviolet to near infrared. Under AM1.5G lighting, the efficiency of the heterojunction is higher than 3.39%.

5.10 HETEROJUNCTION OF OTHER METAL-BASED INORGANIC SEMICONDUCTORS

In addition to the series of common inorganic semiconductors listed above, there are also some inorganic semiconductors used to prepare organic–inorganic heterojunctions, such as SnO_2, MoS_2, WO_3.

Murugan, C., et al. [86] emphasize that p-n heterojunctions formed by p-type organic and n-type inorganic semiconductors have the potential to produce highly efficient sensors. They compare organic–inorganic hybrid composites synthesized by mixing p-PANI and n- SnO_2 with different weight ratios and find that the sensors prepared by mixing PANI with 40 wt% SnO_2 have the best performance for benzene/ toluene analysis that the sensor efficiency is 10%.

Yan, J., et al. [87] fabricate two-dimensional hybrid heterojunction with the p-type polymer PDVT-10 and n-type MoS_2. The PDVT-10/MoS_2 heterojunction devices have bipolar charge transport characteristics that the maximum field effect mobilities of n-type and p-type are 2.45 $cm^2\ V^{-1}\ s^{-1}$ and 0.3 $cm^2\ V^{-1}\ s^{-1}$, respectively. The ideal coefficient is about 6.8. In addition, when exposed to visible light, the PDVT-10/MoS_2 heterojunction exhibits good light response that the light/dark current contrast is clear.

Jeon, D., et al. [88] prepare novel photocatalytic devices with excellent performance and stability by hybridizing organic and inorganic materials. They use different low-cost conductive polymers such as PANi, PEDOT, and PPy to co-deposit on WO_3, the Ru_4POM are deposited together as both water oxidation catalysts and dopants for CPs. Through experimental comparison, WO_3/Ru_4POM: PPy heterojunction exhibits the best performance that the Faradaic efficiency of WO_3 for water oxidation is 21% while of WO_3/PPy: Ru_4POM is 56%.

5.11 CONCLUSIONS

Low-dimensional inorganic–organic hybrid nanomaterials, due to their excellent properties of both organic, inorganic, and nanomaterials, can produce a “1+1> 2” effect in performance, which can form heterojunctions at their interfaces to generate new unique properties, making it a research hotspot. In this chapter, several combinations and applications of organic and inorganic semiconductor nanomaterials are described, as is referred to in Table 5.1. There are many ways and methods of combining inorganic semiconductor materials and organic semiconductor materials. This chapter provides a better description of the choice of inorganic and organic materials to build organic–inorganic semiconductor heterojunctions.

TABLE 5.1
Summary of different combinations of heterojunctions with their applications

Heterojunction	p-type semiconductor	n-type semiconductor	Application	Reference
Pdots/CdS	TPP-doped PFBT	CdS	Photoelectrochemical	[19]
P3HT/CdSe-tBT	P3HT	CdSe-tBT	Solar cell	[20]
P3HT/CdSe	P3HT	CdSe	Solar cell	[21–24]
P3HT/CdSe/PbS	P3HT	CdSe/PbS	Solar cell	[25]
P3HT/CdSe	P3HT	CdSe	Photocathodefabrication	[26]
CdSe/CuPc	CuPc	CdSe	Solar cell	[27]
PDs/QDs	PDs	QDs	Photoelectric sensor	[28]
BE-CdS	BE	CdS	Photocatalyst	[29]
P3HT/CdS	P3HT	CdS	Solar cell	[30]
LC/CdS	LC polymer	CdS	Optoelectronic device	[31]
CdS/PPV	PPV	CdS	Photodetector	[32]
PFO-DBT/ZnO	PFO-DBT	ZnO	Solar cell	[33]
P3HT/ZnO	P3HT	ZnO	Solar cell	[34]
P3HT/ZnO	P3HT	ZnO	Photodetector	[35]
P3HT/ZnO	P3HT	ZnO	Diode	[36]
P3HT/ZnO	P3HT	ZnO	HCHO gas sensor	[37]
PANI/ZnO	PANI	ZnO	Solar cell	[38]
ZnO/spiro-MeOTAD	spiro-MeOTAD	ZnO	Diode/blue light detector.	[39]
PFH/ZnO	PFH	ZnO	Photodetector	[40,41]
PVP/ZnO	PVP	ZnO	Diode	[42]
PEDOT:PSS/ZnO	PEDOT:PSS	ZnO	Photodetector	[43]
ZnO/Tc	Tc	ZnO	Photovoltaic application	[44]
ZnO/CuPc	CuPc	ZnO	Optoelectronic device	[45]
MEH-PPV/P3HT/$CuZn_2AlS_4$	MEH-PPV/ P3HT	$CuZn_2AlS_4$	Solar cell	[46]
ZnSe/MEH-PPV	MEH-PPV	ZnSe	LEDs	[47]
PDINH/TiO_2	PDINH	TiO_2	Photocatalyst	[48]
TiO_2@g-C_3N_4	g-C_3N_4	TiO_2	Photocatalyst	[49]
TNZnPc/TiO_2	TNZnPc	TiO_2	Photocatalyst	[50]
BBT/TiO_2	BBT	TiO_2	Photocatalyst	[51]
TiO_2/P3HT	P3HT	TiO_2	Photovoltaic application	[52]
PEDOT/TiO_2	PEDOT	TiO_2	Photovoltaic application	[53]

(Continued)

TABLE 5.1 (*Continued*)
Summary of different combinations of heterojunctions with their applications

Heterojunction	p-type semiconductor	n-type semiconductor	Application	Reference
NPB/TiO_2	NPB	TiO_2	UV detector	[54]
CSA-PANI/F-Cl-TiO_2	CSA-PANI	F-Cl-TiO_2	Photodetection	[55]
PANI/TiO_2/SnO_2	PANI	TiO_2/SnO_2	Gas microsensor	[56]
PTCDA/TiOPc	PTCDA	TiOPc	Photovoltaic cell	[57]
g-C_3N_4/N-$LaTiO_3$	g-C_3N_4	N-$LaTiO_3$	Photocatalyst	[58]
g-C_3N_4/NiTi	g-C_3N_4	NiTi	Photocatalyst	[59]
g-$C_3N_4/CaTiO_3$	g-C_3N_4	$CaTiO_3$	Photocatalyst	[60]
C-337/n-Si	C-337	n-Si	Solar cell	[61]
n-SO/P-Si	Si	SO	Photovoltaic device	[62]
PEDOT: PSS/Si	PEDOT: PSS	Si	Solar cell	[63,64]
PSi/PPy	PPy	PSi	Electrode	[65]
$MAPbI_3$-TiO_2	TiO_2	$MAPbI_3$	Hydrogen production	[66]
$TiO_2/CH_3NH_3PbI_3$	TiO_2	$CH_3NH_3PbI_3$	Solar cell	[67]
$MASnI_3/TiO_2$	TiO_2	$MASnI_3$	Solar cell	[68]
a-$MAMnI_3/TiO_2$	TiO_2	a-$MAMnI_3$	Photodetector	[69]
$MAPbI_3$/PCBM	PCBM	$MAPbI_3$	Phototransistor	[70]
PCBM/CH3NH3PbI_3	PCBM	CH3NH3PbI_3	Solar cell	[71]
$FAPbI_3$/DNTT	DNTT	$FAPbI_3$	Photodetector	[72]
$MASnI_3/SnO_2$	SnO_2	$MASnI_3$	Sensor	[73]
$CH_3NH_3PbCl_3$/Se	Se	$CH_3NH_3PbCl_3$	Photodetector	[74]
GCN–Ag_3PO_4	GCN	Ag_3PO_4	Photocatalyst	[76]
g-C_3N_4/AgCl	g-C_3N_4	AgCl	Photocatalyst	[77]
PANI/Ag_2MoO_4	PANI	Ag_2MoO_4	Photocatalyst	[78]
Ag/TiO_2@PPy	Ag-Ag_2O	TiO_2@PPy	Photocatalyst	[79]
PPy @Bi_2WO_6	PPy	Bi_2WO_6	Photoelectric detection	[80]
g-C_3N_4/BiOI	g-C_3N_4	BiOI	Photocatalyst	[81]
BiI_3/polymer	PTB7-Th	BiI_3	Solar cell	[82]
PbS/PAE-1	PAE-1	PbS	Optical switch	[84]
PSBTBT/PbS	PSBTBT	PbS	Solar cell	[85]
PANI/SnO_2	PANI	SnO_2	Solar cell	[86]
PDVT-10/MoS_2	PDVT-10	MoS_2	Optoelectronic application	[87]
WO_3/PPy	PPy: Ru_4POM	WO_3	Photoanode	[88]

ABBREVIATIONS AND ACRONYMS

a-$MAMnI_3$	Pb-free and amorphous $MAMnI_3$
BBT	poly (benzothiadiazole)
BPA	bisphenol A
C-337	(2,3,6,7-tetrahydro-11-oxo-1H,5H,11H-[1]-benzopyrano[6,7,8-ij]quinolizine-10-carbonitrile)
CK-MB	creatine kinase-methylene blue
CSA	camphor sulfonic acid
CuPc	copper phthalocyanine
DNTT	dinaphtho[2, 3-b: 2′, 3′-f] thieno-[3, 2-b] thiophene
EDT	1, 2-ethanedithiol
EQE	external quantum efficiency
ETL	electron transporting layer
g-C_3N_4	graphite carbon nitride
GQDs	graphene quantum dots
HOMO	highest occupied molecular orbital
I_{SC}	short-circuit current
k_{obs}	the pseudo first-order rate constant
LC	liquid crystal
LDH	layered double hydroxide
Li-TFSI	Li-bis-(trifluoromethane-sulfonyl)
LUMO	lowest occupied molecular orbital
M-TiO_2	mesoporous TiO_2
Mehcf	metal hexacyanoferrates
MEH-PPV	Poly [2-methoxy-5-(2-ethylhexyloxy)-1, 4-phenylenevinylene]
NPB	N, N′-Bis-(1-naphthalenyl)-N, N′-bis-phenyl-(1, 1′-biphenyl)-4, 4′-diamine
NWs	nanowires
PAE-1	poly [sodium 2-(2-ethynyl-4-methoxyphenoxy) acetate]
PANI	polyaniline
PBTTT	poly (2,5-bis(3-al-kylthiophen-2-yl)thieno[3,2-b]thiophenes)
PCBM	[6,6]-phenyl-C61-butyric acid methyl ester
PCE	Power conversion efficiency
PCPDTBT	poly[2,6-(4,4-bis(2-ethylhexyl)-4H-cyclopenta [2,1-b;3,4-b] dithiophene)-alt-4,7(2,1,3-benzothiadiazole)]
PDINH	perylene-3, 4, 9, 10-tetracarboxylic diimide
PDs	polymer dots
PDVT-10	poly[2,5-bis(2-decyltetradecyl)pyrrolo[3,4-c]pyrrole-1,4(2H,5H) dione-alt-5,5′-di(thiophen-2-yl)-2,2′-(E)-2-(2-(thiophen-2-yl) vinyl)-thiophene]
PEC	photoelectronchemical
PEDOT	poly (3, 4-ethylene dioxythiophene)
PFBT	poly [(9,9-dioctylfluorenyl-2,7-diyl)-co-(1,4-benzo-{2,1′, 3}-thiadazole)]
PFH	poly (9,9-dihexylfluorene)

PFO-DBT	poly [2,7-(9, 9-dioctylfluorene)-alt-4, 7-bis (thiophen-2-yl) benzo-2, 1, 3-thiadiazole]
P3HT	poly (3-hexylthiophene)
PPy	polypyrrole
PPV	poly (p-phenylene vinylene)
PSBTBT	poly [2,6-(4, 4'–bis (2-ethylhexyl) dithieno [3,2-b: 2′, 3′-d] silole)-alt-4, 7 (2,1,3-benzothiadiazole)
Psi	porous silicon
PSS	poly (styrene sulfonate)
PTB7-Th	poly[4,8-bis(5-(2-ethylhexyl)thiophen-2yl) benzo[1,2-b,4,5-b′] dithiophene-2,6-diyl-alt-(4-(2-ethylhexyl)-3-fluorothieno[3,4b] thiophene-)-2-carboxylate-2-6-diyl]
PTCDA	perylenetertracarboxylic dianhydride
spiro-MeOTAD	2,2′,7,7′-tetrakis-(N, N-di-p-methoxyphenylamine)-9, 9′-spirobifluorene
PVP	polyvinyl pyrrolidone
QDs	quantum dots
RRT	reverse recovery transient
Ru_4POM	tetraruthenium polyoxometalate
SO	1,3,3–Trimethy-lindolino-β-naphthopyrylospiran
tBT	tert-butylthiol
Tc	tetracene
TiOPc	titanyl phthalocyanine
TNZnPc	2,9,16,23-tetranitrophthalocyanine zinc
TP	triphenylene
TPP	tetraphenylporphyrin
UV	ultraviolet
V_{OC}	open-circuit voltage

REFERENCES

1. Zhang, J., et al., Evolution of epitaxial semiconductor nanodots and nanowires from supersaturated wetting layers. *Chemical Society Reviews*, 2015. 44(1): p. 26–39.
2. Iveland, J., et al., Direct measurement of Auger electrons emitted from a semiconductor light-emitting diode under electrical injection: Identification of the dominant mechanism for efficiency droop. *Physical Review Letters*, 2013. 110(17): p. 177406.
3. Walker, P.M., et al., Exciton polaritons in semiconductor waveguides. *Applied Physics Letters*, 2013. 102(1).
4. Zimmerman, J., R. Parameswaran, and B.Z. Tian, Nanoscale semiconductor devices as new biomaterials. *Biomaterials Science*, 2014. 2(5): p. 619–626.
5. Yousefi, M., et al., New role for nonlinear dynamics and chaos in integrated semiconductor laser technology. *Physical Review Letters*, 2007. 98(4): p. 044101.
6. Avrutin, V., N. Izyumskaya, and H. Morkoç, Semiconductor solar cells: Recent progress in terrestrial applications. *Superlattices and Microstructures*, 2011. 49(4): p. 337–364.
7. Rim, Y.S., et al., Printable ultrathin metal oxide semiconductor-based conformal biosensors. *ACS Nano*, 2015. 9(12): p. 12174–12181.
8. Suresh, S., Semiconductor nanomaterials, methods and applications: A review. *Nanoscience and Nanotechnology*, 2013.

9. Xu, J., et al., Enhancing visible-light-induced photocatalytic activity by coupling with wide-band-gap semiconductor: A case study on Bi_2WO_6/TiO_2. *Applied Catalysis B: Environmental*, 2012. 111–112: p. 126–132.
10. Li, M., et al., Organic-inorganic heterojunctions toward high-performance ambipolar field-effect transistor applications. *Advanced Electronic Materials*, 2018. 4(9).
11. Umeda, T., et al., Improvement of characteristics on polymer photovoltaic cells composed of conducting polymer - fullerene systems. *Synthetic Metals*, 2005. 152(1–3): p. 93–96.
12. Amrollahi Bioki, H. and M. Borhani Zarandi, Electrical properties of organic–inorganic semiconductor heterojunction. *Indian Journal of Physics*, 2012. 86(6): p. 439–441.
13. Chen, C.-H., Hybrid organic on inorganic semiconductor heterojunction. *Journal of Materials Science Materials in Electronics*, 2006. 17(12): p. 1047–1053.
14. Chen, Y., et al., Self-assembled organic-inorganic hybrid nanocomposite of a perylene-tetracarboxylic diimide derivative and CdS. *Langmuir*, 2010. 26(15): p. 12473–12478.
15. Liu, R., Hybrid organic/inorganic nanocomposites for photovoltaic cells. *Materials (Basel)*, 2014. 7(4): p. 2747–2771.
16. Mir, S.H., et al., Review—organic-inorganic hybrid functional materials: An integrated platform for applied technologies. *Journal of The Electrochemical Society*, 2018. 165(8): p. B3137–B3156.
17. Aleshin, A.N., Organic optoelectronics based on polymer – inorganic nanoparticle composite materials. *Physics-Uspekhi*, 2013. 56(6): p. 627–632.
18. Li, S., et al., Nanocomposites of polymer and inorganic nanoparticles for optical and magnetic applications. *Nano Reviews*, 2010. 1.
19. Mao, L., et al., Organic-inorganic nanoparticles molecularly imprinted photoelectrochemical sensor for α-Solanine based on p-type polymer dots and n-CdS heterojunction. *Analytica Chimica Acta*, 2019. 1059: p. 94–102.
20. Greaney, M.J., et al., Improving open circuit potential in hybrid P3HT:CdSe bulk heterojunction solar cells via colloidal tert-butylthiol ligand exchange. *ACS Nano*, 2012. 6(5): p. 4222–4230.
21. Lim, J., et al., Modular fabrication of hybrid bulk heterojunction solar cells based on breakwater-like CdSe tetrapod nanocrystal network infused with P3HT. *The Journal of Physical Chemistry C*, 2014. 118(8): p. 3942–3952.
22. Xu, W., et al., Efficient Organic/Inorganic Hybrid Solar Cell Integrating Polymer Nanowires and Inorganic Nanotetrapods. *Nanoscale Research Letters*, 2017. 12(1).
23. Zhou, Y., et al., Improved efficiency of hybrid solar cells based on non-ligand-exchanged CdSe quantum dots and poly(3-hexylthiophene). *Applied Physics Letters*, 2010. 96(1): p. 013304.
24. Khan, M.A., U. Farva, and Y.M. Kang, Pyrolytic synthesis of densely packed grown CdSe nanoparticles using a dual function octylamine for hybrid solar cells. *Materials Letters*, 2014. 123: p. 62–65.
25. Tan, F., et al., Core/shell-shaped CdSe/PbS nanotetrapods for efficient organic–inorganic hybrid solar cells. *Journal of Materials Chemistry A*, 2014. 2(35).
26. Lai, L.H., et al., Organic-inorganic hybrid solution-processed H(2)-evolving photocathodes. *ACS Applied Materials & Interfaces*, 2015. 7(34): p. 19083–19090.
27. Saha, S.K., A. Guchhait, and A.J. Pal, Organic/inorganic hybrid pn-junction between copper phthalocyanine and CdSe quantum dot layers as solar cells. *Journal of Applied Physics*, 2012. 112(4): p. 044507.
28. Wang, Q., et al., Semiconducting organic–inorganic nanodots heterojunctions: Platforms for general photoelectrochemical bioanalysis application. *Analytical Chemistry*, 2018. 90(6): p. 3759–3765.
29. Zhang, X.H., et al., Robust visible/near-infrared light driven hydrogen generation over Z-scheme conjugated polymer/CdS hybrid. *Applied Catalysis B-Environmental*, 2018. 224: p. 871–876.

30. Ren, S., et al., Inorganic–organic hybrid solar cell: Bridging quantum dots to conjugated polymer nanowires. *Nano Letters*, 2011. 11(9): p. 3998–4002.
31. Yuan, K., L. Chen, and Y. Chen, Direct anisotropic growth of CdS nanocrystals in thermotropic liquid crystal templates for heterojunction optoelectronics. *Chemistry - A European Journal*, 2014. 20(36): p. 11488–11495.
32. Lin, H.W., et al., Constructing a green light photodetector on inorganic/organic semiconductor homogeneous hybrid nanowire arrays with remarkably enhanced photoelectric response. *ACS Applied Materials & Interfaces*, 2019. 11(10): p. 10146–10152.
33. Sharma, R., et al., Role of zinc oxide and carbonaceous nanomaterials in non-fullerene-based polymer bulk heterojunction solar cells for improved cost-to-performance ratio. *Journal of Materials Chemistry A*, 2015. 3(44): p. 22227–22238.
34. Pei, J., et al., Influence of organic interface modification layer on the photoelectric properties of ZnO-based hybrid solar cells. *Journal of Photochemistry and Photobiology A: Chemistry*, 2018. 364: p. 551–557.
35. Ouyang, B.S., K.W. Zhang, and Y. Yang, Self-powered UV photodetector array based on P3HT/ZnO nanowire array heterojunction. *Advanced Materials Technologies*, 2017. 2(12).
36. Pandey, R.K., et al., Interface engineering for enhancement in performance of organic/inorganic hybrid heterojunction diode. *Organic Electronics*, 2017. 45: p. 26–32.
37. Tai, H.L., et al., The enhanced formaldehyde-sensing properties of P3HT-ZnO hybrid thin film OTFT sensor and further insight into its stability. *Sensors*, 2015. 15(1): p. 2086–2103.
38. Tonpe, D.A., et al., Development of organic/inorganic PANI/ZnO 1D nanostructured hybrid thin film solar cell by soft chemical route. *Journal of Materials Science-Materials in Electronics*, 2019. 30(17): p. 16056–16064.
39. Xie, Y.R., et al., High performance blue light detector based on ZnO nanowire arrays. *Applied Optics*, 2019. 58(5): p. 1242–1245.
40. Li, X.Y., et al., A self-powered nano-photodetector based on PFH/ZnO nanorods organic/inorganic heterojunction. *Results in Physics*, 2018. 8: p. 468–472.
41. Song, J., et al., Poly(9,9-dihexylfluorene)/ZnO nanoparticles based inorganic/organic heterojunction structure: Electrical and photoconductivity properties. *Journal of Nanoscience and Nanotechnology*, 2016. 16(6): p. 6005–6010.
42. Mokhtari, H. and M. Benhaliliba, Organic-inorganic Au/PVP/ZnO/Si/Al semiconductor heterojunction characteristics. *Journal of Semiconductors*, 2017. 38(11).
43. Trung, T.Q., et al., An omnidirectionally stretchable photodetector based on organic-inorganic heterojunctions. *ACS Applied Materials & Interfaces*, 2017. 9(41): p. 35958–35967.
44. Luka, G., et al., Electrical properties of zinc oxide – Tetracene heterostructures with different n-type ZnO films. *Organic Electronics*, 2017. 45: p. 240–246.
45. Padma, N., et al., Performance comparison of p-n junction diodes using zinc oxide and copper phthalocyanine hybrid nanocomposites and bilayer heterostructures. *Nano*, 2014. 9(6).
46. Chaudhary, D.K., et al., Bulk-heterojunction hybrid solar cells with non-toxic, earth abundant stannite phase CuZn2AlS4 nanocrystals. *Thin Solid Films*, 2018. 649: p. 202–209.
47. Chen, G.-H., S.-J. Ho, and H.-S. Chen, Cubic zincblende ZnSe nanowires with an entangling structure grown via oriented attachment and their application in organic–inorganic heterojunction light-emitting diodes. *The Journal of Physical Chemistry C*, 2014. 118(44): p. 25816–25822.
48. Li, X., et al., Self-assembled supramolecular system PDINH on TiO_2 surface enhances hydrogen production. *Journal of Colloid and Interface Science*, 2018. 525: p. 136–142.

49. Xiao, G., et al., Visible-light-driven activity and synergistic mechanism of TiO_2@g-C_3N_4 heterostructured photocatalysts fabricated through a facile and green procedure for various toxic pollutants removal. *Nanotechnology*, 2018. 29(31).
50. Zhang, Q., et al., Tetranitrophthalocyanine Zinc/TiO_2 nanofibers organic-inorganic heterostructures with enhanced visible photocatalytic activity. *Nano*, 2017. 12(10).
51. Hou, H.J., et al., Conjugated microporous poly(benzothiadiazole)/TiO_2 heterojunction for visible-light-driven H-2 production and pollutant removal. *Applied Catalysis B-Environmental*, 2017. 203: p. 563–571.
52. Qin, Y., et al., Top-down strategy toward versatile graphene quantum dots for organic/inorganic hybrid solar cells. *ACS Sustainable Chemistry & Engineering*, 2015. 3(4): p. 637–644.
53. Szkoda, M., et al., Fabrication and photoactivity of organic-inorganic systems based on titania nanotubes and PEDOT containing redox centres formed by different Prussian Blue analogues. *Journal of Alloys and Compounds*, 2017. 723: p. 498–504.
54. Zhang, D.Z., et al., Organics filled one-dimensional TiO_2 nanowires array ultraviolet detector with enhanced photo-conductivity and dark-resistivity. *Nanoscale*, 2017. 9(26): p. 9095–9103.
55. Sharma, S., et al., Fluorine-chlorine co-doped TiO_2/CSA doped polyaniline based high performance inorganic/organic hybrid heterostructure for UV photodetection applications. *Sensors and Actuators a-Physical*, 2017. 261: p. 94–102.
56. Subramanian, E., P. Santhanamari, and C. Murugan, Sensor functionality of conducting polyaniline-metal oxide (TiO_2/SnO_2) hybrid materials films toward benzene and toluene vapors at room temperature. *Journal of Electronic Materials*, 2018. 47(8): p. 4764–4771.
57. Jarka, P., et al., Manufacture of photovoltaic cells with hybrid organic-inorganic bulk heterojunction. *Materials and Manufacturing Processes*, 2018. 33(8): p. 912–922.
58. Rakibuddin, M., H. Kim, and M.E. Khan, Graphite-like carbon nitride (C_3N_4) modified N-doped $LaTiO_3$ nanocomposite for higher visible light photocatalytic and photo-electrochemical performance. *Applied Surface Science*, 2018. 452: p. 400–412.
59. Yazdani, D., A.A. Zinatizadeh, and M. Joshaghani, Organic-inorganic Z-scheme g-C_3N_4-NiTi-layered double hydroxide films for photocatalytic applications in a fixed-bed reactor. *Journal of Industrial and Engineering Chemistry*, 2018. 63: p. 65–72.
60. Kumar, A., et al., Perovskite-structured $CaTiO_3$ coupled with g-C_3N_4 as a heterojunction photocatalyst for organic pollutant degradation. *Beilstein Journal of Nanotechnology*, 2018. 9: p. 671–685.
61. Nawar, A.M. and M.M. Makhlouf, Electrical and photovoltaic responses of an Au/Coumarin 337/n-Si/Sb-Au hybrid organic-inorganic solar cell. *Journal of Electronic Materials*, 2019. 48(9): p. 5771–5784.
62. El-Nahass, M.M. and W.M. Desoky, Fabrication and electrical characterization of n-1,3,3-trimethylindolino-beta-naphthopyrylospiran/p-Si (organic/inorganic) heterojunction for application in photovoltaic device. *Optik*, 2018. 171: p. 44–50.
63. Mahato, S., et al., High efficiency ITO-free hybrid solar cell using highly conductive PEDOT:PSS with co-solvent and surfactant treatments. *Materials Letters*, 2017. 186: p. 165–167.
64. Zhang, J., S.T. Lee, and B.Q. Sun, Effect of series and shunt resistance on organic-inorganic hybrid solar cells performance. *Electrochimica Acta*, 2014. 146: p. 845–849.
65. Harraz, F.A., Electrochemical formation of a novel porous silicon/polypyrrole hybrid structure with enhanced electrical and optical characteristics. *Journal of Electroanalytical Chemistry*, 2014. 729: p. 68–74.
66. Luo, J., et al., Organic-inorganic hybrid perovskite – TiO_2 nanorod arrays for efficient and stable photoelectrochemical hydrogen evolution from HI splitting. *Materials Today Chemistry*, 2019. 12: p. 1–6.

67. Zhou, H.W., et al., Hole-conductor-free, metal-electrode-free $TiO_2/CH_3NH_3PbI_3$ heterojunction solar cells based on a low-temperature carbon electrode. *Journal of Physical Chemistry Letters*, 2014. 5(18): p. 3241–3246.
68. Zhang, W.N., et al., Lead-free organic-inorganic hybrid perovskite heterojunction composites for photocatalytic applications. *Catalysis Science & Technology*, 2017. 7(13): p. 2753–2762.
69. Zhang, X.X., et al., Lead-free and amorphous organic-inorganic hybrid materials for photovoltaic applications: mesoscopic $CH_3NH_3MnI_3/TiO_2$ heterojunction. *RSC Advances*, 2017. 7(59): p. 37419–37425.
70. Jeong, H. and J.K. Lee, Organic–inorganic hybrid ternary bulk heterojunction of nanostructured perovskite–low bandgap polymer–PCBM for improved efficiency of organic solar cells. *ACS Applied Materials & Interfaces*, 2015. 7(51): p. 28459–28465.
71. Rahaq, Y., et al., Highly reproducible perovskite solar cells via controlling the morphologies of the perovskite thin films by the solution-processed two-step method. *Journal of Materials Science-Materials in Electronics*, 2018. 29(19): p. 16426–16436.
72. Luo, L.B., et al., A highly sensitive perovskite/organic semiconductor heterojunction phototransistor and its device optimization utilizing the selective electron trapping effect. *Advanced Optical Materials*, 2019. 7(13).
73. Chen, Y., et al., Light enhanced room temperature resistive NO_2 sensor based on a gold-loaded organic-inorganic hybrid perovskite incorporating tin dioxide. *Microchimica Acta*, 2019. 186(1).
74. Chen, H.-Y., et al., Fabrication and photoelectric properties of organic-inorganic broad-spectrum photodetectors based on Se microwire/perovskite heterojunction. *Chinese Optics*, 2019. 12(5): p. 1057–1063.
75. Huang, K., et al., One-step synthesis of Ag_3PO_4/Ag photocatalyst with visible-light photocatalytic activity. *Materials Research*, 2015. 18(5): p. 939–945.
76. Nagajyothi, P.C., et al., Photocatalytic dye degradation and hydrogen production activity of Ag_3PO_4/g-C3N4 nanocatalyst. *Journal of Materials Science-Materials in Electronics*, 2019. 30(16): p. 14890–14901.
77. Matheswaran, P., P. Thangavelu, and B. Palanivel, Carbon dot sensitized integrative g-C_3N_4/AgCl Hybrids: An synergetic interaction for enhanced visible light driven photocatalytic process. *Advanced Powder Technology*, 2019. 30(8): p. 1715–1723.
78. Mondal, P., et al., Facile fabrication of novel hetero-structured organic-inorganic high-performance nanocatalyst: A smart system for enhanced catalytic activity toward ciprofloxacin degradation and oxygen reduction. *ACS Applied Nano Materials*, 2018. 1(11): p. 6015–6026.
79. Kumar, R., R.M. El-Shishtawy, and M.A. Barakat, Synthesis and characterization of Ag-Ag_2O/TiO_2@polypyrrole heterojunction for enhanced photocatalytic degradation of methylene blue. *Catalysts*, 2016. 6(6).
80. Zhu, L.-B., et al., Enhanced organic–inorganic heterojunction of polypyrrole@Bi_2WO_6: Fabrication and application for sensitive photoelectrochemical immunoassay of creatine kinase-MB. *Biosensors and Bioelectronics*, 2019. 140: p. 111349.
81. Chang, C., et al., Novel mesoporous graphite carbon nitride/BiOI heterojunction for enhancing photocatalytic performance under visible-light irradiation. *ACS Applied Materials & Interfaces*, 2014. 6(7): p. 5083–5093.
82. Ma, S., et al., Vertically oriented BiI3 template featured BiI_3/polymer heterojunction for high photocurrent and long-term stable solar cells. *ACS Applied Materials & Interfaces*, 2019. 11(35): p. 32509–32516.
83. Kumar, R., et al., Preparation of nanocrystalline Sb doped PbS thin films and their structural, optical, and electrical characterization. *Superlattices and Microstructures*, 2014. 75: p. 601–612.

84. Wurst, K.M., et al., Correlated, dual-beam optical gating in coupled organic-inorganic nanostructures. *Angewandte Chemie*, 2018. 130(36): p. 11733–11737.
85. Nam, M., et al., Broadband-absorbing hybrid solar cells with efficiency greater than 3% based on a bulk heterojunction of PbS quantum dots and a low-bandgap polymer. *Journal of Materials Chemistry A*, 2014. 2(11): p. 3978.
86. Murugan, C., E. Subramanian, and D.P. Padiyan, p-n Heterojunction formation in polyaniline-SnO_2 organic-inorganic hybrid composite materials leading to enhancement in sensor functionality toward benzene and toluene vapors at room temperature. *Synthetic Metals*, 2014. 192: p. 106–112.
87. Yan, J., et al., Ambipolar charge transport in an organic/inorganic van der Waals p-n heterojunction. *Journal of Materials Chemistry C*, 2018. 6(47): p. 12976–12980.
88. Jeon, D., et al., WO_3/conducting polymer heterojunction photoanodes for efficient and stable photoelectrochemical water splitting. *ACS Applied Materials & Interfaces*, 2018. 10(9): p. 8036–8044.

6 Organic–Inorganic Heterojunction Nanowires

Yuan Yao and Yanbing Guo

CONTENTS

6.1 INTRODUCTION: BACKGROUND AND DRIVING FORCES

Among crystalline materials, inorganic crystals, such as silicon, were the fundamental building blocks of modern electronics and microelectronics. It was not surprising that the preparation and controlled growth of crystalline nanostructures with various morphologies and orientations was a hot research topic [1,2]. Compared with inorganic materials, organic materials have the advantages of low cost, easy molecular tailoring for property optimization, high flexibility, easy large-scale processing as well as compatibility with lightweight plastic substrates [3–5]. Thus, π-conjugated polymer- and small molecule-based functional organic nanomaterials are considered to be good candidates for the next generation miniature, flexible consumer electronic devices [6–10]. However, the short-range order, amorphous organic materials always share the problem of low-carrier transport ability and low-performance stability [11]. Organic and inorganic materials can be combined by covalent bonds or other interaction forces to form organic–inorganic heterojunction. This strategy not only preserves the original characteristics of each material but also forms new properties through interface interaction, thus realizing the multifunctional materials. On the other hand, one-dimensional nanomaterials have attracted much attention due to their large length-diameter ratio and unique optical and photoelectric properties [12]. The devices made of one-dimensional materials have the characteristics of multifunction and can be used in field-emission diodes, solar cells, photoelectric detection, and other fields [13]. In these devices, the formation of heterojunction has a critical and significant effect on the material properties.

Based on the above analysis, one-dimensional organic–inorganic heterojunction nanowires-based devices have multifunction, unique electrical, optical, and tunable characteristics, and therefore have potential development and application prospects in the fields of electronics, photoelectricity, and catalysis, etc. [11,14]. Strong interactions between organic and inorganic components could result in new or enhanced physical or chemical properties relative to that of a single component, thus achieving synergistic properties [14].

At present, one-dimensional organic–inorganic heterojunction nanowires have made important progress in both synthesis methodology and material properties. There are three research directions: (1) The preparation of novel organic–inorganic heterojunction by the reasonable design of molecular structure and the regulation of interfacial forces; (2) The studies of growth process and formation mechanism; (3) The detection of improved physical or chemical properties. Therefore, we reviewed the recent progress of organic–inorganic heterojunction nanowires, including preparation methods and application fields. Besides, the problems that need to be solved urgently and the future development direction of this field were pointed out to better promote the continuous progress of this field.

6.2 THE SYNTHETIC METHODS OF ORGANIC–INORGANIC HETEROJUNCTION NANOWIRES

Up to now, there are four methods to fabricate organic–inorganic heterojunction nanowires, including chemical vapor deposition, solution phase method and template method. Even though the growth mechanism of nanowires via these methods is different, the growth of nanowires (crystals) is affected by both kinetics and thermodynamics [13]. In this section, the specific synthesis paths of each method are introduced, and the growth mechanism of organic–inorganic heterojunction nanowires is emphasized. This provides theoretical guidance and reference for the design and development of the next generation of higher-quality organic–inorganic heterojunction nanowires and novel synthesis methods.

6.2.1 Solution Phase Method

The solution phase method is the simplest method to synthesize organic–inorganic heterojunction nanowires: the substrate was immersed in the solution of organic precursors, and then it was taken out and cleaned to obtain organic–inorganic heterojunction nanowires after a period of reaction [15].

In 2009, Huibiao Liu et al. used zinc oxide (ZnO) with wide band gap to modify Cu-TCNQ. The ZnO-CuTCNQ hybrid NWs was prepared by impregnating Cu-TCNQ NWs with ZnO colloid, then drying in vacuum [15]. Alejandro L. Briseno et al. also used the solution phase method to build two types of organic/ZnO heterojunction nanowires, including ZnO/P3HT[poly(3-hexylthiophene)] (Figure 6.1a) and ZnO/QT (didodecylquaterthiophene) (Figure 6.1b). As shown in Figure 6.1c and d, the ZnO/P3HT and ZnO/QT core-shell heterojunction nanowires were successfully constructed with shell thicknesses of 7–20 nm and 6–13 nm, respectively. In terms of the ZnO/QT, every two QT molecules form π–π interactions with each other, and phosphonic acids

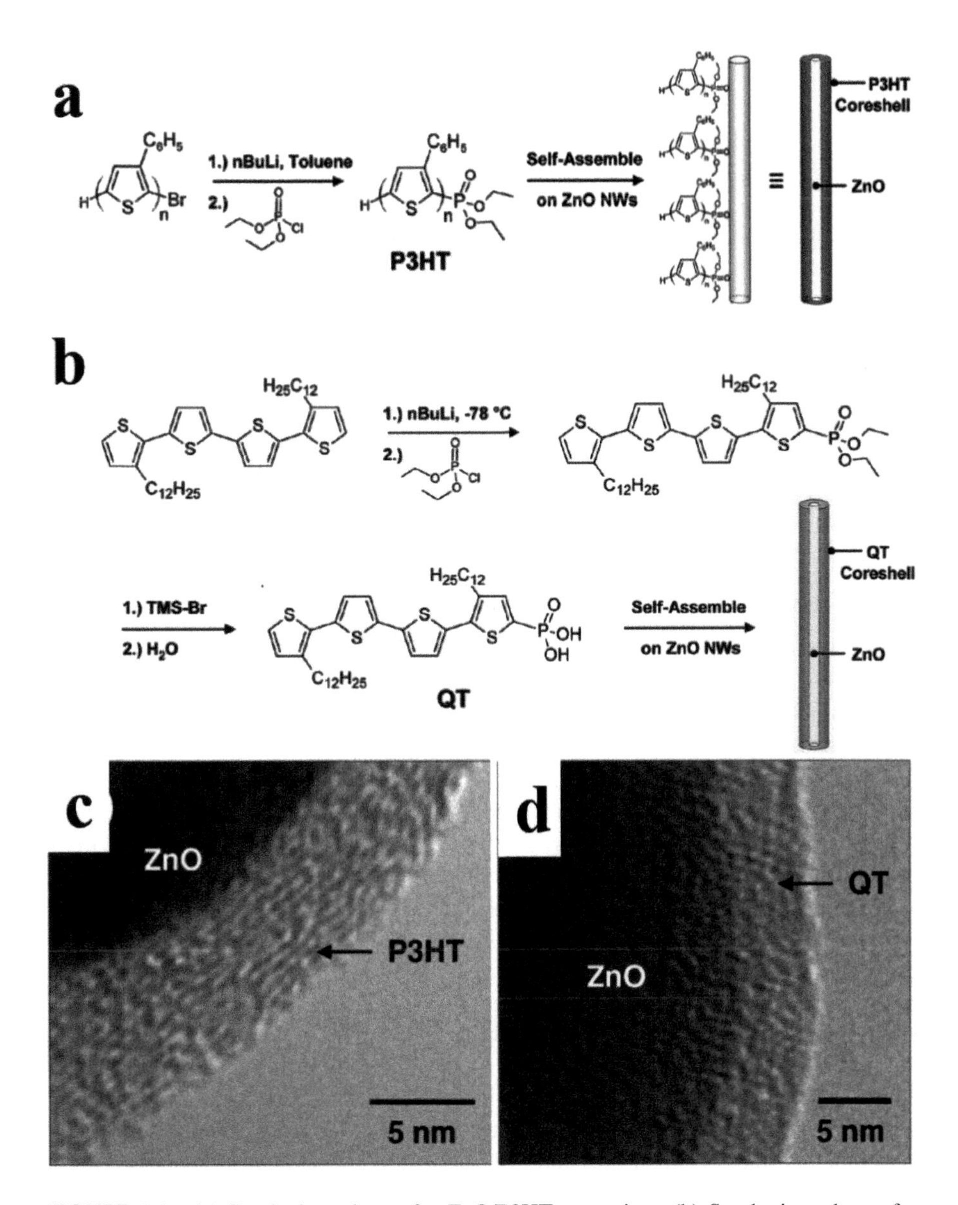

FIGURE 6.1 (a) Synthetic pathway for ZnO/P3HT nanowires; (b) Synthetic pathway for ZnO/QT nanowires; (c) Transmission electron microscopy (TEM) images of the ZnO/P3HT nanowires; (d) TEM images of the ZnO/QT nanowires. (Reprinted with permission from Ref 16. Copyright 2010 American Chemical Society)

group at the ends of different QT molecules form hydrogen bonds (P-O-H···O-P), and alkyl groups of different QT molecules have van der Waals force. These three intermolecular forces stabilize the structure of QT and align it vertically on the surface of ZnO. As for the ZnO/P3HT, P3HT molecules are arranged upside down on the surface of ZnO and there exist partially disordered layers [16]. P3HT:CdSe heterojunction nanowires were also fabricated on silicon by Xianfu Wang et al. Firstly, they prepared high-quality single crystals CdSe growing along [001] orientation with uniform diameter of 100 nm. Then, they mixed the P3HT solution with the CdSe solution and dropped it onto a silicon wafer for vacuum drying to get P3HT:CdSe heterojunction nanowires [17]. ZnO and polythiophene derivatives, (poly[2-(3-thienyl)-ethyloxy-4-butylsulfonate)] (PTEBS) and poly[3-(potassium-6-hexanoate)thiophene-2,5-diyl] (P3KHT))-based core/shell coaxial nanowires, were also reported by Shanju Zhang and co-workers using electrostatic interactions to coat the organic layer on the ZnO nano-crystals [18]. Qing Yang et al. reported the fabrication of Zn/PEDOT:PSS core-shell nanowire and demonstrated effectively enhanced external efficiency of as-prepared ultraviolet light-emitting diode by piezo-phototronic effect [19]. 3D CoO/PPY coaxial nanowire array on nickel foam was also developed by Cheng Zhou et al. through template free solution process [20]. Hyunhyub Ko et al. reported hierarchical fibrillar arrays based on polycarbonate (PC) micropillar (μPLR) arrays decorated with ZnO nanowires (NWs) on flexible substrates. The as-prepared hierarchical PC μPLRs/ZnO NWs exhibited excellent superhydrophobicity which was not observed from the PC μPLRs coated substrate and the blank substrate [21].

6.2.2 Template Method Combined with Electrochemical Polymerization

Anodic aluminum oxide (AAO) is one of the most commonly used templates and is always used to grow organic–inorganic heterojunction nanowires assisted by electrochemical polymerization.

The template method was firstly applied in the synthesis of organic–inorganic heterojunction nanowires by Sungho Park et al. [22]. They prepared three-segment Au-PPY (polypyrrole)-Au heterojunction via the template method associated with electrochemical polymerization of pyrrole. Besides, through analog method, they fabricated four-segment Au-PPY-Cd-Au heterojunction nanowires showing "diode" behavior [22]. Michal Lahav et al. synthesized core-shell and segmented organic–inorganic heterojunction nanowires via this method. By adjusting the pH of the solution (pH = 10.2), PANI (polyaniline)/Au core-shell structure was obtained by the template method. Then, PANI/Au segmented structure was also obtained by forming a self-assembled monolayer (SAM) of thioaniline on the top of Au [23]. Yanbing Guo et al. successfully fabricated CdS-PPY (polypyrrole) heterojunction nanowires using AAO template as shown in Figure 6.2a. The CdS-PPY nanowires were uniform, dense, and smooth with the diameter of 200–400 nm. Figure 6.2b obviously shows the interface between CdS and PPY, indicating the successful construction of heterojunction. As depicted in Figure 6.2c, which shows element distribution of C (green) and Cd (red), the end-to-end structure was revealed, suggesting the successful construction of heterojunction [24]. The schematic diagram of the synthesis of PBPB {Poly[1,4-bis(pyrrol-2-yl)benzene]}/CdS by AAO template is shown in Figure 6.2d. In sequence, PBPB and CdS were deposited into the AAO template

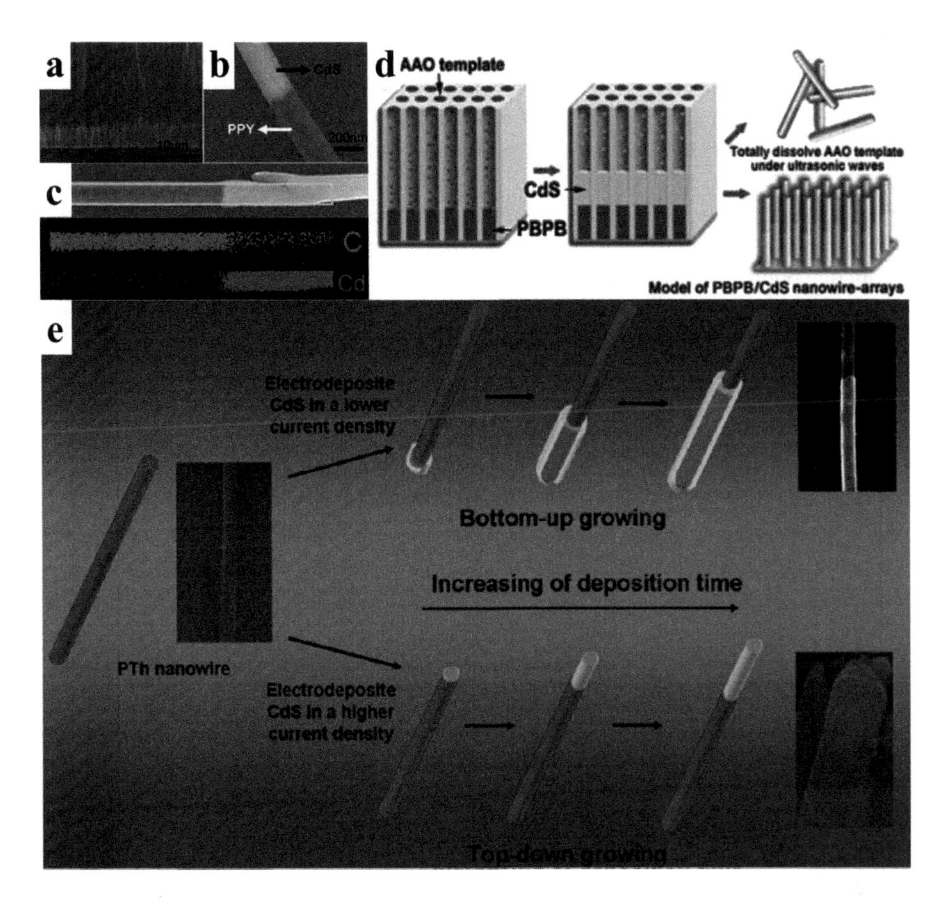

FIGURE 6.2 (a) The side-view SEM image of CdS-PPY nanowires inside the AAO template; (b) The typical SEM image of a single CdS-PPY nanowire; (c) Element mapping of a single CdS-PPY nanowire; (Reprinted with permission from Ref 24. Copyright 2008 American Chemical Society) (d) Synthesis procedure of PBPB/CdS heterojunction nanowire arrays; (Reprinted with permission from Ref 25. Copyright 2011 Royal Society of Chemistry) (e) Schematic illustration of the two different growing processes of PTh-CdS Core-Shell and Segmented Nanorod. (Reprinted with permission from Ref 26. Copyright 2009 American Chemical Society)

assisted by electrochemical polymerization. Finally, the AAO template was etched by NaOH and washed for several times to obtain PBPB/CdS heterojunction nanowires [25]. Besides, Yanbing Guo et al. explored two different growing processes of PTh(polythiophene)-CdS using AAO template method. As depicted in Figure 6.2e, at low current density, CdS would cover PTh nanowires to form core-shell structure (bottom-up growing). While at high current density, CdS would continue to grow at the top of PTh nanowires to form heterojunction (top-down growing) [26]. Haowei Lin et al. also fabricated PANI (polyaniline)/CdS via AAO template method. The diameter of single PANI/CdS nanowire is approximately 200 nm with the length of 10 μm [27]. Zheng Xue et al. prepared GD (graphdiyne)/CuS core-shell structure using AAO template method. Graphdiyne was firstly synthesized into AAO template via self-assembly process, then CuS was electrodeposited on the surface of the graphdiyne to form GD/CuS heterojunction nanowire [28]. Giacomo Mariani et al. fabricated 3D nanostructured core-shell GaAs/PEDOT nanopillar arrays, which were fabricated via a fast, catalyst-free, bottom-up growth approach for GaAs and the following electrochemical deposition process for PEDOT coating [29]. Yuya Oaki et al. fabricated ZnO/PPY co-axial nanowire arrays on an indium tin oxide (ITO) substrate through low-temperature solution processes with irradiation of UV and visible light [30]. Recently, Xinhui Xia et al. demonstrated a few examples of metal oxide/conductive polymer co-axial nanowire including different nanoarray cores (nanowire and nanorod) of metal oxides (Co_3O_4 and TiO_2) and different conducting polymer (PANI and PEDOT) shells through an electrochemical strategy [31].

6.2.2.1 Template Method Combined with Pressure Injection

The template method associated with pressure injection is another successful practice to prepare organic–inorganic heterojunction nanowires.

For example, Yanbing Guo et al. applied AAO template to prepare CdS-OPV3 [oligo(*p*-phenylenevinylene)] heterojunction. OPV3 (Figure 6.3a) and CdS nanocrystals (Figure 6.3b) were firstly synthesized, then mixed to form colloid. A porous AAO membrane was immersed into the colloid and kept for 5–8 min under reduced pressure. Due to the differential pressure, the solution was immediately injected into the pores of the membrane. Finally, the AAO membrane was etched by NaOH solution and dried (Figure 6.3c). As depicted in Figure 6.3d, the diameter of CdS-OPV3 nanowires was 150–200 nm with the length of 2–5 μm. A clear interface between CdS and OPV3 was observed by HRTEM and is shown in Figure 6.3e, and the lattice spacing observed is about 0.204 nm, which is in good agreement with the plane spacing of the direction parallel (110) in wurtzite phase of CdS [32].

6.2.3 Vapor–Liquid–Solid Method

The vapor–liquid–solid (VLS) method was used to grow GDY (graphdiyne)/ZnO organic–inorganic heterojunction nanowires by Xuemin Qian et al. for the first time. As shown in Figure 6.4a, they first grew ZnO nanowires on silicon substrate, then low molecular weight GDY vaporized and reacted with ZnO on the surface of ZnO to form metallic Zn. Small Zn molten droplets (melting point of Zn: 415 °C) formed on the tips of ZnO, then adsorbed GDY vapor. Subsequently, GDY molecules segregated to form GDY nuclei, and then GDY nanowires. During the growth, GDY pushed the Zn droplets up. Finally, Zn was oxidized to ZnO to form GDY/ZnO nanowires. TEM images (Figure 6.4b–d) showed the successful synthesis of the GDY/ZnO nanowires with uniform morphology [33].

6.3 THE APPLICATIONS OF ORGANIC–INORGANIC HETEROJUNCTION NANOWIRES

The construction of organic–inorganic heterojunction nanowires not only retains the characteristics of each material but also produces novel electrical and catalytic properties through the interfacial interaction, which showed great potential in field emission, diode rectification, solar cells, photoelectric detection, and logic gates.

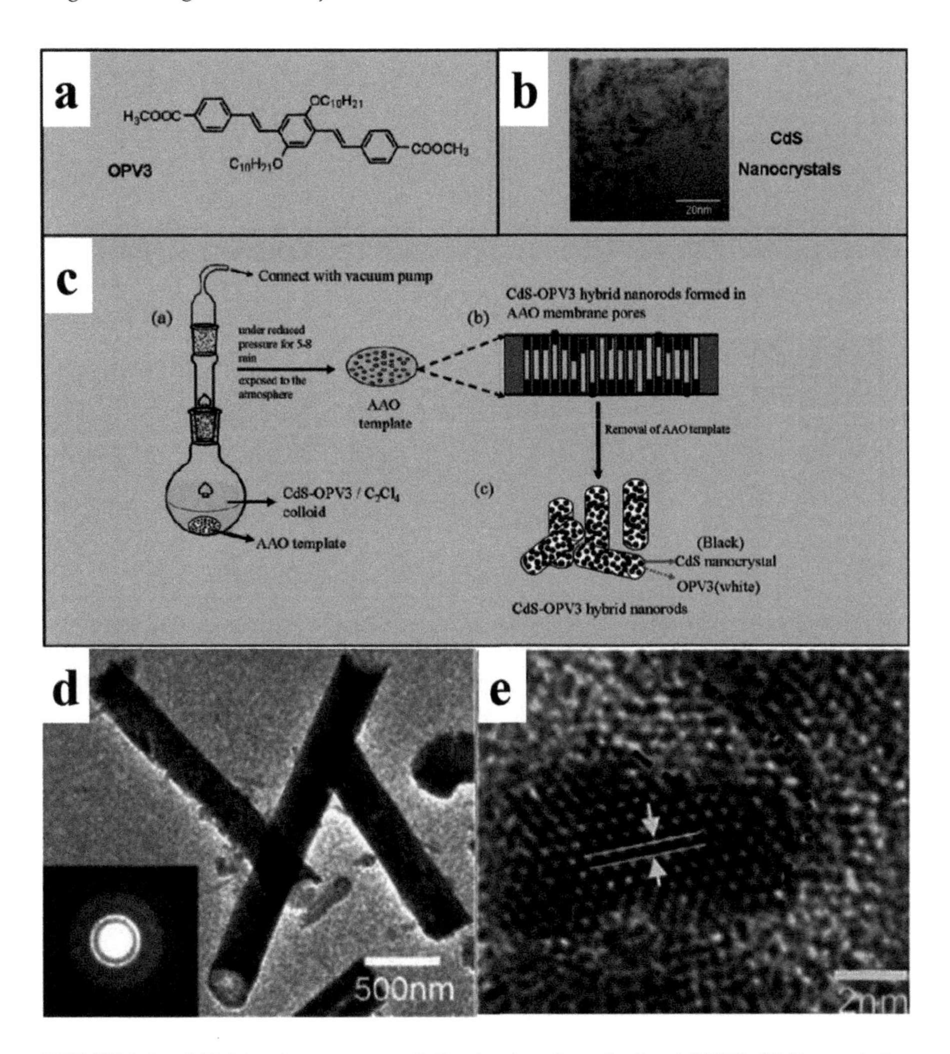

FIGURE 6.3 (a) Molecular structure of oligo(p-phenylenevinylene) OPV3; (b) Transmission electron microscope (TEM) image of CdS nano-crystals; (c) Schematic illustration of the template synthesis of CdS-OPV3 hybrid nanorods; (d) TEM image of a few CdS-OPV3 nanorods. The inset in (d) is the selective area electron diffraction pattern (SAED) taken from the nanorods; (e) HRTEM image of CdS-OPV3 interface under a higher magnification. (Reprinted with permission from Ref 32. Copyright 2008 American Chemical Society)

6.3.1 Field Emission

Field emission refers to the phenomenon that electrons are released from the surface of cathode under a strong electric field, which could be used in field-emission microscope, microwave power amplifier, and electron beam etch, etc. [11].

Huibiao Liu et al. loaded ZnO nanoparticles onto the Cu-TCNQ nanowires to form ZnO-CuTCNQ hybrid NWs. The turn-on field of ZnO-CuTCNQ (6.5 V/μm) was 3V lower than that of CuTCNQ (9.5 V/μm). They also synthesized In_2O_3-CuTCNQ

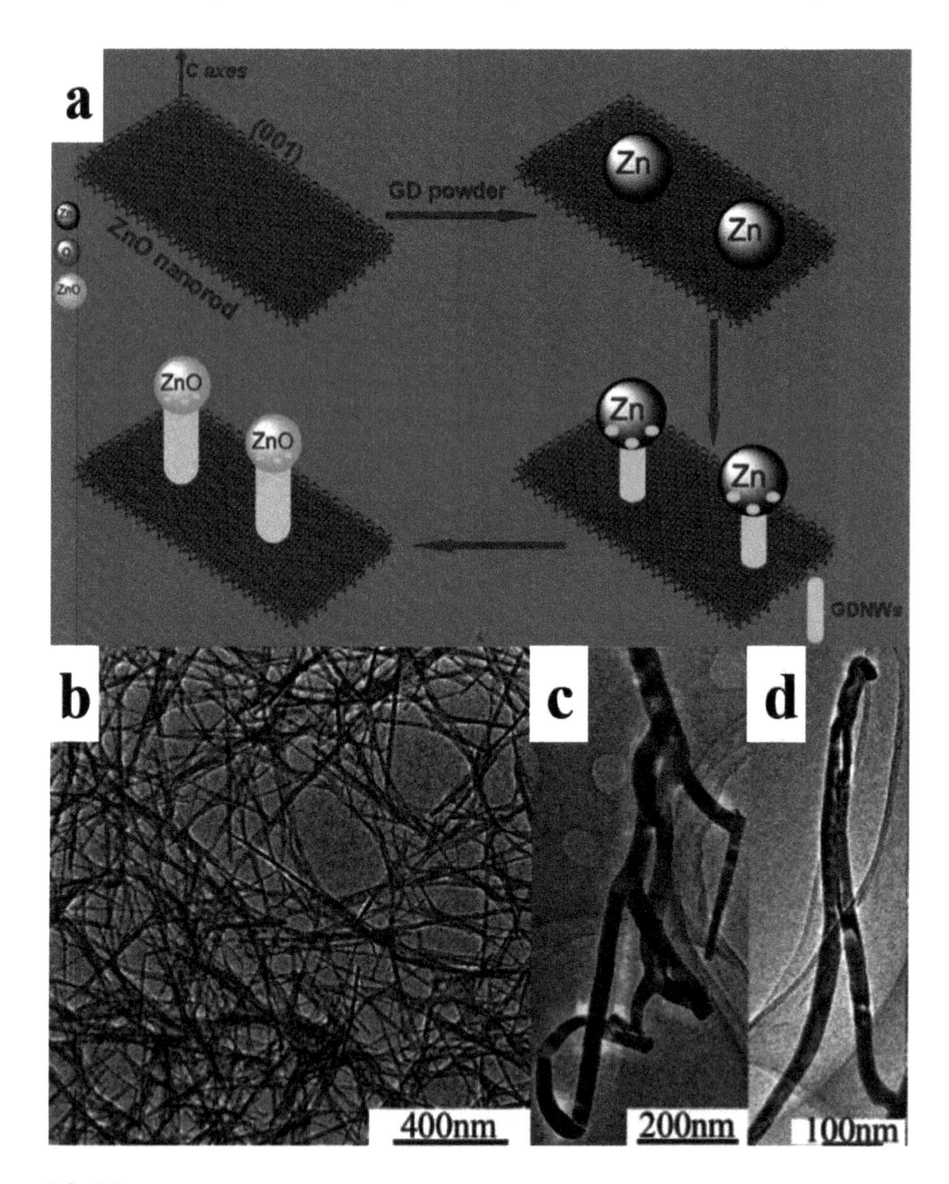

FIGURE 6.4 (a) Schematic illustration of the VLS process in the growth of GDY/ZnO nanowires; (b) Low magnification TEM image of GDY/ZnO nanowires; (c) and (d) High-magnification TEM images of GDY/ZnO nanowires. (Reprinted with permission from Ref 33. Copyright 2012 Royal Society of Chemistry)

hybrid NWs by the same liquid phase method, which proved the universality of the method [15]. Yanbing Guo et al. prepared PTh-CdS core-shell and segmented structure using the AAO template method. Figure 6.5a shows that large-area and dense PTh-CdS core-shell was successfully fabricated. The half part of the PTh rounded by a shell of CdS is observed in Figure 6.5b. And the length of the shell can be controlled

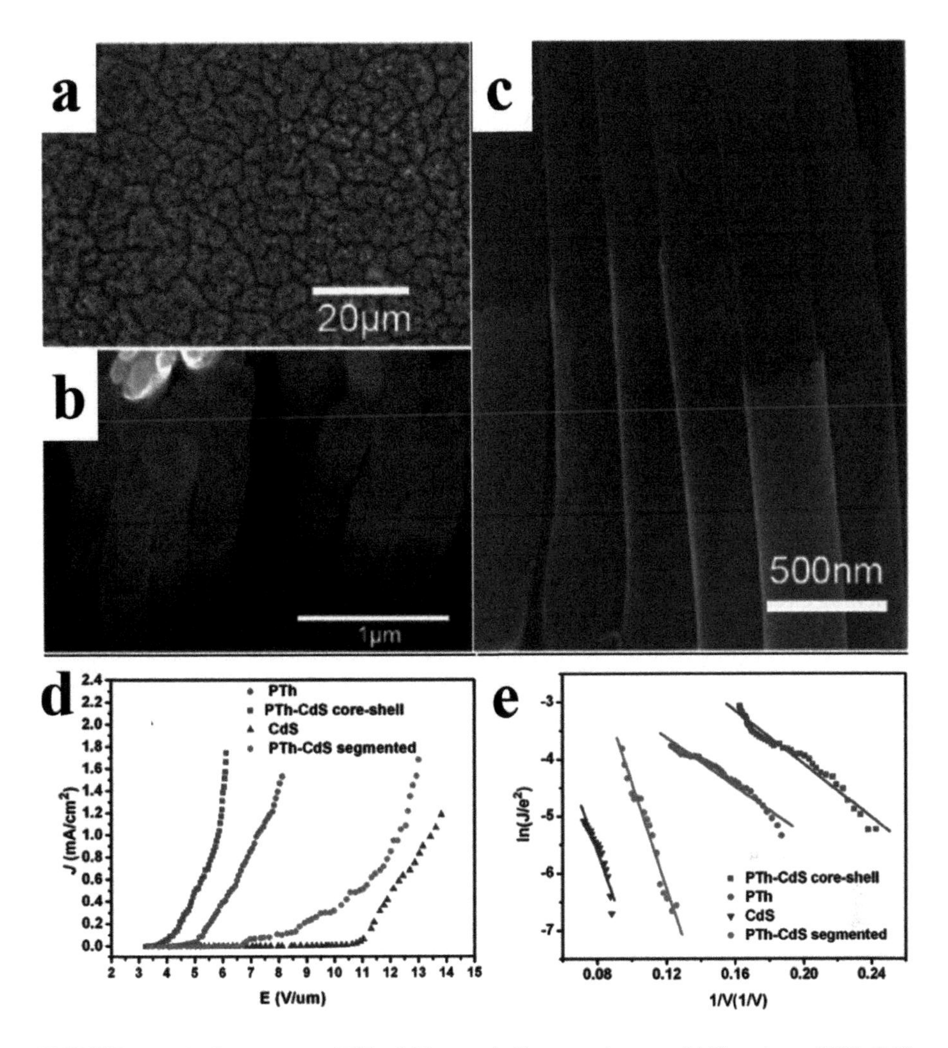

FIGURE 6.5 (a) Large area of PTh-CdS core-shell nanorod array; (b) Top view of PTh-CdS core-shell nanorod array under a higher magnification; (c) Side view of PTh-CdS core-shell nanorod array; (d) J-E plot of CdS nanorods, PTh nanorods, PTh-CdS segmented nanorods, and PTh-CdS core-shell nanorods; and (e) Corresponding F-N plot of those nanorods above. (Reprinted with permission from Ref 26. Copyright 2009 American Chemical Society)

by changing the time of electrodeposition. Figure 6.5c further indicates that the shell length of CdS was 2 μm. The field-emission performance of the PTh-CdS core-shell and segmented structure were tested and are shown in Figure 6.5d. The minimum turn-on field was achieved by the PTh-CdS core-shell structure (3.62 V/μm), which was lower than PTh (4.58 V/μm), the PTh-CdS segmented structure (6.72 V/μm) and CdS (10.41 V/μm). According to Figure 6.5e, the linear results showed that the field emission of the PTh-CdS core-shell and segmented structure was generated by the quantum tunneling mechanism [26]. To the PTh-CdS core-shell array, the triple junctions are almost vertical to the anode, and electrons emitted from such triple junctions arrive at the anode easily, which resulted in the low turn-on field and low threshold field. However, as for the PTh-CdS segmented structure, triple junctions are almost parallel with the anode because the end-to-end structure, and the top fraction of CdS will shield the triple junction emissions which led to a higher turn-on field and threshold field. This proved that the structure of heterojunction can also regulate the field-emission performance.

In summary, the field-emission performance of organic–inorganic heterojunction nanowires can be adjusted by changing the organic ligand, morphology, and heterostructure. Among them, triple junction formed by metal-semiconductor-vacuum and conductive polymer-carbon nanotube-vacuum plays an important role in the field-emission properties (such as turn-on field).

6.3.2 Diode Rectification

The diode (except for Schottky barrier diode) is mainly composed of P-N junction and can control the direction of the current in the circuit [11]. Diodes can be used for switching circuits, automotive lighting, light sources of electronic products, etc. [34].

Yanbing Guo et al. firstly prepared CdS-PPY organic–inorganic P-N junction nanowires and tested it as a diode rectification device. As shown in Figure 6.6c, TEM image showed the obvious interface, indicating that the CdS-PPY organic–inorganic P-N junction was successfully constructed. The diode performance of the CdS-PPY was measured (Figure 6.6d), and the light-response behavior of the CdS-PPY was similar to that of the PBPB/CdS mentioned above [14]. When light was introduced, the CdS-PPY showed strong light-response current, suggesting that it could be used as a potential diode material [24]. Haowei Lin et al. fabricated PTCM [poly(3-thiophene carboxylic acid methyl ester)]/PbS organic–inorganic heterojunction nanowires, which exhibited a rectification ratio of 15.7. Besides, distinct electrical switching properties were also observed for the PTCM/PbS heterojunction nanowires with a high ON/OFF ratio of 83.5 [35]. Nan Chen prepared and studied PbS/PPY heterojunction nanowires via the template method. Due to the intact, large-area contact between PbS and PPY, an obvious improvement was found for the PbS/PPY with a high rectification ratio ($\geq$ 100). The PbS-PPY showed tight contact between organic and inorganic polymers and the larger heterogeneous junction area led to stronger interaction on the interface, which improved the property of diode rectification [36]. Then, Nan Chen et al. also utilized organic semiconductor polymer PBPB (P-type) and inorganic semiconductor CdS (N-type) to prepare PBPB/CdS organic–inorganic P-N junction nanowires and applied it in diode rectification (Figure 6.6a). As shown in Figure 6.6b, the diode rectification characteristics of PBPB/CdS heterojunction nanowires were investigated under different illumination conditions. In the dark, the light-response current of the PBPB/CdS was very low under the forward bias voltage. While white light was applied, the light-response signal increased obviously and improved with the increased light intensity. This phenomenon is due to the sensitivity of CdS to light and the tight contact with organic surface on the interface, resulting in a unique photoelectric response signal. Due to the different electron affinity between CdS and PBPB, the electrons and holes generated in the photoexcitation process

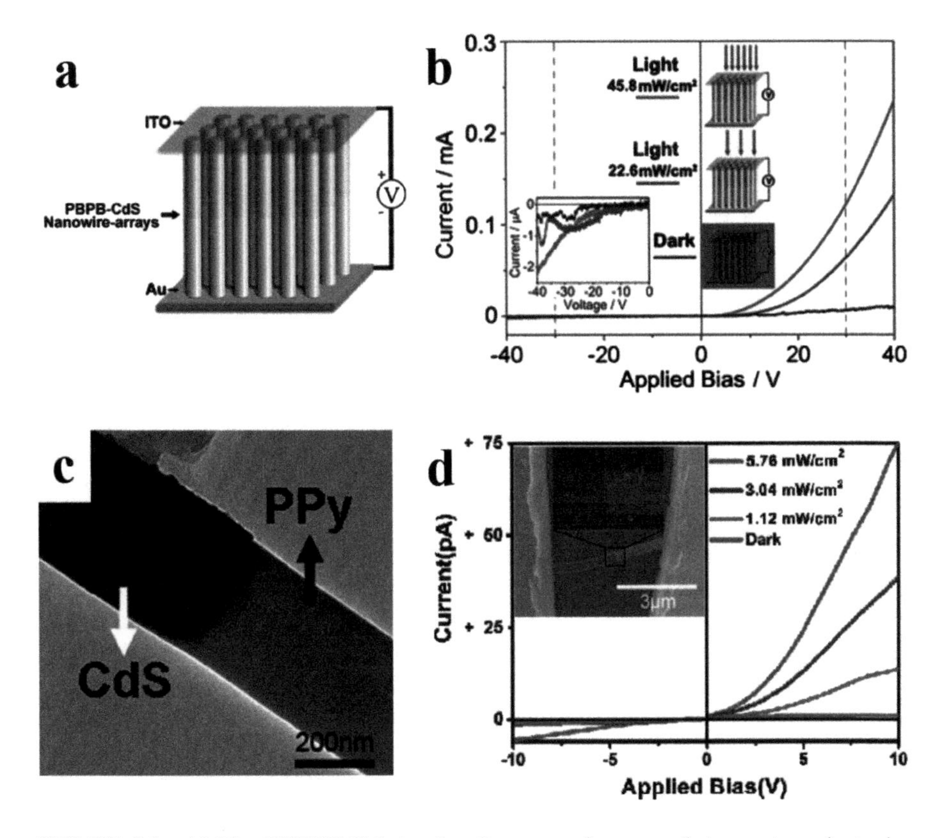

FIGURE 6.6 (a) The PBPB/CdS heterojunction nanowire arrays between two electrodes under an applied electric field; (b) Typical current–voltage (I-V) curves for PBPB/CdS heterojunction nanowire arrays under light illumination with different intensities; (Reprinted with permission from Ref 25. Copyright 2011 Royal Society of Chemistry) (c) Typical TEM image of a single CdS-PPY heterojunction nanowire; (d) Typical current-voltage (I-V) curves for a single CdS-PPY heterojunction nanowire under light illumination with different intensities at room temperature (inset: SEM image and EDS line analysis of the measured nanodevice). (Reprinted with permission from Ref 24. Copyright 2008 American Chemical Society)

would be transferred through CdS and PBPB, respectively. As the light intensity increases, the number of electrons and holes increases, so the diode effect is better when the light intensity is stronger [25]. Based on this interesting phenomenon, they believed that the photoelectric response of CdS must be selective. Therefore, they further fabricated PANI/CdS nanowires, and the light-response behavior of the PANI/CdS was tested under different wavelength illumination (254–610 nm). They found that the light-response current of the PANI/CdS irradiated with a blue light of 420 nm was much greater than that of the sample irradiated with other wavelengths, demonstrating the selectivity of the PANI/CdS. Besides, they proposed that the PANI/CdS can be used as a photoelectric detector to detect the intensity of blue light [27].

To summarize, organic–inorganic P-N junction nanowires can be constructed by selecting suitable N-type and P-type semiconductor materials, including organic semiconductor polymer and inorganic semiconductor. Excellent light-controlled diode rectification performance can be realized via the use of organic–inorganic heterojunction nanowires. However, diode performance, such as the rectifier ratio and the simplification of preparation process, needs to be further improved.

6.3.3 Solar Cells

Solar cells are electronic devices that absorb sunlight and convert it into electricity, making them more environmentally friendly than conventional batteries and charge–discharge batteries. Due to its low cost, simple preparation and high stability, organic–inorganic hybrid thin-film solar cells have attracted much attention [11].

Alejandro L. Briseno et al. fabricated ZnO/P3HT and ZnO/QT organic–inorganic heterojunction nanowires and tested their photovoltaic performance. Figure 6.7a shows a schematic diagram of the photovoltaic performance test of a single ZnO/oligothiophene nanowire. The SEM image and the typical I-V curve of the ZnO/P3HT are shown in Figure 6.7b. The ZnO/P3HT produced a short circuit (J_{sc}) of 0.32 mA/cm^2, an open circuit voltage (V_{oc}) of 0.40 V, and an efficiency of 0.036%. Similar to the ZnO/P3HT, the ZnO/QT nanowires yield a J_{sc} of 0.29 mA/cm^2, a V_{oc} of 0.35 V, and an efficiency of 0.033% (Figure 6.7c). They proposed that single-nanowire devices improved the shunt resistance by removing the short-circuit path in bulk ZnO array devices [16]. Yanbing Guo et al. utilized a single CdS-PPY organic–inorganic P-N junction nanowire to convert light energy into electricity. The construction of heterojunction significantly enhanced the range and intensity of light absorption, which is beneficial for the working condition device under visible light. The CdS-PPY delivered an efficiency of 0.018% under low light intensity of 6.05 mW/cm^2, while photovoltaic characteristics were not found in individual CdS and PPY nanowires. Additional tests were carried out on individual CdS and individual PPY nanowires to demonstrate that the photovoltaic properties are caused by unique organic/inorganic p-n junctions formed in a single nanowire [37]. Recently, Sanghoon Yoo et al. also reported the fabrication of Au-PPY-CdSe axial nanorod and its application as a solar cell. They found that unadorned Au-PPY-CdSe-Au nanorods with up to 1.1% power conversion efficiency could be obtained by using a porous Au nanorod electrode in the core of the PPY-CdSe nanorod. Due to the presence of electrophilic nitrogen atoms in the pyrrole ring unit, PPY can be chemically adsorbed to Au to generate tight contact. Nitrogen atom allowed electrons to jump from the p-conjugated molecular chain to the Au, thereby reducing band bending at the Au-PPY interface. Thus, the dissociation efficiency of excitons was enhanced [38]. Hao Xin et al. reported that the poly(3-butylthiophene) nanowires/[6,6]-phenyl-C61-butyric acid methyl ester (P3BT-nw/C61-PCBM) nanocomposite solar cell with P3BT nanowire (8–10 nm in width and up to 5–10 μm in length) as the donor component embedded in a sea of C_{61}-PCBM acceptors showed a dramatically improved energy conversion efficiency compared with its counterpart P3BT:C_{61}-PCBM blend. It was concluded that the substantially better hole transport ability of P3BT nanowire network in the P3BT-nw/C_{61}-PCBM solar cells contributed to the improved performance [39]. Yajie Zhang and co-workers reported organic single-crystalline P-N junction nanoribbons of CuPc and F_{16}CuPc, which showed a relatively lower energy conversion efficiency (η) of ~0.007% under 100 mW/cm^2 light intensity [40].

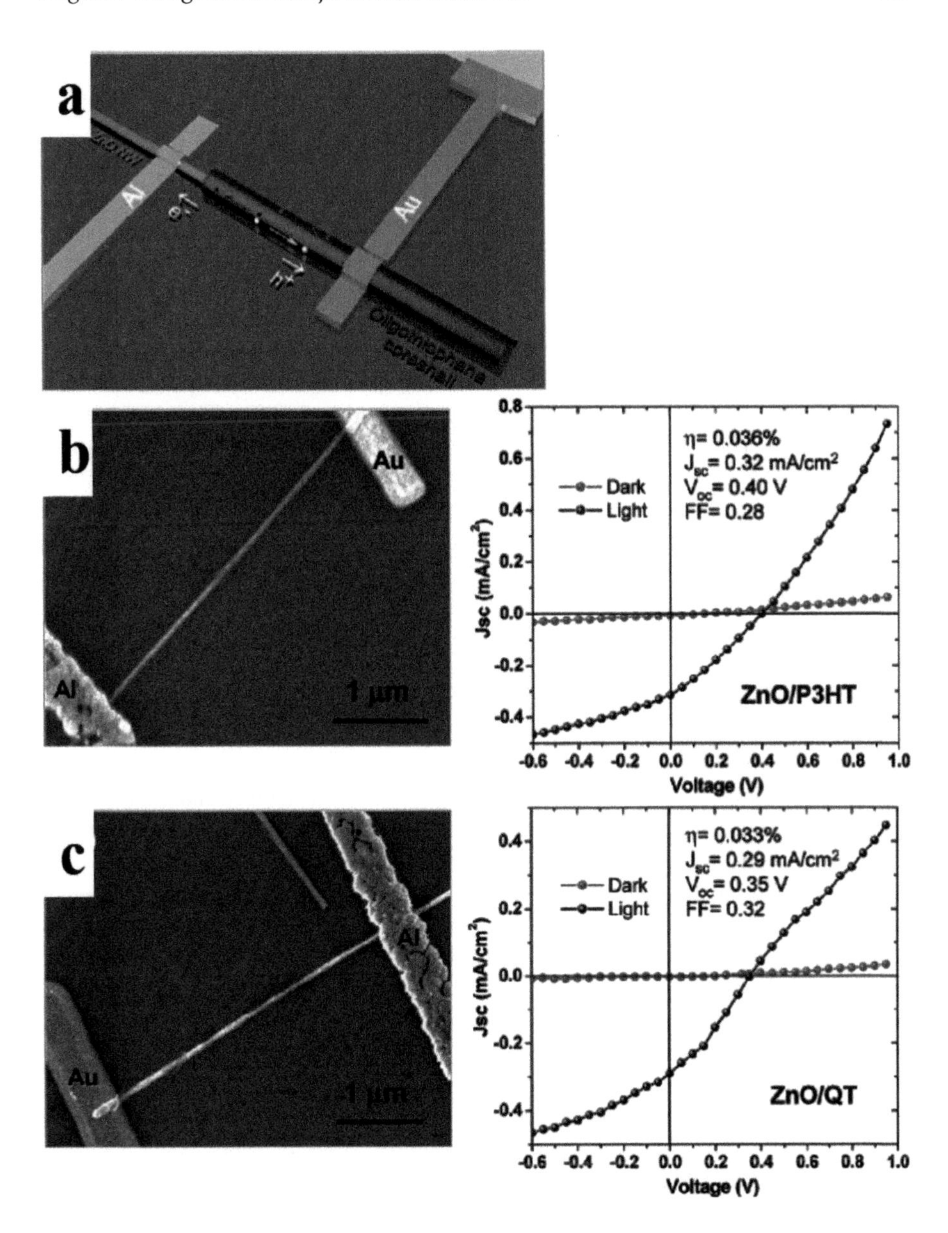

FIGURE 6.7 (a) Schematic configuration of a discrete ZnO/oligothiophene nanowire solar cell; (b) SEM image of a discrete ZnO/P3HT nanowire device fabricated by EBL and the corresponding current-voltage characteristics; (c) SEM image of a discrete ZnO/QT nanowire device and the current-voltage characteristics. The devices were measured under AM 1.5 irradiation (100 mW/cm^2). (Reprinted with permission from Ref 16. Copyright 2010 American Chemical Society)

In summary, organic–inorganic heterojunction nanowires showed certain photovoltaic performance, proving their feasibility as solar cell material. But their photovoltaic performance, including the short circuit, the open circuit voltage, and the efficiency, needs to be improved.

6.3.4 Photoelectric Detection

Owing to incident radiation, the conductivity of semiconductors would change a lot leading to a light-dependent behavior. This phenomenon can be used for photoelectric detection which has wide applications, such as optical communication, sensors [17].

Guozhen Shen et al. prepared P3HT:CdSe heterojunction nanowire photodetectors showing a improved photocurrent with a short recovery time ($\leq$0.1 s). The photocurrent would enhance with the light wavelength (350–650 nm) increased, while it would decrease with the light wavelength (700–850 nm) increased. This demonstrated a strong light-dependent behavior of the P3HT:CdSe heterojunction nanowires. The high hole transport rate of P3HT, the high conductivity of CdSe, and the synergistic absorption spectra of each component in the visible spectrum were responsible for the enhanced optical response and stability of the P3HT:CdSe [17]. Haowei Lin et al. successfully fabricated uniform PANI/CdS nanowire photodetectors as shown in Figure 6.8a and b. The electrical properties of the PANI/CdS were measured as shown in Figure 6.8c. The experimental results (Figure 6.8d) showed that the PANI/CdS was more sensitive to blue light (420 nm) than other tested wavelengths. The rectification ratio of the PANI/CdS improved (from 11.4 to 34.1) with enhanced light intensity (Figure 6.8e). Besides, as depicted in Figure 6.8f, the PANI/CdS exhibited excellent stability under the illumination of 420 nm. Some unique properties of the PANI/CdS were thought to be beneficial for high response speeds: (1) firstly, the CdS had high crystallinity and the trap density caused by defects was greatly increased; (2) secondly, high surface-to-volume ratio of the PANI/CdS allows surface defects and dangling bonds act as recombination centers, enhancing the recombination of carriers and thus reducing the decay time; (3) finally, the lowering of the recombination barrier is caused by Fermi level pinning [27]. Then, Haowei Lin et al. demonstrated the feasibility of CdS/PPV (*p*-phenylene vinylene) as a photodetector. The results showed that the photocurrent of the CdS/PPV enhanced dramatically under the illumination of 545 nm while the photocurrent of the CdS was almost 0. The significant enhancement of the photoelectric performance of CdS/PPV hybrid nanowire arrays is mainly due to the matching energy level between PPV and CdS. PPV can be used as a bridge to consume or transfer photogenerated holes and electrons, therefore restraining their recombination [41].

6.3.5 Logic Gates

Logic gates are the basic components in an integrated circuit that perform logic operations through high and low potentials. In general, a logic gate carries out logical operations on one or more inputs and produces an output [42].

Nan Chen et al. firstly reported and proposed a novel principle device for logic gates fabricated by using a three-segment (organic–inorganic–organic) heterojunction nanowire and the synthesis process is shown in Figure 6.9a. The component of

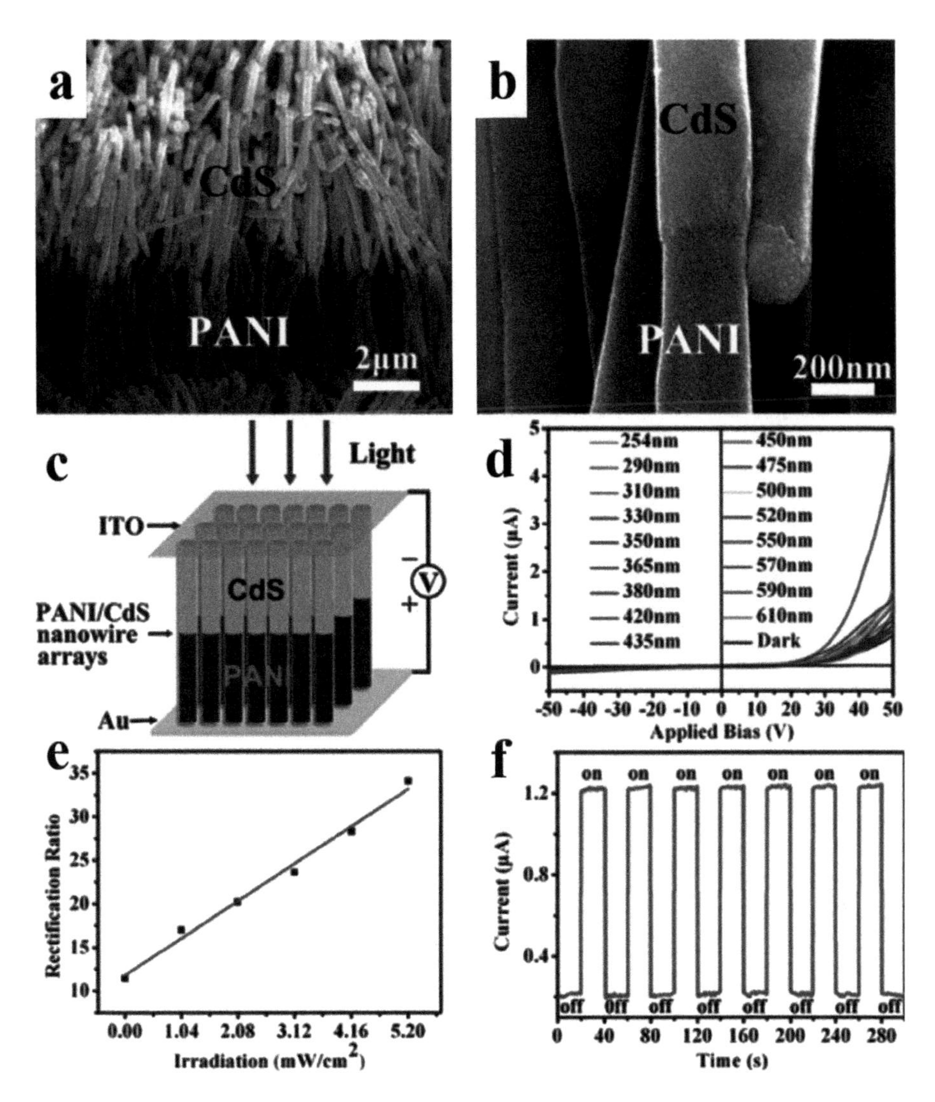

FIGURE 6.8 (a) Cross-view SEM image of PANI/CdS heterojunction nanowires; (b) High-resolution SEM image of PANI/CdS heterojunction nanowires; (c) Working model of PANI/CdS heterojunction nanowire array device; (d) Typical I-V curves of PANI/CdS heterojunction nanowire arrays in dark and under illumination of different wavelength light; (e) Irradiance dependence of the rectification ratio of PANI/CdS heterojunction nanowire arrays in the dark and under 420 nm light illumination; (f) On/off switching of PANI/CdS heterojunction nanowire arrays upon pulsed illumination from 420 nm wavelength light with a power density of 5.21 mW/cm^2. (Reprinted with permission from Ref 27. Copyright 2011 American Chemical Society)

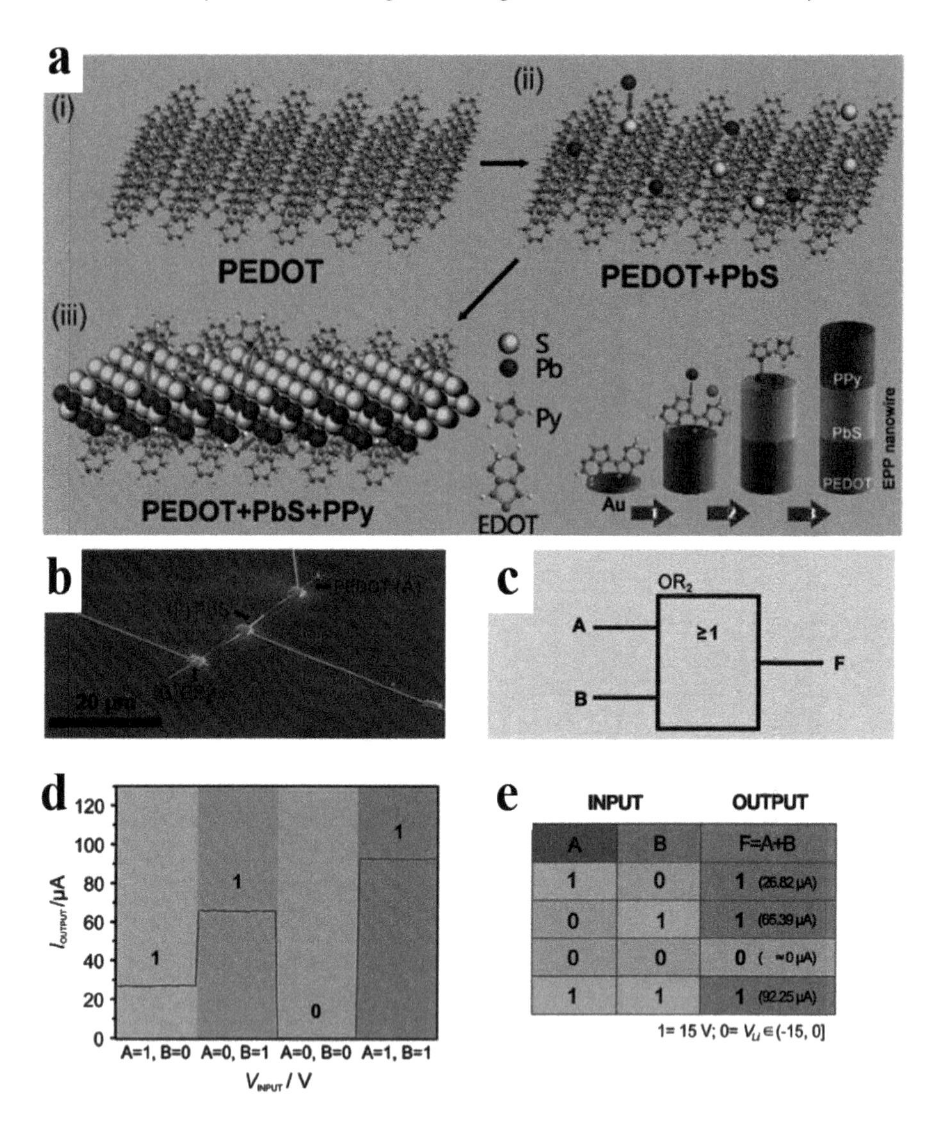

A	B	F=A+B
1	0	1 (26.82 μA)
0	1	1 (65.39 μA)
0	0	0 (≈0 μA)
1	1	1 (92.25 μA)

FIGURE 6.9 (a) Schematic diagram of the synthesis process of the EPP heterojunction nanowire arrays; (b) SEM image of a single EPP heterojunction nanowire device made by focus ion beam; (c) Standard symbols for logic OR_2 gate (two-input OR gate); (d) The signal output of the logic two-input OR gate constructed using EPP nanowire; and (e) The output data of the logic two-input OR gate constructed using EPP nanowire. (Reprinted with permission from Ref 42. Copyright 2013 Springer Nature)

three-segment heterojunction nanowire was poly(3,4-ethylenedioxythiophene) (PEDOT), PbS, and polypyrrole (PPY), respectively. The device was prepared by EPP (PEDOT/PbS/PPY) nanowires via ion beam (Figure 6.9b). Figure 6.9c showed a typical equivalent circuit (logic OR_2 gate) which was consistent with the behavior of the EPP nanowires. The experimental results (Figure 6.9d) revealed a "logical gates" behavior: when the input was higher than "1" (voltage ≥ 5 V), there was an output; while the input was lower than "1," there was no output. Detailed data results are summarized in Figure 6.9e [42].

6.4 SUMMARY AND PERSPECTIVE

This chapter introduces the research progress of organic–inorganic heterojunction nanomaterials. Organic–inorganic heterojunction nanowires have been successfully fabricated by four methods (solution phase method, template method combined with electrochemical polymerization, template method combined with pressure injection, and VLS method) and showed excellent performance in field emission, diode rectification, solar cells, photoelectric detection, and logic gates. Among these researchers, Yuliang Li et al. extensively prepared and studied all kinds of organic–inorganic heterojunction nanowires, explored the growth mechanism of them, and explained the structure–activity relationship between structure and performance. To sum up, the interface structure of organic–inorganic heterojunction nanowires can be adjusted by changing the organic ligand, growth mode, and morphology to adjust the electrical and catalytic properties.

In the future, novel synthesis methods (such as VLS method) need to be proposed and practiced to extend the materials diversity of organic–inorganic heterojunction nanowires. For the materials reported so far, they may also be used in new electronic devices (such as logic gates). At present, organic–inorganic heterojunction nanowires have been widely studied and applied in electric and photoelectric field. However, novel physical or chemical properties brought by the construction of heterojunction need to be further explored. For example, Yanbing Guo et al. fabricated Cu-TCNQ nanowires showing excellent performance for CO catalytic oxidation for the first time, which opens up a novel application field [43]. Further improvement would be achieved if Cu-TCNQ nanowires can be further compounded with other metals (such as Pt) or metal oxides (such as CuO) to form organic–inorganic heterojunction nanowires. It is a reminder that organic–inorganic heterojunction nanowires still have a lot of potential of application to be discovered. On the other hand, the conductivity of the material would be greatly improved for organic–inorganic heterojunction nanowires under the illumination or the dark [15,17]. This suggests that organic–inorganic heterojunction nanowires synthesized by reasonable design may be applied in electrochemical water splitting or photoelectrochemical water splitting.

REFERENCES

1. Meng, F. E. I., Morin, S. A., Forticaux, A., et al. Screw dislocation driven growth of nanomaterials. *Acc. Chem. Res.*, 2013, 46, 1616–1626.
2. Zang, L., Che, Y., Moore, J. S. One-dimensional self-assembly of planar π-conjugated molecules: adaptable building blocks for organic nanodevices. *Acc. Chem. Res.*, 2008, 41, 1596–1608.
3. Mei, J., Diao, Y., Appleton, A. L., et al. Integrated materials design of organic semiconductors for field-effect transistors. *J. Am. Chem. Soc.*, 2013, 135, 6724–6746.
4. Figueira-Duarte, T. M., Mullen, K. Pyrene-based materials for organic electronics. *Chem. Rev.*, 2011, 111, 7260–7314.

5. Virkar, A. A., Mannsfeld, S., Bao, Z., et al. Organic semiconductor growth and morphology considerations for organic thin-film transistors. *Adv. Mater.*, 2010, 22, 3857–3875.
6. Schwartz, G., Tee, B. C. K., Mei, J., et al. Flexible polymer transistors with high pressure sensitivity for application in electronic skin and health monitoring. *Nat. Commun.*, 2013, 4, 1–8.
7. Facchetti, A. Organic semiconductors: Made to order. *Nat. Mater.*, 2013, 12, 598–600.
8. Pramod, P., Thomas, K. G., George, M. V. Organic nanomaterials: morphological control for charge stabilization and charge transport. *Chemistry-An Asian Journal*, 2009, 4, 806–823.
9. Henson, Z. B., Müllen, K., Bazan, G. C. Design strategies for organic semiconductors beyond the molecular formula. *Nat. Chem.*, 2012, 4, 699–704.
10. An, B. K., Gierschner, J., Park, S. Y. π-Conjugated cyanostilbene derivatives: a unique self-assembly motif for molecular nanostructures with enhanced emission and transport. *Acc. Chem. Res.*, 2012, 45, 544–554.
11. Qian, X., Liu H., Li, Y. Self-assembly low dimensional inorganic/organic heterojunction nanomaterials. *Chin Sci Bull*, 2013, 58, 2686–2697.
12. Wang, Z. Characterizing the structure and properties of individual wire-like nanoentities. *Adv. Mater.*, 2000, 12, 1295–1298.
13. Guo, Y., Xu, L., Liu, H., et al. Self-assembly of functional molecules into 1D crystalline nanostructures. *Adv. Mater.*, 2015, 27, 985–1013.
14. Zheng, H., Li, Y., Liu, H., et al. Construction of heterostructure materials toward functionality. *Chem. Soc. Rev.*, 2011, 40, 4506–4524.
15. Liu, H., Cui, S., Guo, Y., et al. Fabrication of large-area hybrid nanowires arrays as novel field emitters. *J. Mater. Chem.*, 2009, 19, 1031–1036.
16. Briseno, A. L., Holcombe, T. W., Boukai, A. I., et al. Oligo- and Polythiophene/ZnO hybrid nanowire solar cells. *Nano Lett.*, 2010, 10, 334–340.
17. Wang, X., Song, W., Liu, B., et al. High-performance organic-inorganic hybrid photodetectors based on P3HT:CdSe nanowire heterojunctions on rigid and flexible substrates. *Adv. Funct. Mater.*, 2013, 23, 1202–1209.
18. Zhang, S., Pelligra, C. I., Keskar, G., et al. Directed self-assembly of hybrid oxide/polymer core/shell nanowires with transport optimized morphology for photovoltaics. *Adv. Mater.*, 2012, 24, 82–87.
19. Yang, Q., Liu, Y., Pan, C., et al. Largely enhanced efficiency in ZnO nanowire/p-polymer hybridized inorganic/organic ultraviolet light-emitting diode by piezo-phototronic effect. *Nano Lett.*, 2013, 13, 607–613.
20. Zhou, C., Zhang, Y., Li, Y., et al. Construction of high-capacitance 3D CoO@ polypyrrole nanowire array electrode for aqueous asymmetric supercapacitor. *Nano Lett.*, 2013, 13, 2078–2085.
21. Ko, H., Zhang, Z., Takei, K., et al. Hierarchical polymer micropillar arrays decorated with ZnO nanowires. *Nanotechnology*, 2010, 21, 295305.
22. Park, S., Chung, S. W., Mirkin, C. A. Hybrid organic-inorganic, rod-shaped nanoresistors and diodes. *J. Am. Chem. Soc.*, 2004, 126, 11772–11773.
23. Lahav, M., Weiss, E. A., Xu, Q., et al. Core-shell and Segmented polymer-metal composite nanostructures. *Nano Lett.*, 2006, 6, 2166–2171.
24. Guo, Y., Tang, Q., Liu, H., et al. Light-controlled organic/inorganic P-N junction nanowires. *J. Am. Chem. Soc.*, 2008, 130, 9198–9199.
25. Chen, N., Qian, X., Lin, H., et al. Synthesis and characterization of axial heterojunction inorganic–organic semiconductor nanowire arrays. *Dalton Trans.*, 2011, 40, 10804–10808.
26. Guo, Y., Liu, H., Li, Y., et al. Controlled core-shell structure for efficiently enhancing field-emission properties of organic-inorganic hybrid nanorods. *J. Phys. Chem. C*, 2009, 113, 12669–12673.

27. Lin, H., Liu, H., Qian, X., et al. Constructing a blue light photodetector on inorganic/organic p-n heterojunction nanowire arrays. *Inorg. Chem.*, 2011, 50, 7749–7753.
28. Xue, Z., Yang, H., Gao, J., et al. Controlling the interface areas of organic/inorganic semiconductor heterojunction nanowires for high-performance diodes. *ACS Appl. Mater. Interfaces*, 2016, 8, 21563–21569.
29. Mariani, G., Wang, Y., Wong, P. S., et al. Three-dimensional core-shell hybrid solar cells via controlled in situ materials engineering. *Nano Lett.*, 2012, 12, 3581–3586.
30. Oaki, Y., Oki, T., Imai, H. Enhanced photoconductive properties of a simple composite coaxial nanostructure of zinc oxide and polypyrrole. *J. Mater. Chem.*, 2012, 22(39), 21195–21200.
31. Xia, X., Chao, D., Qi, X., et al. Controllable growth of conducting polymers shell for constructing high-quality organic/inorganic core/shell nanostructures and their optical-electrochemical properties. *Nano Lett.*, 2013, 13, 4562–4568.
32. Guo, Y., Li, Y., Xu, J., et al. Fabrication of homogeneous hybrid nanorod of organic/inorganic semiconductor materials. *J. Phys. Chem. C*, 2008, 112, 8223–8228.
33. Qian, X., Ning, Z., Li, Y., et al. Construction of graphdiyne nanowires with high-conductivity and mobility. *Dalton Trans.*, 2012, 41, 730–733.
34. Zhao, J., Li, S., Tao, M. The working principle and main application of the diode. *Public Commun. Sci. Technol.*, 2010, 10, 45–46.
35. Lin, H., Liu, H., Qian, X., et al. Synthesizing axial inserting p-n heterojunction nanowire arrays for realizing synergistic performance. *Inorg. Chem.*, 2013, 52(12), 6969–6974.
36. Chen, N., Xue, Z., Yang, H., et al. Growth of axial nested P-N heterojunction nanowires for high performance diodes. *Phys. Chem. Chem. Phys.*, 2015, 17, 1785–1789.
37. Guo, Y., Zhang, Y., Liu, H., et al. Assembled organic/inorganic p-n junction interface and photovoltaic cell on a single nanowire. *J. Phys. Chem. Lett.*, 2010, 1, 327–330.
38. Yoo, S. H., Liu, L., Ku, T. W., et al. Single inorganic-organic hybrid photovoltaic nanorod. *Appl. Phys. Lett.*, 2013, 103, 1–4.
39. Xin, H., Kim, F. S., Jenekhe, S. A. Highly efficient solar cells based on poly (3-butylthiophene) nanowires. *J. Am. Chem. Soc.*, 2008, 130, 5424–5425.
40. Zhang, Y., Dong, H., Tang, Q., et al. Organic single-crystalline p-n junction nanoribbons. *J. Am. Chem. Soc.*, 2010, 132, 11580–11584.
41. Lin, H., Chen, K., Li, M., et al. Constructing a green light photodetector on inorganic/organic semiconductor homogeneous hybrid nanowire arrays with remarkably enhanced photoelectric response. *ACS Appl. Mater. Interfaces*, 2019, 11, 10146–10152.
42. Chen, N., Chen, S., Ouyang, C., et al. Electronic logic gates from three-segment nanowires featuring two p–n heterojunctions. *NPG Asia Mater.*, 2013, 5, 59.
43. Hu, S., Xiao, W., Yang, W., et al. Molecular O_2 activation over Cu(I)-mediated C≡N bond for low temperature CO oxidation. *ACS Appl. Mater. Interfaces*, 2018, 10, 17167–17174.

7 Electroluminescence of Organic Molecular Junction in Scanning Tunneling Microscope

Xiaoguang Li

CONTENTS

7.1 INTRODUCTION: MOLECULAR JUNCTIONS AND DEVICES

In this chapter, we focus on the optoelectronic properties of molecular devices. Over the past few decades, advancements in nanofabrication techniques and quantum theory of electronic transport and optics have enabled us to explore and to understand the basic characteristics of rudimentary electronic circuits in which single molecules or molecular assemblies are used as functional building blocks. The molecular devices discussed here should not be confused with organic electronics, where molecular materials are investigated as possible components of various macroscopic electronic devices.

Compared with conventional functional devices composed of bulk semiconductor materials or even 2D semiconductors, a single molecule or molecular assembly is naturally nanoscale and can achieve multiple functions due to its rich and mutually coupled internal degrees of freedom. The study of single-molecule devices is now no longer limited to the early electronic transport properties [1,2], but involves a series of novel physical phenomena, including mechanics, thermoelectricity, optoelectronics, and spintronics, as depicted in Figure 7.1a [3]. Interesting new phenomena have been found to emerge due to competition between those different transition processes

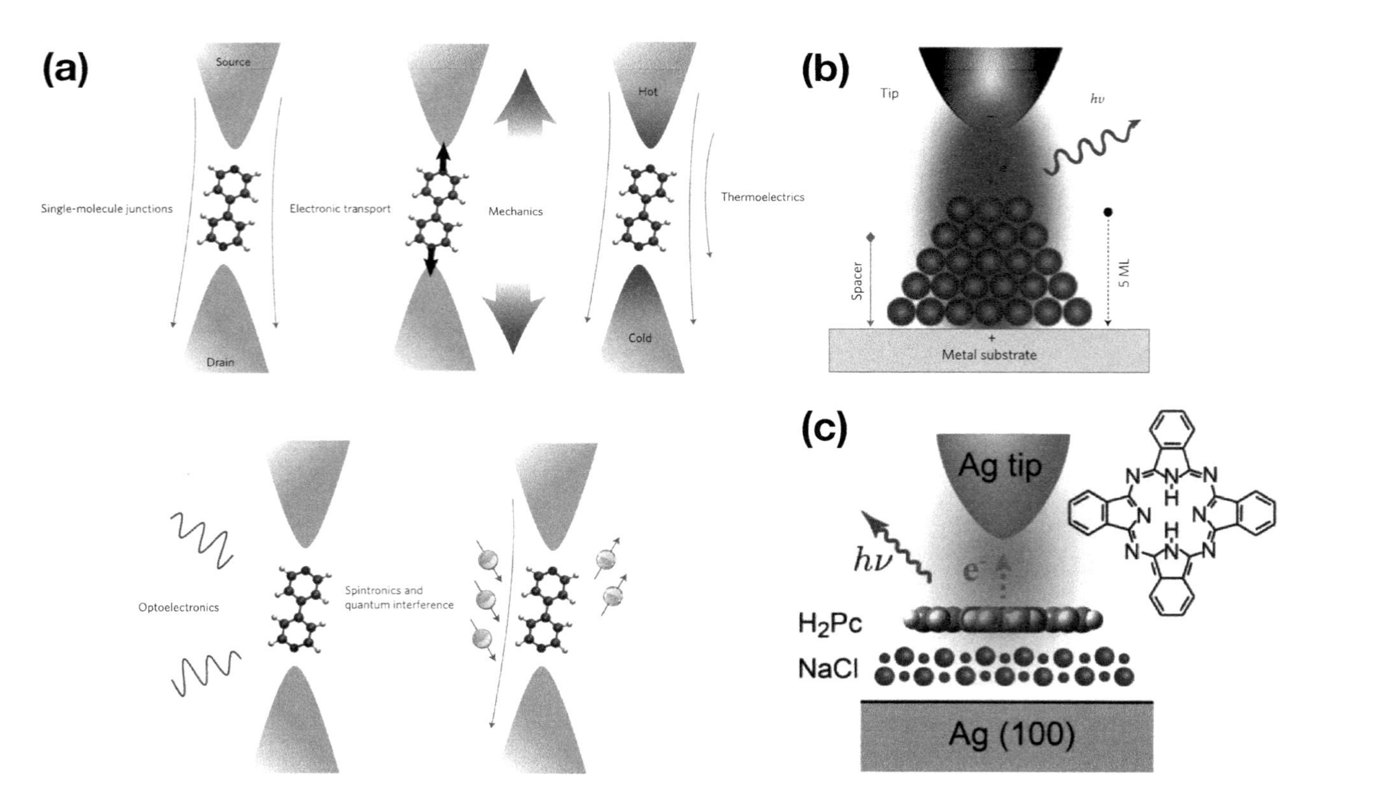

FIGURE 7.1 (a) Schematic of the electronic transport, mechanic response, thermoelectric phenomena, optical effects, and spin-dependent transport of single-molecule junctions. (Source: Aradhya and Venkataraman 2013 [3]) (b) Schematic junction geometry with multimonolayer stacking and localized electrical excitation from a nanotip. (Source: Dong et al. 2010 [5]) (c) Schematic of the STM-induced single-molecule emission from a single H_2Pc molecule on two monolayer NaCl. (Source: Chen et al. 2019 [8])

with different energy and time scales. These findings are highlighting single-molecule devices broad application prospects and important scientific significance.

Molecular luminescence is an elemental and crucial process that forms the basis of various organic optoelectronic devices [4]. Despite decades of in-depth research, many fundamentally important issues remain unresolved, partly because the experimental probes available in this field are limited. It has been demonstrated recently that the molecular luminescence induced by scanning tunneling microscope (STM) provides an unprecedented opportunity to explore many important optoelectronic phenomena, including hot luminescence [5–7], upconversion electroluminescence [5,8], Fano resonance [9–11], superradiance [12], and Electrofluorochromism [13]. The typical experimental setup for those observations is shown in Figure 7.1b and c. The observed phenomena are not only fundamentally intriguing but also may play an important role in optimizing the performance of optoelectronic devices.

STM utilizes the sensitivity of the tunneling rate of electrons to the width of the potential barrier between metal tip and substrate and can obtain extremely high spatial resolution. When molecules are inserted into STM, the appearance of the molecular electronic orbitals could drastically change the transport characteristics of the tunneling electrons. In turn, the optical properties of the molecule itself could be greatly modulated by the STM due to the surface plasmon mode of the nanocavity formed by metal tip and substrate. The most interesting thing is that the metal nanostructure of STM could help to solve the issue that light-matter coupling is too weak for single-molecule devices due to the huge size mismatch between the wavelength of visible light and the size of single molecules. Consequently, the STM system with molecular junction can combine the high resolution of space, time, and energy spectra, which can help us systematically analyze the observed optoelectronic phenomena [14–17].

In this chapter, we will focus on two peculiar electroluminescence phenomena in metal-molecule-metal junctions formed by a molecule inserted in STM: the hot luminescence and upconversion electroluminescence in Section 7.4. To elucidate the underlying physics of both phenomena, we will focus on some necessary fundamental knowledge about transport mechanisms of the metal-molecule-metal junction in Section 7.2, and the optical properties of the molecule and the metal nanocavity formed in an STM in Section 7.3. As we will gradually introduce in the following sections, our main emphasis for metal-molecule-metal junctions is on the energy-level alignment, energy scales between different pieces, and time scales of different processes in the system. Unfortunately, this chapter does not provide a comprehensive review of optoelectronics in a molecular junction. Readers are encouraged to look at the other chapters of this book as well as the many excellent reviews and treatises that are now available on this topic [18–25].

7.2 TRANSPORT MECHANISM IN MOLECULAR JUNCTIONS

In this section, we discuss the transport mechanism in the metal-molecule-metal junction, covering both the coherent and incoherent transports [26]. We first give an example of coherent transport, the so-called resonant tunneling model, where the electron tunnels through the molecule by forming a virtual charged molecule state.

Then we will introduce the incoherent transport, which is important for the electroluminescence. The discussion is only based on simple toy models or handwaving arguments, trying to convey the core physical ideas instead of the tedious calculation details.

7.2.1 Coherent Transport

Different from the macroscopic electronic conduction, where the resistance mainly comes from the electron–phonon interaction, the conductance G of the atomic scale STM junction should be described by the Landauer formula as

$$G = \frac{e^2}{h} \sum_i T_i, \tag{7.1}$$

where e^2/h is the quantum unit of the conductance determined by electron charge e and Planck constant h, and T_i is the transmission of an individual transition mode. Essentially, the quantum nature of electrons should be considered in such a small spatial distance, and the conductance of the system is closely related to the potential barrier tunneling problem in quantum mechanics, namely, the resistance mainly comes from the interface scattering.

When a molecule is inserted into the STM metal–metal junction, the transmission of an electron can be evaluated by using Breit–Wigner formula as

$$T(E,V) = \frac{4\Gamma_S \Gamma_T}{\left[E - \epsilon(V)\right]^2 + \left(\Gamma_S + \Gamma_T\right)^2}, \tag{7.2}$$

where Γ_S (Γ_T) indicates the coupling between the substrate (tip) and molecule, and E and $\epsilon(V)$ are the energies of tunneling electrons and molecular electronic orbitals, respectively, as shown in Figure 7.2a. A tunneling process of an electron (hole) through this metal-molecule-metal junction is depicted by the red dot (circle) and arrows in Figure 7.2b. When the Fermi surface of the tip is higher than that of the substrate, an electron can tunnel through the molecule LUMO (Lowest Unoccupied Molecular Orbital) with a probability determined by Γ_T, and then a virtual/transient (or real as long as $E > \epsilon$) charged molecular anion state M* is formed, and finally the electron in the LUMO of state M* can tunnel to the substrate with a probability determined by Γ_S. Conversely, we can also consider a hole state transport through the molecule by forming a transient cation state with a hole in HOMO (Highest Occupied Molecular Orbital). Both processes contribute a positive-charge current from the substrate to tip. So, the coherent transport through molecular junctions is mainly determined by the strength of the metal-molecule coupling as well as by the alignment between the molecular electronic orbital (usually HOMO and LUMO) and the metal Fermi level. Another way to understand this tunneling process is that due to the hybridization of the molecular orbitals and the electronic state in metal, the molecular orbitals acquire a finite broadening that provides a nonzero density of state aligning with the energy of the tunneling electron.

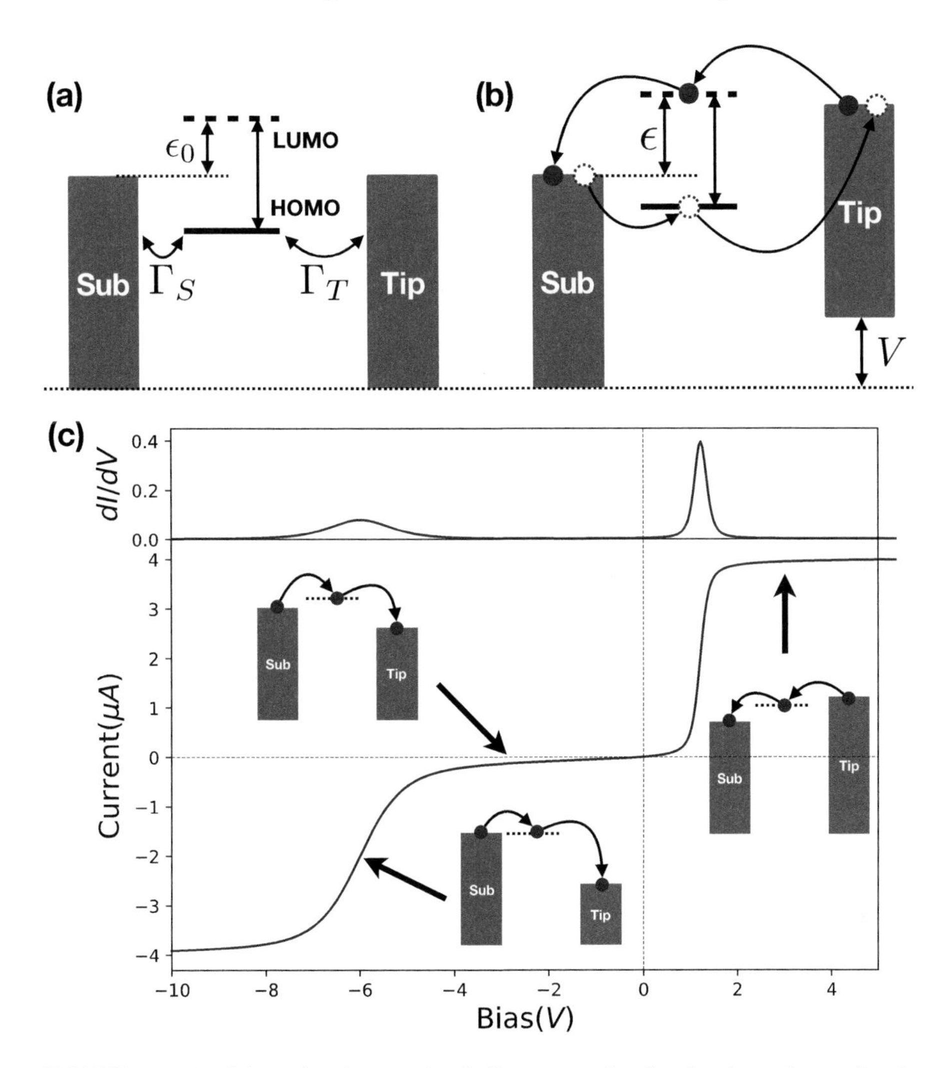

FIGURE 7.2 (a) Schematic of energy-level alignment and molecule–electrode coupling in STM system. (b) Schematic of the resonant tunneling process through the HOMO and LUMO of the molecule. (c) Current and dI/dV vs. bias voltage in the resonant tunneling model for $\Gamma_T = 0.2$ eV, $\Gamma_S = 0.01$, and $\epsilon_0 = 0.1$.

By further obtaining a quantitative result for the current-bias relation in the STM junction, the tunneling current I at a bias voltage V can be evaluated by adopting the Landauer formula as

$$I(V) = \frac{2e^2}{h} \int_{-\infty}^{\infty} dE\, T(E,V)\left[f(E,V_S) - f(E,V_T)\right] \tag{7.3}$$

where

$$f(E,\mu) = \frac{n}{e^{(E-\mu)/k_B T} + 1} \tag{7.4}$$

is the state occupation at the energy E for the metal electrode with the chemical potential μ, density of state n, temperature T, and Boltzmann constant k_B. For the transmission $T(E, V)$, it could be assumed that

$$\epsilon\left(V\right)=\epsilon_0+V\frac{\Gamma_T}{\left(\Gamma_S+\Gamma_T\right)}, \tag{7.5}$$

indicating the bias voltage dependence of the molecular orbital with its energy measured from the Fermi level of the metal substrate. Clearly, the voltage drop is considered only to happen at the tunneling barrier between the two metal electrodes and the molecule. In the STM system discussed here, we have $\Gamma_S \gg \Gamma_T$. The expression for $\epsilon(V)$ simply reflects the fact that if one of the coupling strength is much greater than the other, the molecular orbital follows the shift of the chemical potential of the electrode that is better coupled.

Figure 7.2b shows the current and dI/dV under different bias evaluated from the Landauer formula (Equation 7.3). The result clearly exhibits the so-called Coulomb blockade transport behavior of the molecular junction. Essentially, at a low-bias voltage and weak-coupling regime, the electron tunneling rate is extremely small because tunneling electrons cannot "classically" run into the molecule. In this case, the Fermi levels of the electrodes lie somewhere within the HOMO–LUMO gap of the molecule and are far from the HOMO and LUMO levels. The transmission of the electron is low, and the increase of the current with bias is due to the increase of the available tunneling electrons in the bias window between the Fermi level of the tip and substrate. As the bias further increases until the Fermi level of the tip approaches the molecular LUMO level, the conductance drastically climbs up due to the increasing transmission rate. At the resonant condition with $E - \epsilon(V) = 0$, we obtain the largest transmission $\frac{4\Gamma_S\Gamma_T}{\left(\Gamma_S+\Gamma_T\right)^2}$. As the bias further increases with the LUMO entering into the bias window, the conductance of the molecular junction jumps to a higher value with one quantum unit increasing. In this case, the tunneling electron can really run into the molecule and stay at the LUMO level, and the original high-order tunneling process relying on the two simultaneous tunneling events could now be done by two independent so-called charge injection processes. The tunneling rate is thus largely increased.

By now, we have seen that the transport property of the metal-molecule-metal junction is mainly determined by the coupling between the molecule and metal electrodes, and the energy-level alignment between molecular orbital and Fermi level of metal. The different parameters of the model, including the energy levels and the coupling constants, could be considered as phenomenological parameters, while they could in principle be obtained from a fit to the experimental results or from theoretical ab initio calculation.

7.2.2 Incoherent Transport

Electroluminescence typically originates from the generation of electron–hole pairs by incoherent transport through the metal-molecule-metal junction. In this subsection, we present an introductory description of incoherent transport.

Probably, the simplest picture of the incoherent transport can be understood when both HOMO and LUMO levels lie in the bias window. In this case, as shown in Figure 7.3a, after one electron from the tip tunneling to the molecular LUMO, the next tunneling procedure could be another electron in HOMO tunneling from the molecule to the substrate, instead of the corresponding coherent case with the electron further tunneling from LUMO to the substrate. So, the whole incoherent process includes two different charge injection steps and leaves the molecule at a neutral yet excited state.

Here, we explicitly emphasize on the "neutral", because the transport process involves the ground and excited states of the molecules not only for the neutral species but also for the cation and anion states. For example, if we consider the molecule as a two-level electronic system with level 0 occupied and level 1 empty for the neutral ground state, i.e., HOMO, the charge injection for the neutral molecule requires the Fermi level of the tip rising to be higher than LUMO. However, for a cation state with only one electron at level 0, the charge injection happens when the Fermi level of the tip reaches the spin-singlet state S1 as shown in Figure 7.3a. So, for a charge injection process, the net charge of the molecule is varied, and the position of the molecular levels participating in the carrier injection is different from what is referred to in the one-electron molecular orbital picture.

With the above understanding in mind, you may immediately realize that in the above-mentioned incoherent transport, after the electron tunneling into LUMO, the

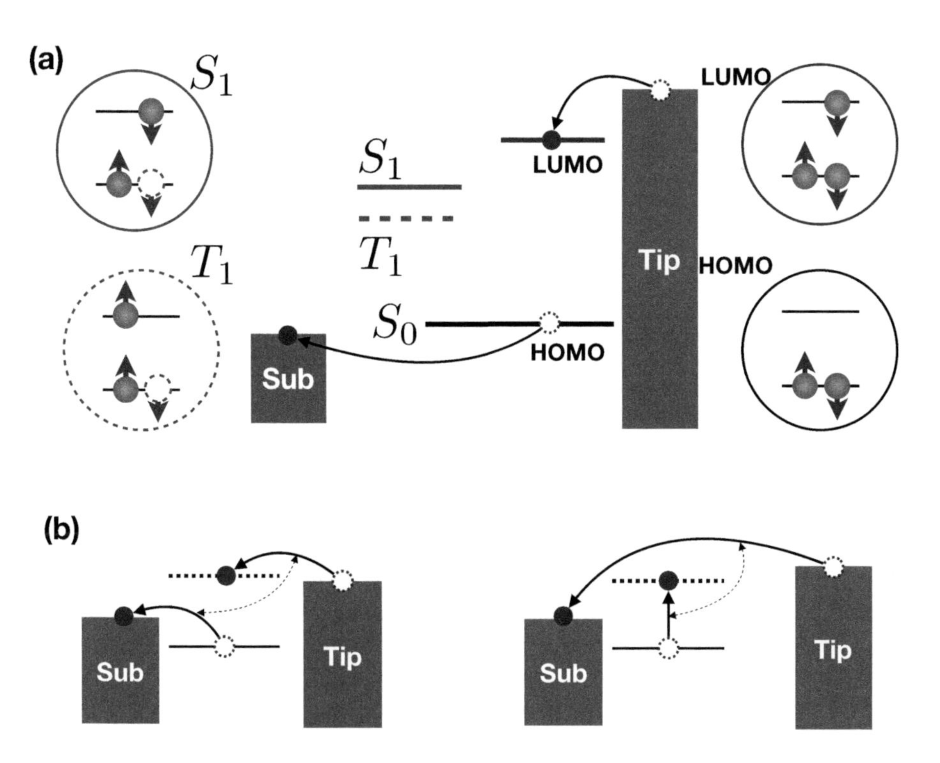

FIGURE 7.3 (a) Schematic of different electronic state configurations of a single molecule. (b) Schematic of inelastic tunneling processes.

second charge injection from the molecule to substrate does not require the "neutral" HOMO higher than the Fermi level of the substrate. Actually, since the molecule is now at the anion state, the second charge injection only requires that the Fermi level of the substrate lies E_{T1} below the LUMO level, and then the injection is energetically allowed. Or conversely, as the bias increases, if the HOMO becomes aligned with the Fermi level of the substrate, one electron at level 0 can then tunnel into the substrate and leave the molecule at a cation state. The subsequent charge injection from the tip to molecule can happen as long as the corresponding Fermi level is higher than the T_1 energy level.

Here, it is worth to emphasize that as long as the spin of the tunneling electrons of the two charge injection processes is different, the spin state of the molecule can be changed, namely, both the triplet and singlet states can be excited by the inelastic transport. This is clearly different from the excitation by an external light source.

Another possible situation is that the bias voltage could reach the excitation energy of the T1 state with neither HOMO nor LUMO aligning with the Fermi level of metal electrodes, as shown in Figure 7.3b. In this case, the two-step charge injection transport cannot happen. However, similar as in the coherent transport, the excitation could happen mediated by a virtual cation state, with one electron tunneling from the tip to molecular HOMO, and simultaneously one electron tunneling from the molecular LUMO to the substrate. This process can be considered as electron–electron scattering, which may also be described as shown in Figure 7.3c with an electron directly tunneling from the tip to substrate and scattering the electron in the molecule to a higher energy level. These two processes cannot be distinguished as long as the spin state of the molecule is not changed.

So far, we have introduced the fundamental pictures of coherent and incoherent transports in the metal-molecule-metal junction. It would be very challenged to calculate the occurrence probability of different processes in actual systems, but the understanding of the above pictures can help us to explain the observed electroluminescence phenomena by combining the rate equation method and the different transition rates obtained from experimental fitting.

7.3 OPTICAL PROPERTIES OF MOLECULAR JUNCTIONS

Different from the conventional semiconductor functional devices, in single-molecule devices, the light-emission property of the metal-molecule-metal junction depends not only on the intrinsic electronic structures of the molecule but also on its surrounding electromagnetic environment dominated by the nanocavity, formed by the metallic tip and substrate. This nanocavity plays an important role in the optical response of the junction due to the corresponding surface plasmon mode. In metal materials, surface plasmons describe the collective oscillation mode of electrons at the surface of the material, and their corresponding highly localized enhanced electric fields give surface plasmons a wide range of applications [27–43]. Extremely strong electromagnetic near-field allows surface plasmon to be used as a nano-antenna to improve the coupling of various optoelectronic devices with the external light field and thus enhance absorption and emission [32,36,38,43]. In addition, the highly localized characteristics also allow it to break the diffraction limit of the optical field, which is essential for the miniaturization of optoelectronic devices.

In a vacuum, the emission of excited molecules due to the spontaneous radiation is a pure quantum phenomenon, which originates from the interaction between the transition dipole of the molecule and the vacuum fluctuations of the electromagnetic field. It is well known that the presence of dielectric nanocavity surrounding the molecule could modify the photon density of states, and therefore change the spontaneous decay rate, known as the Purcell effect. The strength of the Purcell effect is described by the Purcell factor

$$F = \frac{6}{\pi^2}\frac{Q}{V}\left(\frac{\lambda}{2}\right)^3, \tag{7.6}$$

where $Q = \omega/\Gamma$ with resonant frequency ω and linewidth Γ is the quality factor of the nanocavity, V is the volume of the cavity, and λ is the wavelength of the emitted light. For a dielectric cavity, the diffraction limit gives $V > (\lambda/2)^3$, and therefore conduct the upper limit for the Purcell factor $F \leq \frac{6}{\pi^2} Q$ [44].

A metallic plasmonic nanocavity can also change the photon density of states and thus utilize the Purcell effect to dramatically enhance the spontaneous decay rates of the molecule. Since the surface plasmon can break the diffraction limit of the optical field, the fundamental limitation for its Purcell factor comes from the quality factor Q, more specifically, the linewidth Γ of the nanocavity. Essentially, a large part of the radiative energy transferring from the molecule to metal nanocavity will be dissipated through the non-radiative decay in metals due to the electron–electron scattering.

In addition, the de-excitation of the molecule close to metals also suffers from the dissipation due to the possible charge transfer and non-radiative energy transfer, so the distance between the molecule and metal tip and substrate should be carefully controlled. Different insulating films, such as Al_2O_3, NaCl, and several layers of molecules themselves, have been used as spacers to avoid those issues, as shown in Figure 7.1b and c.

An example of the role of the metal nanocavity in the de-excitation of the molecule is shown in Figure 7.4, where the total and radiative decay rates of a molecule between two spherical metal nanoparticles are evaluated by using the generalized Mie theory [6]. In comparison with the result in a vacuum, the radiative decay rate of the molecule is increased by 3–5 orders of magnitude in optical wavelengths, driving the radiative lifetime for a nanosecond to picosecond regime. By comparing the radiative and total decay rates, we also see the presence of a large portion of non-radiative decay. In addition, we can find that the gold dimer exhibits a large non-radiative decay rate in the high-frequency region, which is due to the interband transition loss in gold above ~2 eV. Therefore, to achieve a large radiative enhancement for the molecular junction, we need to carefully select proper plasmonic materials and control the size and shape for the tip and substrate in the STM.

7.4 SPECIAL PHENOMENA: HOT LUMINESCENCE AND UPCONVERSION

As we have mentioned in the previous two sections, the energy alignment largely determines the transport behaviors, while the coupling between the metallic nanocavity formed by the tip and substrate strongly affects the emission of the molecule.

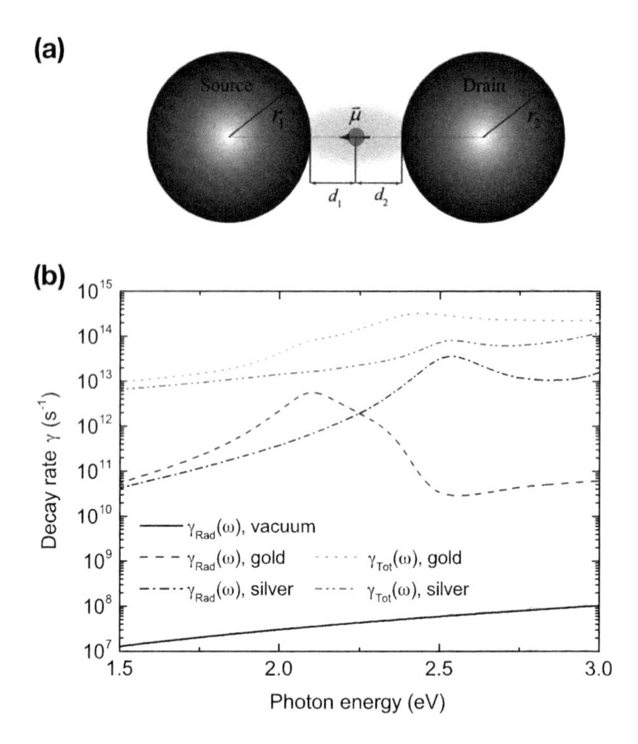

FIGURE 7.4 (a) Schematic picture of the system setup. The tunneling junction is modeled with two metal spheres and the molecule is treated as an electric dipole locating on the axis of the dimer. (b) Radiative and total decay rates for an electric dipole placed in a vacuum, or within a gold dimer and silver dimer, respectively. (Source: Chen et al. [6])

When we bring the transport and optics together in this metal-molecule-metal junction, some representative electroluminescence phenomena can be observed, including the hot luminescence [5–7], upconversion [5,8], and the Fano phenomenon [9–11], superradiance [12], and Electrofluorochromism [13]. In this section, we discuss two of them, i.e., hot luminescence and upconversion electroluminescence of the molecular junction in an STM. The main emphasis is on the competition between different transport and optical processes due to their different energy and time scales.

7.4.1 Hot Luminescence

The excitation of the molecule usually combines both the electronic and vibrational parts, namely, the electron jumps to the high energy state and atoms start to oscillate. The corresponding excited state can be denoted as $|e, v\rangle$. In a vacuum, light emission from an excited molecule follows Kasha's rule, which states that only emission from the lowest vibrational level of the excited state is possible. This is because, for free molecules, the vibrational damping rate is typically 10^{12} s^{-1}, while the spontaneous emission rate determined by the transition dipole is typically $10^7 \sim 10^8$ s^{-1}. Consequently, the molecule will first cool down to the vibrational ground state $|e, 0\rangle$, and then make a radiative transition to electronic ground state $|g, v\rangle$. The emission final state may not be $|g, v = 0\rangle$, because the equilibrium position of a vibrational mode upon the excited electronic state is usually not coincident with that of the ground electronic state.

The Kasha's rule, however, is founded to be broken in some molecular junctions in a STM. In a work done by Dong et al. [5], people find the modulation of molecular emission profile as well as the presence of high energy emission, which seems coming from the directly radiative transition from a "hot" electronic excited state $|e, v\rangle$ to ground state $|g, v'\rangle$ without going through a picosecond cooling process. A schematic plot of the excitation and emission processes in the experiment has been shown in Figure 7.5a. The molecule is excited by a charge injection process with one electron tunneling from $|g, 0\rangle$ (HOMO) to drain and another electron tunneling from source to higher energy level in the molecule, exciting both electronic and vibrational motions. The normal luminescence starts from a cooling process labeled by γ_{vib}, following by the transition labeled by γ_s from $|e, 0\rangle$ to the ground state, while the hot luminescence labeled by γ_h will not go through the cooling process.

Within the framework of the quantum master equation proposed earlier by Johansson et al., and by considering the spontaneous emission enhancement due to the metallic nanocavity, Chen et al. [6] show that the electroluminescence spectral profile of the metal-molecule-metal junction can be largely modulated. A hot luminescence could be observed, when the radiative decay rate of the molecule is enhanced to be comparable with the vibration relaxation rates.

To obtain the luminescence spectra, two essential quantities of the molecule, the radiative decay rate and the state population at the dynamic equilibrium, should be evaluated. An example of the enhancement of the radiative decay rate has been shown in Figure 7.4 in Section 7.3. Here, the results of the spectral profiles at different system configurations have been shown in Figure 7.5, which shows that the spontaneous decay rate of the molecule can be strongly modulated by tuning the geometry and material of the nanocavity. Figure 7.5b shows that the spectra can be selectively modulated depending on the quantum efficiency of the cavity, which is related to the geometrical structure of the nanocavity. In turn, the intrinsic properties of the molecule will also greatly affect the luminescence spectrum, as shown in Figure 7.5d and e, in which the intensity and direction of the transition dipole can strongly affect the thermal luminescence peak with an energy higher than 2 eV. In essence, the presence of hot luminescence depends on the delicate competition between the different transition processes. As shown in Figure 7.5c, the theoretical study also indicates that some hot luminescence results may be overlooked before, because the corresponding peak could overlap with a norm luminescence peak.

7.4.2 Upconversion Electroluminescence

Upconversion electroluminescence in an STM is a good example to show how the energy and time scales of different transition processes and the energy-level alignment affect the optoelectronic properties of a system. Upconversion luminescence is a conceptually counterintuitive quantum phenomenon in which a single emitted photon has higher energy than that of the excitation sources. In the STM system, it means that the energy of an emitted photon is higher than energy loss by a single tunneling

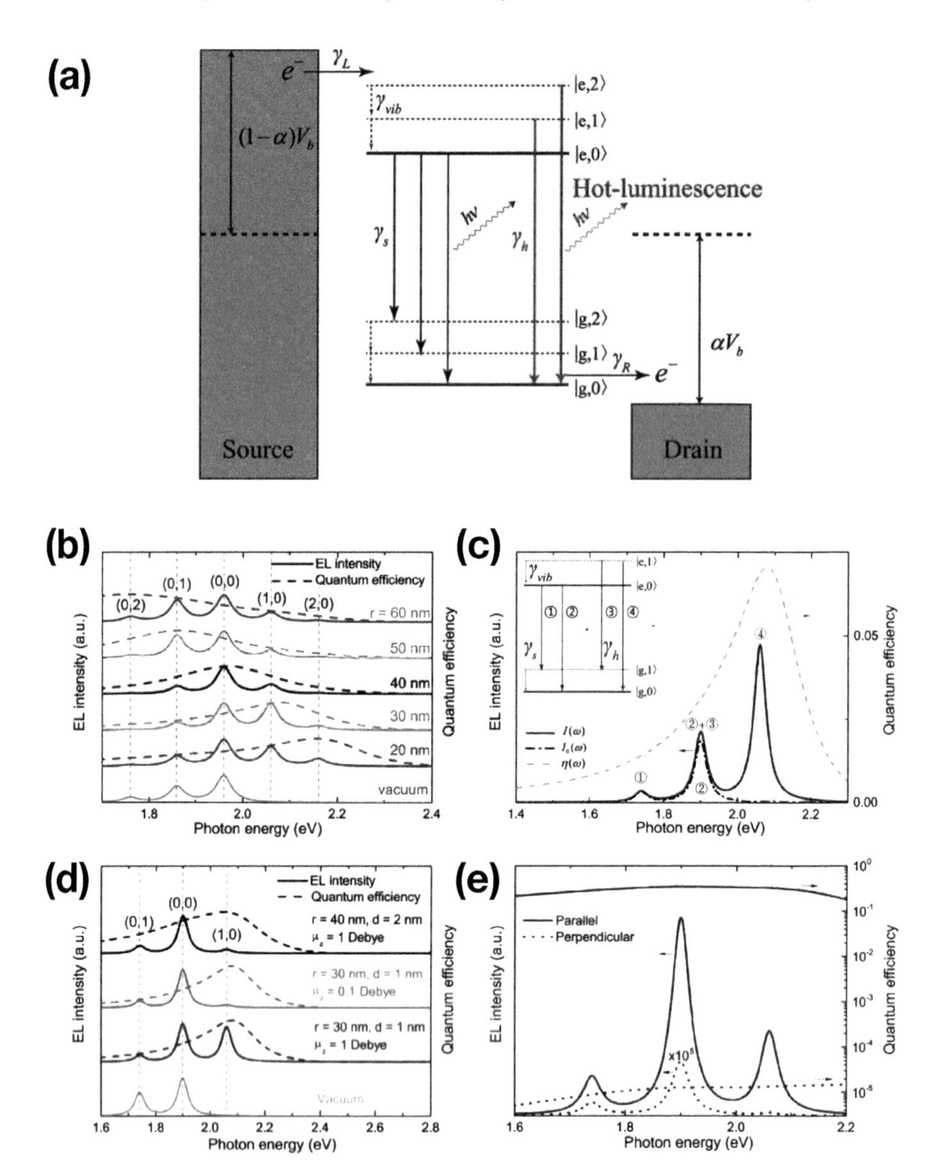

FIGURE 7.5 (a) Schematic diagram of energy levels and various transition processes within the system formed by a molecule in a tunnel junction. (b) Molecular fluorescence spectra in vacuum and in a tunnel junction for gold dimers. (c) Illustration of contributions from different channels for molecular emission. (d) Molecular fluorescence spectra in the junction of a gold dimer for different dipole moment and gap distance. (e) Molecular fluorescence spectra and quantum efficiency for electric dipoles parallel and perpendicular to the axis of the gold dimer. (Source: Chen et al. [6])

electron, provided by the bias voltage. Clearly, this phenomenon requires a multi-electron excitation process of fluorescent radiation. Therefore, all the different mechanisms proposed in earlier studies of upconversion electroluminescence, such as intermolecular triplet–triplet annihilation (TTA) [45,46] and molecular vibration-assisted plasmonic pumping [5,47], involve intermediate states to capture and store multiples of the energy quanta from the excitation source.

In a recent experiment done by Chen et al. [8], it has been shown that the upconversion electroluminescence could happen at the single-molecule and low-tunneling current level. Figure 7.1c exhibits the geometrical configuration of the system with isolated H_2Pc molecules between tip and substrate, and electronically decoupled by a two-monolayer thick NaCl spacer. The main experiment findings include two aspects. One is the upconversion phenomenon, which is interpreted to be mediated by the spin-triplet excited state in the single molecule. The other is the observation of the three distinct efficiency-bias regions, implying the different transport and excitation mechanisms. The conclusion is based on the detailed analysis of the energy and time scales of different processes in the system, and also the energy alignment between the molecule and substrate.

Considering the experiment setup with the low tunneling current, the observed upconversion luminescence cannot be explained by most of the previously proposed intermediate states, including the surface plasmon of the nanocavity and vibrational mode of the molecule. In this experiment, the tunneling current is as low as ~100 pA, corresponding to an average time interval of ~1 ns between the electron-tunneling events. However, the surface plasmon of the nanocavity has a typical femtosecond lifetime [48], and the molecular vibrational mode has a typical picosecond lifetime [49]. Consequently, the spin-triplet state of a free H_2Pc molecule with a lifetime ~100 μs [50] seems the only possible intermediate state in this system.

The bias voltage threshold for the upconversion also implies the spin-triplet state mediating the upconversion. Figure 7.6a shows the optical spectra from the single molecule at different negative bias voltages. The normal electroluminescence happens at $V_b = -2.5$ V with the energy of a tunneling electron higher than that of the spin-singlet state at ~1.81 eV. The upconversion electroluminescence could be observed at $V_b = -1.7$ V with a clear emission peak at 1.81 eV. The voltage threshold for the upconversion can be seen in Figure 7.6b at around –1.2 V, which coincides with the energy of the spin-triplet state of H2Pc molecule.

As shown in Figure 7.6c, the emission intensity shows three distinct stepwise increases in different voltage regions with constant current, implying the different excitation efficiency or say different inelastic excitation rate in three bias regions: (I) $V_b < -2.25$, (II) $-2.25\ \mathrm{V} < V_b < -1.8$ V, and (III) $V_b > -1.8$ V. The schematic pictures of the electronic transport and excitation mechanisms have been shown in Figure 7.6d.

In region (I) with $V_b < -2.25$ V, similar as we have exhibited in Section 7.2, the molecular HOMO lies above the Fermi level of the tip and the bias voltage is larger than the energy of S_1, and therefore the excitation of the molecule is dominant by the charge injection process, with the first electron tunneling from HOMO to the tip following by the second electron tunneling from the substrate to S_1.

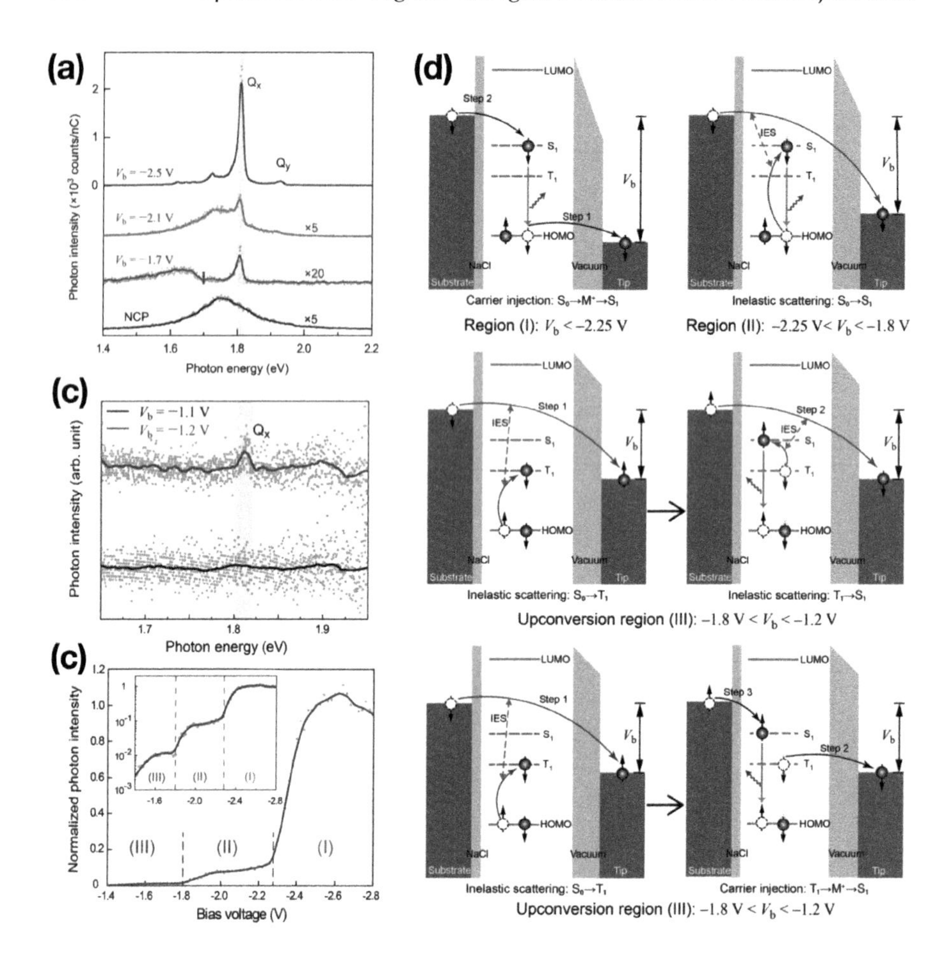

FIGURE 7.6 (a, b) Electroluminescence spectra from a single H2Pc molecule at different bias voltages. (c) Normalized bias-dependent intensity of the spin-singlet emission peak at a constant current of 100 pA, with the logarithmic plot shown in the inset. (d) Schematics on competing molecular excitation mechanisms. (Source: Chen et al. [8])

In region (II) with −2.25 V < V_b < −1.8 V, the above charge injection picture is forbidden because both HOMO and LUMO are out of the bias window. However, the bias voltage is still larger than the energy of S_1, and therefore the excitation is dominated by the resonant tunneling with the two charge injection processes described above that simultaneously occur. It explains the much lower emission intensity in this bias region as shown in Figure 7.6c.

In the upconversion region (III) V_b > −1.8 V, the excitation of T_1 state could also be done by the resonant tunneling as long as V_b < −1.2 V. However, it is very intriguing to notice that since the energy level T_1 is higher than the Fermi level of the tip in this bias region, the final upconversion process from T_1 to S_1 could be done by a charge injection process as shown in the bottom panel of Figure 7.6d. This transition of the inelastic transport picture from resonant tunneling to charge injection due to the different alignment of T1 and HOMO with the Fermi level of the tip is vital. It means that after the molecule is excited to the T1 state, it does not have to wait a too long time to be upconverted to S1, thus giving the upconversion electroluminescence an observable intensity. Actually, a quantitative simulation based on quantum master equations in the study has found that upconversion due to the resonant tunneling plus charge injection processes is ~16 times more efficient than that of two resonant tunneling pictures, as shown in the middle panel of Figure 7.6d.

In conclusion, the proper energy and lifetime, together with the proper energy alignment between T1 state and the Fermi level of the metal, justify the role of T1 as the intermediate state of the upconversion electroluminescence in the experiment.

7.5 SUMMARY AND OUTLOOK

This chapter has tried to briefly introduce the optoelectronics in the metal-molecule-metal junction. The main focus is on the hot luminescence and upconversion electroluminescence from a molecular junction in an STM, and the corresponding backgrounds regarding the quantum transport and nano-optics. We believe that through the above simple examples, readers can already see a promising research direction.

We can see that when molecules are inserted into the junction, the interaction between the tunneling electrons, molecular electronic and vibrational states, and the nanocavity plasmon in the system will bring extremely rich physical phenomena. In this extreme scale, the metal electrodes have to be considered as an important part of the whole system, because they play a critical role in both the transport and optical properties. Consequently, the study on the single molecular optoelectronics is also partly motivated by the tremendous interest in plasmonics and nano-optics.

This field has seen remarkable progress in the past two decades, and the growing availability of scanning tip configurations that can combine optical and electronic probes allows to simultaneously obtain high temporal, spatial and spectral resolution in the system. We believe that STM electroluminescence technology can reveal more important fundamental understanding and may prove to be instrumental in the future design of organic optoelectronic devices with optimized performances.

REFERENCES

1. A. S. Martin, J. R. Sambles, and G. J. Ashwell, *Physical Review Letters* 70, 218 (1993).
2. M. A. Reed, C. Zhou, C. J. Muller, T. P. Burgin, and J. M. Tour, *Science* 278, 252 (1997).
3. S. V. Aradhya and L. Venkataraman, *Nature Nanotechnology* 8, 399 (2013).
4. E. Flaxer, O. Sneh, and O. Cheshnovsky, *Science* 262, 2012 (1993).
5. Z. C. Dong et al., *Nature Photonics* 4, 50 (2010).
6. G. Chen, X. G. Li, Z. Y. Zhang, and Z. C. Dong, *Nanoscale* 7, 2442 (2015).
7. M. C. Chong, L. Sosa-Vargas, H. Bulou, A. Boeglin, F. Scheurer, F. Mathevet, and G. Schull, *Nano Letters* 16, 6480 (2016).
8. G. Chen et al., *Physical Review Letters* 122, 177401 (2019).
9. L. Zhang et al., *Nature Communications* 8, 1 (2017).
10. H. Imada, K. Miwa, M. Imai-Imada, S. Kawahara, K. Kimura, and Y. Kim, *Physical Review Letters* 119, 1 (2017).

11. J. Kröger, B. Doppagne, F. Scheurer, and G. Schull, *Nano Letters* 18, 3407 (2018).
12. Y. Luo et al., *Physical Review Letters* 122, 233901 (2019).
13. B. Doppagne, M. C. Chong, H. Bulou, A. Boeglin, F. Scheurer, and G. Schull, *Science* 361, 251 (2018).
14. X. H. Qiu, G. V. Nazin, W. Ho, and W. Hot, *Science* 299, 542 (2003).
15. S. W. Wu, N. Ogawa, and W. Ho, *Science* 312, 1362 (2006).
16. Y. Y. Zhang et al., *Nature* 531, 623 (2016).
17. H. Imada, K. Miwa, M. Imai-Imada, S. Kawahara, K. Kimura, and Y. Kim, *Nature* 538, 364 (2016).
18. W.-D. Schneider, *Surface Science* 514, 74 (2002).
19. K. Watanabe, D. Menzel, N. Nilius, and H. J. Freund, *Chemical Reviews* 106, 4301 (2006).
20. F. F. Rossel, M. Pivetta, and W.-D. D. Schneider, *Surface Science Reports* 65, 129 (2010).
21. T. Shamai and Y. Selzer, *Chemical Society Reviews* 40, 2293 (2011).
22. M. Galperin and A. Nitzan, *Physical Chemistry Chemical Physics* 14, 9421 (2012).
23. L.-G. Chen, C. Zhang, R. Zhang, M. Zhen-Chao Dong, A. Kar, Z.-Q. Zou, L. Chen, and Z. Dong, *Japanese Journal of Applied Physics* 54, 8 (2015).
24. T. Wang and C. A. Nijhuis, *Applied Materials Today* 3, 73 (2016).
25. K. Okamoto, M. Funato, Y. Kawakami, and K. Tamada, *Journal of Photochemistry and Photobiology C: Photochemistry Reviews* 32, 58 (2017).
26. J. C. Cuevas and E. Scheer, *Molecular Electronics* (World Scientific, 2010), Vol. 1, World Scientific Series in Nanoscience and Nanotechnology.
27. K. Kneipp, Y. Wang, H. Kneipp, L. T. Perelman, I. Itzkan, R. R. Dasari, and M. S. Feld, *Physical Review Letters* 78, 1667 (1997).
28. S. Nie, *Science* 275, 1102 (1997).
29. H. Xu, E. J. Bjerneld, M. Käll, and L. Börjesson, *Physical Review Letters* 83, 4357 (1999).
30. W. Srituravanich, N. Fang, C. Sun, Q. Luo, and X. Zhang, *Nano Letters* 4, 1085 (2004).
31. E. Ćavar, M.-C. Blüm, M. Pivetta, F. Patthey, M. Chergui, and W.-D. Schneider, *Physical Review Letters* 95, 196102 (2005).
32. S. Kühn, U. Håkanson, L. Rogobete, and V. Sandoghdar, *Physical Review Letters* 97, 017402 (2006).
33. N. Engheta, *Science* 317, 1698 (2007).
34. M. A. Noginov et al., *Nature* 460, 1110 (2009).
35. R. F. Oulton, V. J. Sorger, T. Zentgraf, R.-M. Ma, C. Gladden, L. Dai, G. Bartal, and X. Zhang, *Nature* 461, 629 (2009).
36. H. A. Atwater and A. Polman, *Nature Materials* 9, 205 (2010).
37. A. Aubry, D. Y. Lei, S. A. Maier, and J. B. Pendry, *Physical Review Letters* 105, 233901 (2010).
38. A. L. Feng, M. L. You, L. Tian, S. Singamaneni, M. Liu, Z. Duan, T. J. Lu, F. Xu, and M. Lin, *Scientific Reports* 5, 1 (2015).
39. P. Törmö, W. L. Barnes, P. Törmä, W. L. Barnes, P. Törmö, and W. L. Barnes, *Reports on Progress in Physics* 78, 13901 (2015).
40. A. I. Fernández-Domínguez, F. J. García-Vidal, and L. Martín-Moreno, *Nature Photonics* 11, 8 (2017).
41. X. Li, L. Zhou, Z. Hao, and Q.-Q. Wang, *Advanced Optical Materials* 6, 1800275 (2018).
42. J. C. W. Song, *Nature* 557, 501 (2018).
43. J. Luan et al., *Light: Science and Applications* 7 (2018).
44. S. I. Bozhevolnyi and J. B. Khurgin, *Nature Photonics* 11, 398 (2017).

45. T. Uemura, M. Furumoto, T. Nakano, M. Akai-Kasaya, A. Salto, M. Aono, and Y. Kuwahara, *Chemical Physics Letters* 448, 232 (2007).
46. T. Uemura, M. Akai-Kasaya, A. Saito, M. Aono, and Y. Kuwahara, *Surface and Interface Analysis* 40, 1050 (2008).
47. K. Miwa, M. Sakaue, and H. Kasai, *Nanoscale Research Letters* 8, 204 (2013).
48. T. J. Y. Derrien, J. Kruger, and J. Bonse, *Journal of Optics* 18, 115007 (2016).
49. V. Krishna and J. C. Tully, *The Journal of Chemical Physics* 125, 054706 (2006).
50. J. Mcvie, R. S. Sinclair, and T. G. Truscott, *Journal of the Chemical Society, Faraday Transactions* 74, 1870 (1978).

8 Recent Research Progress on Organic–Inorganic Hybrid Solar Cells

Wenjie Zhao, Na Li, Xin Jin, Shengnan Duan, Baoning Wang, Aijun Li, and Xiao-Feng Wang

CONTENTS

8.1 INTRODUCTION

Nowadays, organic–inorganic solar cells have recently gained remarkable attention from the photovoltaic research community because of their remarkable device performance and high power-conversion efficiencies (PCEs). The crystalline Si and other alternative inorganic materials are more commonly used in the manufacturing of solar cells, but their implementation in modern electronic devices is restricted as a result of the inflexibility of the solar cells and the aggravated air pollution problem. One of the possible alternatives is organic-based solar cells, where the cell can be fabricated at low temperature over a large area in an inexpensive way. Now the organic materials have been developed greatly as eco-friendly solar energy to replace non-renewable sources.[1–3] However, the low electron mobility or the inefficient charge transport due to the presence of the traps in the organic active film leads to a decrease of the efficiency of the organic photovoltaic cell structures. Later research provides the evidence that the limitation of the organic photovoltaic structures linked to the charge transport can be overcome by combining organic materials with inorganic semiconductors nanoparticles (NPs), which improve the charge transport based on the low ionization potential of the organic part and the high electron affinity of the inorganic component.[4,5] The combined characteristics of the organic and inorganic semiconductors are implemented into the hybrid solar cells, which exhibit the good charge-transfer characteristics of inorganic semiconductors and the easy processing and tunability of organic semiconductors.

Among the inorganic compounds applied on hybrid photovoltaic structures, ZnO[6] and TiO_2[7–10] with different kinds of morphologies have been proposed as promising materials in hybrid organic–inorganic photovoltaic device applications because of their optical and electrical properties. ZnO is an n-type semiconductor characterized by its unique physical and chemical properties, such as high chemical stability, high electrochemical coupling coefficient, broad range of radiation absorption, and high photostability. TiO_2 as another electron acceptor for use in photovoltaic devices has the advantages of resistance to acid and base, good chemical and photochemical stability, non-toxicity, low cost, and better charge separation properties. These advantages have attracted intensive interest focusing on organic–inorganic hybrid solar cells with ZnO or TiO_2 as inorganic compounds and many efforts have been made to further improve their performance. Recent studies show that a better control of the organic–inorganic heterojunction morphology can develop the performance of solar cells by enhancing the generation and transfer efficiency of excitons. The grain size and morphology of inorganic compounds are proved to be critical for the performance of the solar cells, because the charge generation and transport highly rely on percolating pathways. Employing inorganic semiconductor NPs could effectively increase charge diffusion and effective carrier lifetime, thereby promoting charge transfer at the interface and bulk materials. Besides the bulk properties of organic–inorganic heterojunction, the interfacial properties of semiconductor oxide could also influence the relevant photophysical processes of it. Due to the surface-induced defect states, the doping could effectively adjust the bandgap of the inorganic semiconductor and the electrical properties, thus suppressing the change of its bulk property. The device performance of the organic–inorganic hybrid solar cells can be improved by optimizing of the donor–acceptor interface to improve the charge separation efficiency and suppress the charge recombination. An efficient way to enhance the interfacial properties is the surface modification of inorganic materials with semiconductor quantum dots and small organic molecules. Controllable nanomorphology, well-structured interfaces, and superior optoelectronic interactions are proved to be essential to develop highly efficient hybrid solar cells.[11]

Further efforts are made by applying both ZnO and TiO_2 together on hybrid bulk heterojunction to improve the efficiency solar cells. Moreover, as an abundant natural pigment with low costs and non-toxic property, a series of chlorophyll (Chl) derivatives are applied into organic–inorganic-based solar cells (OISCs) to achieve the goal of using "green" photovoltaics with high efficiency. Their outstanding photoelectron performance and strong absorption ability around near-infrared region provide new research ideas and methods for using clean and renewable energy.

8.2 ZNO ORGANIC HYBRID SOLAR CELLS

In organic–inorganic hybrid solar cells, ZnO is widely used to replace the electron acceptor organic semiconductors of organic solar cells. And the efficiency of organic–inorganic solar cells critically not only relies on the compactness of mixing of the donor and acceptor semiconductors, but also depends on the presence of unobstructed

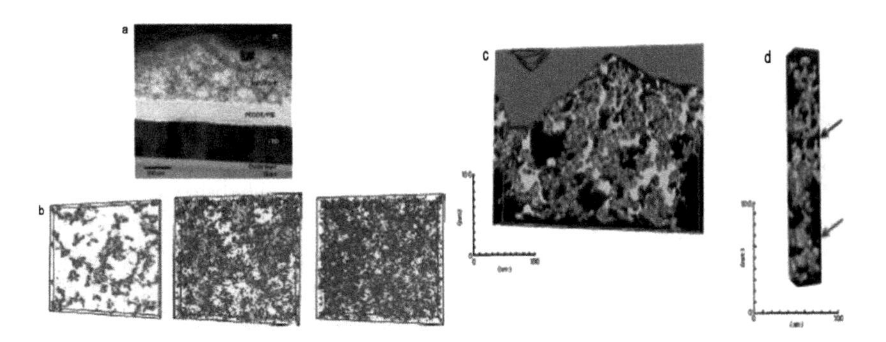

FIGURE 8.1 Electron tomography of P3HT/ZnO solar cells (a) transmission electron micrograph of a cross-section of P3HT/ZnO photovoltaic cell, (b) reconstructed volumes of P3HT/ZnO layers obtained by electron tomography, (c) reconstructed volume of a cross-section of the active layer of a completed P3HT/ZnO device, (d) the green arrow indicates an isolated ZnO domain, the red arrow indicates a ZnO domain that is connected to the top, but not through a strictly rising path. Reprinted with permission from Ref. 1, Copyright 2009, Springer Nature.

transport pathways of electrons and holes to achieve efficient charge transfer and collection.

In 2009, Oosterhout *et al.* first spatially resolve the morphology of 2%-efficient organic–inorganic hybrid solar cells consisting of poly(3-hexylthiophene) as the donor and ZnO as the acceptor in the nanometer range by electron tomography. And via solving the three-dimensional exciton-diffusion equation, a consistent and quantitative correlation between solar cells performance, photophysical data and the three-dimensional morphology could be obtained for solar cells with different layer thicknesses. Figure 8.1 shows the electron tomography of P3HT/ZnO hybrid solar cells.

The relatively poor performance of organic–inorganic solar cells is related to inefficient charge generation as a result of the low thickness of inorganic materials and the coarse phase separation, as well as the exciton losses impaired by the electrodes. However, the solar cells with thicker photoactive layers (organic–inorganic heterojunction), charge generation is much more efficient, owing to a much more favorable phase separation. Therefore, as expected, a better control of the organic–inorganic heterojunction morphology could improve performance of the solar cells through enhancing the generation and transfer efficiency of excitons.

8.2.1 ZnO-NP Organic Hybrid Solar Cells

Because the charge generation and transport highly rely on percolating pathways to ensure that the charge carriers can arrive at their respective electrodes without recombination due to trapping in dead ends on the isolated ZnO domain, the grain size and morphology of ZnO are essentially critical for the performance of the solar cells.

In 2004, Beek *et al.* reported nano-crystalline ZnO as n-type semiconductor in organic–inorganic hybrid solar cells, which is a cheap and environmentally-friendly material that could be synthesized in high purity and crystallinity at low temperatures.[12]

Figure 8.2 displays the layout of organic–inorganic hybrid solar cells employed in nano-crystalline ZnO. The nano-crystalline ZnO could be disappeared in relatively apolar solvent mixture, thus it could blend with poly (2-methoxy-5-(3′,7′-dimethyloctyloxy)-1,4-phenylenevinylene](MDMO-PPV). Upon excitation, ultra-fast charge transfer could occur between the interface of nano-crystalline ZnO–MDMO-PPV, which is utilized to generate an efficient organic–inorganic hybrid solar cell with a high fill factor and open-circuit voltage.

This demonstrates that the effectiveness of using a combination of organic materials and inorganic materials by using nano-crystalline inorganic semiconductor would be benefited to the charge separation and transfer at the interface of the donor–acceptor and photovoltaic performance of the organic–inorganic hybrid solar cells.

In 2012, Wu *et al.* had systematically investigated poly (2-methoxy-5-(2-ethylhexyloxy)-1,4-phenylenevinylene) (MEH-PPV) and vertically aligned ZnO nanorod array (ZnO-NA).[13] As shown in Figure 8.3, they had prepared the organic–inorganic solar cells with three kinds of MEH-PPV/ZnO-NA layouts and found the device layout and illuminated photoactive area impose significant effects on the steady-state and dynamic performances of the devices even though the device architecture, the materials property, and the Au electrode in the devices are not changed.[13]

ZnO nanorod arrays could grow from the substrate vertically and have a great potential for organic–inorganic heterojunction solar cells due to their ease of synthesis[14] and high electron mobility (~102 $cm^2 \cdot V^{-1} \cdot s^{-1}$)[15] with a direct transport pathway to the corresponding electrode.[16] However, the inefficient charge generation and

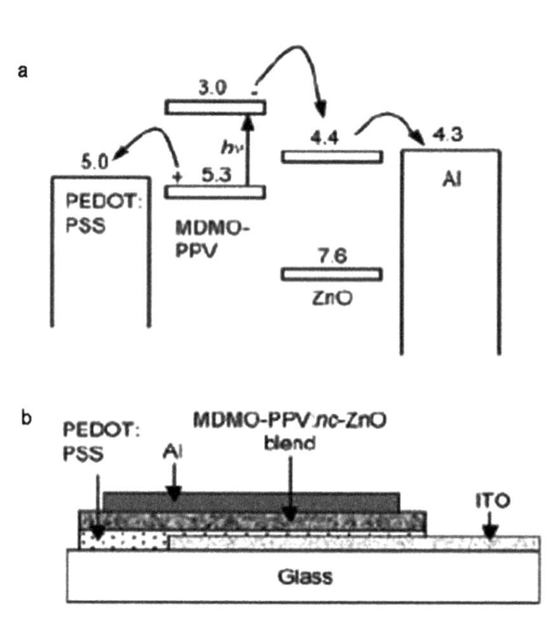

FIGURE 8.2. (a) The schematic energy level diagram and (b) the device structure of the organic–inorganic hybrid solar cells employed in nano-crystalline ZnO. Reprinted with permission from Ref. 2, Copyright 2004, WLEY-VCH Verlaine GmbH & Co. KGaA, Weinheim.

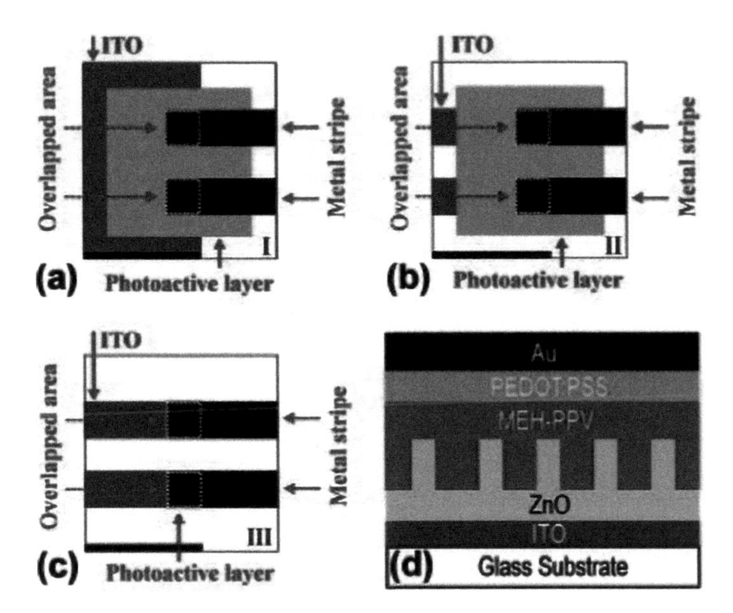

FIGURE 8.3. The device layouts (a–c) and architecture (d) used in the study. The black bare (b and c) identifies the ITO stripe length, which is the same to the width of the continuous ITO layer in (a). The area enclosed by dotted lines (a–c) identifies the OLA region in each device. Reprinted with permission from Ref. 3, Copyright 2012, Elsevier Ltd.

charge transfer, which are influenced by the morphology of heterojunction, affected the PCE of the solar cells effectively. In order to further improve the organic–inorganic hybrid (ZnO nanorods/polymer) solar cells, Ruankham *et al.* had improved the photovoltaic performances by controlling their charge dynamics via addition of ZnO NPs into poly(3-hexylthiophene) (P3HT) photoactive layer.[17] The inter-rod space of ZnO nanorod substrates is completely filled with the solution-processed ZnO NPs–P3HT blends, forming homogeneous junction among the components. And the PCE of the solar cells has been achieved to 1.02% with 13 vol % ZnO NPs employed in ZnO nanorods/polymer. Figure 8.4 shows the device architecture and possible charge transport diagram of the solar cells studied in this work. And their research demonstrates that formation of ZnO NP domain extending across the active layer provides larger interfacial area of ZnO–P3HT interface and more effective percolation path for the charge carriers.[17] Therefore, employing inorganic semiconductor NPs could effectively increase charge diffusion and effective carrier lifetime, thereby promoting charge transfer at the interface and bulk materials.

8.2.2 Modified ZnO Organic Hybrid Solar Cells

While the bulk properties of organic–inorganic heterojunction are used to described its' interface, it is known that the interfacial properties of semiconductor oxide could also influence the relevant photophysical processes of it. The doping could effectively adjust the bandgap of the inorganic semiconductor and the electrical

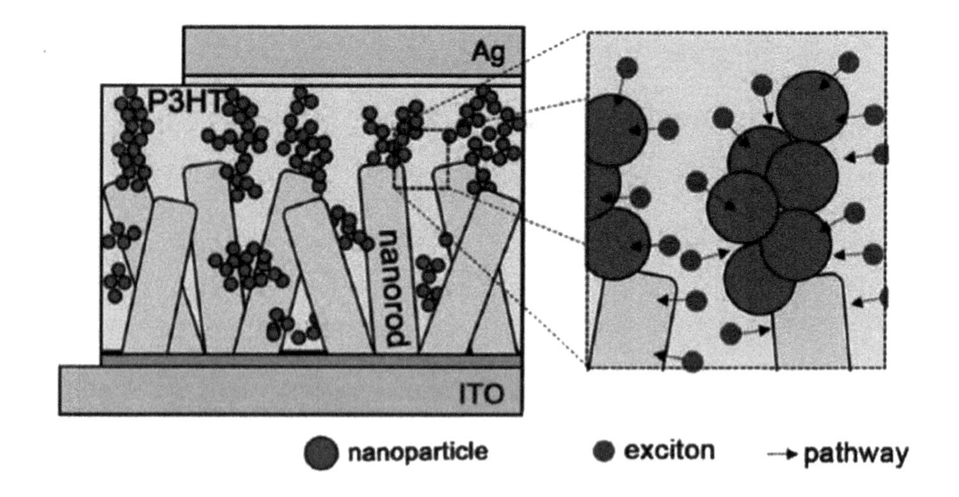

FIGURE 8.4. Device architecture and possible charge transport diagram of the ITO/dense ZnO/ZnO nanorods/ZnO nanoparticles/P3HT/VO_x/Ag solar cells. Arrow lines show possible charge transport pathway. Reprinted with permission from Ref. 4, Copyright 2015, Springer Nature.

properties, thereby suppressing the change of its bulk property due to the surface-induced defect states. Musselman *et al.* had doped zinc oxide with nitrogen (ZnO:N) to tune its electron concentration, reducing it by approximately two orders of magnitude in 2014.[18] And they further studied the effect of bulk electron concentration on the surface properties of ZnO and the exciton dissociation on the double-layer ZnO–P3HT interface model. As illustrated in Figure 8.5, the formation of a space-charge region could be generated by electron trapping and oxygen chemisorption of ZnO at its surface and grain boundaries. Under illumination, photo-generated holes created in the ZnO near its surface could recombine with trapped surface electrons, potentially releasing chemisorbed oxygen molecules.[18] However, more efficient light-induced de-trapping of electrons is observed form the ZnO:N surface, which enhances exciton dissociation and electron transfer from the P3HT to the ZnO.

Therefore, doping ZnO with a small amount of nitrogen could reduce its electron concentration in dramatically improved surface properties and enhance the ability of excitons dissociation.

The performance efficiency of organic–inorganic hybrid solar cells is not high yet, only about (η = 2–3%).[19] Among the limiting factors in the hybrid devices, the poor compatibility between inorganic acceptor materials and organic donor materials components and the serious charge recombination at the interface of acceptor–donor are the major interfacial difficulties for efficient devices. The incompatibility between hydrophilic NPs and hydrophobic polymers frequently causes a macroscopic phase separation and a bad interfacial contact between acceptor and donor components, resulting in a low efficiency of the charge transfer from donor to acceptor materials. Moreover, the normally low carrier mobility (ca. 10^{-1} to 10^{-9} $cm^2 \cdot V^{-1} \cdot s^{-1}$)[20] in conjugated polymers often results in a poor hole transfer from the donor to acceptor, and

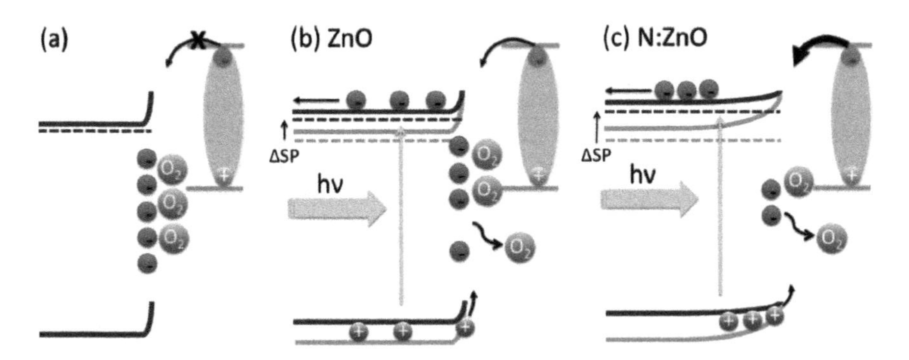

FIGURE 8.5. (a) The trapping of electrons at the surface of ZnO results in the formation of a space-charge region, (b) photogenerated holes in the ZnO can recombine with trapped electrons, releasing adsorbed species and reducing the space charge and associated band bending at the surface, (c) a thicker space-charge region is expected for the ZnO:N, which should enhance the de-trapping process and result in a greater reduction in the surface's work function upon UV illumination (as indicated by the change in surface potential, ΔSP). Reprinted with permission from Ref. 5, Copyright 2014, Published by WLEY-VCH Verlaine GmbH & Co. KGaA, Weinheim.

then leading to a poor spatial separation of photo-generated charge carriers (electron and hole), yielding an easy interfacial charge recombination. Therefore, optimization of the donor–acceptor interface to improve the charge separation efficiency and suppress the charge recombination is an essential strategy to enhance the performance of the organic–inorganic hybrid solar cells.

Semiconductor quantum dots, such as CdS, CdSe, and PbS, with a tunable bandgap in the visible region can serve as sensitizers for wide bandgap semiconductor materials.[21] Furthermore, the semiconductor quantum dots could also provide new chances to utilize hot electrons or generate multiple charge carriers with a single photon.[22] Thus, combination of organic–inorganic solar cells with semiconductor quantum dots could enhance the charge transfer processes between organic and inorganic materials. Hao *et al.* had employed a simple and low cost electro-deposition method to fabricate organic–inorganic solar cell with CdSe-modified ZnO and P3HT as active layer.[23] They demonstrated that the CdSe-ZnO core-shell nanorod arrays with P3HT to form p-n heterojunctions that could largely improves the photovoltaic performance of solar cell, exhibiting a PCE of 0.88%.[23]

Organic sensitizers, which are available in the form of dyes, have been largely used for photovoltaic works in the form of dye-sensitized solar cells because of their high absorption coefficients that benefit the light absorption spectra of the devices. Organic sensitizers could be used to modify the inorganic semiconductor materials based on their advantages. And ZnO nanorod arrays modified with organic sensitizers display specific light-harvesting and charge-collecting properties, which are promising for enhancing the characteristic performance of organic–inorganic hybrid solar cells based on ZnO/poly(3-hexylthiophene). Ruankham *et al.* had used common organic sensitizers (N719 (Ru-based complex), NKX2677 (coumarin dye), and D205 (indoline dye) and synthesized derivative of square molecules)-modified ZnO nanorod arrays to improve the performance of the solar cells in 2010.[24] The mechanisms of modified solar cells are summarized in Figure 8.6. For unmodified devices and D205- and N719-modified devices, the excitons in P3HT could separate and the generated free electrons could transfer through the interface (red line in Figure 8.3), while this process does not occur for NKX2677-modified devices.[24] The induced space-charge layer for D205- and NKX2677-modified devices could suppress a flow of charge leakage[25] (gray dotted line in Figure 8.6); however, this does not take place in unmodified devices and N719- and square-modified devices. In a work, Ruankham *et al.* had present a workable approach for improving the performance of organic–inorganic solar cells via organic sensitizers-modified ZnO nanorods, and square-modified devices gave the best PCE of about 0.82%, with improved structure and performance could reach as high as 1.02%.

Surface modification of inorganic materials with semiconductor quantum dots and small organic molecules has been an efficient way to enhance the interfacial properties and the device performance in the organic–inorganic hybrid solar cells. For further enhancement of the photovoltaic performance of the solar cells, Bi *et al.* had employed amphiphilic and carboxylated dye molecules (Z907) to modify vertically aligned ZnO nanorod arrays via grafting Z907 onto ZnO surfaces at different soaking times.[26] The modification effects on the performance of the organic–inorganic hybrid solar cells with poly(2-methoxy-5-(2-ethylhexyloxy)-1,4-phenylenevinylene) (MEH-PPV) as the donor materials had been investigated systematically. Amphiphilic sensitization with Z907 at aligned ZnO nanorod arrays improves the compatibility between ZnO nanorod arrays and MEH-PPV and enhances the charge separation efficiency at the interface of MEH-PPV–ZnO as a result of the enhanced electronic coupling property at the interface for charge transfer.[26] However, the presence of Z907 could reduce the surface defect concentration of ZnO nanorod arrays, but increase the defects at the heterojunction interface. This phenomenon is attributed to electron-rich carbonyl groups of Z907. Therefore, the content of Z907 on the ZnO

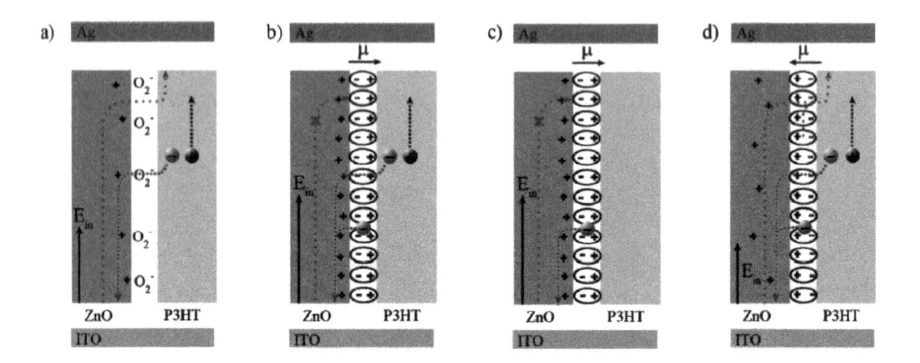

FIGURE 8.6. Device mechanisms along the ZnO nanorods for (a) unmodified devices, (b) D205-modified devices, (c) NKX2677-modified devices, and (d) N719- and Sq-modified devices. E_{in} is the internal electric field of the ZnO nanorods. Red and blue dotted lines are the electron and hole transport pathways, and the gray dotted line is the charge leakage. Reprinted with permission from Ref. 6, Copyright 2011, American Chemical Society.

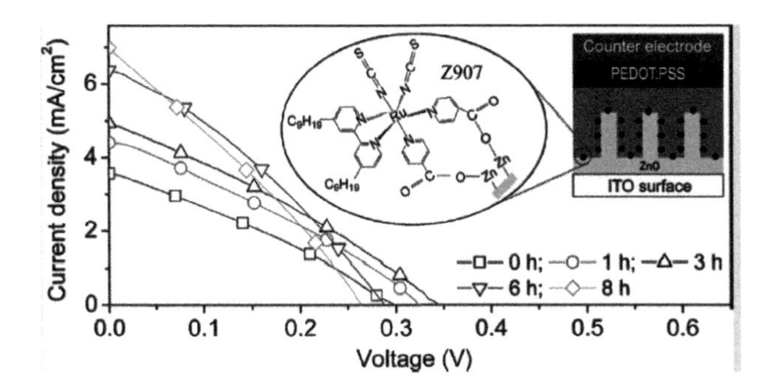

FIGURE 8.7. The device structure and photovoltaic performance of Z907 modified organic–inorganic hybrid solar cells. Reprinted with permission from Ref. 7, Copyright 2011, American Chemical Society.

nanorod arrays surface induced by modification time could generate a great effect on the performance of the solar cells. Figure 8.7 shows the device structure and photovoltaic performance of the Z907-modified organic–inorganic hybrid solar cells. Their study had demonstrated that trapping electrons induced by surface defects could promote the charge transfer at the interface of donor–acceptor for efficient charge separation in the organic–inorganic hybrid solar cells, and both open-circuit voltage and charge recombination rate in the solar cell are correlated to the occupation of injected electrons in conduction band and surface defects.

8.3 TIO_2 ORGANIC HYBRID SOLAR CELLS

Organic–inorganic hybrid materials based on TiO_2 and p-conjugated polymers have attracted great attentions in optoelectronic applications over the past decades. Similar to ZnO, the performance of TiO_2-based organic–inorganic solar cells relies on several factors, such as the particle size, doping, and modifications of TiO_2.

In 1991, the Lausanne group first reported the regenerative photo electrochemical solar cells base on photosensitization of nano-crystalline TiO_2. Their photovoltaic cell created from medium-purity materials through low-cost processes exhibits a commercially realistic energy-conversion efficiency of 7.1–7.9% in simulated solar light. Figure 8.8 shows the schematic representation of the principle of the dye-sensitized photovoltaic cell that indicates the electron energy level in the different phases.[27]

8.3.1 TiO_2-NP Organic Hybrid Solar Cells

Efficient charge separation relies upon molecular interfaces to separate charges, and the domain size of the organic donor and inorganic acceptor materials should be comparable to or smaller than the exciton diffusion length to increase the probability of exciton dissociation across the donor–acceptor interfaces. However, coagulation problems with inorganic nano-crystals often occur during the device fabrication process. The blended organic–inorganic components may undergo macro-phase separation that reduces the interfacial interactions between the polymer and inorganic

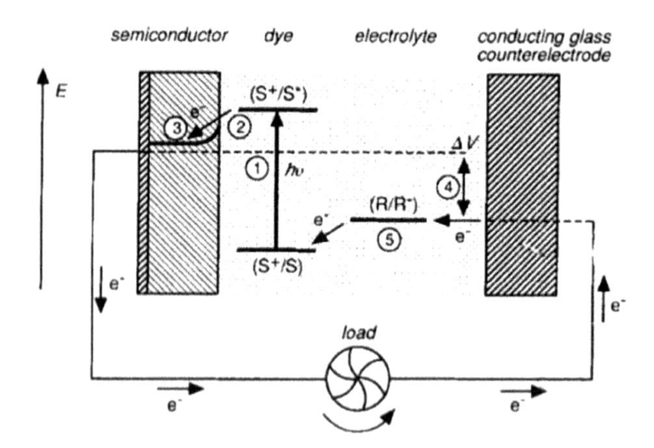

FIGURE 8.8. Schematic representation of the principle of the dye-sensitized photovoltaic cell to indicate the electron energy level in the different phases. The cell voltage observed under illumination corresponds to the difference, ΔV, between the quasi-Fermi level of TiO_2 under illumination and the electrochemical potential of the electrolyte. The latter is equal to the Nernst potential of the redox couple (R/R^-) used to mediate charge transfer between the electrodes. S, sensitizer; S^*, electronically excited sensitizer; S^+, oxidized sensitizer. Reprinted with permission from Ref. 8, Copyright 1991,Springer Nature.

materials, resulting in a decrease in the quantum efficiency, thereby degrading the device performance. So that, controllable nanomorphology, well-structured interfaces, and, superior optoelectronic interactions are the critical factors for developing highly efficient hybrid solar cells.

In 2008, Su *et al.* had systematically investigated the microscopic mechanisms of charge separation and charge transport in the poly(3-hexylthiophene)-TiO_2 nanorod nanocomposites.[28] They found charge separation and transport efficiency could be improved by adding an adequate amount of TiO_2 nanorods in polymer. Figure 8.9a shows the UV–vis absorption spectra of the nanocomposites consisting of 0, 30, 50, and 60 wt% TiO_2 nanorods, respectively. The pristine P3HT exhibits a broad absorption spectrum ranged from 400 to 650 nm, and TiO_2 nanorods have an absorption edge at about 350 nm. The optical density of the absorption spectra in the hybrid is simply the sum of the absorption spectra of the constituent parts. In contrast, the yield of the PL emission decreases with the increasing TiO_2 nanorod content, suggesting the occurrence of photo-induced electron transfer from P3HT to TiO_2 nanorods. The presence of photo-induced charge transfer at the P3HT–TiO_2 nanorod interfaces can be further evident from time-resolved photoluminescence spectroscopy. Figure 8.9b shows the PL decay curves for the pristine P3HT and the hybrid films, respectively. The addition of TiO_2 nanorods in polymer results in a new relaxation process, which provides the donor a further non-radiative process and leads to the enhancement of the non-radiative decay rate. The measured PL decay lifetime for the pristine P3HT and hybrids consisting of 30, 50, and 60 wt% TiO_2 nanorods are 720 ps, 580 ps, 520 ps, and 470 ps, respectively. This produces a charge separated

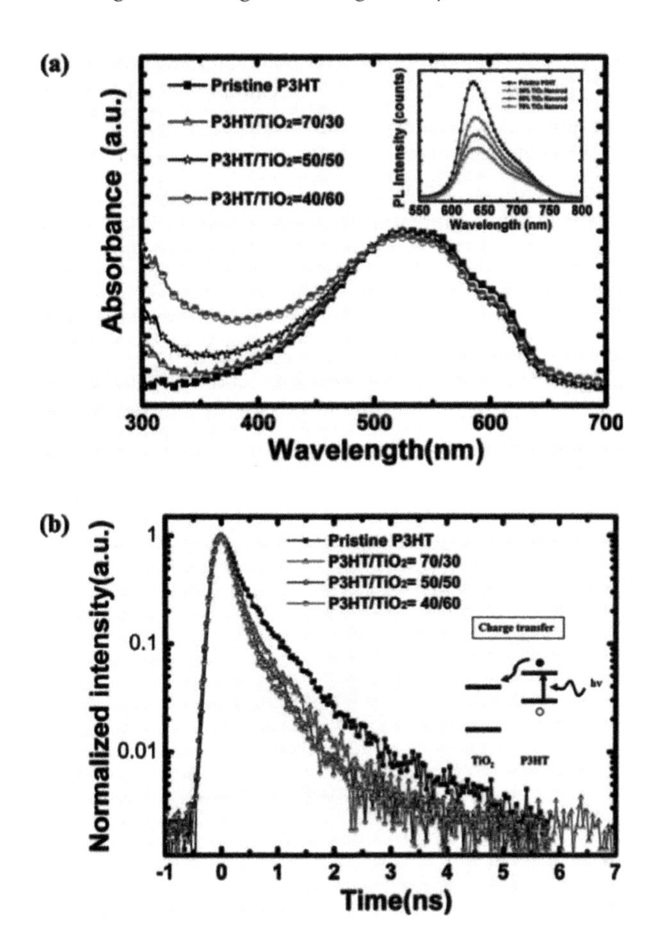

FIGURE 8.9. (a) UV–vis absorption spectra of the P3HT–TiO_2 nanorod hybrid materials consisting of 0 wt%, 30 wt%, 50 wt%, and 60 wt% TiO_2 nanorods respectively. The inset shows the corresponding PL yield of the hybrids. (b) Time-resolved PL spectroscopy of the hybrid materials. Reprinted with permission from Ref. 9, Copyright 2008, Royal Society of Chemistry.

state with an electron on the TiO_2 nanorods and a hole on the polymer. As the content of TiO_2 nanorods in the hybrid is increased, the efficiency of charge separation at the polymer nano-crystals is enhanced, resulting in shortening of the measured lifetime τ PL and quenching of the PL efficiency.

In 2010, Goubard*et al.* prepared poly(3,4-ethylenedioxythiophene) (PEDOT) as electronic conducting polymer and nano-crystalline TiO_2 as host matrix by the template method.[29] They applied an original in situ photopolymerization technique to synthesize PEDOT inside the TiO_2 pore and characterized the polymer and pore filling by different analysis. In situ generation of PEDOT by photopolymerization was observed to be higher and self-limiting after 22% filling of the mesoporous TiO_2 network. Hybrid materials were used to fabricate an indium-tin oxide/nano-crystalline

TiO_2/PEDOT/Au device. The current-voltage characteristics indicate that a built-in electrical field has been created at the nano-crystalline TiO_2–PEDOT interface with energy conversion efficiency of 0.09% without dye. Figure 8.10 shows the structure of solar cell device and the J-V curves.

Reeja-Jayan *et al.* applied a bias-dependent fluorescence modulation technique to probe the hybrid organic–inorganic interface between pristine and carboxylate P3HT(P3HT-COOH) and a TiO_2 acceptor layer.[30] Charge traps were found in the case of P3HT and the open circuit voltage and fill factor was improved in devices containing P3HT-COOH compared to P3HT. A challenge with this work was that the carboxylate polymeric materials were not completely functionalized due to synthetic limitations, which led to a mixture of both P3HT and P3HT-COOH in the organic/polymer layer. In order to overcome this limitation, the researchers furthered their study by synthesizing well-defined oligothiophenes and using them as the interfacial modifier between P3HT and TiO_2.[31] The oligothiophenes offered a more well-defined system due to their discrete, monodisperse molecules than the corresponding polymer. Another advantage was that carboxylate oligothiophenes can be isolated using

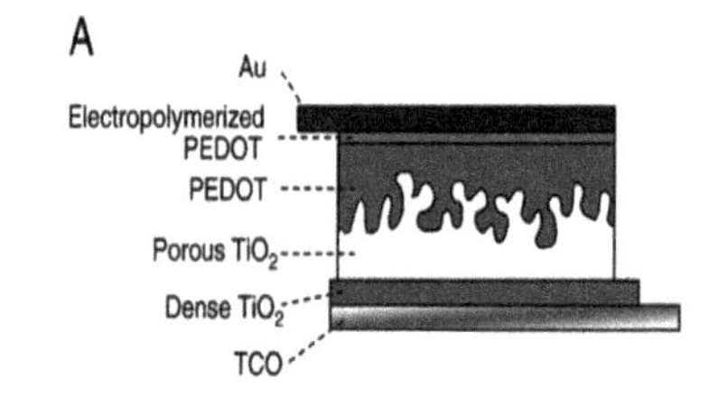

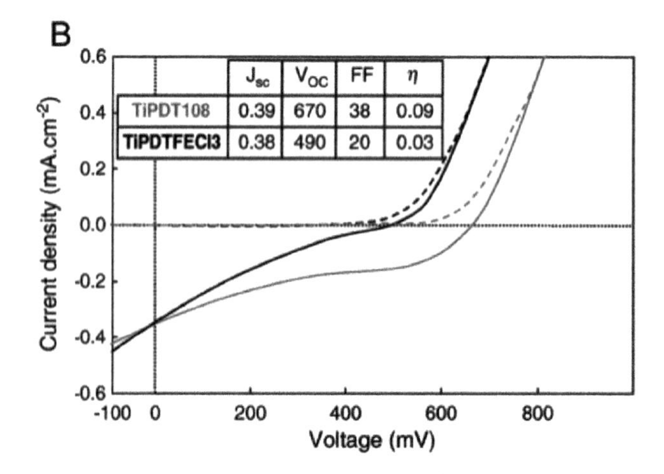

FIGURE 8.10. (a) Device architecture, (b) current-density-voltage characteristics under simulated standard solar irradiation (AM1.5, 100 mW cm^2, solid lines) and in the dark (dashed lines) for TiPDT-108 (green curve) and $TiPDTFECl_3$ (red curve) J_{sc}, short-circuit current density (mA cm^{-2}), V_{OC}, open-circuit voltage (mV); FF, fill factor (%); and η, energy-conversion efficiency (%). Reprinted with permission from Ref. 10, Copyright 2010, Elsevier B.V.

conventional purification techniques resulting in pure, monodisperse molecules with 100% carboxylation. Device prototypes using carboxylate oligothiophenes as interfacial modifiers showed improved performance in the open-circuit voltage and fill factor over devices using unmodified oligothiophenes as interfacial modifiers. It was found that interface layer adhesion was improved by functionalizing oligothiophenes with a carboxyl moiety, and devices made using carboxylate oligothiophenes had fewer aggregates in the P3HT layers atop the modified TiO_2 surface. The best interfaces were found using oligothiophenes functionalized with carboxylates, which created smooth layers on TiO_2, and showed no hysteresis, suggesting elimination of interfacial charge traps. However, hysteresis could be reintroduced by varying the bias, suggesting that even with a good interface the phenomenon of interfacial trapping was not solved. These results provide insight into the use of modifiers to enhance interfacial interactions between the diverse materials employed in organic solar cells, which are expected to lead to the rational design of novel organic photovoltaic devices with long-term stability.

Study of the hybrid BHJ solar cells shows that structure control with the fine adjustment of electron donor/acceptor distribution is still a challenge for people to improve the PCE. The key factors for developing highly efficient hybrid solar cells are controllable morphology of nanohybrids, well-defined interface, and superior optoelectronic interactions. Efficient exciton dissociation requires excitons generated at a distance no longer than exciton diffusion lengths, which are very short in organic semiconductors, at a few tens of nanometers. However, coagulation problems during the device fabrication process may reduce the interfacial interactions between the polymer and inorganic materials and degrade the device performance. In order to achieve a high miscibility of organic–inorganic components and prepare a favorable dispersion of nano-crystals in hybrid photovoltaic devices, a strategy has been developed that involves a π-conjugated polymer matrix such as P3HT containing a precursor of an inorganic component required for the synthesis process.

Chiang *et al.* put forward an in situ method to fabricate highly elongated P3HT nanowires with continuous ZnO nano-crystal pathways by simultaneously organizing organic P3HT molecules and inorganic zinc precursors into highly ordered nanowires with micrometer-scale lengths, followed by a thermal oxidation treatment to directly grow discrete ZnO nano-crystals on the existing nanofibrillary template.[32] However, the prepared nano-crystals owns poor quality and less crystalline than those produced by calcination or colloidal synthesis due to the low annealing temperature to prevent degradation of polymer components. A hydrothermal technique of TiO_2 preparation was developed to fabricate TiO_2-conjugated polymer nanohybrids with a controllable nanostructure by an in situ synthetic method, which provides a prospective route to prepare a well-crystalline and phase-pure oxide in an one-step approach in a tightly closed stainless steel autoclave at controlled temperatures and/or pressures. Chiang *et al.* fabricated P3HT–TiO_2 nanohybrids by a novel in situ growth strategy, where an one-step approach was used to simultaneously organize organic P3HT chains and inorganic titanium precursors into highly elongated nanofibrils, followed by the hydrothermal synthesis process to directly grow highly crystallized TiO_2 NPs on the existing P3HT nanofibrils (Figure 8.11).[33] The P3HT–TiO_2 hybrid system showed a superior structural stability, which provided intimate

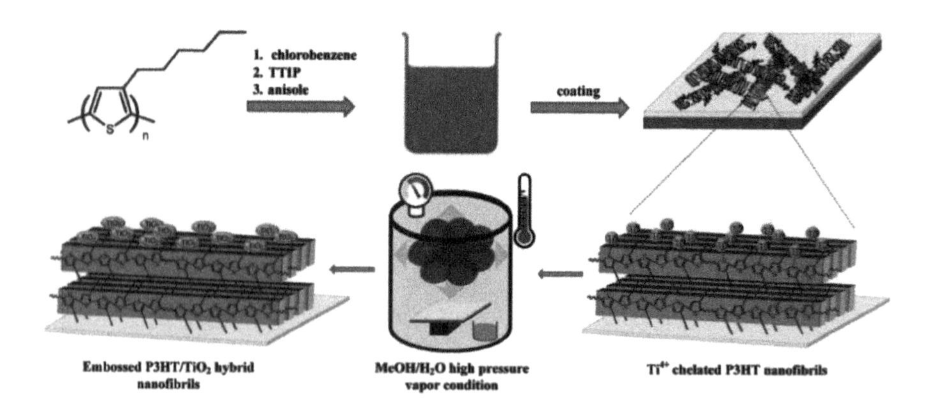

FIGURE 8.11. Reaction scheme for the formation of *in situ* nanohybrids containing TiO_2 nanoparticle-decorated P3HT nanofibrils via the high-pressure hydrothermal crystallization process. Reprinted with permission from Ref. 11, Copyright 2016, Royal Society of Chemistry.

contacts between p-type P3HT and n-type TiO_2 nano-crystals, and created a large number of interfaces for charge dissociation. The enhanced vibronic absorption, PL quenching, and shorter exciton lifetime indicated that excitons in the P3HT–TiO_2 embossed hybrid samples were efficiently dissociated into charge carriers. The enhanced interfacial area for charge separation as well as efficient pathways for charge transport improved optoelectronic properties and device performance of the system. Compared with the ex situ P3HT–TiO_2 hybrid device, the efficiency achieved for the P3HT–TiO_2 embossed hybrid device was approximately a 4.67 times improvement. This novel in situ approach shows a feasible way to fabricate organic–inorganic nanohybrid materials of conjugated copolymers with different inorganic NPs for the applications of future optoelectronic devices.

8.3.2 Modified TiO_2 Organic Hybrid Solar Cells

The hybrid BHJ solar cells were introduced in the study of dye sensitized solar cells (DSSCs) in order to address a few major challenges such as instability caused by corrosive liquid electrolytes and inefficient electron injection from organic dyes. Hybrid BHJ solar cells are potential alternatives for DSSCs as any appropriate visible/infrared light absorbing materials could be adopted to enhance the photo-absorption through which photo-carrier generation is significantly improved. It was reported that using perylene derivative in TiO_2/P3HT-based BHJ solar cells resulted in much improved photon absorption and hence enhancement in J_{sc} of the solar cell. Surface-modified TiO_2 NPs with P3HT were used in BHJ solar cell, yet it was concluded that further optimization would be required to boost the photovoltaic performance.

In 2016, Dai*et et al.* demonstrated an approach to increase performance of P3HT:TiO_2 solar cell either by electron deficient boron or electron-rich bismuth doping into TiO_2 nanorods.[34] The B doping increases the absorption, crystallinity, and electron mobility of TiO_2 nanorods. The Bi-doped TiO_2 has higher J_{sc} as compared with B-doped TiO_2, mainly due to the improvement of electron density and increased absorption of TiO_2 nanorods. Figure 8.12 shows the J-V curves of P3HT:TiO_2 solar cells based on different modified TiO_2 nanorods.

The devices were fabricated from TiO_2 nanorods being surface modified by organic dye W-4. The dye facilitates the bandgap alignment and compatibility between TiO_2 and P3HT. The PCE of solar cell has been increased by 1.33 times and 1.30 times for Bi-doped TiO_2 and B-doped TiO_2, respectively, as compared with that of as-synthesized TiO_2. The results suggest the optical and electronic properties of TiO_2 can be tuned by various dopants to enhance the device performance.

In 2011, Frey *et al.* pointed out that the chemical compositions and structures of organic–inorganic interfaces in mesostructurally ordered conjugated polymer-TiO_2 nanocomposites are shown to have a predominant influence on their photovoltaic properties.[35] Such interfaces can be controlled by using surfactant structure-directing agents (SDAs) with different architectures and molecular weights to promote contact between the highly hydrophobic electron-donating conjugated polymer species and hydrophilic electron-accepting TiO_2 frameworks. As shown in Figure 8.13, MEHPPV polymer chains incorporated within the SDA-directed cubic-like TiO_2 mesostructured films maintain their photoluminescence (PL) properties.

SDAs circumvent macrophase separation between highly hydrophobic conjugated polymers and hydrophilic metal oxide donor–acceptor pairs and additionally allow the compositions, structures, and interactions of active components to be controlled in mesostructured hybrid materials. For cubic mesostructured TiO_2 films containing conjugated polymers, molecular-level differences in the interactions among

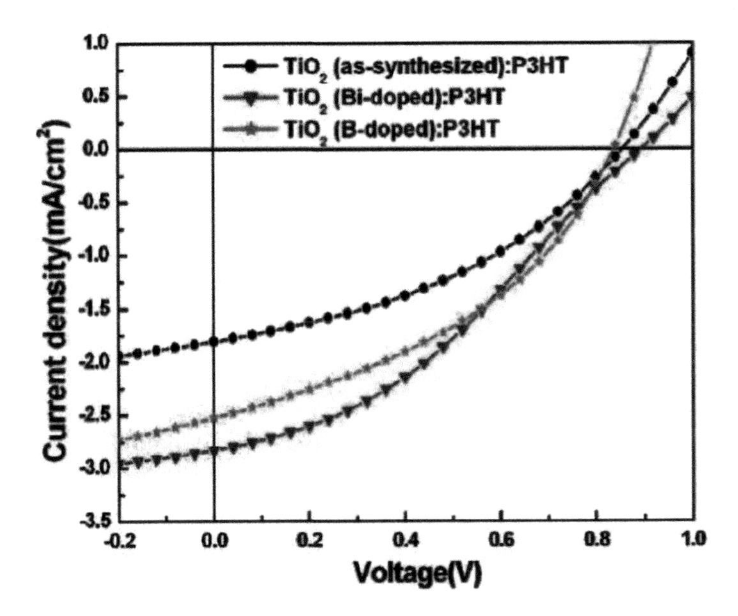

FIGURE 8.12. J-V curve of P3HT:TiO_2 solar cells fabricated from different W4-dyemodified TiO_2 nanorods. Reprinted with permission from Ref. 12, Copyright 2015, Elsevier Inc.

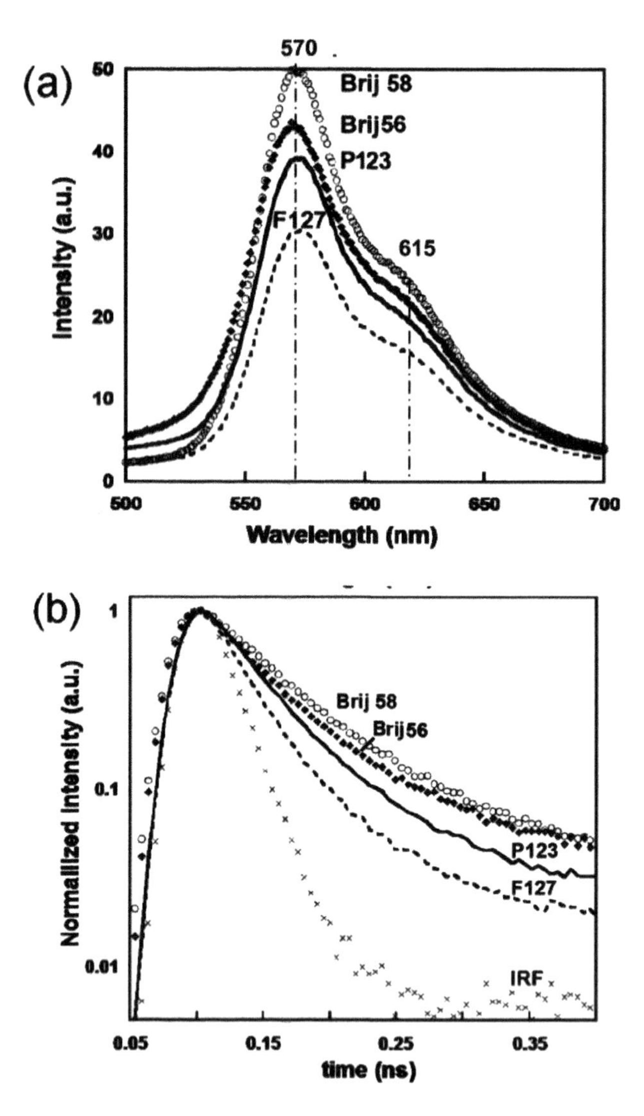

FIGURE 8.13. (a) Photoluminescence (PL) spectra (excitation wavelength 420 nm), and (b) PL decay curves (acquired at 575 nm) of 100–200-nm thick mesostructured TiO_2/SDA/MEHPPV thin films synthesized with different SDAs: F127 (dashed line), P123 (solid line), Brij56 (diamonds), Brij58 (circles). Signals (a) centered at *ca.* 570 and 615 nm correspond to the (0, 0) and (0, 1) vibronic transitions, respectively. The instrument response function is denoted by the "×" symbols in (b). Reprinted with permission from Ref. 13, Copyright 2011, American Chemical Society.

the SDAs, TiO_2 frame works, and conjugated polymer guest species are shown to be correlated with the macroscopic PL and photovoltaic device properties of otherwise identical materials synthesized with different structure-directing species. More specifically, ethylene oxide-based SDAs with different lengths of hydrophilic EO segments and different types and lengths of hydrophobic segments, for example,

Pluronic P123 and F127 triblock copolymers or low molecular weight Brij56 and Brij58 surfactants, result in materials with significantly different interactions among the SDAs, titania, and conjugated polymer guests.

In 2012, Yu *et al.* demonstrated of hybrid BHJ solar photovoltaic cell employing molybdenum disulfide (MoS_2)/TiO_2 nanocomposite (~151 μm thick) and P3HT active layers.[36] As shown in Figure 8.14, the dominant Raman peak at 146 cm^{-1} confirmed TiO_2, while two other peaks observed at 383 cm^{-1} and 407 cm^{-1} asserted MoS_2 in the nanocomposite film.

The demonstrated BHJ solar cell, having a stacked structure of indium tin oxide/TiO_2/MoS_2/P3HT/gold, exhibits a short-circuit current density of 4.7 mA/cm^2, open-circuit voltage of 560 mV, and photo conversion efficiency of 1.3% under standard AM1.5 illumination condition. The quality of TiO_2–MoS_2–P3HT interfaces, as reflected in the dark saturation current in low- and medium-forward-bias region, plays a key role in affecting solar cell performance due to interfacial recombination effect as shown in Figure 8.15.

Interface control is an important approach in polymer-based solar cells because the interface properties on bulk heterojunction can govern the device performance. In 2015, Guo *et al.* selected an organic triphenylamine-type sensitizer to tune the interfacial characters in TiO_2 nanorod array–P3HT hybrid solar cell device (Figure 8.16).[37] In addition to physically improving the compatibility between TiO_2 nanorod and polymer contact junction, the introduction of modifier reduces the charge recombination, prolongs the electron lifetime, and thus optimizes the device performance.

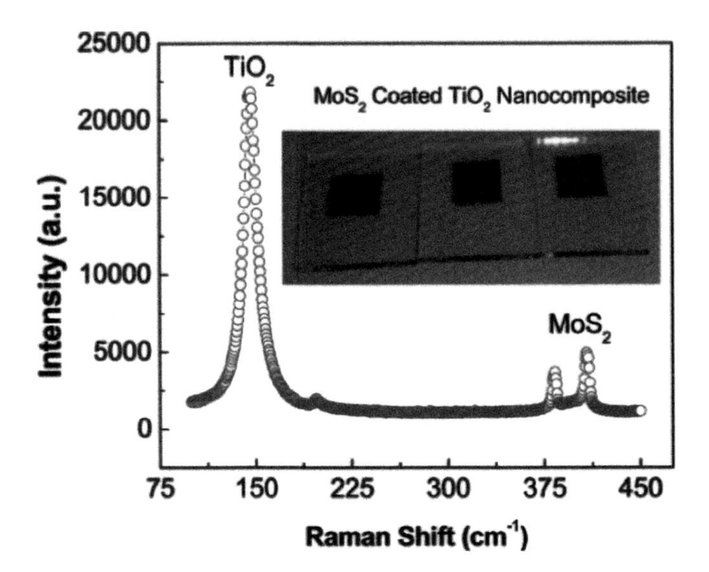

FIGURE 8.14. Raman spectrum obtained from the TiO_2/MoS_2 nanocomposite film. The inset shows the photograph of three prepared nanocomposite films. One of the representative peaks of MoS2 is 407 cm^{-1}, sitting between the signature of MoS_2 monolayer (405 cm^{-1}) and that of bulk MoS_2 film (409 cm^{-1}), suggesting the co-existence of both monolayer and multi-layer flakes. Reprinted with permission from Ref. 14, Copyright 2012, AIP Publishing.

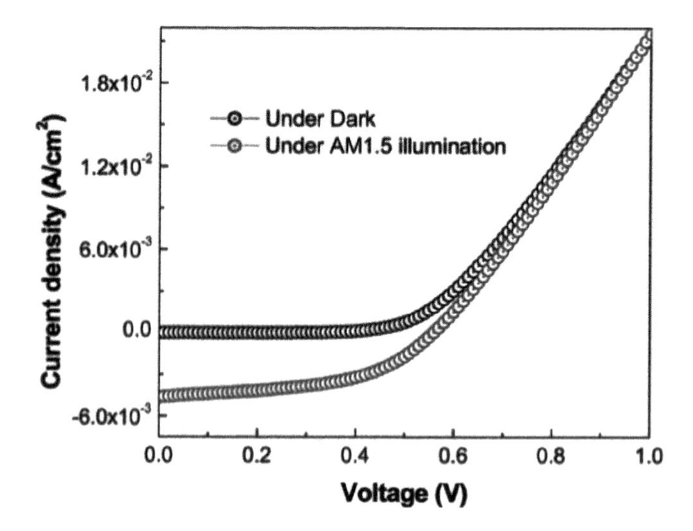

FIGURE 8.15. Dark and illuminated J-V characteristics of the fabricated BHJ solar cell with TiO_2/MoS_2 nanocomposite film. Reprinted with permission from Ref. 14, Copyright 2012, AIP Publishing.

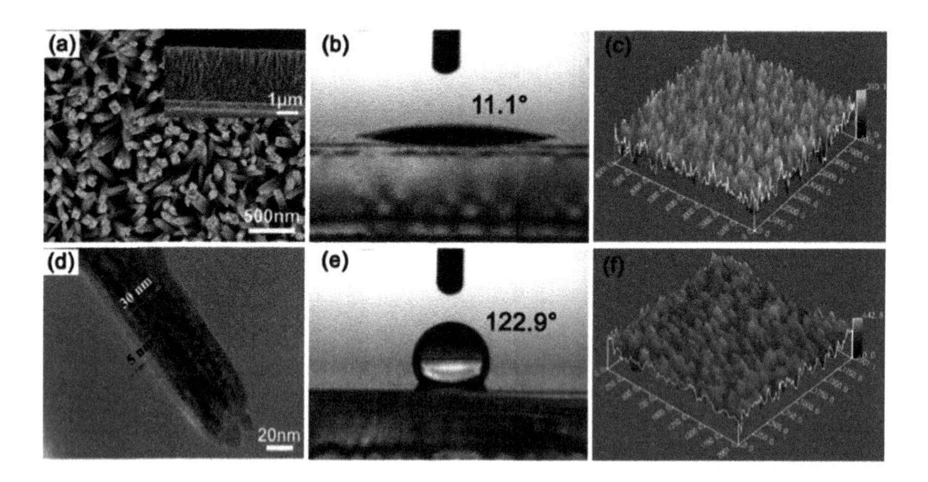

FIGURE 8.16. Top and cross-section view (insert) SEM images of the as-prepared TiO_2 nanorod array on FTO substrate (a) and HR-TEM images of dye modified TiO_2 nanorod (d); images with water contact angles onto films of FTO/TiO_2 (b) and FTO/TiO_2/Modifier surface (e); AFM topographic images of the P3HT surface with TiO_2-nanorods underneath before (c) and after (f) triphenylamine-dye solution treatment. Reprinted with permission from Ref. 15, Copyright 2015, Elsevier B.V.

Furthermore, dyes can also be used to modify $TiO_{2,}$ especially when the combination of regionally regular P3HT and nano-TiO_2 becomes an important prototype of hybrid solar cells, the PCE of TiO_2–P3HT hybrid solar cells is still very low. The key problems in this system are the incomplete polymer infiltration into the mesoporous

TiO_2 and the poor electronic communication between the organic and inorganic materials. In order to improve the device performance, various approaches have been used. For example, optimizing the TiO_2 nanostructure[38] can facilitate infiltration and direct charge transport. Besides, surface modification of TiO_2 by chemisorption of dyes containing carboxylate, phosphonate, or sulfonate groups has also become one of the effective methods today.[39] These dyes show efficient electron injection into TiO_2, as in the well-known DSSCs.[40] In addition, if the absorption of the selected dye is complementary to the P3HT, the photocurrent of the TiO_2/P3HT solar cells can be enhanced by obtaining a more panchromatic light capture response.

In 2008, Wang *et al.* had prepared BHJ solar cells using TiO_2 nanotubes/P3HT as active layer to promote electron transport, in which TiO_2 acts as the acceptor in the blend.[41] Compared with the original polymer P3HT solar cells, the performance of TiO_2 nanotubes–P3HT hybrid solar cells has been significantly improved. To reduce the surface defects in the TiO_2 nanotubes and improve the efficiency of solar cells, standard dye N719 was used in the hybrid bulk heterojunction solar cells. The strong affinity of the carboxylic group in dye N719 to the surface of the metal oxide (TiO_2) facilitates the dispersion of the hybrid material in organic solvents.[2] And it also improves the charge separation.[2] Figure 8.17 shows the UV–vis–NIR absorption spectra of different active layers. Because the absorption edge of TiO_2 is around 350 nm, the addition of TiO_2 nanotubes cannot enhance the absorption in the near-infrared spectrum. However, dye-sensitized TiO_2 nanotube hybrid polymers present stronger near-infrared absorption, and then through the active role of dye N719 in light absorption and charge dissociation, the PCE of dye-sensitized TiO_2 nanotubes–P3HT bulk heterojunction solar cells also improves.

At the same time, Zhicheng Zhong *et al.* reported copper phthalocyanine (CuPc) as a sensitizer in organic–inorganic hybrid solar cells to enhance photon absorption.[42]

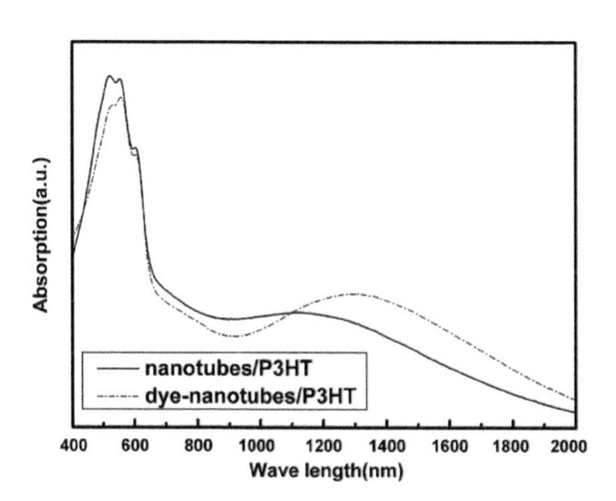

FIGURE 8.17. UV–vis–NIR absorption spectra of different active layers used in solar cells. The solid curve and the dot one represent the TiO_2 nanotubes hybrid with polymer and dye-sensitized TiO_2 nanotubes hybrid with polymer as active layer respectively. Reprinted with permission from Ref. 16, Copyright 2008, Elsevier B.V.

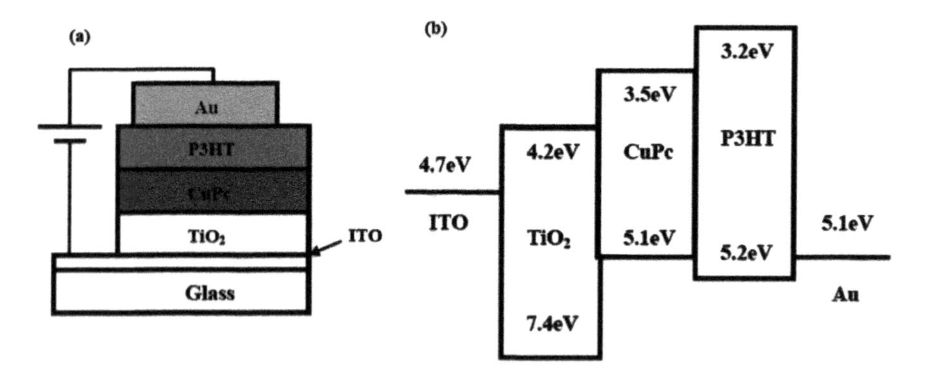

FIGURE 8.18. (a) Schematic structure and (b) schematic energy level diagram of the photovoltaic device. Reprinted with permission from Ref. 17, Copyright 2008, AIP Publishing.

Figure 8.18 displays the device structure and the energy levels of materials used in this work. Here, TiO_2 acts as electron acceptor, CuPc and P3HT work synergistically to improve the solar cells efficiency as electron donors. Besides, it had been confirmed that TiO_2/CuPc has strong absorption between 400 nm and 500 nm and 600 nm and 700 nm, where the absorption of TiO_2/P3HT is poor. Because the absorption spectrum of the final device covers the visible spectral range, more excitons are generated under the illumination using CuPc layer, which greatly increases the photocurrent, and the increase in J_{sc} is the main contribution to improve the PCE of devices.

Subsequently, in 2011, Kevin Sivula *et al.* expended the selection of the dyes to modify TiO_2 using a porphyrin sensitizer (called as YD2).[43] And for the first time, a TiO_2/Dye/P3HT-based solar cells using porphyrin-based YD2 dye was demonstrated, which can enhance light absorption in the near infrared and near ultraviolet regions. In addition, a TiO_2 layer with larger particles and larger pores (*ca.* 75 nm) was prepared to solve the problem that P3HT is hard to infiltrate the 20 nm pores usually used in TiO_2-based solar cells. In order to compare the TiO_2/YD2/P3HT-based devices with the devices without panchromatic light absorption, TiO_2/P3HT devices without YD2 dye and YD2-sensitized TiO_2 devices with transparent hole-transporting material, Spiro-OMeTAD were prepared. The results show that under standard illumination conditions due to the panchromatic light collection provided by the two complementing absorbers, the hybrid solar cells based on TiO_2/YD2/P3HT generated a maximum photocurrent and achieved a PCE of 3.13%. And the continued development of these dye-modified TiO_2 hybrid systems will eventually be able to manufacture highly efficient and inexpensive solution-processing photovoltaic devices for solar conversion.

8.4 ZNO/TIO$_2$ ORGANIC HYBRID SOLAR CELLS

The PCE of solar cells with planar junctions is limited because the exciton diffusion length of the donor material is shorter than its absorption length. The bulk heterojunction topology was proposed to address this problem by mixing the donor and acceptor phases intimately to generate the majority of excitons within a diffusion length of the interface. In bulk heterojunction cells, polymer–inorganic hybrid solar cells have attracted more attention because of the solution processability of polymers and the high electron mobility of inorganic semiconductors. However, inefficient charge transport causes a charge transport bottleneck in isotropic bulk heterojunctions and then many attempts have been done to solve this problem by creating ordered film architectures. The optimal design of polymer-inorganic device topology till now is a perfect vertical array of single-crystalline nanorods of the appropriate dimensions and pitch, encased in a film of the polymer.

Over past several years, ZnO nanorod arrays suitable for both polymer-inorganic cells or dye-sensitized cells have been investigated, but the device performance needs to be further improved. Many attempts have been done, such as changing conditions nanorods grow, and the best isotropic polymer-metal oxide bulk heterojunctions outperform the ordered architectures by a factor of 2–3. Further study showed that though bulk ZnO and TiO_2 have similar bandgaps and band edge energies, the interfacial structures and charge-transfer dynamics of ZnO–P3HT and TiO_2–P3HT interfaces are quite different. By coating ZnO nanorod arrays with a thin shell of TiO_2, as shown in Figure 8.19, Greene *et al.* improved the efficiency of nanorod/P3HT solar cells by a factor of five-fold.[44] In this study, both ZnO-TiO_2 core-shell geometry and air exposure are required to produce cells with higher efficiencies. The efficiency of cells stored in air for 1 month reached 0.29%. It is concluded that both the bandgap of the high-density nanorod material and the organic–inorganic interface should be optimized for effective charge separation to achieve a high-efficiency nanorod polymer solar cell.

In order to realize environmental-friendly and energy renewable devices, conjugated polymers were applied during the manufacture of solar cells. Due to the high charger carrier mobility, one of conjugated polymers, P3HT, was applied both in the BHJ's active layer or as a sensitizer of TiO_2 in DSSCs. However, the performance of the corresponding solar cells was limited because of the loss of reflected sunlight due to the photon scattering. Then ZnO was proposed as anti-reflecting material to solve this problem. Moreover, surface modification such as the 1D nanostructures including nanorods, nanowires, and nanotubes was another way used to increase performance. The synthesis methods based on vapor phase deposition for 1D nanostructures, such as Metal–Organic Chemical Vapor Deposition, Vapor–Liquid–Solid, Vapor–Solid, and pulsed laser deposition, require high temperature and expensive equipment. More researchers used aqueous solution method such as hydrothermal route to synthesis 1D nanostructures for its low cost, simplicity, and suitable for mass production. A study of ZnO NR/TiO_2 NR-P3HT as an active layer based on hybrid bulk heterojunction was reported to optimize DSSC performs.[45] ZnO and TiO_2 nanorods were synthesized by hydrothermal method, and ZnO nanorods were used as anti-reflecting to increase light harvesting of DSSCs. It was found that DSSC with metal inorganic and conjugated organic polymer mixture showed a better efficiency than that with only TiO_2 as an electrode. The dependence of efficiencies on different hybrid TiO_2-P3HT concentration was further investigated, which exhibited optimum condition with the concentration of 10 mg/ml.

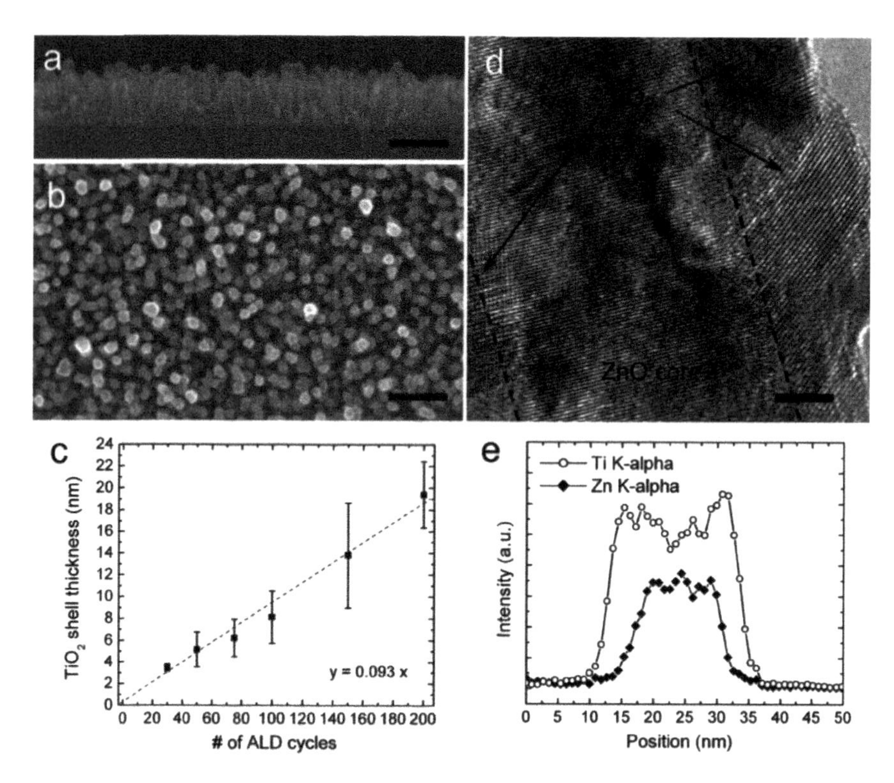

FIGURE 8.19. Structural characterization of the ZnO-TiO_2 core-shell nanorod arrays used in this study. (a) Cross-sectional and (b) plan-view SEM images of a ZnO nanorod array on silicon coated with 6 nm of TiO_2 by ALD. Scale bars, 200 nm. (c) TiO_2 shell thickness against number of ALD cycles. Error bars represent one standard deviation of 30 data points. The dashed line is a least-squares fit, intercept set at origin. (d) High-resolution TEM image of a ZnO-TiO_2 nanorod showing a single-crystalline ZnO core coated with polycrystalline anatase TiO_2 (as determined by XRD (not shown)). Scale bar, 5 nm. The ZnO core is outlined with dashed lines. (e) Corresponding energy dispersive spectroscopy (EDS) elemental line scan across the nanorod. Reprinted with permission from Ref. 18, Copyright 2007, American Chemical Society.

8.5 NEW TYPE ORGANIC–INORGANIC SOLAR CELLS BASED ON ALL CHL DERIVATIVE

Developing and utilizing renewable and clean energy have become a top priority for society today due to the energy and environmental crisis. Solar cells, which could convert solar energy efficiently into heat and electrical energy, have been developed greatly.[46–48] However, the high costs and uneasy degradation have limited the further development of traditional inorganic- or organic materials-based photovoltaics. Photosynthesis is the most basic channel for converting renewable solar energy into chemical energy and biological energy on the earth.[49] As an essential natural pigment in photosynthesis, Chl plays an important role in light capturing, energy/charge transferring in natural photosynthetic complex.[50]

Chl derivatives are widely used in DSSCs, perovskite solar cells, and organic solar cells due to their main merits of easy synthesis, non-toxics, and low costs.[51–53] Furthermore, their outstanding photoelectron performance and strong absorption ability around near-infrared region provide new research ideas and methods for the usage of clean and renewable energy.[54] In addition, previous studies about DSSCs have proved that charge dissociation could take place at the Chl sensitizer and TiO_2 interface.[55,56] Considering above-mentioned advantages, a series of Chl derivatives are semi-synthesized and applied into OISCs to achieve the goal of using "green" photovoltaics with high efficiency.

Three kinds of carboxyl-functioned Chl derivatives methyl trans-3^2-carboxypyropheophorbide *a* (Chl-1) or its zinc complex (ZnChl-1) as sensitizers and zinc methyl 3-devinyl-3-hydroxymethyl-pyropheophorbide *a* (named as ZnChl-2) as hole transporter materials (HTM) were employed to fabricate the mesoporous OISC with an architecture of FTO/compact TiO_2/mesoporous TiO_2/Chl-1 or ZnChl-1/ZnChl-2/Ag (see their molecular structure in Figure 8.20 and device architecture in Figure 8.21a).[57] The photoexcited excitons are dissociated at the TiO_2 and Chl sensitizer interface due to the strong driving force between their TiO_2 and Chl heterojunction. The electrons

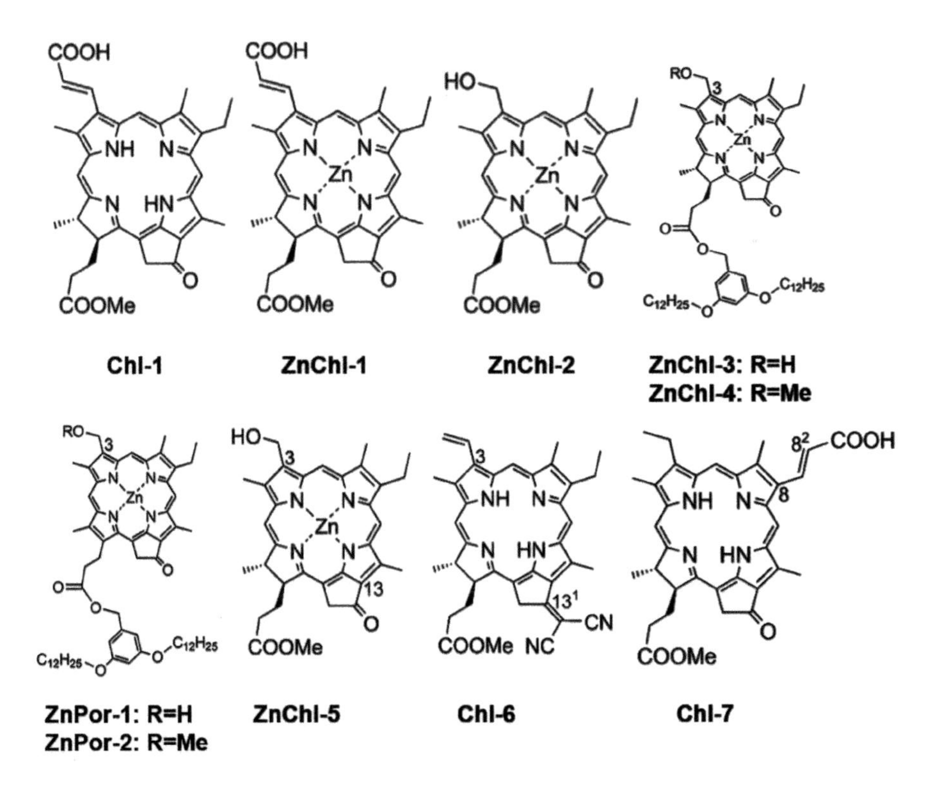

FIGURE 8.20. Molecular structures of the Chl and porphyrin derivatives used in the organic–inorganic heterojunction based BSCs. Reprinted with permission from Ref. 19, Copyright 2019, American Chemical Society.

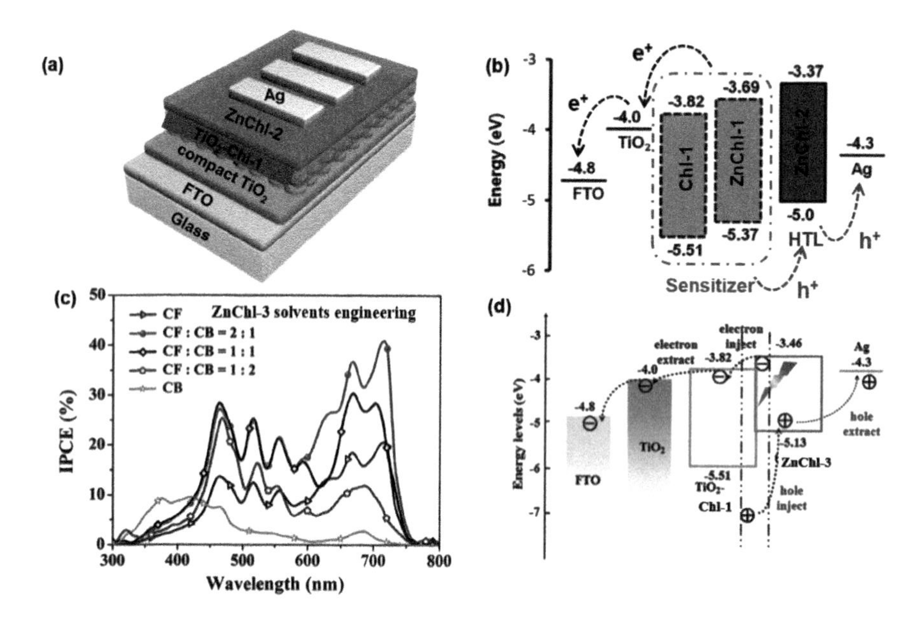

FIGURE 8.21. (a) Device structure and (b) energy alignments of the whole Chl derivative as photoactive material-based OISCs, (c) IPCE spectra of the Chl-1/ZnChl-3-based OISCs via solvents engineering, and (d) possible charge transfer pathway for the Chl-1/ZnChl-3-based OISCs with CF:CB=2:1 as solvents for ZnChl-3. Reprinted with permission from Ref. 20, Copyright 2015, Elsevier B.V.

are extracted by the TiO_2 and are collected at FTO side. In the meanwhile, the holes are transferred to Ag side through ZnChl-2 (see their working principle in Figure 8.21b).

It is worthy to mention that the hole transferring ability among ZnChl-2 is greatly increased due to their formed aggregates. As an initial attempt of fabricating the OISCs with Chl derivatives as sensitizer and hole transporter, such an interesting and unique device, could work normally but only gave an efficiency of 0.11%. Here, the ZnChl-2 had no contribution to the photocurrent since there was no incident photon to converted electron (IPCE) single from ZnChl-2, which means that ZnChl-2 only functioned as a hole transporter instead of photoactive layer. The photocurrents were generated only at the Chl-1 and TiO_2 such an organic–inorganic interface. Although the current efficiency was relatively low, this study offered new possibility of using bio-resources as the whole photoactive materials for next generation of photovoltaics and also deepen our recognition of Chl derivatives.

Considering the relative low efficiency for the previous OISC devices, molecular engineering is employed to improve the hole extraction ability of hole transporter. Four kinds of hole transport materials were synthesized with either Chl or porphyrin skeleton together with different substitutes at C^3 position named as ZnChl-3/4 and ZnPor-1/2, respectively.[58] All the HTMs showed suitable energy levels and well-aggregated state for a favorable hole transfer from the sensitizer to Ag. In the meanwhile, it proved that the larger (highest occupied molecular orbital) HOMO energy gap between sensitizer and HTM contributes to a larger photovoltage. As a result, with an improved carrier mobility as maximum as 4.11×10^{-3} $cm^2 \cdot V^{-1} \cdot s^{-1}$ and highest HOMO energy level for ZnChl-3, a better PCE of 0.86% was achieved for this ZnChl-3-based OISC devices with the same working principle as previous one. And it is noted that the fabrication process of this device is without employing of any additives, which is favorable to realize a true "green energy" utilization.

Although the PCE for this whole Chl-based OISC device has improved through optimizing the molecular structure of the Chl HTM, their overall efficiency still needs to be further improved. Previous study found that the final photovoltaic performance is greatly influenced by the intermolecular arrangement.[59,60] And solvents engineering could influence the molecular arrangement. Thus, the aggregation of ZnChl-3 HTM was controlled by solvents engineering. Different solvents (chloroform (CF), chlorobenzene (CB), CF:CB=2:1, CF:CB=1:1, and CF:CB=1:2) were employed as the chosen solvents for ZnChl-3 HTM.[61] The spin-coated ZnChl-3 films prepared by different solvents differ from their absorption ability, film morphology, and carrier mobility. When controlling the CF and CB as the ratio as 2:1, the ZnChl-3 film showed the highest carrier mobility and the most uniform film morphology, which means the ZnChl-3 showed the best aggregation state in such a situation, leading to the most favorable charge transfer from Chl-1 to Ag. Here, the device architecture of this device is FTO/compact TiO_2/mesoporous TiO_2/Chl-1/ZnChl-3/Ag. And the PCE of this system was further enhanced to 2.13% through solvent engineering.

The most surprising thing was that, for the first time, the ZnChl-3 HTM contributes to the photocurrent generation after solvents optimization, which means both Chl sensitizer and HTC can be excited to generate photocurrent in this solar cell, as proved by the IPCE spectra in Figure 8.21c. However, the photocurrent of the device based on CB as solvents showed a major contribution from sensitizer, which is similar to the previous study. And for the device with CB as the solvents for ZnChl-3, the ZnChl-3 only could act as HTL. Furthermore, the heterojunction between the Chl-1 sensitized TiO_2 and HTC may form, thus leading to an enhanced photocurrent generation through analyzing the spectral analysis of the devices and films (Figure 8.21d). The excitons of ZnChl-3 are dissociated at the ZnChl-3 and Chl-1-sensitized TiO_2 interface. And then electrons from ZnChl-3 are transferred to TiO_2 and finally collected by FTO. In the meanwhile, the holes from both Chl-1 and ZnChl-3 are gathered at Ag side.

In order to clear out the excited state dynamics at the Chl-1 sensitizer and ZnChl-3 HTL interface, sub-picosecond time-resolved absorption spectroscopy (TAS) was employed to investigate above-mentioned system.[62] After pumping the Chl-1-sensitized TiO_2 and ZnChl-3 bilayer at both 680 nm and 720 nm, there are two bleaching signals at *soret* and Q_y bands together with an positive absorption signal at 675 nm (Figure 8.22a and b). A charge transfer state between Chl-1sensitizer and ZnChl-3 was observed at 640 nm after exciting at the 680 nm and 720 nm. A fast electron injection process from Chl-1 to TiO_2 was observed followed by the process of radical cation transferring from Chl-1 to ZnChl-3. Such a new charge transfer state was observed when exciting both at 680 nm and 720 nm, indicating that charge dissociation and transfer could take place at the Chl-1 to ZnChl-3 interface.

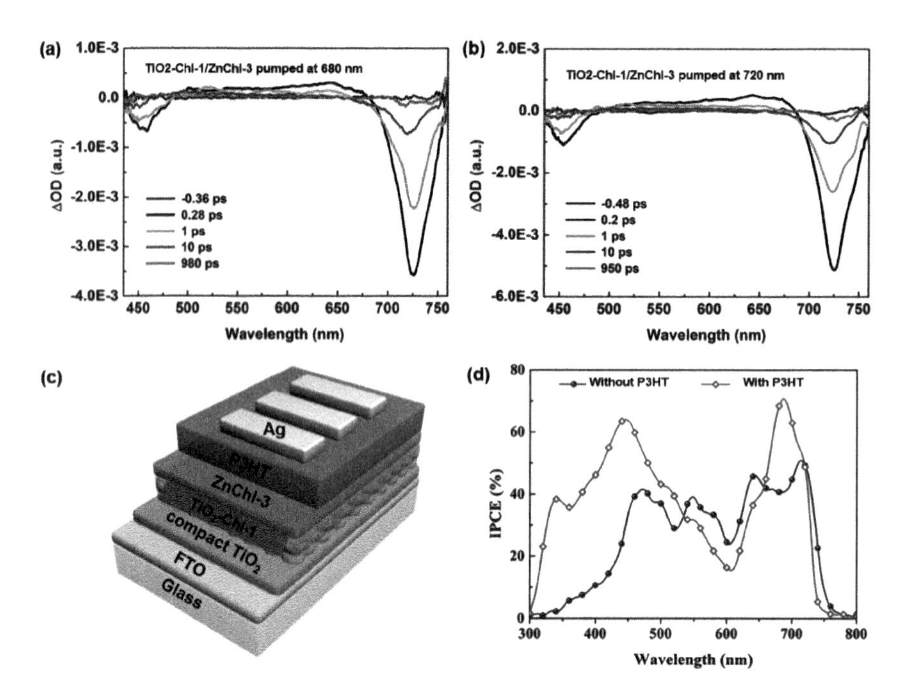

FIGURE 8.22. The TAS of TiO_2-Chl-1/ZnChl-3 films pumped at (a) 680 nm and (b) 720 nm, (c) device architecture of the Chl derivative-based OISC devices with P3HT as HTL, and (d) IPCE spectra of the OISC devices with or without P3HT as HTL. Reprinted with permission from Ref. 21, Copyright 2019, Royal Society of Chemistry.

Considering the formation of heterojunction between Chl-1 sensitized TiO_2 and ZnChl-3, thus there may be no electron-blocking layer or hole-extracting layer between ZnChl-3 and Ag interface. A traditional polymer P3HT was employed as hole transporter to improve the charge collection efficiency of the whole Chl derivative-based organic–inorganic heterojunction solar cells[63] (see device structure in Figure 8.22c). After adding the P3HT layer, the charge transfer resistance of the device is reduced while the charge recombination resistance is increased, leading to a smoother and more efficient charge transfer and collection from ZnChl-3 to Ag. Therefore, the photocurrents and fill factor of this OISCs device (FTO/compact TiO_2/mesoporous TiO_2/Chl-1/ZnChl-3/P3HT/Ag) were enhanced greatly. In addition, the photon-to-electron conversion in both 300–540 nm and 660–725 nm wavelength regions have enhanced significantly (Figure 8.22d). Therefore, the P3HT enhanced the charge transport ability and suppressed the charge recombination simultaneously, thus achieving a higher efficiency for the OISC device.

Recently, our group has successfully fabricated a kind of bilayer, Chl-based BSCs, to simulate the electron transfer process of the natural Z-scheme oxygenic photosynthesis.[64] However, charge extraction between the planar ZnO and Chl derivative interface might be insufficient, which origins from the relatively small contact area. In order to improve the charge extraction ability and also further mimic photosynthesis systems, we proposed a trilayer Chl-based OISC device by employing TiO_2-sensitized Chl-1 as primary electron acceptor, ZnChl-5 as PSI simulator as Chl-6 as PSII simulator[65] (see molecular structure in Figure 8.20). The device structure we fabricated here is FTO/compact TiO_2/mesoporous TiO_2/Chl-1/Chl-5/Chl-6/Ag (Figure 8.23a). Compared the bilayer Chl based devices with the trilayer Chl based ones, the planar ETL ZnO is replaced by Chl-1-sensitized TiO_2, which owns larger interface contact areas and could capture more complimentary light than previous ZnO. Such an enhanced light absorption ability and charge extraction ability will lead to a higher photocurrent and lower charge recombination.

The possible charge transfer pathway is shown in Figure 8.23b. The TiO_2-sensitized Chl-1, ZnChl-5, and Chl-6 could be photoexcited simultaneously. The photoexcited electrons in ZnChl-5 (PSI simulator) is extracted by Chl-1-sensitized TiO_2 (primary electron acceptor simulator) and collected at FTO side while the holes of Chl-6 (PSII simulator) are gathered at Ag side. The rest holes of ZnChl-5 (PSI simulator) are recombined with the electrons of Chl-6 (PSII simulator), this charge transfer step is similar to the charge transfer of natural Z-scheme photosynthesis. The PCE for such a unique device could reach to 3.21%. Considering that the electron injection efficiency could be further improved by co-sensitizing different sensitizer. In order to get a best electron injection efficiency at Chl-1-sensitized TiO_2 and ZnChl-5 interface, Chl-7 is introduced as co-sensitizer together with Chl-1. After optimizing the ratio of Chl-1 and Chl-7 as 100:1 wt%, the highest PCE of 4.14% was reached by this co-sensitized trilayer Chl-based OISC system, which is due to the decreased charge transfer resistance. And this study shows endless possibility to get a high efficiency for the Chl-based OISC devices by mimicking the natural photosynthesis process.

The above-mentioned series of exploration by using natural Chl derivatives as photoactive materials for OISCs prove that Chl derivatives, as an abundant natural pigment with low costs and non-toxic property together with biodegradable ability, are promising for the next generation photovoltaics. Improving the efficiencies of Chl derivatives-based OISCs with organic–inorganic heterojunction also makes them

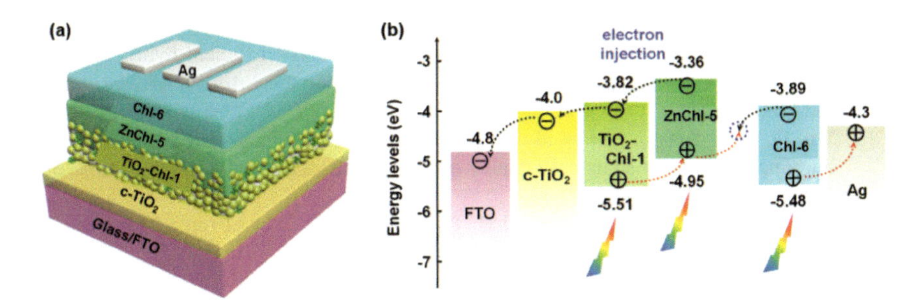

FIGURE 8.23. (a) Device architecture of the trilayer Chl derivative-based OISCs and (b) energy alignments together with possible electron transfer pathway for this natural Z-scheme photosynthesis-simulated OISC devices. Reprinted with permission from Ref. 19, Copyright 2019, American Chemical Society.

charming and unique, thus gathering more and more attentions. We believe in deep that Chl derivative-based OISCs will be further developed and ultimately applied commercially.

REFERENCES

1. Beek, W. J. E., M. M. Wienk, and R. A. J. Janssen. 2004. Efficient hybrid solar cells from zinc oxide nanoparticles and a conjugated polymer. *Advanced Materials* 16 (12): 1009–1013.
2. Bouclé, Johann, Sabina Chyla, Milo S. P. Shaffer, James R. Durrant, Donal D. C. Bradley, and Jenny Nelson. 2008. Hybrid solar cells from a blend of poly(3-hexylthiophene) and ligand-capped TiO_2 nanorods. *Advanced Functional Materials* 18 (4):622–633.
3. Burgi, L., T. J. Richards, R. H. Friend, and H. Sirringhaus. 2003. Close look at charge carrier injection in polymer field-effect transistors. *Journal of Applied Physics* 94 (9):6129–6137.
4. Caricato, A. P., M. Cesaria, G. Gigli, et al. 2012. Poly-(3-hexylthiophene)/6,6-phenyl-C-61-butyric-acid-methyl-ester bilayer deposition by matrix-assisted pulsed laser evaporation for organic photovoltaic applications. *Applied Physics Letters* 100 (7).
5. Choi, Hyung Woo, Kyu-Sung Lee, and T. L. Alford. 2012. Optimization of antireflective zinc oxide nanorod arrays on seedless substrate for bulk-heterojunction organic solar cells. *Applied Physics Letters* 101 (15).
6. Friend, R. H., G. J. Denton, J. J. M. Halls, et al. 1997. Electronic excitations in luminescent conjugated polymers. *Solid State Communications* 102 (2):249–258.
7. Coakley, K. M., and M. D. McGehee. 2003. Photovoltaic cells made from conjugated polymers infiltrated into mesoporous titania. *Applied Physics Letters* 83 (16): 3380–3382.
8. Coffey, D. C., and D. S. Ginger. 2006. Time-resolved electrostatic force microscopy of polymer solar cells. *Nature Materials* 5 (9):735–740.
9. Cozzoli, P. D., A. Kornowski, and H. Weller. 2003. Low-temperature synthesis of soluble and processable organic-capped anatase TiO_2 nanorods. *Journal of the American Chemical Society* 125 (47):14539–14548.
10. Ferber, Jörg, Rolf Stangl, and Joachim Luther. 1998. An electrical model of the dye-sensitized solar cell. *Solar Energy Materials and Solar Cells* 53 (1):29–54.
11. Oosterhout, S. D., M. M. Wienk, S. S. van Bavel, et al. 2009. The effect of three-dimensional morphology on the efficiency of hybrid polymer solar cells. *Nature Materials* 8 (10):818–824.
12. Beek, W. J. E., M. M. Wienk, and R. A. J. Janssen. 2004. Efficient hybrid solar cells from zinc oxide nanoparticles and a conjugated polymer. *Advanced Materials* 16 (12):1009-+.
13. Wu, F., W. J. Yue, Q. Cui, et al. 2012. Performance correlated with device layout and illumination area in solar cells based on polymer and aligned ZnO nanorods. *Solar Energy* 86 (5):1459–1469.
14. Lori E. Greene, Matt Law, Dawud H. Tan, et al. 2005. General route to vertical ZnO nanowire arrays using textured ZnO seeds. *Nano Letters* 5 (7):1231–1236.
15. Jason B. Baxter, and Charles A. Schmuttenmaer. 2006. Conductivity of ZnO nanowires, nanoparticles, and thin films using time-resolved terahertz spectroscopy. *The Journal of Physical Chemistry B* 110:25229–25239.
16. Baoquan Sun, Eike Marx, and Neil C. Greenham. 2003. Photovoltaic devices using blends of branched CdSe nanoparticles and conjugated polymers. *Nano Letters* 3 (7):961–963.

17. Ruankham, Pipat, Supab Choopun, and Takashi Sagawa. 2015. Control of charge dynamics by blending ZnO nanoparticles with poly(3-hexylthiophene) for efficient hybrid ZnO nanorods/polymer solar cells. *Applied Physics A* 121 (1):301–310.
18. Musselman, K. P., S. Albert-Seifried, R. L. Hoye, et al. 2014. Improved exciton dissociation at semiconducting polymer:ZnO donor:acceptor interfaces via nitrogen doping of ZnO. *Advanced Functional Materials* 24 (23):3562–3570.
19. Yun-Yue Lin, Tsung-Hung Chu, Shao-Sian Li, et al. 2009. Interfacial nanostructuring on the performance of polymer/TiO_2 nanorod bulk heterojunction solar cells. *Journal of American Chemical Society* 131:3644–3649.
20. Skompska, Magdalena. 2010. Hybrid conjugated polymer/semiconductor photovoltaic cells. *Synthetic Metals* 160 (1–2):1–15.
21. Istvan Robel, Vaidyanathan Subramanian, Masaru Kuno, and Prashant V. Kamat. 2006. Quantum Dot Solar Cells. Harvesting light energy with CdSe nanocrystals molecularly linked to mesoscopic TiO_2 Films. *Journal of American Chemical Society* 128:2385–2393.
22. Schaller, Richard D., Vladimir M. Agranovich, and Victor I. Klimov. 2005. High-efficiency carrier multiplication through direct photogeneration of multi-excitons via virtual single-exciton states. *Nature Physics* 1 (3):189–194.
23. Hao, Yanzhong, Juan Pei, Yao Wei, et al. 2010. Efficient semiconductor-sensitized solar cells based on poly(3-hexylthiophene)-CdSe-ZnO core-shell nanorod arrays. *The Journal of Physical Chemistry C* 114:8622–8625.
24. Ruankham, Pipat, Lea Macaraig, Takashi Sagawa, Hiroyuki Nakazumi, and Susumu Yoshikawa. 2011. Surface modification of ZnO nanorods with small organic molecular dyes for polymer–inorganic hybrid solar cells. *The Journal of Physical Chemistry C* 115 (48):23809–23816.
25. Chiatzun Goh, Shawn R. Scully, and Michael D. McGehee. 2007. Effects of molecular interface modification in hybrid organic-inorganic photovoltaic cells. *Journal of Applied Physics* 101:114503.
26. Bi, Dongqin, Fan Wu, Qiyun Qu, et al. 2011. Device performance related to amphiphilic modification at charge separation interface in hybrid solar cells with vertically aligned ZnO nanorod arrays. *The Journal of Physical Chemistry C* 115 (9):3745–3752.
27. O'Regan, Brian, and Michael Gratzel. 1991. A low-cost, high-efficiency solar cell based on dye-sensitized colloidal TiO_2 films. *Nature* 353:737–740.
28. Chang, Chia-Hao, Tse-Kai Huang, Yu-Ting Lin, et al. 2008. Improved charge separation and transport efficiency in poly(3-hexylthiophene)–TiO_2 nanorod bulk heterojunction solar cells. *Journal of Materials Chemistry* 18 (19):2201.
29. Dehaudt, J., L. Beouch, S. Peralta, et al. 2011. Facile route to prepare film of poly(3,4-ethylene dioxythiophene)-TiO_2 nanohybrid for solar cell application. *Thin Solid Films* 519 (6):1876–1881.
30. Reeja-Jayan, B., Takuji Adachi, Robert J. Ono, David A. Vanden Bout, Christopher W. Bielawski, and Arumugam Manthiram. 2013. Effect of interfacial dipoles on charge traps in organic–inorganic hybrid solar cells. *Journal of Materials Chemistry A* 1 (10):3258.
31. Reeja-Jayan, B., K. A. Koen, R. J. Ono, D. A. Vanden Bout, C. W. Bielawski, and A. Manthiram. 2015. Oligomeric interface modifiers in hybrid polymer solar cell prototypes investigated by fluorescence voltage spectroscopy. *Physical Chemistry Chemical Physics* 17 (16):10640–10647.
32. Lee, Yi-Huan, Yu-Ping Lee, Chi-Ju Chiang, et al. 2014. In situ fabrication of poly(3-hexylthiophene)/ZnO hybrid nanowires with D/A parallel-lane structure and their application in photovoltaic devices. *Macromolecules* 47 (16):5551–5557.
33. Chiang, Chi-Ju, Yi-Huan Lee, Yu-Ping Lee, et al. 2016. One-step in situ hydrothermal fabrication of D/A poly(3-hexylthiophene)/TiO_2 hybrid nanowires and its application in photovoltaic devices. *Journal of Materials Chemistry A* 4 (3):908–919.

34. Tu, Y. C., H. Lim, C. Y. Chang, J. J. Shyue, and W. F. Su. 2015. Enhancing performance of P3HT:TiO(2) solar cells using doped and surface modified TiO(2) nanorods. *Journal of Colloid and Interface Science* 448:315–319.
35. Neyshtadt, S., J. P. Jahnke, R. J. Messinger, et al. 2011. Understanding and controlling organic-inorganic interfaces in mesostructured hybrid photovoltaic materials. *Journal of the American Chemical Society* 133 (26):10119–10133.
36. Shanmugam, Mariyappan, Tanesh Bansal, Chris A. Durcan, and Bin Yu. 2012. Molybdenum disulphide/titanium dioxide nanocomposite-poly 3-hexylthiophene bulk heterojunction solar cell. *Applied Physics Letters* 100 (15):153901.
37. Pei, Juan, Yan Zhong Hao, Hai Jun Lv, Bao Sun, Ying Pin Li, and Zhi Min Guo. 2016. Optimizing the performance of TiO_2/P3HT hybrid solar cell by effective interfacial modification. *Chemical Physics Letters* 644:127–131.
38. Shankar, K., G. K. Mor, H. E. Prakasam, O. K. Varghese, and C. A. Grimes. 2007. Self-assembled hybrid polymer-TiO_2 nanotube array heterojunction solar cells. *Langmuir* 23 (24):12445–12449.
39. Liu, Y. X., S. R. Scully, M. D. McGehee, et al. 2006. Dependence of band offset and open-circuit voltage on the interfacial interaction between TiO_2 and carboxylated polythiophenes. *Journal of Physical Chemistry B* 110 (7):3257–3261.
40. Grätzel, Michael. 2005. Mesoscopic solar cells for electricity and hydrogen production from sunlight. *Chemistry Letters* 34 (1):8–13.
41. Wang, Z. J., S. C. Qu, X. B. Zeng, et al. 2008. Hybrid bulk heterojunction solar cells from a blend of poly(3-hexylthiophene) and TiO_2 nanotubes. *Applied Surface Science* 255 (5):1916–1920.
42. Shen, Liang, Guohui Zhu, Wenbin Guo, et al. 2008. Performance improvement of TiO_2/P3HT solar cells using CuPc as a sensitizer. *Applied Physics Letters* 92 (7):073307.
43. Moon, S. J., E. Baranoff, S. M. Zakeeruddin, et al. 2011. Enhanced light harvesting in mesoporous TiO_2/P3HT hybrid solar cells using a porphyrin dye. *Chemical Communications (Cambridge)* 47 (29):8244–8246.
44. Greene, L. E., M. Law, B. D. Yuhas, and P. D. Yang. 2007. ZnO-TiO_2 core-shell nanorod/P3HT solar cells. *Journal of Physical Chemistry C* 111 (50):18451–18456.
45. Saputri, Liya Nikmatul Maula Zulfa, Ari Handono Ramelan, Qonita Awliya Hanif, Yesi Ihdina Fityatal Hasanah, Lau Bekti Prajanira, and Sayekti Wahyuningsih. 2016. Optimalization activity of ZnO NR/TiO_2 NR-P3HT as an active layer based on hybrid bulk heterojunction on Dye Sensitized Solar Cell (DSSC). In *3rd International Conference on Advanced Materials Science and Technology*, edited by Sutikno, Khairurrijal, H. Susanto, R. Suryana, K. Triyana, and Markusdiantoro.
46. Mazzarella, Luana, Yen-Hung Lin, Simon Kirner, et al. 2019. Infrared light management using a nanocrystalline silicon oxide interlayer in monolithic perovskite/silicon heterojunction tandem solar cells with efficiency above 25%. *Advanced Energy Materials* 9 (14).
47. Albrecht, Steve, and Bernd Rech. 2017. Perovskite solar cells: On top of commercial photovoltaics. *Nature Energy* 2 (1).
48. Meng, Lingxian, Yamin Zhang, Xiangjian Wan, et al. 2018. Organic and solution-processed tandem solar cells with 17.3% efficiency. *Science* 361:1094–1098.
49. Barber, James. 2009. Photosynthetic energy conversion: natural and artificial. *Chemical Society Reviews* 38:185–196.
50. Kosumi, D., T. Horibe, M. Sugisaki, R. J. Cogdell, and H. Hashimoto. 2016. Photoprotection mechanism of light-harvesting antenna complex from purple bacteria. *The Journal of Physical Chemistry B* 120 (5):951–956.
51. Andreas Kay, Michael Grätzel. 1993. Artificial Photosynthesis. 1. Photosensitization of TiO_2 solar cells with chlorophyll derivatives and related natural porphyrins. *Journal of Physics and Chemistry* 93:6272–6277.

52. Li, Mengzhen, Na Li, Gang Chen, et al. 2019. Perovskite solar cells based on chlorophyll hole transporters: Dependence of aggregation and photovoltaic performance on aliphatic chains at C17-propionate residue. *Dyes and Pigments* 162:763–770.
53. Duan, Shengnan, Qiang Zhou, Chunxiang Dall'Agnese, et al. 2019. Organic solar cells based on the aggregate of synthetic chlorophyll derivative with over 5% efficiency. *Solar RRL.*
54. Duan, Shengnan, Guo Chen, Mengzhen Li, et al. 2017. Near-infrared absorption bacteriochlorophyll derivatives as biomaterial electron donor for organic solar cells. *Journal of Photochemistry and Photobiology A: Chemistry* 347:49–54.
55. Wang, X. F., H. Tamiaki, L. Wang, et al. 2010. Chlorophyll-a derivatives with various hydrocarbon ester groups for efficient dye-sensitized solar cells: static and ultrafast evaluations on electron injection and charge collection processes. *Langmuir* 26 (9):6320–6327.
56. Wang, Xiao-Feng, Li Wang, Naoto Tamai, Osamu Kitao, Hitoshi Tamiaki, and Shin-ichi Sasaki. 2011. Development of solar cells based on synthetic near-infrared absorbing purpurins: Observation of multiple electron injection pathways at cyclic tetrapyrrole–semiconductor interface. *The Journal of Physical Chemistry C* 115 (49):24394–24402.
57. Li, Yue, Shin-ichi Sasaki, Hitoshi Tamiaki, et al. 2015. Zinc chlorophyll aggregates as hole transporters for biocompatible, natural-photosynthesis-inspired solar cells. *Journal of Power Sources* 297:519–524.
58. Li, Yue, Wenjie Zhao, Mengzhen Li, et al. 2017. Chlorophyll-Based Organic-Inorganic Heterojunction Solar Cells. *Chemistry - A European Journal* 23 (45):10886–10892.
59. Wang, J., and Z. Liang. 2016. Synergetic solvent engineering of film nanomorphology to enhance planar perylene diimide-based organic photovoltaics. *ACS Applied Materials & Interfaces* 8 (34):22418–22424.
60. Duan, Sheng-Nan, Chunxiang Dall'Agnese, Haruhiko Ojima, and Xiao-Feng Wang. 2018. Effect of solvent-induced phase separation on performance of carboxylic indoline-based small-molecule organic solar cells. *Dyes and Pigments* 151:110-115.
61. Zhao, Wenjie, Shin-ichi Sasaki, Hitoshi Tamiaki, et al. 2018. Enhancement of performance in chlorophyll-based bulk-heterojunction organic-inorganic solar cells upon aggregate management via solvent engineering. *Organic Electronics* 59:419–426.
62. Zhao, W., L. Wang, L. Pan, et al. 2019. Charge transfer dynamics in chlorophyll-based biosolar cells. *Physical Chemistry Chemical Physics* 21 (40):22563–22568.
63. Zhao, Wenjie, Xiao-Feng Wang, Chunxiang Dall'Agnese, et al. 2019. P-type P3HT interfacial layer induced performance improvement in chlorophyll-based solid-state solar cells. *Journal of Photochemistry and Photobiology A: Chemistry* 371:349–354.
64. Duan, Shengnan, Chunxiang Dall'Agnese, Gang Chen, et al. 2018. Bilayer chlorophyll-based biosolar cells inspired from the Z-scheme process of oxygenic photosynthesis. *ACS Energy Letters* 3 (7):1708–1712.
65. Zhao, Wenjie, Chunxiang Dall'Agnese, Shengnan Duan, et al. 2019. Trilayer chlorophyll-based cascade biosolar cells. *ACS Energy Letters*: 384–389.

9 Nanogenerators Based on Organic–Inorganic Heterojunction Materials

Md Masud Rana, Asif Abdullah Khan, and Dayan Ban

CONTENTS

9.1 INTRODUCTION

Scavenging sustainable power by converting ubiquitously available unutilized energy to usable electrical energy holds promise to meet ever-expanding energy demands as the conventional fossil energy sources are being quickly exhausted. Energy harvesting based on nanotechnology [1] is attracting intensive interest and attention for two primary reasons: (i) the potential to realize self-powered electronics [1–4] as portable devices, sensors, and implantable biomedical devices which typically consume very low electrical power [5] and (ii) the potential to reduce global dependency on energy sources based on fossil fuels [6]. Over the past decades, researchers have been investigating different ambient energy harvesting technologies based on electromagnetic [7], electrostatic [8–9], and piezoelectric methods [10–11]. The contact electrification effect was first recorded as early as 2600 years ago, and the first demonstrated piezoelectric effect was reported by Pierre and Jacques Curie in 1880. Nevertheless, the unprecedented potentialities of these effects in energy harvesting applications had not been fully revealed for a very long time. Following the first demonstration of the PENG in 2006 [12] and the triboelectric (contact-electrification) nanogenerator in 2012 [13], significant efforts have been devoted toward accomplishing a brand new era of self-powered electronics by using organic–inorganic heterojunction nanomaterials to build sophisticated micro/nano systems [14–23].

Triboelectric nanogenerators (TENGs) based on the coupling effect of contact electrification between two different materials and electrostatic induction have emerged as a viable technology to convert ambient mechanical energy into electrical energy. TENGs has numerous advantages, including large power density, high energy conversion efficiency, versatile options for material selection, lightweight, low cost, etc. They have been successfully used as self-powered sensors in wind speed sensing, micro liquid biological and chemical sensing, vibration monitoring, transportation and traffic management, motion tracking, powering biomedical microsystems, among others [24–26]. The contact electrification-induced surface charge density (SCD) is defined as the key figure of merit for TENGs, which originates from the different work functions between two materials. Therefore, increasing surface area by creating nanostructures, designing nanomaterials with high energy-storing capabilities, or having dense surface states are the efficient routes toward a highly-efficient TENG. A state-of-the-art TENG can produce a power density of up to 500 W/m^2 [27]. However, in contact-separation-triggered TENGs, the air breakdown effect can significantly limit the SCD on triboelectric surfaces [28].

On the other hand, PENGs, which have compact and flexible working modes, are a very promising alternative solution. When mechanical stress is applied on a piezoelectric material, the centers of positive and negative charges are separated, thus creating polarization-induced piezo-potential. PENG research is focused on manipulating material structures (porous, nanowires, etc.) by lithography, etching, or others to improve stress distribution profiles or growing materials with very high inherent spontaneous polarization. PENG device performance has been improved by a series of structure-driven techniques, such as adopting nanowires (NWs) [29], aspect ratio tuning, film porosity modulation through a multi-stage etching process [30], cascading multiple devices [31], and reducing charge-screening effects [32]. However, PENGs suffer from comparatively lower output performance than other existing harvesters (e.g., TENGs, electromagnetic generators) and operate preferably in higher frequency regimes.

The energy conversion efficiency and power output can be further improved by combining harvesters of different types. Hybrid NGs integrate different harvesters in a single unit, where several energy sources can be leveraged either simultaneously or individually. This approach provides a continuous supply of power through harvesting renewable and green energy resources and helps maximize energy utilization.

Integration of a PENG and a TENG, or a PENG and a solar cell (SC), or a PENG, a TENG, and a SC yields some of the notable hybrid energy harvesters. In this chapter, different organic–inorganic heterojunction-based NGs are introduced. The device working principle, design and fabrication, and some promising applications are discussed in detail.

9.2 FUNDAMENTALS OF NANOGENERATOR

NGs efficiently transform mechanical energy into electrical power/signal, which has broad applications in energy science, environmental protection, wearable electronics, self-powered sensors, medical science, robotics, and artificial intelligence [33]. TENGs are generally based on contact electrification. When two dissimilar materials are brought into contact, electrostatic charges are created on the material surfaces due to the different electron affinities of the materials. When the two materials are subsequently separated, the developed voltage forces the electrons to flow between two electrodes, generating an alternating current in the TENG. When mechanical stress is applied, a piezoelectric material is polarized, creating a piezo-potential. The physics behind NGs can be explained using Maxwell's equations.

Ampere's circuital law with Maxwell's addition is

$$\nabla \times H = J + \frac{\partial D}{\partial t} \tag{9.1}$$

where H is the magnetic field and D is the displacement field.

$$D = \epsilon_o E + P \tag{9.2}$$

Here P is the polarization field and E is the electric field. Therefore, Maxwell's displacement current can be defined as:

$$J_D = \frac{\partial D}{\partial t} = \epsilon_o \frac{\partial E}{\partial t} + \frac{\partial P}{\partial t} \tag{9.3}$$

The first part on the right of Eq. 9.3 gives the birth of electromagnetic waves. The second part relates to the output of the NG. If the SCD of a PENG is σ_p, and there is no external electric field, the displacement current is reduced to [34]

$$\frac{\partial D}{\partial t} = \frac{\partial P}{\partial t} = \frac{\partial \sigma_P}{\partial t} \tag{9.4}$$

Equation 9.4 denotes the observed output current in PENGs. In TENGs, the electrostatic field built by the triboelectric charges (with a SCD of σ_c) drives the electrons to flow through an external load, resulting in an accumulation of free electrons in the electrode $\sigma_1(z, t)$. $\sigma_1(z, t)$ is a function of the gap distance $z(t)$ between the two dielectrics. The corresponding displacement current is [34]

$$\frac{\partial D}{\partial t} = \frac{\partial \sigma_1(z,t)}{\partial t} = \sigma_c \frac{\partial z}{\partial t} \tag{9.5}$$

This is the observed current for TENG. $\partial z / \partial t$ depicts the speed at which two triboelectric layers contact each other. This basic theoretical understanding of NG

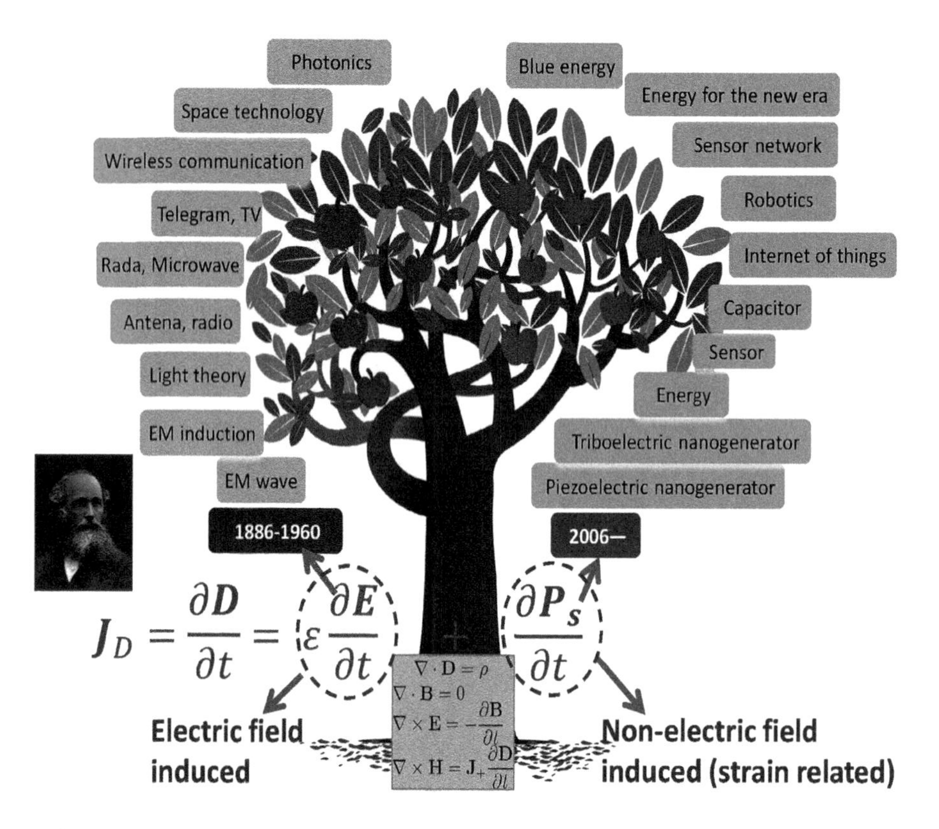

FIGURE 9.1 A tree idea to illustrate Maxwell's displacement current: the first term $\varepsilon\partial E/\partial t$ is responsible for the electromagnetic waves theory, and the second term $\partial \boldsymbol{P}_s/\partial t$ is related to energy and sensor applications, such as NGs. Reproduced with permission [33]. Copyright 2020, Elsevier.

operation is vital to further modeling and analysis, which unveils enormous potentialities for meeting future energy demands as shown in Figure 9.1.

9.3 PIEZOELECTRIC NANOGENERATORS BASED ON ORGANIC–INORGANIC HYBRID NANOMATERIAL

9.3.1 Basic Concept of PENGs and Its Operating Principle

Originating from Maxwell's displacement current, the PENG concept was coined by Prof. Wang in 2006 [12], where an array of zinc oxide (ZnO) nanowires (NWs) grown on a metallic substrate was bent by a conductive atomic force microscope (AFM) cantilever probe, as shown in Figure 9.2.

The tetrahedrally coordinated Zn^{2+} and O^{2-} were accumulated layer-by-layer along the *c*-axis (Figure 9.2a). At its original state, the charge center of the anions and cations coincide with each other. Once an external force is applied, the ZnO NW structure is deformed and stretched on one side while compressed on the other side, which accumulates negative and positive charges on the respective sides. Therefore, the

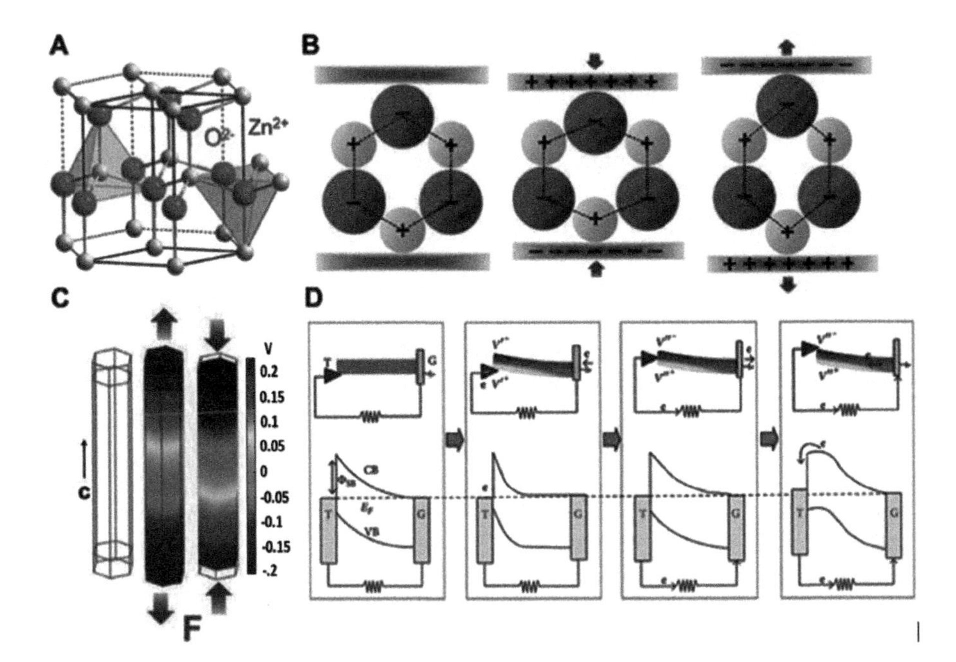

FIGURE 9.2 Mechanism of piezoelectricity. (a) Atomic model of the wurtzite-structured ZnO. (b) Different piezopotential in tension and compression modes of the PENG. (c) Numerical calculation of the piezoelectric potential distribution in a ZnO NW under axial strain. Reproduced with permission [35]. Copyright 2017, Wiley-VCH. (d) Band diagram for the charge outputting and flowing processes in the PENG. Reproduced with permission [36]. Copyright 2017, Wiley-VCH.

negative and positive charge centers are separated and form an electric dipole leading to a piezoelectric potential (Figure 9.2b). If an external load is connected to the deformed material, the free electrons are driven to partially screen the piezoelectric potential and flow through the external circuit to realize a new equilibrium state [35–36]. The resultant piezopotential is observed through the formed Schottky barrier between the AFM probe tip and semiconducting ZnO NW, which forces the electrons to flow between the electrodes, through an external circuit (Figure 9.2c–d).

Since this pioneering work, intensive research efforts were made to enhance output power generation from ZnO NW-based PENGs as it is environmentally friendly, easy to grow at low temperatures, and is self-poled. In contrast, most of the ferroelectrics require high-temperature processing conditions [16].

9.3.2 Material Design Criteria and Techniques for Performance Enhancement

The key points for better device performance have been attributed to the higher aspect ratios of NWs: Schottky barrier formation between the top electrodes and ZnO due to the higher work function of a PdAu electrode than other commonly used metals and reduced charge-screening effects. As the tensile stress and compressive stress induce negative and positive piezopotential, respectively, an approach stems from creating pn-junction ZnO NWs rather than intrinsic ones to reduce local charge-screening effects inside the NWs [37]. However, the lower piezoelectric coefficient (d_{33} ~ 12 pC/N) and the fragile nature of ZnO NWs, nanobelts, or nanorods are limiting their applications as a high-performance renewable power source. On the contrary, polymer materials like polyvinylidene fluoride (PVDF) and polyvinylidene fluoride trifluoro-ethylene (PVDF-TrFE) are promising piezoelectric materials that have higher flexibility, piezoelectric coefficients, and long-term reliability. Altering the microstructures of such piezoelectric films to enhance strain-dependent piezoelectric polarization has proven to be an effective energy-harnessing mechanism. Film porosity modulation through a multi-stage etching process, aspect ratio tuning, and cascading multiple devices are remarkable structure-driven techniques, further pushing the piezoelectricity limit. For example, by using random and highly porous (50%) PVDF structures (through etching process), Mao *et al.* enhanced the output voltage and current of PENG to ~11.1 V and 9.7 μA, respectively, which is higher than lithography-assisted porous PVDF NW array [38]. Recently, Yuan *et al.* presented a cascade-type six-layer rugby-ball-shaped PENG and improved the output performance to 88.62 $V_{P\text{-}P}$ and 353 μA, setting a record value for multilayer PENGs [39]. Although piezoelectricity can be enhanced by these strategies, optimally unifying appropriate mechanical and electrical properties in a single piezoelectric film remains a challenge. Growing organic–inorganic molecular perovskite solution or synthesizing an organic–inorganic perovskite single-crystal has recently achieved record d_{33} coefficients (~185 pC/N) [40] by surpassing their inorganic counterparts, for example lead zirconium titanate (PZT), lead magnesium niobate lead titanate (PMN-PT), and barium titanate (BTO). By dispersing highly piezoelectric nanoparticles (NPs) in a flexible polymer, composite films can be developed, which has been proven as an attractive, easier approach in terms of fabrication scalability, device flexibility, improved mechanical strength, and enhanced electrical output and stability.

9.3.3 InN Nanowire-Based High-Performance PENGs

NW arrays' unique advantages, such as enhanced surface area, relatively high mechanical flexibility, and high sensitivity to small forces, make them an ideal candidate for PENG applications [41–43]. Such NGs produce piezoelectric potential (piezopotential) under external dynamic strain and drive electrons to flow in an external load [44–45]. III-nitrides NWs such as AlN, AlGaN, GaN, and InN are noted for their tunability, direct bandgap, high chemical stability and strong resistance to atmospheric moisture, and their unique piezoelectric property arising from their non-centrosymmetric wurtzite crystal structures [46–47]. The large dislocation density that is often observed in InN planar structures grown on lattice-mismatched substrates can be substantially minimized in InN NW structures. This is attributed to highly efficient strain (and thermal) relaxation in NW lateral surfaces [48–51]. In addition, by adding *p*-dopants into InN NWs, the piezoelectric device performance is significantly improved compared to intrinsic InN NW-based devices. A systematic study on the *p*-type InN NW material and devices for piezoelectric NG applications has been conducted, and the experimental results demonstrate excellent reproducibility and reliability of *p*-type InN NW-based NG devices [52].

Using radiofrequency plasma-assisted molecular beam epitaxy, intrinsic and magnesium (Mg)-doped (*p*-type) InN NWs have grown on Si (111) substrates under nitrogen-rich conditions, respectively, using the InN NW-growth procedure detailed by others [53–54]. To fabricate the NG devices, an insulating layer (polymethyl methacrylate (PMMA), Micro Chem 950k A11) is spin-coated to first encapsulate the NWs. The PMMA layer is then cured at 90 °C for 3 hours. The encapsulation prevents electrical shorting between the NGs' top and the bottom electrodes [55] but does not affect the application of external strain to the NWs due to the PMMA's softness [56]. Two Cu wire leads are glued using silver paste to the bottom high-doped Si substrate and top Au electrode, respectively, so that electrical measurements could be taken. To minimize electromagnetic interference with the measurements, the two copper wires connected to the devices under test are twisted together. All measurements are performed at ambient room temperature.

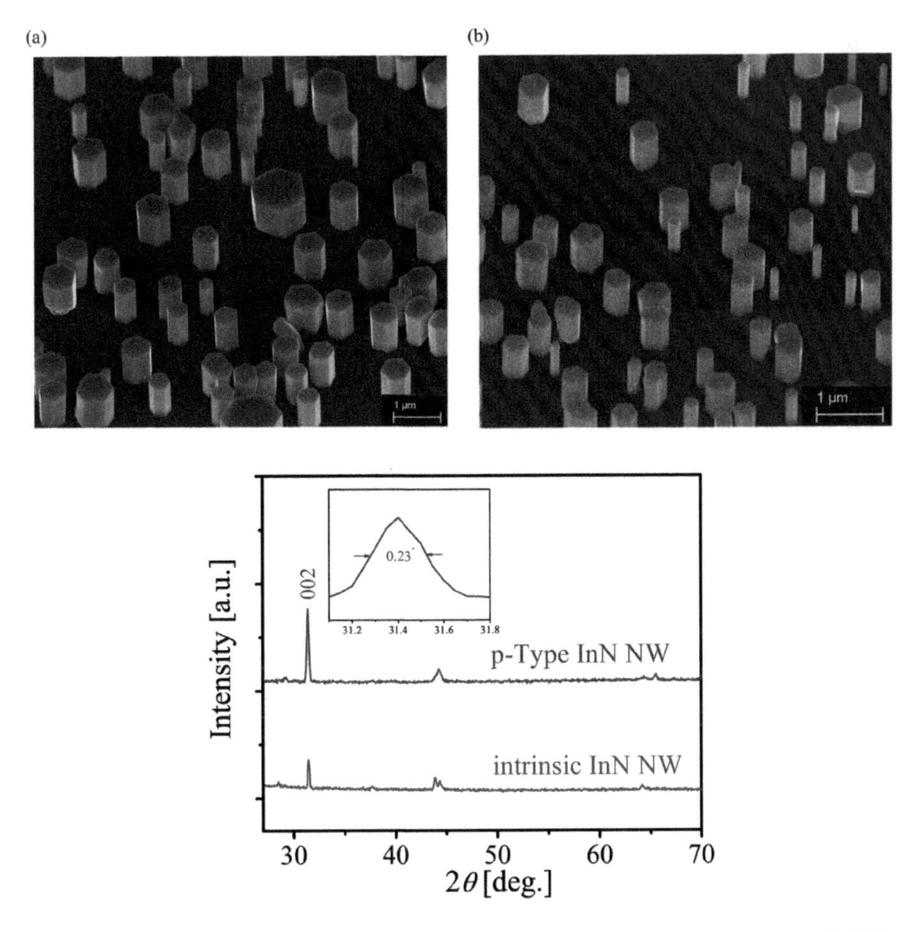

FIGURE 9.3 SEM images of (a) ***p***-type and (b) intrinsic InN NWs have grown on Si (111) substrate, taken at a 45° angle. (c) The 2θ-scan spectra of the XRD intensity for ***p***-type and intrinsic InN NWs, demonstrating the high crystalline quality of the as-grown NWs on the silicon substrate. Reproduced with permission [52]. Copyright 2016, RSC-Pub.

Figure 9.3a–b shows a 45° tilted SEM image of Mg-doped (p-type) InN NWs and intrinsic InN NWs, which have grown on a Si (111) substrate, exhibiting a non-tapered surface morphology with well-defined hexagonal structure [52]. Typically, the NWs have a diameter of approximately 200 ± 10 nm, an area density of 1.6×10^9 cm^{-2}, a length of ~1.0 μm, and are nearly perfectly straight, with smooth surfaces along the c-axis. Figure 9.3c shows the 2θ-scan spectra of XRD measurements. The sharp peak at 31.4° (corresponding to the (002) planes) reveals the high crystalline quality of InN NWs and their c-axis-preferring orientation. The piezoelectric energy harvesting performance of compression/release p-type InN NWs is first investigated using a conductive AFM (C-AFM, Figure 9.4a).

A Pt-coated Si cantilever probe is scanned across an array of as-grown NWs in a contact mode with a scan speed of 10 μm/s and under a constant normal force of 66.7 nN. Both the topography image of the InN NWs on the substrate and the corresponding piezoelectric current signals due to mechanical deformation are detected simultaneously from the scan. One typical result is presented in Figure 9.4b, showing that electric current spikes of more than 200 pA can be observed at the leading edge of each NW along the AFM probe scanning direction (left to right).

The two consecutive electric current spikes on the right hand side should be attributed to two neighboring NWs in a row. The two NWs are in such a close proximity that the line profile of topography cannot resolve them as two individual features. They instead appear as a broader peak with a small shoulder structure on the left side. The good correlation between the location of electric current spikes and the leading edge of NWs indicates that the measured electric current indeed comes from the bending InN NWs dragged by the scanning AFM probe. In summary, all experimental observations confirm that the measured electric current signals on p-InN NWs originate from the material's piezoelectric effect. Figure 9.4c–d are the three-dimensional (3D) images of measured electric current from the p-type and intrinsic InN NWs over a scan area of 10×10 μm^2, respectively. The results show that most of the current spikes from the p-type InN NWs are positive (flowing from NWs to the tip), with a maximum output current of 331 pA. Four p-type InN NWs samples are tested and show similar results. Quantitative analyses reveal that these NWs can produce an average output current density of 84.3 pA over an area of 100 μm^2. The output current is generally governed by [57]:

$$I_s \approx V_s / \left(r_0 + r_c\right) \tag{9.6}$$

where I_S is the output current, V_S is the piezopotential generated by strain, r_0 is the inner resistance of the NW including the surface depletion effect [58], r_c is the contact resistance of the metal-semiconductor interface [59]. A series of C-AFM scans are performed on the p-InN NWs under identical experimental conditions except the force applied to the AFM probe is progressively changed. The measured piezoelectric current is roughly proportional to the applied force as expected (Figure 9.4e). At a force of 83 nN, the mean piezoelectric current from the p-InN NWs is around 93 pA.

The sweeping up/down curves in Figure 9.4f show a clear asymmetric and rectifying behavior, which can be ascribed to the Schottky contact formed between the Pt-coated

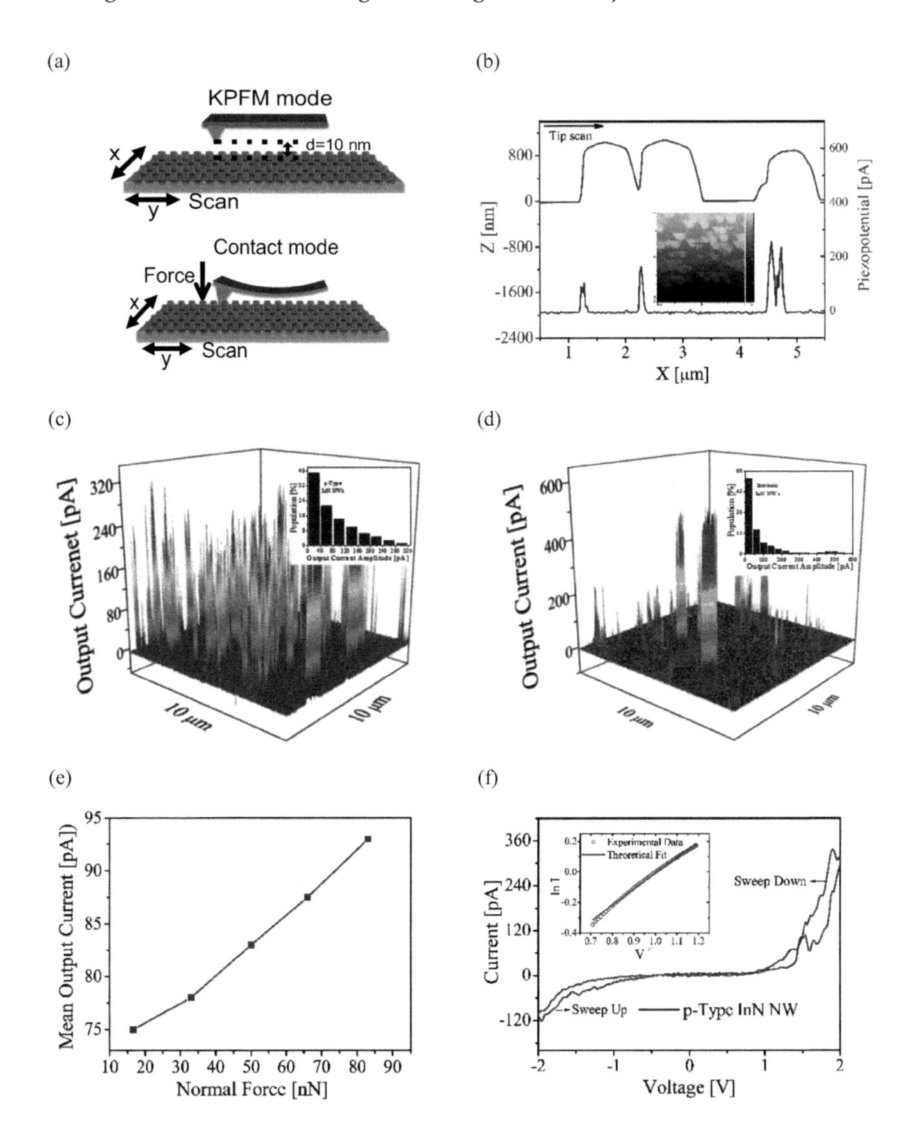

FIGURE 9.4 (a) Schematic illustration of the experimental setup showing the NWs deposited on Si substrate and measured by C-AFM using a Pt/Ir tip. (b) Line profile of topography and output current output; the inset AFM images reveal the surface of Mg-doped InN NWs. (c)–(d) 3D positive and negative current output signals when scanning over an area of $10 \times 10\ \mu m^2$ with a scan speed of 10 μm/s and under a constant normal force of 66.7 nN; the inset is the statistical distribution of the amplitude of output current for ***p***-type and intrinsic InN NWs, (e) Measured current output as a function of applied force under the same experimental conditions. (f) I–V curves measured fitting with thermionic emission–diffusion model (inset). Reproduced with permission [52]. Copyright 2016, RSC-Pub.

AFM tip and the InN NWs. The transportation behavior through a Schottky junction has been well described by the thermionic-emission-diffusion (TED) model [60].

$$E_m = \sqrt{\frac{2qN}{\varepsilon_s}(V + V_{\text{bi}} - \frac{kT}{q}} \tag{9.7}$$

where A is the area of the Schottky barrier, A^{**} is the effective Richardson constant, T is the temperature, ϕ_B is the Schottky barrier height (SBH), k_B is the Boltzmann constant, q is the electron charge, V_f is the voltage drop on the forward-biased Schottky diode, and nn is the ideality factor. In the inset of Figure 9.4f, the experimental I–V curve and the theoretical curve based on Eq. 9.7 are presented for comparison, showing an excellent agreement in the bias range of 0.7–1.2 V. The long-term stability of the InN NW-based NGs is investigated. The p-type NG is tested for 1105 consecutive cycles. A strain is applied at 3 Hz frequency, and the piezoelectric response is recorded. The results show fairly consistent open-circuit voltage and/or short-circuit current output. The device performance exhibits no perceivable degradation due to the tests or storage time over the cycles. The 3D integrated p-type NG shows that the maximum output voltage and current reach 0.055 V and 211 nA at 3 Hz, respectively. Due to the increased area density of active NWs, a substantially reduced contact resistance between the metal electrode and the NW, and the highly energetically stable surface of p-type InN NWs, the p-type NGs outperform the intrinsic InN NW-based NGs by 70% under the same mechanical excitation condition.

9.3.4 1D/2D ZnO Nanostructure-Based PENGs

After the first demonstration of PENGs based on ZnO NW arrays [12], researchers continue to design and fabricate new device structures to improve device performance. Many PENGs are based on ZnO NWs or nanorods because one-dimensional (1D) ZnO nanostructures with a high aspect ratio exhibit better piezoelectric performance. However, 1D ZnO nanostructures suffer from mechanical fragility and instability. On the contrary, two-dimensional (2D) ZnO nanosheets can generate direct current-type piezoelectric output, which is attributed to their buckling behavior and formation of a self-formed anionic nano clay layer. A PENG based on the integration of 1D and 2D ZnO nanostructures on the same substrate was demonstrated for the first time, which was synthesized using a simple, low-temperature, and low-cost hydrothermal method [61]. Figure 9.5a shows the 3D schematic of the PENG, and Figure 9.5b shows the illustration of the fabrication steps. The structure is fabricated on a 1.2 × 1.2 cm^2 shim substrate. The shim substrate is an alloy with the following material compositions: aluminum (99.29%), zinc (0.04%), manganese (0.04%), silicon (0.13%), iron (0.48%), and others (0.02%). First, the substrate is cleaned using acetone, isopropyl alcohol, and deionized water. After that a 400-nm aluminum doped zinc oxide (AZO) layer is deposited on the substrate using plasma-enhanced chemical vapor deposition (PECVD) at 250 °C and 5 mT. The AZO layer serves as the seed layer for the hydrothermal growth of subsequent ZnO nanostructures and

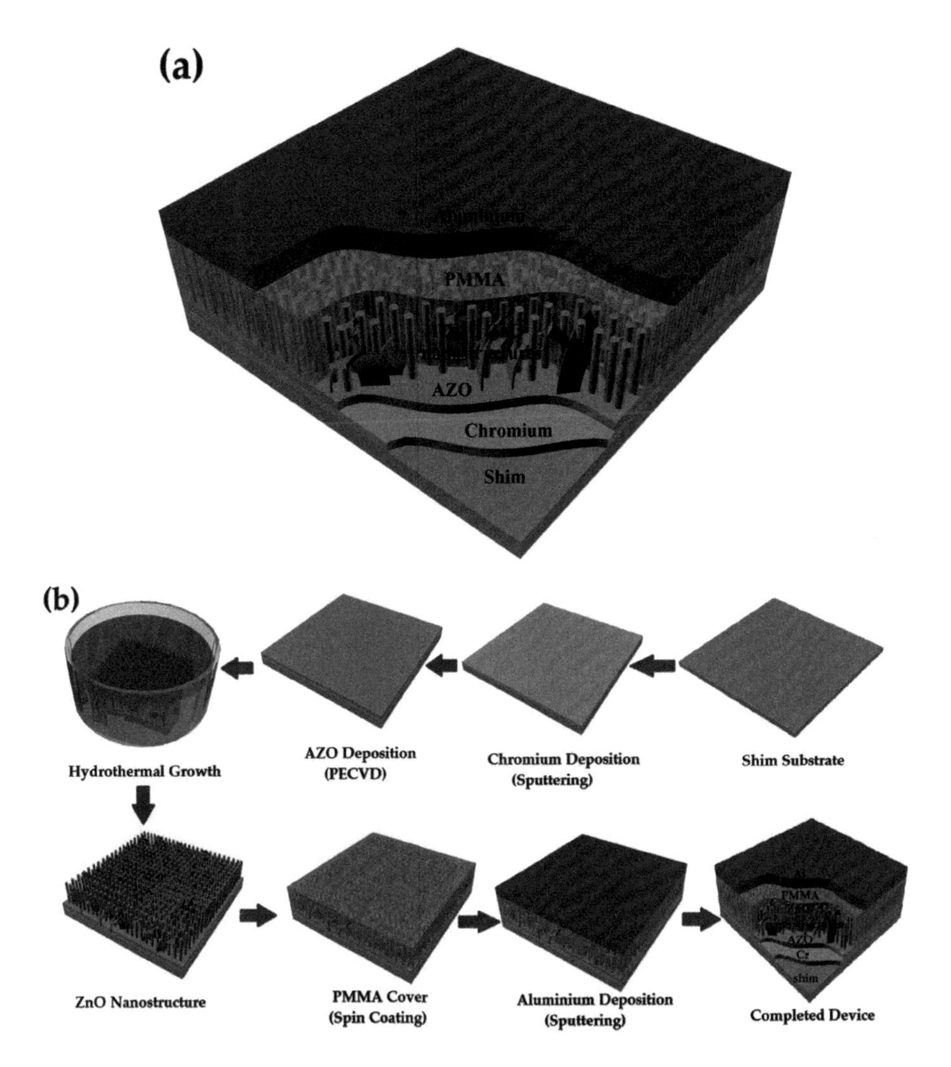

FIGURE 9.5 (a) 3D schematic of the PENG. (b) The step-by-step fabrication process of the PENG. Reproduced with permission [61]. Copyright 2018, Wiley-VCH.

plays a critical role in the growth process. To grow the ZnO nanostructures, the substrate is attached to a pre-cleaned glass substrate and immersed into a mixture of solutions: zinc nitrate hexahydrate (25×10^{-3} M), hexamethylenetetramine (HMTA) (25×10^{-3} M), and aluminum nitrate nonahydrate (25×10^{-3} M) using a substrate holder in such a way that the AZO-deposited layer faces downward to avoid the accumulation of any debris on the substrate during the hydrothermal process [62].

The solution is kept at a constant 88 °C during the hydrothermal growth, which is optimal for the growth of the desired ZnO nanostructures instead of growing nanoballs (below 75 °C) and nanorods (above 95 °C). The structure height is simply controlled by the growth time. In this study, HMTA is used to synthesize the ZnO nanostructures that react with water to produce ammonia, which provides a slow and controlled supply of OH^- anions. Due to the lowest surface energy of (002) facet, the wurtzite ZnO crystal grows preferentially along [001] direction and the growth velocity along ⟨100⟩ directions is slower than that along [001] direction, which leads to the formation of 1D ZnO NWs and nanorods. However, as the substrate is a shim, ZnO crystals with sheet-like morphology are formed and consequently it is reasonable to assume that Al should be responsible for the 2D growth of ZnO and the suppression effect along [001] direction also likely originates from Al. After the hydrothermal growth of ZnO nanostructures, the sample is cleaned using a standard process.

Then a 2% PMMA solution in toluene is spin-coated onto the sample and then cured at a temperature of 120 °C for 3 hours to cover the ZnO nanostructures with a 100 nm PMMA layer. Finally, a 100-nm aluminum layer is deposited on top of the PMMA layer as the top electrode using magnetron sputtering. Then, two copper wires are connected to the top aluminum and the shim substrate, which serve as the external electrodes for the testing of the device. Figure 9.6a–b shows the 45° tilted view SEM image of as-grown ZnO nanostructures at different scales, and the growth is distinctly uniform for a large area. From the magnified image, in Figure 9.6b, it is clear that both 1D and 2D ZnO nanostructures are grown on the substrate.

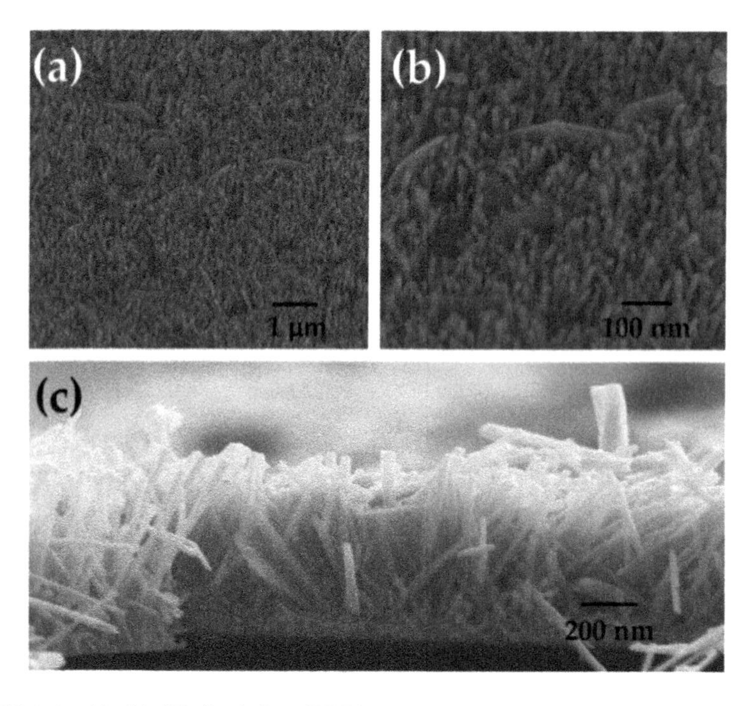

FIGURE 9.6 (a)–(b) 45° tilted view SEM images of the hydrothermally grown ZnO nanostructure that shows the average diameter of the 1D ZnO NWs is ≈70 nm and the thickness of 2D ZnO nanosheets is ≈50 nm. (c) SEM image of the cross-section of the ZnO. Reproduced with permission [61]. Copyright 2018, Wiley-VCH.

The diameter of the 1D ZnO NWs varies between 50 nm and 120 nm with an average diameter of 70 nm. They are closely packed on the substrate surface, which is reasonable due to the thick AZO seed layer that not only facilitates to grow NWs with higher aspect ratio but also increases the NW density. The SEM images also show that the nanoplates are buckled and have a self-assembled interlaced configuration with a thickness of ≈50 nm. Figure 9.6c shows an SEM image of the cross-section of the ZnO nanostructures showing an average height of 1.3 μm. Figure 9.7a–b shows the output open-circuit voltages and short-circuit currents, respectively, at different frequencies when the force is kept fixed at 5 N and the hammer peak-to-peak displacement is 5 mm. Since the strain rate on the device increases as the frequency increases at a constant force, the output piezopotential increases and eventually enhances the output open-circuit voltage and current. When the excitation frequency approaches the resonance frequency, the output gets saturated and beyond the resonant frequency, the output drops quickly. However, as the frequencies of environmental vibration sources are relatively low, this device will not reach its resonant frequency and is capable of producing enhanced piezopotential. Figure 9.7c–d shows the output open-circuit voltages and short-circuit currents at different acceleration

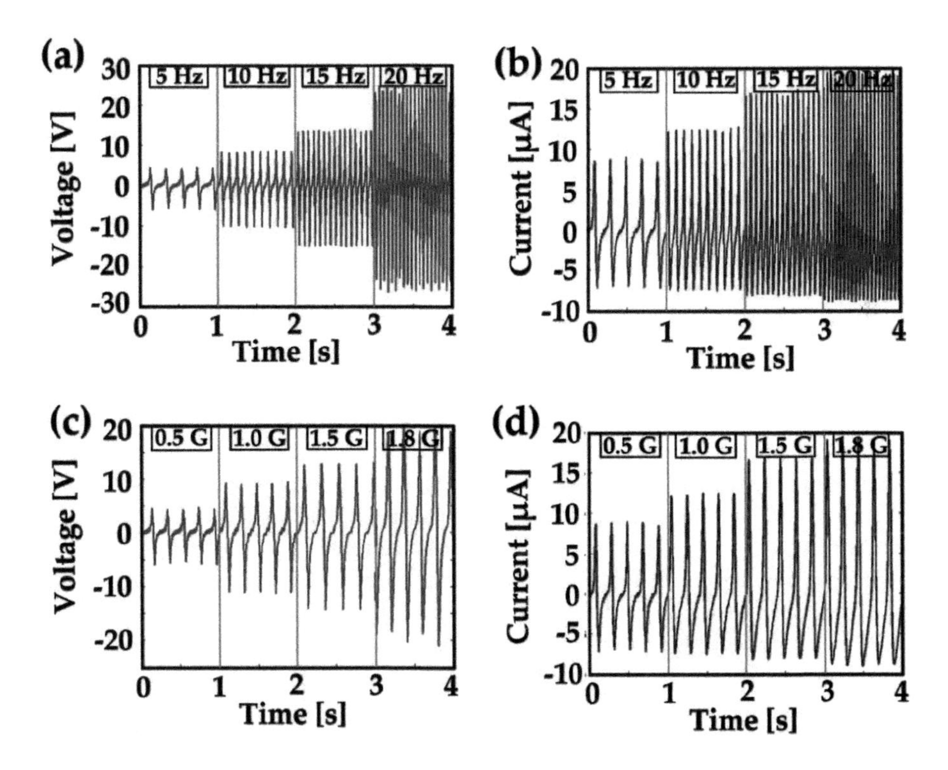

FIGURE 9.7 Output performance of the PENG at different frequency and acceleration (a) Open-circuit voltages and (b) Short-circuit currents at different frequencies. (c) Open-circuit voltages, and (d) Short-circuit currents at different acceleration. Reproduced with permission [61]. Copyright 2018, Wiley-VCH.

levels when the force and the frequency are kept fixed at 5 N and 5 Hz, respectively. The average peak output voltages and currents also increase linearly with varying acceleration, as the impact on the device is increasing proportionally. At a force of 5 N and a frequency of 5 Hz, the average peak-to-peak output open-circuit voltage and short-circuit current reach up to 10.18 V and 15.9 μA, respectively, which allows the device to power up an array of sensors in an aircraft SHM system, making it self-powered.

Even though PENGs exhibit numerous advantages, they are limited by comparatively lower output performance and inefficiency in harvesting vibration energy at low frequencies. TENGs emerge as an alternative promising choice for energy harvesting with complementary properties since triboelectrification is a universal and ubiquitous effect with an abundant choice of materials. In the following section, different organic–inorganic heterojunction-based TENGs will be discussed.

9.4 TRIBOELECTRIC NANOGENERATORS BASED ON ORGANIC–INORGANIC HYBRID NANOMATERIAL

9.4.1 Basic Concept of TENGs and Its Operating Principle

Founded on the omnipresent but detrimental contact-electrification effect TENGs are originally developed for scavenging the mechanical energy from impacts, sliding, and rotations [63–64], producing higher output voltages than other harvesters. Triboelectrification/contact electrification creates static polarized charges on two material surfaces, whereas electrostatic induction on the electrodes by the tribo-charges converts the applied mechanical energy to electrical energy by changing the separating distance and hence creating potential differences [65]. According to the lump circuit model of TENG, it is a variable capacitor-type voltage generator in which the device output is associated with the separation distance between the electrodes (Figure 9.8c). TENGs offer advantages such as more flexible material choices, easy fabrication, lightweight, and low cost. In 2012, Prof. Wang's group first introduced the TENG model in energy-harvesting device fabrication, exhibiting unprecedented energy generation potentialities at low frequency. As demonstrated in Figure 9.8a, the TENG device structure is composed of Au/Kapton/Air gap/PET/Au. Periodic contact-separation between PET and Kapton can generate sufficient output power. From there onward, significant research progress has been made and a maximum output power density of 500 W/m^2 has been reported recenty [27].

Depending on the of the polarization change and electrode configuration, four different operation modes of the TENG have been proposed [13] including vertical contact-separation (CS) mode, lateral-sliding (LS) mode, single-electrode (SE) mode, and freestanding triboelectric-layer (FT) mode, as shown in Figure 9.9. These can scavenge almost all types of mechanical energy from the environment. The vertical CS mode uses relative motion perpendicular to the interface and the potential change between electrodes, and thus external current flow is dictated by the gap distance between material surfaces. The LS mode uses the relative displacement in the direction parallel to the interface, and it can be implemented in a compact package via rotation-induced sliding.

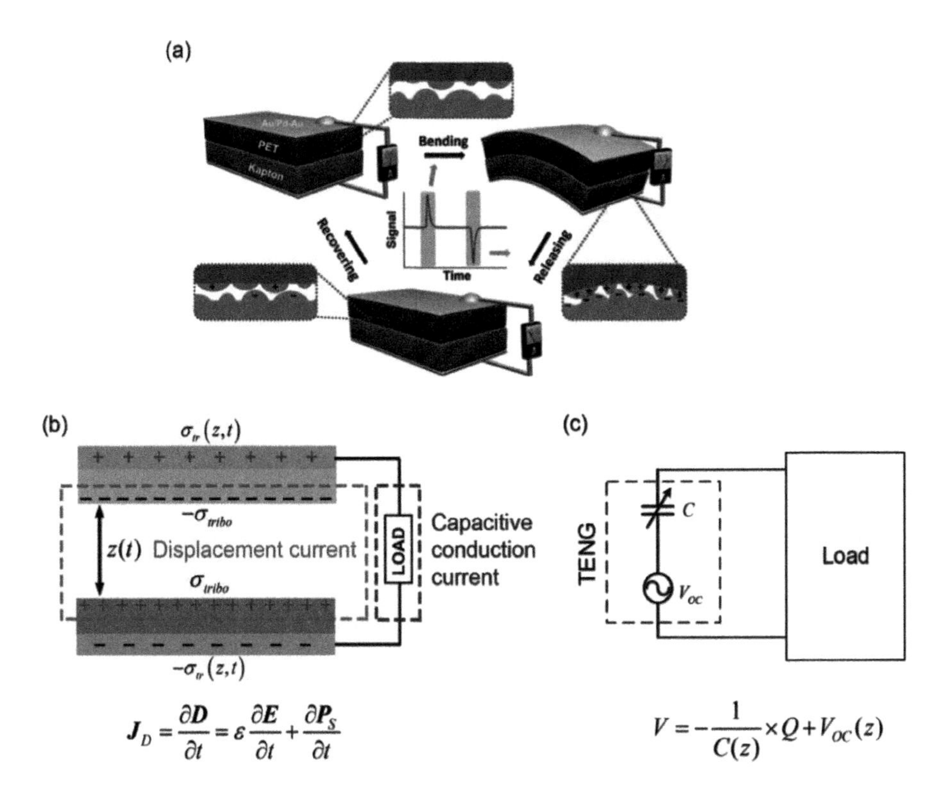

FIGURE 9.8 Theoretical models of TENG. (a) Schematic illustration of the first TENG and its operating cycle. Reproduced with permission [13]. Copyright 2012, Elsevier. (b) The displacement current model of a contact-separation-mode TENG. (c) The equivalent electrical circuit model of TENG. Reproduced with permission [65]. Copyright 2019, Wiley-VCH.

The SE mode takes the ground as the reference electrode and is versatile in harvesting energy from a freely moving object without attaching an electric conductor, such as a hand typing, human walking, and moving transportation. The FT-layer mode is developed upon the SE mode, but instead of using the ground as the reference electrode, it uses a pair of symmetric electrodes and electrical output is induced from asymmetric charge distribution as the freely moving object changes its position [65]. One thing worth noting is that the practical application of TENG is not limited to one single mode but relies more on the conjunction or hybridization of different modes to harness their full advantages.

9.4.2 Material Design Criteria and Techniques for Performance Enhancement

Based on the electron affinity of different materials, the triboelectric series has been created to quantify the figure of merits of TENG design. The displacement current model, the capacitive model, and the figure of merits of TENG all suggest that its

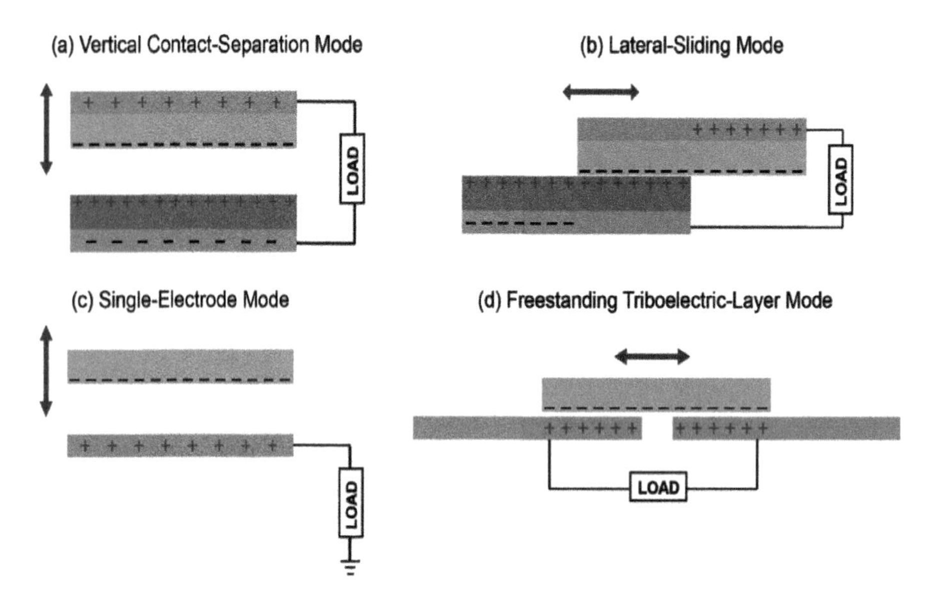

FIGURE 9.9 Four working modes of TENG. (a) Vertical contact-separation (CS) mode. (b) Lateral-sliding (LS) mode. (c) Single-electrode (SE) mode. (d) Freestanding triboelectric layer (FT) mode. Reproduced with permission [65]. Copyright 2019, Wiley-VCH.

output current and voltage are proportional to the triboelectric SCD and that the output power is proportional to the square of the SCD [65]. That is why the underlying mechanism concerning the origin of the triboelectric charge still requires further investigation.

The improvement of SCD can be classified into three major approaches: material composition modification, enhancement of effective contact area, and adjustment of environmental conditions. The material modification strategy can be further divided into chemical surface functionalization and bulk composition manipulation. In chemical surface functionalization, the triboelectric material is modified by changing the functional groups exposed on the surface so that its charge capture capability is enhanced [66–68]. For example, Wang *et al.* demonstrated the use of self-assembled monolayers, thiols, and silanes to modify the surfaces of the conductive material Au and dielectric material SiO2, respectively [67]. The results show that the output of the Au-based TENG is enhanced by the largest scale when the more triboelectrically positive function group, amine ($-NH_2$), is introduced on the Au surface, while its performance deteriorates when the triboelectrically negative group (–Cl) is used. This approach eludes the change in bulk material and their properties and still possesses long-term stability with experimental validation.

Secondly, the SCD can be improved by increasing the surface contact area through surface engineering. The active contact area of two solid materials is generally small due to surface roughness, and thus by simply improving the contact effectiveness, the total amount of triboelectric charges will increase. Some forthright and widespread approaches such as surface texturing and nanostructure preparation can be adopted through lithography-assisted nanofabrication techniques, which elevate the SCD to several times higher for the same material.

The third approach is the control and tuning of environmental conditions such as temperature and pressure. Lu *et al.* showed that the performance of a PTFE-based TENG and its electrical output decreased with increasing temperature in the range of −20 to 20 °C, remained stable from 20 °C to 100 °C, and then dropped subsequently [69]. This temperature-dependent change in output charge density is attributed to the change in material permittivity and temperature-induced surface defects such as surface oxidation or defluorination. Wang *et al.* enhanced the triboelectric charge density of a basic Cu-PTFE-based TENG to a record-high value of 660 μC/m^2 by simply operating the device in a high vacuum to prevent air breakdown, which is the biggest performance-limiting factor of TENG [68–70]. Besides the aforementioned three approaches, high dielectric constant materials such as $BaTiO_3$ and $SrTiO_3$ have been widely used as fillers (P(VDF-TrFE)) as the matrix material for charge-attracting, and high-dielectric barium titanate (BTO) NPs for charge-trapping to enhance TENG performance [71–73]. Adding a carbon nanotube (CNT) charge transport layer between the triboelectric and dielectric material can effectively increase triboelectric charge density by facilitating the charge accumulation process [74].

Therefore, the quest for new device design, material innovation, and underlying fundamental physics investigation by considering the ambient parameters, such as temperature, humidity, ambient pressure, is important to realize an era of self-powered TENG-based micro/nanosystems.

9.4.3 High-Performance TENGs

An all-in-one or multifunctional triboelectric nanogenerator (MTENG), which can simultaneously act as a sensor and as an energy harvester to operate a whole radio-frequency (RF) transmitter and signal processor unit, is developed and integrated with a self-powered wireless sensing and monitoring system [75]. The long-term reliability of the MTENG output and the RF transmission capability are also tested without any interruption for ~38,000 cycles. Each TENG unit consists of a nanostructured Al foil and polytetrafluoroethylene (PTFE) as triboelectrically positive and negative layers, respectively.

For producing a nanostructured PTFE surface, 10-nm gold (Au) is deposited on the PTFE surface by e-beam evaporation and the shadowing effect of the thin Au NPs is employed in the following dry etch process. PTFE nanostructures on the surface are shown in Figure 9.10i–ii.

The self-powered wireless sensing and monitoring system contains two units; the top unit contains the TENG units and the bottom unit contains energy and control circuits. As shown in Figure 9.10a, the mechanical structure is made of three aluminum plates of 6.5 cm × 6.5 cm × 0.5 cm. The device is sandwiched between the top and middle plate, while the bottom unit remains immovable due to the fixed aluminum blocks in order to carry and protect a current rectification unit, an energy management module (EMM), and an RF module. The top TENG unit is used for sensing purposes and the rest of the units are used for harvesting mechanical energy. Then the whole device is encapsulated to the top unit of the spring-assisted structure as illustrated in Figure 9.10b–c.

The working principle of each TENG unit is demonstrated in Figure 9.11a. Herein, at first, the contact between the top Al electrode and the PTFE surface creates positive triboelectric charges on the top electrode and negative charges on the PTFE surface (state i). Then the separation between the top electrode and the PTFE film produces a difference in electric potential between the two electrodes, which drives the flow of free electrons from the bottom Al electrode to the top one (state ii). The current continues until the physical separation reaches the maximum (state iii).

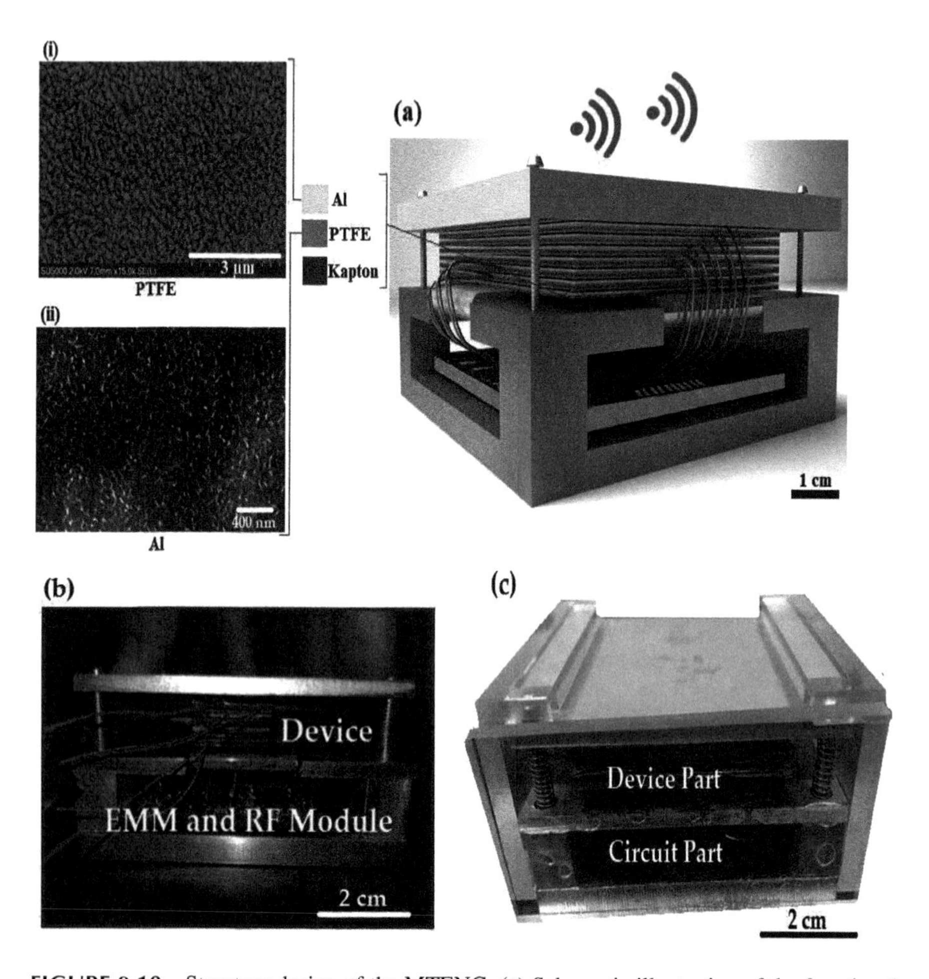

FIGURE 9.10 Structure design of the MTENG. (a) Schematic illustration of the functional components of MTENG, which is mainly composed of a TENG unit and an integrated circuit unit. (i–ii) SEM images of the nanostructured PTFE and Al surfaces. Photographs of (b) an as-fabricated MTENG before encapsulation with the acrylic and (c) an as-fabricated MTENG after encapsulation with the acrylic. Reproduced with permission [75]. Copyright 2019, Elsevier.

When the top Al electrode and the PTFE surface get closer to each other, the free electrons flow from the top electrode back to the bottom one, thus generating a reverse current (state iv). As shown in Figure 9.11b the peak-to-peak output voltage from the top TENG unit is ~700 V and the maximum peak output voltage reaches ~400 V. To theoretically validate the result, finite element simulations are performed using COMSOL (Figure 9.11c). Based on the electron affinity of PTFE (−190 nC/J) and the applied mechanical force (7 N corresponding to the potential energy of ~0.035 J), a maximum surface charge density (MSCD) ~6.65 $\mu C/m^2$ is expected.

The MTENG device exhibits a peak output voltage of ~400 V, corresponding to a SCD of ~3.75 $\mu C/m^2$, which is ~56% of the theoretical MSCD (~6.65 $\mu C/m^2$). It was previously reported that triboelectric materials cannot attain the MSCD due to the limitations imposed by air breakdown, thermal fluctuations, and humidity in the environment [68].

Then the output current from the device is measured by connecting all the TENG units in parallel, and after rectification, the average output current reaches ~300 μA with normal hand pressing (Figure 9.11d). It can be seen from the output current signal that the rectified output current displays a higher peak followed by a lower peak in each cycle. The higher peak is from the pressing motion while the lower one is from the releasing motion. Subsequently, different resistors are used to investigate the reliance of the output electric power of MTENG on the external load. The corresponding instantaneous output power as a function of the load resistance ($P = I^2R$) is presented in Figure 9.11e. The maximum output power of ~10 mW and the corresponding power density of ~4 W/m^2 are achieved at a load resistance of 1 MΩ and with a hand-tapping frequency of ~5 Hz, which is sufficient for sustainably powering up the whole RF module. The decrease in the matched resistance of the TENG with eight units compared to the TENG with a single unit is attributed to the increase in total capacitance, according to the matched resistance expression of $1/\omega C$ [76–78].

The effect of the vibration frequency of the linear shaker on the output performance of the MTENG is also investigated. An iron mass of 0.5 kg is attached to the spring-supported top plate of the MTENG and the combined output current is measured with a constant acceleration of 1 G. With 5-mm peak-to-peak vibration displacement from the linear shaker at 10 Hz, the combined output current from the devices is ~30 μA. The output current drops as the frequency increases from 10 Hz to 60 Hz (Figure 9.11f). The displacement profile of the linear shaker with the same acceleration condition is shown in the inset (i) of Figure 9.11f. The correlation between the short-circuit current and the displacement implies that the amplitude of vibration plays a critical role in TENG output performance.

Through the innovative structural design of the MTENG and the correlation of output with the ambient vibration frequency, it can be utilized as a vibration sensor and an energy harvester unit. The EMM unit collects, stores, and manages the generated electrical energy. An RF wireless module is then powered to transmit the vibration signal to multiple receivers simultaneously, which holds many promising applications, especially in structural health monitoring, automobile engine vibration monitoring, and biomechanical applications.

Though TENGs have higher output energy compared to PENGs, the harvested energy by TENGs may be insufficient for some high-power applications under

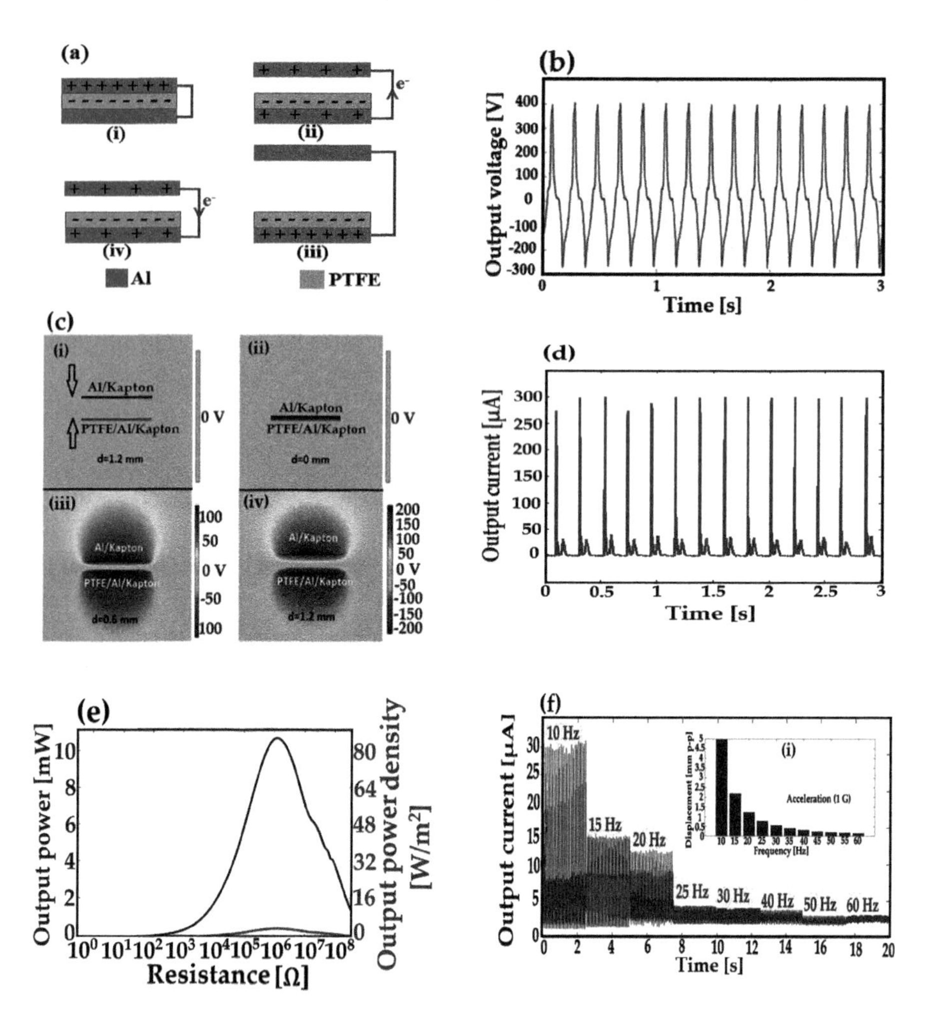

FIGURE 9.11 Theoretical simulation and output performance of the multifunctional TENG (a) Schematic diagram showing the working principle of the MTENG. (b) Simulated potential distribution of the MTENG at four different displacement conditions (i–iv) by COMSOL software. (c)–(d) Measured output voltage and rectified short-circuit current of the MTENG with a frequency of ~5 Hz. (e) The measured output power of the MTENG with a frequency of ~5 Hz and applied force of ~7 N. (f) Comparison of the rectified output current at different frequency excitation of a linear motor (Inset (i) showing the displacement variation of the linear motor with different frequencies. Reproduced with permission [75]. Copyright 2019, Elsevier.

certain scenarios. The output power of the state-of-art energy harvesters based on a single mechanism (i.e., piezoelectric or triboelectric) is still quite low, limiting the range of their applications. Therefore, it is highly desirable to employ multiple mechanisms not only to miniaturize NGs but also at the same time to obtain high output performance for powering the devices. Some such examples are PENG and SC-based hybrid energy cells, PENG and biochemical cell-based hybrid cells, PENG/TENG-based hybrid cells, and a combination of all. In the following section, different combinations of hybrid NGs will be discussed.

9.5 HYBRID NANOGENERATORS BASED ON ORGANIC–INORGANIC HYBRID NANOMATERIAL

9.5.1 Basic Concept of HNGs and Its Operating Principle

Hybrid NGs integrate different types of harvesters in a single unit, where several energy sources can be leveraged either simultaneously or individually, making it possible to use whatever energy is avaliable at the time. This approach provides a continuous supply of power through renewable and green energy resources and helps maximize energy utilization to achieve a stable electrical output. Within a hybrid cell, constituent units can be connected in series or in parallel to enhance the voltage or current, respectively.

Extending from the basic understanding of NG operation from Maxwell's equations, a specific theoretical model for hybrid NGs was developed in 2017. Song *et al.* [79] presented the first theoretical model for piezo-tribo-based hybrid NGs, and later extended the analytical approach based on the following three conditions: (a) PENG and TENG are separate units, (b) tribo-charges are uniformly distributed on the electrode, and (c) electric potentials are regarded as a closed loop in a circuit. Considering the above three conditions, it is possible to write from Figure 9.12b that [80]

$$V_a + V_{PT} + V_1(t) = 0 \; and \, V_{PV} + V_2(t) = 0 \tag{9.8}$$

where V_a, V_{PT} and V_{PV} are the voltages across the air gap, the PTFE layer, and the PVDF layer, respectively, and V_1 and V_2 are the voltages across the external resistors. The test cases for this hybrid NG can be assumed from Figure 9.12 that after applying the external impact from time $t = 0$, the top electrode ED1 makes contact with PTFE at $t = t1$ (after Δt_1 time from $t = 0$), then the whole device starts deforming and reaches its highest deformed stage at $t = t2$ (after Δt_2 time from $t = t1$).

So, after time $t = t2$ the device starts releasing and at $t = t3$ (after Δt_3 time from $t = t2$) the device reaches its same position again like $t = t1$, and finally at $t = t4$ (after

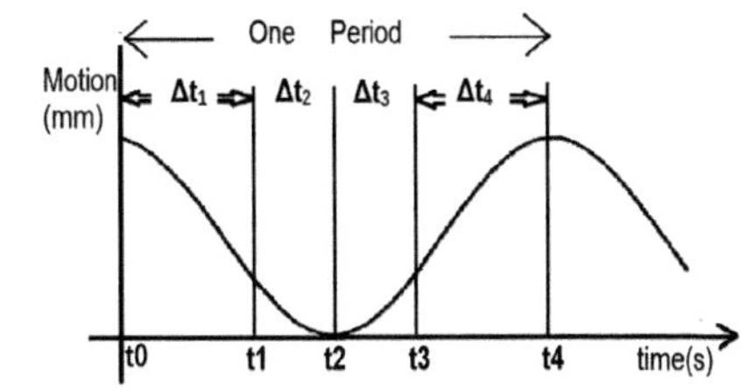

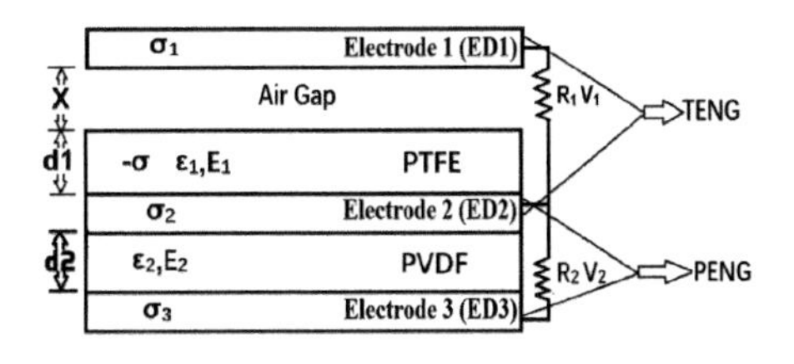

FIGURE 9.12 Theoretical analysis approach for piezo-tribo hybrid NG. (a) Motion characteristics of external load and (b) PENG–TENG hybrid device. The dynamics is used for ED1 and ED2 are $\boldsymbol{x}_1$ and $\boldsymbol{x}_2$ respectively. Reproduced with permission [80]. Copyright 2019, IEEE.

Δt_4 time from $t = t3$) ED1 starts breaking contact with PTFE and reaches the initial position, thus completing one cycle of operation. In this approach, x_1 and x_2 will be used to express the dynamics of ED1 and ED2 (top electrode of PVDF-based PENG). If the transferred charges in ED_1, ED_2, and ED_3 are Q_1, Q_2, and Q_3 respectively, and assuming the initial condition ($Q_1 = Q_3 = 0$ and $Q_2 = \sigma S$) at ($t = t0 = 0$), then the V-Q-x relationship for the top TENG unit can be expressed as [81]

$$R_1 \frac{dQ_1}{dt} + \left(d_{r1} + x_1(t)\right)\frac{Q_1}{S \in_0} - \frac{\sigma}{\in_0} x_1(t) = 0 \tag{9.9}$$

where s is the surface area, d_{r1} is the effective thickness constant for PTFE ($d_{r1} = d/\in_r$), $\in_r$ is relative permittivity, and σ is the SCD. Assuming the initial condition ($Q_1 = Q_3 = 0$ and $Q_2 = \sigma s$) at ($t = t_0 = 0$), Eq. 9.9 can be solved as

$$Q_1(t) = S\sigma - e^{\int_0^t \frac{(d_r + x_1(\tau))}{R_1 S \in_0}} \left(S\sigma + \int_0^t \frac{\sigma x_1(\tau)}{R_1 \in_0} e^{\int_0^\tau \frac{(d_r + x_1(\gamma))}{R_1 S \in_0} d\gamma} d\tau \right) \tag{9.10}$$

This is the charge transfer equation for the TENG unit and is valid within the time range of Δt_1 and Δt_4.

Similarly, by using the same approach and boundary condition, the charge transfer equation for the PENG unit is found as [81]

$$Q_2 = Q_0 e^{-\int_0^t \frac{(d_0 - x_2(\tau))}{R_2 S \in_{r2} \in_0} d\tau} \tag{9.11}$$

It is clear from Figure 9.12a that Eq. 9.11 is valid for $t = 0$ and $t = \Delta t_2 + \Delta t_3$ because the PENG unit operates in that period. From the charge transfer equation for the PENG and TENG units, it is possible to obtain the current and voltage expressions for this hybrid NG. This theoretical approach for hybrid NGs can be used to qualitatively [81] describe the PENG and TENG-based hybrid NG; however, it should be used with caution as it could yield substantial errors when predicting PENG characteristics. Another important limitation is that the output of the two units is simply superimposed in this model, and the potential synergetic effect in hybrid NGs is ignored.

9.5.2 Various Approaches Taken to Design High-Performance HNGs

9.5.2.1 Cascade-Type Hybrid Nanogenerator

Solar [82] and vibration [83] energies are most commonly available in the ambient environment. However, vibrations generate power only while motion persists and solar energy is significant only when optical illumination is sufficient. The nanotechnology-based compact hybrid energy cell (CHEC) can individually and concurrently harvest vibrations and/or solar energies [84–85]. In typical hybrid energy harvesters, the components that scavenge different types of energy are designed and fabricated independently, following distinct physical principles. Due to their different output characteristics, each energy harvesting modality requires its own power conversion and management circuitry. For example, PENGs have large output impedance and can produce high voltage but low current, while SCs have small output impedance, with high current but low voltage [86]. Designing compact cells that can effectively and simultaneously harvest energy from multiple types of sources will increase their applicability and levels of output power.

A cascade-type transparent vibration/solar energy cell synthesized on a polyethylene naphthalate (PEN) flexible substrate is presented [87]. The cascade-type CHEC's monolithically integrated two-terminal structure substantially suppresses the large interfacial electrical losses typically encountered in mechanically stacked devices. Furthermore, integrating the SC on top of the PENG significantly enhances the output power density, through effective, simultaneous, and complementary harvesting of ambient strain and solar energies. The cell consists of a vertically aligned *n-p* ZnO homojunction NW-based NG and a hydrogenated nanocrystalline/amorphous silicon (nc/a-Si:H) n^+-i-p^+ junction SC. The device's full inorganic heterostructure improves chemical stability and mechanical durability. It can function as a sensor, a SC, or a NG.

The ZnO homojunction NWs are grown hydrothermally [88]. A SiN buffer layer and aluminum-doped ZnO (AZO, 2 wt.% Al_2O_3 + 98 wt.% ZnO) layer are deposited onto a pre-cleaned polyethylene naphthalate (PEN) substrate using radio-frequency (RF) magnetron sputtering at 150 °C. To obtain *p*-type ZnO NWs, a doping reagent, lithium (Li) nitrate (75 mM), is added to the solution (heavily *p*-type). Additionally, the *n-n* homojunction NWs are prepared with an intrinsic (effectively *n*-type) NW growth procedure for use as control samples in the experiments [88]. The solar component of the CHECs consists of a stack of n^+-i-p^+ nc/a-Si:H thin-film layers, deposited on top of the synthesized *n-p* and *n-n* homojunction ZnO NWs by PECVD at a substrate growth temperature of 150 °C. To minimize electromagnetic interference, the two copper wires connected to the device under test are twisted together. All measurements are conducted at ambient room temperature.

Figure 9.13a shows a schematic diagram of a fabricated CHEC and its architecture. An equivalent circuit of the CHEC, showing the NG and SC connected in series, appears in Figure 9.13b. The nc/a-Si:H n^+-i-p^+ layers are integrated directly on top of the underlying lithium-doped ZnO NW layer. Two types of ZnO NWs are employed in the device fabrication: ZnO *n-p* homojunction NWs and ZnO *n-n* homojunction NWs. Figure 9.13c shows a photograph of the patterned array of CHECs with varying side lengths (from 1 mm to 1 cm) and insulation separation. This array configuration provides the basis for effectively comparing the output for a range of CHECs. Figure 9.13d shows a cross-sectional helium ion microscope (HIM) image of a typical CHEC and confirms the monolithic and seamless integration between the nc/a-Si:H n^+-i-p^+ layers and the underlying ZnO NW layer. The ZnO NWs are functioning as the piezoelectric material for mechanical energy conversion and as the electron transport layer for photocurrent collection of the SC component. Figure 9.13e shows top-view HIM images of the as-grown *n-p* and *n-n* homojunction ZnO NWs, revealing uniform growth of high-density and vertically aligned NWs. The average length

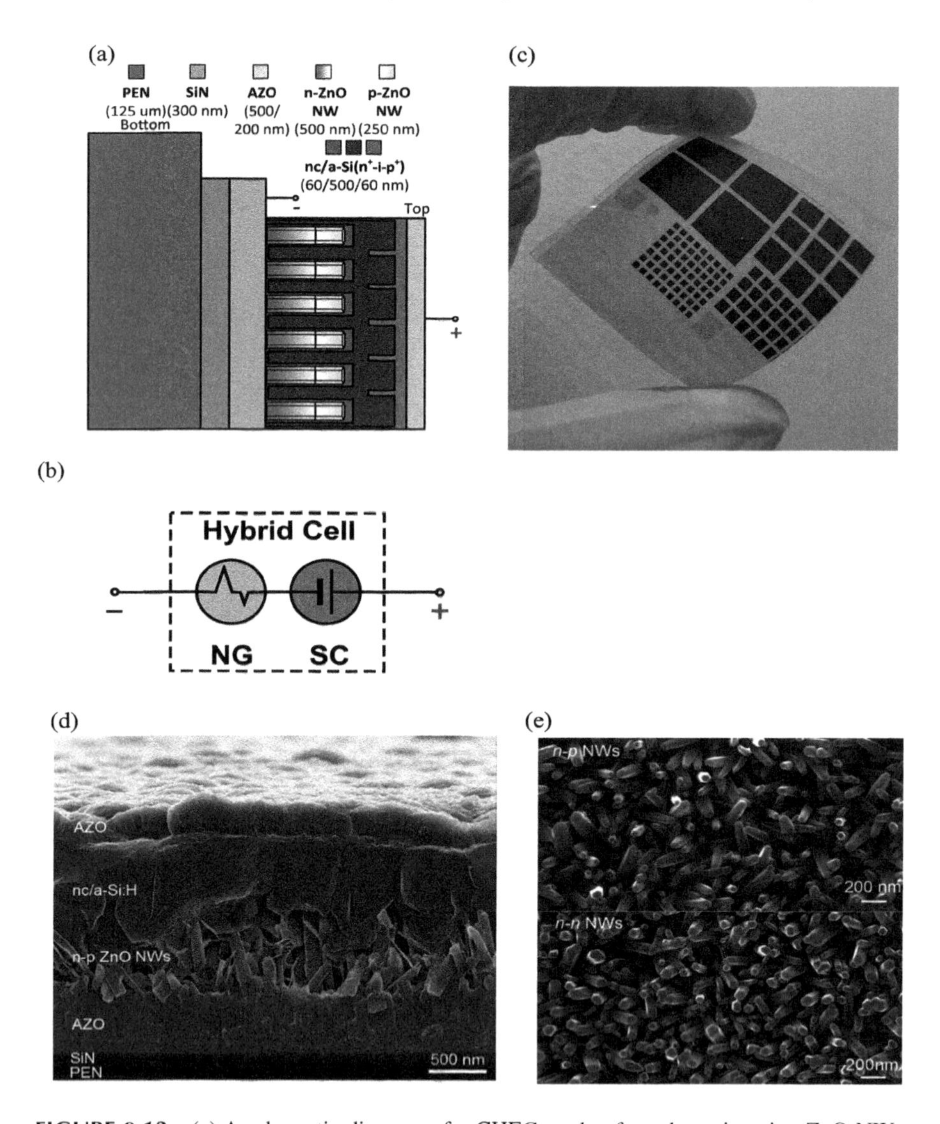

FIGURE 9.13 (a) A schematic diagram of a CHEC made of ***n-p*** homojunction ZnO NWs grown on a flexible substrate (cross-sectional view). (b) A schematic showing an equivalent circuit of the hybrid energy cell. (c) A photograph of patterned CHEC arrays. (d) A cross-sectional helium ion microscopy (HIM) image of a fabricated CHEC. (e) HIM images of the ***n-p*** (top) and ***n-n*** (bottom) homojunction ZnO NW arrays. Reproduced with permission [87]. Copyright 2016, Elsevier.

and diameter of these NWs are ~750 nm and 80 nm, respectively. This monolithic CHEC can exploit piezo-potential under compressive strain and photovoltaic potential under ambient optical illumination, to generate electrical power. The CHECs, when placed solely under optical illumination, function as traditional SCs and produce continuous photocurrent output. The photocurrent flows from the n^+-nc-Si layer to the p^+-nc-Si layer, or from the left (the bottom) to the right (the top), as illustrated in Figure 9.13a. The hybrid energy cell's potential to charge capacitors, power LEDs, and drive wireless sensor nodes is illustrated using the *n-p* CHECs under ~10 mW/cm^2 illumination and an acceleration amplitude of 3 m/s^2 at 3 Hz frequency. Their pulsed voltage output is rectified using a full-wave bridge.

Figure 9.14a shows the charging curves of a 10 μF capacitor charged by a 1 cm-sized CHEC. Under optical illumination only, the capacitor can be charged from 0 V to 0.61 V in less than 0.3 seconds. Voltage remains constant afterward (left inset, Figure 9.14a). Under mechanical excitation only, the voltage across the capacitor increases slowly and almost linearly, reaching ~1.27 V in 580 seconds (right inset, Figure 9.14a).

Under combined optical and mechanical input, the CHEC charges the same capacitor to a voltage of 2.0 V in 920 seconds. The comparison indicates that the hybrid cell can effectively compensate for the lower voltage output of the SC component. To enhance the CHEC's output, six cells are integrated with series to charge a 1000 μF capacitor. The capacitor is then deployed to power eight blue and three white LEDs connected in parallel. The emitted light lasted for 0.5–1.0 seconds and is captured against the background, in Figure 9.14b.

The CHECs' capacity to sustainably drive a wireless sensor node is tested on a commercial EH-LINK wireless sensor (strain gauge) node (LORD Corporation). On this node, the output of six CHECs connected in series is first rectified by the full-wave bridge. The charge is stored in the 1000 μF capacitor. A custom-made full Wheatstone bridge is implemented using four 350 Ω commercial strain gauge sensors (Vishay precision group) (Figure 9.14c) to measure the strain at the instrumented root of a cantilever beam.

The wireless strain sensor node is used to transmit the measured strain signal to a USB base station connected to a computer that acquire and record data. Figure 9.14d shows the recorded strain signals obtained from this experimental setup. The strain in the beam is measured by the wireless sensor node powered by an electronic circuit consisting of the CHECs, the capacitor, and the full-wave bridge. Depending on whether the beam is under mechanical excitation or not, measurable strain signals are recorded (lower graph) or not (upper graph, Figure 9.14d). When the excitation frequency of the beam is set to 3 Hz and the acceleration amplitude to 3 m/s^2, the intermittently measured strain is about 1600 με. These results demonstrate that the CHECs are capable of powering commercial electronics.

This work presents a compact hybrid energy cell (CHEC) made of an inorganic SC monolithically integrated with a ZnO PENG. Employing *n-p* junction-based ZnO NWs in the NG component improves the piezoelectric voltage output of the CHECs by more than two orders of magnitude (138 times). This cascade-type ZnO *n-p* homojunction NW CHEC represents a significant step toward effectively combined energy harvesting from the ambient environment, offering a flexible power supply for self-powered electronics.

9.5.2.2 Organic–Inorganic Hybrid NG

A PENG and a TENG can be integrated to synergistically harvest mechanical energy and convert it into electricity. The NG device performance is therefore further improved.

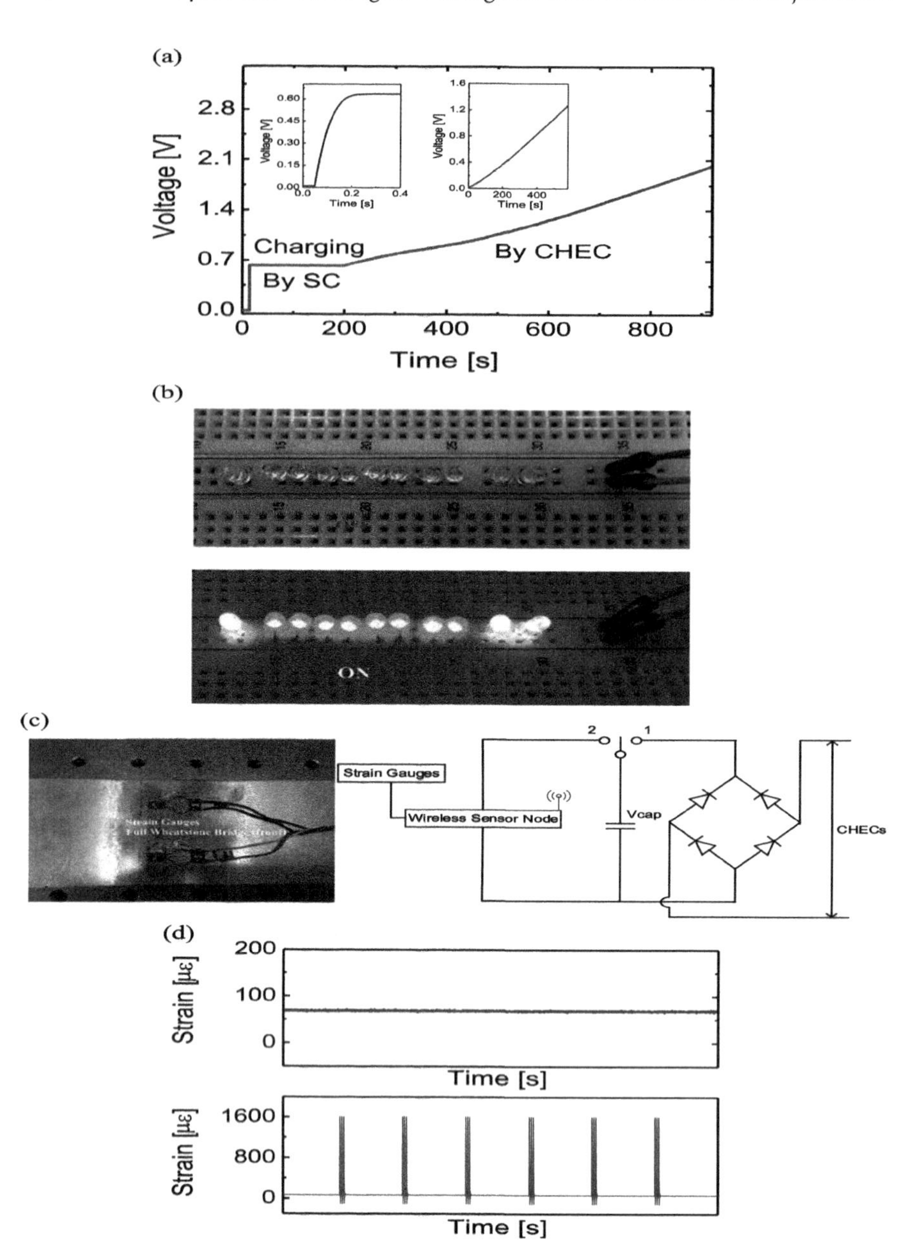

FIGURE 9.14 (a) The charging curves of a 10 μF capacitor being charged by an ***n-p*** individual CHEC. The insets are the curves for the NG and SC components, separately. (b) A photograph of eight blue and three white LEDs before and after being powered by a charged 1000 μF capacitor. (c) A photograph of two commercial strain gauges (the front-side of a Wheatstone bridge) incorporated into the wireless sensor node. (d) The measured strain signals (top) without vibration and (bottom) with vibration from the wireless strain gauge sensor. Reproduced with permission [87]. Copyright 2016, Elsevier.

This synergistic phenomenon was recently demonstrated in a hybrid NG on a shim substrate, which integrates both piezoelectric and triboelectric components based on inorganic *p-n* junction ZnO nanostructures and nanostructured organic PTFE film, respectively [88]. In this design, individual components can be operated independently or concurrently. Moreover, when operated concurrently, component performance is mutually enhanced, enabling more efficient conversion of mechanical energy into electrical energy in a single press-and-release cycle. When triggered with 25 Hz frequency and 1 G acceleration of external force, the PENG component generates a peak-to-peak output voltage of 34.8 V, which is ~3 times higher than its output when it acts alone. Similarly, the TENG component generates a peak-to-peak output voltage of 356 V under the same conditions, which is higher than its initial output of 280 V when acting alone. The NG unit produces an average peak output voltage of 186 V, a current density of 10.02 $\mu A/cm^2$, and an average peak power density of 1.864 mW/cm^2 when operated in the hybrid configuration. The device can even produce an average peak-to-peak voltage of ~160 V from normal hand movement when placed under a wristband fitness tracker, and ~670 V from human walking when placed within a walker's shoe. Figure 9.15 shows the step-by-step fabrication of the hybrid NG.

The fabrication of the entire device can be divided into two components: the PENG component and the TENG component. Figure 9.16a presents a 3D schematic of the curve-shaped TENG integrated on top of a flat-shaped piezoelectric device. The piezoelectric part consists of the Al/*p-n* junction-type ZnO nanostructures covered with PMMA/AZO/Cr/shim substrate from the bottom of the structure, and the

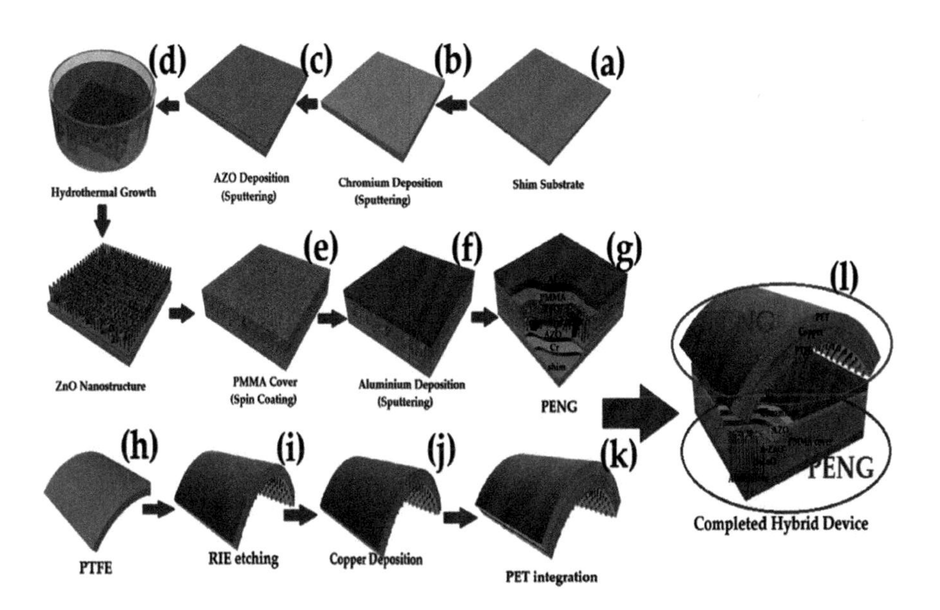

FIGURE 9.15 The fabrication process of the hybrid NG. (a)–(g) Step-by-step progress toward the piezoelectric component of the device. (h)–(k) Fabrication steps for the triboelectric component. (l) Integration of piezoelectric and triboelectric components to form the hybrid structure. Reproduced with permission [88]. Copyright 2019, Elsevier.

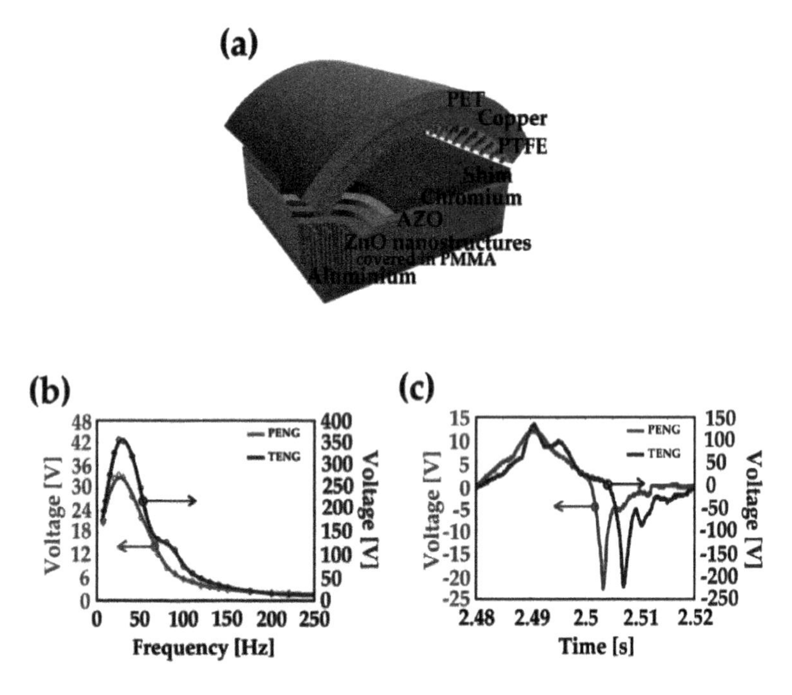

FIGURE 9.16 (a) 3D schematic of the curved-shaped hybrid device. (b) The average peak-to-peak output voltage from the PENG and the TENG components of the device at different frequencies, respectively. (c) The output voltage from the PENG and the TENG components in a single press-and-release cycle, respectively. Reproduced with permission [88]. Copyright 2019, Elsevier.

triboelectric part consists of Polyethylene terephthalate (PET)/Cu/PTFE (with nanostructures on the surface)/nanoporous shim from the top of the structure. Since the *p-n* junction-type ZnO nanostructures have demonstrated better piezoelectric performance compared to pristine ZnO NWs [37,89], a low temperature, cost-efficient, and straight forward hydrothermal method is used to grow the *p-n* junction-type ZnO nanostructures as reported earlier [61,87]. Hence, both the piezoelectric and triboelectric output signals can be leveraged simultaneously within one press-and-release cycle as shown in Figure 9.16c, enhancing the energy conversion efficiency of the device. Before starting the device characterization, that is, output voltage, short-circuit current, output power, charging capability, etc., the optimized operating frequency is determined by sweeping the frequency of the applied mechanical vibration from 1 Hz to 250 Hz as shown in Figure 9.16b. Both of the piezoelectric and triboelectric components yield their highest output voltage at 25Hz, thus subsequent measurements are taken at this frequency unless otherwise mentioned. When the output performance is measured separately, the piezo-tribo hybrid energy harvester unit (2×2.5 cm^2 in dimension) produces piezoelectric and triboelectric peak-to-peak output voltages of ~34.8 V and ~356 V, respectively, as shown in Figure 9.17a–b. However, when the two units are combined in parallel using hybrid operation mode, the unit produces a peak-to-peak output voltage of ~106V as shown in Figure 9.17c, which is higher than that of the

PENG unit alone but lower than that the TENG unit alone. This can be attributed to the mismatch between the internal resistances of the piezoelectric and TENG units. It is well established that when two power sources are connected in parallel, the power source with a lower internal impedance dominates the output voltage [87]. In addition, at the testing frequency, the phase difference between the voltages from the two components results in voltage cancelation and thus degrades the output voltage. To mitigate the foregoing issues, two bridge rectifier units are utilized to collect the electrical signals from the two components separately (Figure 9.17d) as well as their combined output during hybrid operation mode (Figure 9.17e). The full-wave bridge rectifiers do not allow voltage degradation due to the different internal resistances and eliminate the voltage cancellation effect from the phase mismatch [30–31,44].

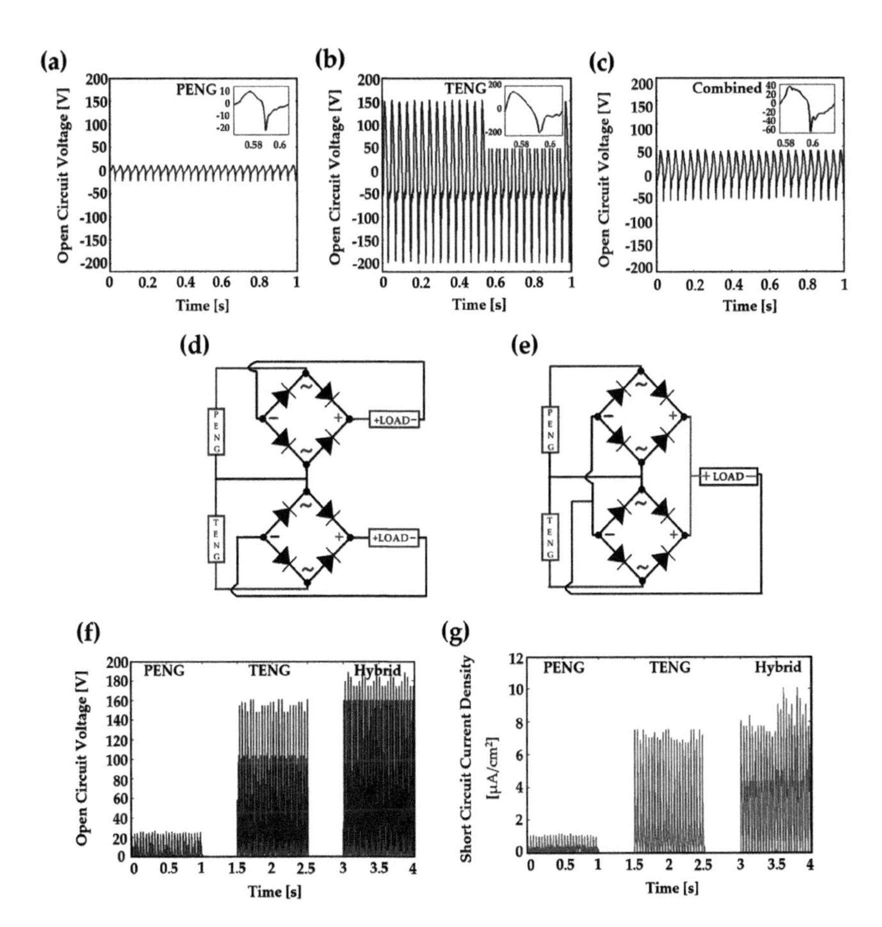

FIGURE 9.17 Output voltage of (a) piezoelectric, (b) triboelectric, and (c) hybrid NG without using rectifiers. (d) Concurrent measurement method for piezo and triboelectric outputs. (e) The piezoelectric and triboelectric outputs are combined in parallel for the hybrid output measurement. Measured piezoelectric, triboelectric, and hybrid output voltages (f) and short-circuit currents (g) using the rectifier circuits. Reproduced with permission [88]. Copyright 2019, Elsevier.

Figure 9.17f–g shows the rectified piezoelectric, triboelectric, and hybrid output voltages and short-circuit currents under a periodic mechanical vibration at 25 Hz frequency, 1 G acceleration, and 5 mm peak-to-peak hammer displacement. Finally, to demonstrate its long-term mechanical stability and reliability, the device is tested over 200,000 cycles over 4 consecutive weeks. The consistent output voltage waveforms from both the TENG and PENG components of the hybrid device, with no perceivable performance degradation over time, clearly demonstrates the long-term stability and robustness of the device. The high output as well as the fast charging characteristics confirm that the hybrid energy harvester can be used as a power source for storing electrical energy from mechanical vibrations in the surroundings, demonstrating great potential in the field of self-powered systems or sensor networks.

9.6 CONCLUSION

This chapter summarizes contemporary studies of organic–inorganic heterojunction nanomaterial-based NGs. Due to the inherent synergistic enrichment of their functional properties and diversity of combinations, organic–inorganic heterojunction nanomaterials hold great potentials for NG applications. NGs based on flexible organic–inorganic nanomaterials can be placed on many surfaces, including bones, human skins, and wearable devices with curved surfaces. Significant efforts have been made by numerous research groups to enhance the performance of organic–inorganic nanomaterial-based NGs through advanced device fabrication, novel device design, and new nanomaterials. For example, PENG–TENG hybrid NGs based on organic–inorganic nanomaterials exhibit one order of magnitude enhancement in peak-to-peak output voltage and current. However, studies on organic–inorganic nanomaterials-based NGs are still in its premature stage. Nevertheless, a giant-shift in the output performance of nanogenerator could be achieved by changing the nano-microscopic morphologies of piezoelectric materials, doping with chemical compounds, and preparing composite films, stabilizing the inorganic NPs by suitable host polymer and choosing an effective electrical poling method. At the same time, unfolding the underlying physics of triboelectrification, conjoining its different working modes, developing environment-specific durable TENG structure design, and overcoming the air-breakdown phenomena are vital areas of future research. Along with innovating advanced energy materials to keep pace with the burgeoning demand for the self-powered wearable and portable electronic devices, focus should also be pointed toward system integration, prototype testing, and product development to reach commercialization. It is anticipated that the NG field will continue its rapid growth in the next decade and some killer applications will play a key role in solving the power source problems for the next generation of nano/micro devices.

ACKNOWLEDGMENTS

The main content of this chapter is based on research that was financially supported by Natural Sciences and Engineering Research Council of Canada (NSERC), Ontario Centers of Excellence (OCE), University of Waterloo, and Shimco North America Inc., Cambridge, Ontario, Canada.

REFERENCES

1. Z. L. Wang, W. Wu, *Angew. Chem. Int. Ed.* 51 (2012) 11700–11721.
2. F. R. Fan, W. Tang, Z. L. Wang, *Adv. Mater.* 28 (2016) 4283–4305.
3. H. Askaria, E. Hashemia, A. Khajepoura, M. B. Khameseea, Z. L. Wang, *Nano Energy* 53 (2018) 1003–1019.
4. Y. Suab, X. Wena, G. Zhua, J. Yang, J. Chen, P. Bai, Z. Wu, Y. Jiang, Z. L. Wang, *Nano Energy* 9 (2014) 186–195.
5. J. A. Paradiso, T. Starner, *IEEE Pervasive Computing*, 4 (2005) 18–27.
6. C. Thomas, M. Greenstone, C. R. Knittel, *J. Econ. Persp.* 30 (2016) 117–138.
7. S. J. Park, S. H. Lee, M. L. Seol, S. B. Jeon, H. Bae, D. Kim, G. H. Cho, Y. K. Choi, *Nano Energy* 55 (2019) 115–122.
8. H. Tian, S. Ma, H. M. Zhao, C. Wu, J. Ge, D. Xie, Y. Yang, T. L. Ren, *Nanoscale* 5 (2013) 8951.
9. P. D. Mitcheson, P. Miao, B. H. Stark, E. M. Yeatman, A. S. Holmes, T. C. Green, *Sens. Actuators A Phys.* 115 (2004) 523.
10. M. P. Lu, J. Song, M. Y. Lu, M. T. Chen, Y. Gao, L. J. Chen, Z. L. Wang, *Nano Lett.* 9 (2009) 1223.
11. M. L. Seol, H. Im, D. I. Moon, J. H. Woo, D. Kim, S. J. Choi, Y. K. Choi, *ACS Nano* 7 (2013) 10773.
12. Z. L. Wang, J. Song, *Science* 312 (2006) 24.
13. F. Fan, Z. Tian, Z. L. Wang, *Nano Energy* 1 (2012) 328.
14. S. R. Anton, H. A. Sodano, *Smart Mater. Struct.* 16 (2007) 1–21.
15. X. Wang, *Nano Energy* 1 (2012) 13–24.
16. J. Briscoe, S. Dunn, *Nano Energy* 14 (2015) 15–29.
17. Z. L. Wang, G. Zhu, Y. Yang, S. Wang, C. Pan, *Mater. Today* 15 (2012) 532–543.
18. Z. L. Wang, J. Chen, L. Lin, *Energy Environ. Sci.* 8 (2015) 2250–2282.
19. R. Hinchet, W. Seung, S. W. Kim, *Chem. Sus. Chem.* 8 (2015) 2327–2344.
20. Z. L. Wang, *Faraday Discuss.* 176 (2014) 447–458.
21. M. Ma, Z. Kang, Q. Liao, Q. Zhang, F. Gao, X. Zhao, Z. Zhang, Y. Zhang, *Nano Res.* 11 (2018) 2951–2969.
22. Z. L. Wang, T. Jiang, L. Xu, *Nano Energy* 39 (2017) 9–23.
23. X. Cao, Y. Jie, N. Wang, Z. L. Wang, *Adv. Energy Mater.* 6 (2016) 1600665.
24. Y. Qi, M. C. Mcalpine, *Energy Environ. Sci.* 3 (2010) 1275.
25. R. S. Yang, Y. Qin, L. M. Dai, Z. L. Wang, *Nat. Nanotechnol.* 4 (2009) 34.
26. M. Y. Choi, D. Choi, M. J. Jin, I. Kim, S. H. Kim, J. Y. Choi, S. Y. Lee, J. M. Kim, S. W. Kim, *Adv. Mater.* 21 (2009) 2185.
27. Z. Zhu, et al., *Adv. Mater.* 26 (2014) 3788.
28. Y. Zi, C. Wu, W. Ding, Z. L. Wang, *Adv. Funct. Mater.* 27 (2017) 1700049.
29. S. Lu, Q. Liao, J. Qi, S. Liu, Y Liu, G. Zhang, Y. Zhang, *Nano Res.* 9 (2016) 372.
30. Y. Su, K. Gupta, Y. Hsiao, R. Wang, C. Liu, *Energy Environ. Sci.* 12 (2019) 410.
31. X. Yuan, X. Gao, J. Yang, X. Shen, Z. Li, S. You, Z. Wang, S. Dong, *Energy Environ. Sci.* 13 (2020) 152.
32. K. Zhang, S. Wang, Y. Yang, *Adv. Energy Mater.* 7 (2017) 1601852.
33. Z. L. Wang, *Nano Energy* 68 (2020) 104272.
34. Z. L. Wang, *Mater. Today* 20 (2017) 74.
35. Q. Zheng, B. Shi, Z. Li, Z. L. Wang, *Adv. Sci.* 4 (2017) 1700029.
36. Z. L. Wang, *Adv. Mater.* 21 (2009) 1311–1315.
37. G. Liu, E. A. Rahman, D. Ban, *J. Appl. Phys.* 118 (2015) 094307.
38. Y. Mao, P. Zhao, G. McConohy, H. Yang, Y. Tong, X. Wang, *Adv. Energy Mater.* 4 (2014) 1301624.
39. Z. Yang, S. Zhou, J. Zu, D. Inman, *Joule* 2 (2018) 642–697.
40. Y. You, W. Liao, D. Zhao, H. Ye, Y. Zhang, Q. Zhou, X. Niu, J. Wan, *Science* 357 (2017) 306.

41. M. T. Todaro, F. Guido, L. Algieri, V. M. Mastronardi, D. Desmaele, G. Epifani, M. D. Vittorio, *IEEE Trans. Nanotechnol.*, 17 (2018) 2.
42. Z. L. Wang, *Nano Today* 5 (2010) 540.
43. A. I. Hochbaum, P. Yang, *Chem. Rev.* 110 (2010) 527–546.
44. T. I. Lee, S. Lee, E. Lee, S. Sohn, Y. Lee, S. Lee, et al., *Adv. Mater.* 25 (2013) 2920–2925.
45. Z. L. Wang, *Adv. Mater.* 24 (2012) 4632–4646.
46. B. Gil, *Oxford University Press*, 18 (2013).
47. X. Wang, J. Song, F. Zhang, C. He, Z. Hu, Z. Wang, *Adv. Mater.* 22 (2010) 2155–2158.
48. V. Polyakov, F. Schwierz, F. Fuchs, J. Furthmüller, F. Bechstedt, *Appl. Phys. Lett.* 94 (2009) 022102.
49. X. Wang, S. Liu, N. Ma, L. Feng, G. Chen, F. Xu, et al., *Appl. Phys. Expr.* 5 (2012) 015502.
50. S. Wang, H. Liu, B. Gao, H. Cai, *Appl Phys. Lett.* 100 (2012) 142105.
51. J. Wu, *J. Appl. Phys.* 106 (2009) 011101.
52. G. Liu, S. Zhao, R. D. E. Henderson, Z. Leonenko, E. A. Rahman, Z. Mi, D. Ban, *Nanoscale* 8 (2016) 2097.
53. S. Zhao, S. Fathololoumi, K. Bevan, D. Liu, M. Kibria, Q. Li, *et al.*, *Nano Lett.* 12 (2012) 2877–2882.
54. S. Zhao, B. Le, D. Liu, X. Liu, M. Kibria, T. Szkopek, et al., *Nano Lett.* 13 (2013) 5509–5513.
55. S. Lee, S. H. Bae, L. Lin, Y. Yang, C. Park, S. W. Kim, et al., *Adv. Funct. Mater.* 23 (2013) 2445–2449.
56. G. Zhu, A. C. Wang, Y. Liu, Y. Zhou, Z. L. Wang, *Nano Lett.* 12 (2012) 3086–3090.
57. J. Liu, P. Fei, J. Zhou, R. Tummala, Z. L. Wang, *Appl. Phys. Lett.* 92 (2008) 173105.
58. B. Simpkins, M. Mastro, C. Eddy Jr, P. Pehrsson, *J. Appl. Phys.* 103 (2008) 104313.
59. C. H. Wang, W. S. Liao, N. J. Ku, Y. C. Li, Y. C. Chen, L. W. Tu, et al., *Small* 10 (2014) 4718–4725.
60. D. K. Schroder, *John Wiley & Sons*, 2006. ISBN: 978-0-471-73906-7.
61. A. Mahmud, A. A. Khan, P. Voss, T. Das, E. Abdel-Rahman, D. Ban, *Adv. Mater. Interfaces* 5 (2018) 1801167.
62. Z. L. Wang, X. Y. Kong, Y. Ding, P. Gao, W. L. Hughes, R. Yang, Y. Zhang, *Adv. Funct. Mater.* 14 (2004) 943–956.
63. S. Wang, L. Lin, Z. L. Wang, *Nano Lett.* 12 (2012) 6339.
64. X. Zhang, M. Han, R. Wang, F. Zhu, Z. Li, W. Wang, H. Zhang, *Nano Lett.* 13 (2013) 1168.
65. C. Wu, A. C. Wang, W. Ding, H. Guo, Z. L. Wang, *Adv. Energy Mater.* 9 (2019) 1802906.
66. W. C. Lin, S. H. Lee, M. Karakachian, B. Y. Yu, Y. Y. Chen, Y. C. Lin, C. H. Kuo, J. J. Shyue, *Phys. Chem. Chem. Phys.* 11 (2009) 6199.
67. S. Wang, Y. Zi, Y. S. Zhou, S. Li, F. Fan, L. Lin, Z. L. Wang, *J. Mater. Chem. A* 4 (2016) 3728.
68. S. Wang, Y. Xie, S. Niu, L. Lin, C. Liu, Y. S. Zhou, Z. L. Wang, *Adv. Mater.* 26 (2014) 6720.
69. C. X. Lu, C. B. Han, G. Q. Gu, J. Chen, Z. W. Yang, T. Jiang, C. He, Z. L. Wang, *Adv. Energy Mater.* 19 (2017) 1700275.
70. J. Wang, C. Wu, Y. Dai, Z. Zhao, A. Wang, T. Zhang, Z. L. Wang, *Nat. Commun.* 8 (2017) 88.
71. X. Chen, K. Parida, J. Wang, J. Xiong, M. F. Lin, J. Shao, P. S. Lee, *ACS Appl. Mater. Interfaces* 9 (2017) 42200.
72. A. Danish, Y. Bin, D. Xiaochao, Y. Hao, Z. Meifang, *Nanotechnology* 28 (2017) 075203.
73. J. Chen, H. Guo, X. He, G. Liu, Y. Xi, H. Shi, C. Hu, *ACS Appl. Mater. Interfaces* 8 (2016) 736.

74. N. Cui, L. Gu, Y. Lei, J. Liu, Y. Qin, X. Ma, Y. Hao, Z. L. Wang, *ACS Nano* 10 (2016) 6131.
75. A. A. Khan, A. Mahmud, S. Zhang, S. Islam, P. Voss, D. Ban, *Nano Energy* 62 (2019) 691–699.
76. T. Jiang, X. Chen, C. B. Han, W. Tang, Z.L. Wang, *Adv. Funct. Mater.* 25 (2015) 2928–2938.
77. X. Liang, T. Jiang, G. Liu, T. Xiao, L. Xu, W. Li, F. Xi, C. Zhang, Z. L. Wang, *Adv. Funct. Mater.* 29 (2019) 1807241.
78. C. Xu, Y. Zi, A. C. Wang, H. Zou, Y. Dai, X. He, P. Wang, Y. C. Wang, P. Feng, D. Li, Z. L. Wang, *Adv. Mater.* 30 (2018) 1706790.
79. C. Song, T. Xiaoming, Z. Wei, Y. Bao, S. Songmin, *Adv. Energy Mater.* 7 (2016) 1614–6832.
80. A. A. Khan, A. Mahmud, D. Ban, *IEEE Trans. Nanotechnol.* 18 (2019) 21.
81. S. Jian, Y. Bao, Z. Wei, P. Zehua, L. Shuping, L. Jun, T. Xiaoming, *Adv. Mater. Technol* 3 (2018) 1800016.
82. M. Law, L. E. Greene, J. C. Johnson, R. Saykally, P. Yang, *Nat. Mater.* 4 (2005) 455.
83. S. Xu, B. J. Hansen, Z. L. Wang, *Nat. Commun.* 1 (2010) 93.
84. C. Xu, Z. L. Wang, *Adv. Mater.* 23 (2011) 873.
85. D. Choi, K. Y. Lee, M. J. Jin, S. G. Ihn, S. Yun, X. Bulliard, W. Choi, S. Y. Lee, S. W. Kim, J. Y. Choi, *Energy Environ. Sci.* 4 (2011) 4607.
86. C. Pan, W. Guo, L. Dong, G. Zhu, Z. L. Wang, *Adv. Mater.* 24 (2012) 3356.
87. G. Liu, N. Mrad, E. A. Rahman, D. Ban, *Nano Energy* 26 (2016) 641–647.
88. A. Mahmud, A. A. Khan, S. Islam, P. Voss, D. Ban, *Nano Energy* 58 (2019) 112–120.
89. K. C. Pradel, W. Wu, Y. Ding, Z. L. Wang, *Nano Lett.* 14 (2014) 6897–6905.

10 Organic–Inorganic Semiconductor Heterojunctions for Hybrid Light-Emitting Diodes

J. Bruckbauer and N. J. Findlay

CONTENTS

10.1 INTRODUCTION

The beginning of the twenty-first century saw a transformation from conventional light sources to solid-state lighting (SSL) [1,2]. Conventional light sources such as incandescent light bulbs, almost unchanged since its invention in the 1800s, and fluorescent tubes are either highly inefficient in converting electricity into light or can often contain toxic materials. In the early 1990s, crucial breakthroughs in the improvements of nitride semiconductors by three Japanese scientists enabled the realization of highly efficient blue light-emitting diodes (LEDs) [3]. This was recognized with the 2014 Nobel Prize in Physics *"for the invention of efficient blue light-emitting diodes which has enabled bright and energy-saving white light sources."* [4]

White light emission using LEDs can be achieved by different methods [2,5]: Combining inorganic LEDs of multiple wavelengths; combining an inorganic LED with a suitable color converter or using purely organic electroluminescent materials. The most commonly used and commercially established approach is the combination of a blue LED and a yellow-emitting phosphor. Issues with phosphors generated a strong drive toward phosphor-free LEDs. These issues include challenging color tuning and rendering, availability of green and red phosphors, self-absorption and use of expensive rare-earth materials [6,7].

Organic LEDs (OLEDs) are made entirely of luminescent organic or organometallic molecules [8]. Their thin-film deposition allows the fabrication of large area and flexible devices. The absorption and emission properties of these organic materials can be directly influenced and tuned by modifying and adjusting their chemical structure [9].

In this chapter, a different approach is presented which utilizes the 'best of both worlds' through combining the ultra-high efficiency of inorganic nitride-based LEDs with the flexibility and low-cost of organic energy-down converters. The first part (Section 10.2) will cover the basic principles of inorganic LEDs and give an introduction to colorimetry, radiometry, and efficiency of light sources. This is followed by a short overview on the chemical design considerations for the organic compound in Section 10.3. In Section 10.4, examples of light-emitting polymers will be given. Finally, the main Section 10.5 illustrates the versatility of small molecules as organic color converters.

10.2 BASIC INTRODUCTION TO WHITE LEDS

10.2.1 III-Nitride Semiconductors and Inorganic LEDs

The III-nitride semiconductor material system consists of the materials like gallium nitride (GaN), indium nitride (InN), and aluminium nitride (AlN) along with their alloys. This material family has gained a lot of interest due to the bandgap tunability of the ternary (e.g., InGaN and AlGaN) and quaternary alloys (e.g., InAlGaN) from the deep ultraviolet (UV) to the near infrared wavelength regions [10,11]. By alloying GaN with InN, it is possible to cover the entire visible spectrum, whereas alloying GaN with AlN allows the bandgap to be shifted into the UV spectral region. Figure 10.1a illustrates the spectral range covered by the III-nitrides showing the room temperature bandgap of wurtzite AlN, GaN, and InN plotted against their

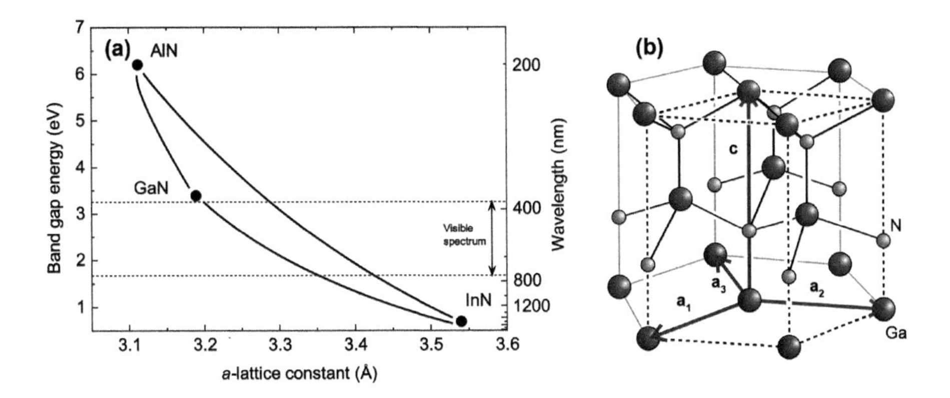

FIGURE 10.1 (a) Bandgap energy versus ***a***-lattice constant of wurtzite AlN, GaN, and InN at room temperature. (b) Thermodynamically stable hexagonal wurtzite structure of GaN.

a-lattice constant [12]. The most common crystal structure for the III-nitrides is the thermodynamically stable hexagonal wurtzite structure as shown in Figure 10.1b, which is defined by the *a*- and *c*-lattice constants [13].

The basic building block of an LED is a p-n junction consisting of a p- and n-type doped semiconductor [14]. In an unbiased p-n junction in thermal equilibrium, electrons originating from donors on the n-side diffuse toward the p-side where they recombine with holes. The same process happens with holes diffusing to the n-side. This creates a depletion region in the vicinity of the junction devoid of free carriers. Ionized donors and acceptors (the dopants), however, stay behind forming a space charge region, which creates an electric field across the junction. This electric field counteracts the diffusion current and leads to a drift current of electrons (holes) from the p-side (n-side) to the n-side (p-side). In thermal equilibrium, both currents cancel each other, meaning there is no total current flowing across the junction. Figure 10.2a shows the bending of the conduction and valance bands caused by the internal electric field, which represents a barrier for free carriers. In the case of a positively biased p-n junction (forward bias), the externally applied electric field opposes the internal field and lowers the barrier created by it. Injected free carriers diffuse across the junction into the region of opposite conductivity and recombine there by emitting light as shown in Figure 10.2b. In the case of a negatively biased p-n junction, the potential barrier is increased and no current flows across the junction.

The area in a p-n junction where recombination occurs is defined by the diffusion length, which is the mean distance a carrier diffuses until recombination. This reduces the electron (n) and hole densities (p) and hence the radiative recombination rate. To make the recombination process more efficient, most modern LEDs possess a quantum well (QW) structure (also referred to as the active region), which is a semiconductor with lower bandgap (well), positioned between the p- and n-type semiconductors of larger bandgaps (barrier) as displayed in Figure 10.2c. In the case of a blue LED, the well layer consists of InGaN with an alloy composition giving a bandgap in the blue region and GaN barrier layers. Here injected carriers are trapped or

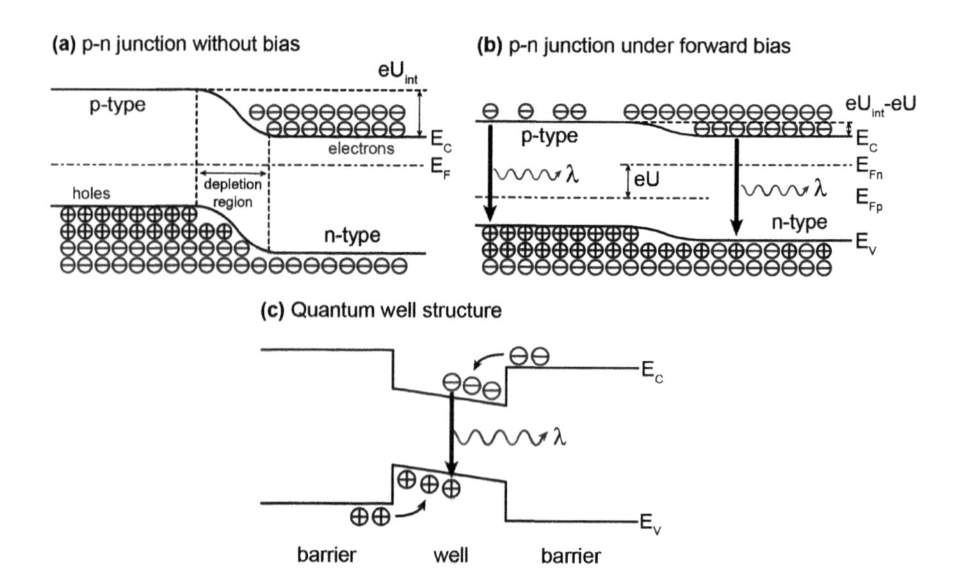

FIGURE 10.2 (a) *p-n* Junction without bias with U_{int} caused by the internal electric field, (b) p-n junction under forward bias *U* and (c) quantum well structure. E_C and E_V denote the conduction and valance bands, respectively.

confined to the well region because of the lower potential. The QW width is generally around a few nanometers and therefore much smaller than the diffusion length, which is 50–60 nm in GaN and could be up to a few tens of micrometer in the case of GaAs [15–18]. The result is a higher carrier density, which is proportional to the radiative recombination rate $R \propto np$ (n and p are the electron and hole densities, respectively) in comparison with a p-n junction where the area is defined by the diffusion length. LEDs commonly consist of multiple wells to further increase confinement, which is referred to as a multiple quantum well (MQW) structure.

10.2.2 Colorimetry, Radiometry, Photometry, and Efficacy

The "science of color," or *colorimetry*, describes the quantification and perception of light by the human eye. Unlike other physical quantities, the perception of light is a subjective quantity, depending strongly on each individual. To standardize the measurement of colors, the *Commission Internationale de l'Eclairage* (CIE, International Commission for Illumination) has introduced the chromaticity diagram [19], which essentially describes the quality of color. It uses three color-matching functions to calculate the chromaticity coordinates x and y, which span the chromaticity diagram. The color-matching functions correspond to the response of the three cone cells in the human eye, which are sensitive in the red, green and blue (RBG) spectral ranges. Additionally, the response of the human eye is taken into account by defining the green color-matching function to be identical to the *eye sensitivity function* $V(\lambda)$ (CIE 1978), which has its maximum at 555 nm [20]. The red and blue color-matching functions are mathematically transformed into a new set with the green color-matching function fixed. Any spectrum can be described by these color-matching functions, with three *tristimulus values X, Y,* and *Z* specifying the contribution of each color-matching function. The chromaticity coordinates *x, y,* and *z* are calculated by normalizing the tristimulus values according to:

$$x = \frac{X}{X+Y+Z} \tag{10.1}$$

$$y = \frac{Y}{X+Y+Z} \tag{10.2}$$

$$z = \frac{Z}{X+Y+Z} = 1 - x - y \tag{10.3}$$

The *z* chromaticity coordinate is redundant, because it can be calculated from the *x* and *y* chromaticity coordinates and does not provide any additional information; therefore, color is defined by a simple (*x, y*) coordinate system.

It was shown that the CIE 1931 chromaticity diagram (*x, y*), Figure 10.3a, possesses small elliptical areas of color, which appear identical to the human eye [21]. Taking these geometrical features into account, the modified CIE 1976 (*u*', *v*') uniform chromaticity diagram was introduced, which is shown in Figure 10.3b [22]. Monochromatic light (as indicated) can be found on the perimeter, whereas white light is located in a region close to the center of the diagram. Every color can be described by chromaticity coordinates, that is, its location in the chromaticity diagram. Although the light from inorganic LEDs is almost monochromatic; it has a spectral linewidth. The coordinates of LEDs, therefore, can be found in close

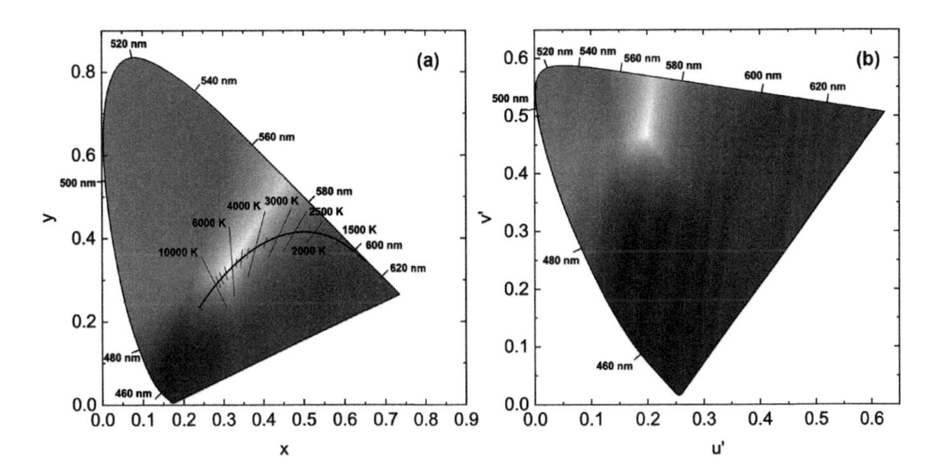

FIGURE 10.3 (a) CIE 1931 (*x, y*) chromaticity diagram and (b) CIE 1976 (*u*′, *v*′) uniform chromaticity diagram. Monochromatic color is located on the perimeter, whereas white light is close to the center of the diagram. The Planckian locus for color temperatures between 1500 K and 10,000 K is also shown.

proximity to the perimeter of the chromaticity diagram. If the linewidth becomes broader, the location moves toward the center of the diagram, such as for white light, which covers most of the visible spectrum.

White light can be generated from a combination of numerous potential spectra. To characterize white light, defined standards are used. One unique standard is the *Planckian black body radiator*, because it can be described with only one parameter, namely the color temperature T. The black body spectrum $I(\lambda)$ is described by *Planck's law* [23,24]:

$$I(\lambda) = \frac{2hc^2}{\lambda^5 \left(e^{\frac{hc}{\lambda k_B T}} - 1 \right)} \tag{10.4}$$

where h is the Planck constant, c the speed of light, k_b the Boltzmann constant, T the temperature, and λ the wavelength of the body. Figure 10.4 shows Planckian black body spectra for various temperatures. The maximum of the peak can be described by *Wien's displacement law* [25].

$$\lambda_{\max} = \frac{2889\,\mu\text{m K}}{T} \tag{10.5}$$

With increasing temperature, the maximum of the peak shifts toward shorter wavelengths. The temperature T, also referred to as *color temperature*, can be used to describe the emission spectra of a Planckian black body radiator.

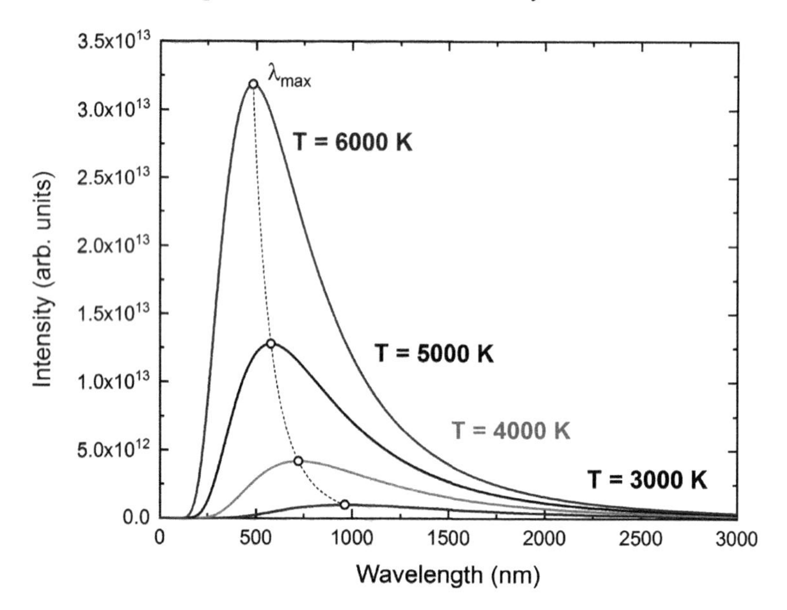

FIGURE 10.4 Intensity distribution of a Planckian black body radiator as a function of wavelength λ for different color temperatures T. The position of the maximum (λ_{max}) is described by Wien's law.

Each Planckian black body spectrum for different color temperatures can be described with chromaticity coordinates, spanning a curve in the chromaticity diagram. This curve is known as the *Planckian locus*, displayed in Figure 10.3a. The spectrum of a black body with very low color temperatures contains more infrared light, and its location in the chromaticity diagram can be found near the perimeter of red light. With increasing color temperature, the spectrum shifts into the range of the visible light and the position in the chromaticity diagram moves to the region where white light is located.

The term color temperature stems from the fact that, with increasing temperature, a black body glows red, orange, yellow, white, and then blueish. Therefore, the *color temperature* (unit: K) of a white light source corresponds to the temperature of a Planckian black body radiator of the same chromaticity coordinates. Not all white light sources, however, can be described with a Planckian black body radiator, that is, they are not located exactly on the Planckian locus. Hence, the *correlated color temperature* (CCT, unit: K) of a white light source is the temperature of a black body radiator whose color has the best resemblance to the white light source [26]. The choice of white light with a certain CCT depends strongly on individual preference. In Japan, white light with higher CCTs (around 5000 K) describing cold white light is preferred, whereas warm white light (CCTs below 3000 K) is generally desired in the UK and US [27].

A true white light source needs to have the ability to reproduce the real color of the illuminated object accurately. The *color rendering index* (CRI) is a measure of the ability of a light source to render or reproduce the color of an object faithfully in comparison with the light of an ideal or natural source [22,28]. The CRI is dimensionless, and its maximum value is 100, representing an ideal color rendering source. The CRI is an important parameter to describe the quality of a light source. Light sources in homes or offices should have high color rendering capabilities (high CRI), whereas color rendering is less crucial (lower CRI) for street lights or lights for general illumination. The CRI of a Planckian black body radiator is defined to have the highest color rendering capabilities (CRI of 100), because it most closely resembles natural daylight. The closest light source to a black body radiator is the incandescent light bulb, which (as a black body emitter) has the highest possible CRI of 100, as a consequence all other light sources have a lower CRI.

Radiometry describes the science of the measurement of electromagnetic radiation, which includes visible light. For LED characterization, the radiant flux $\Phi_{rad}(\lambda)$ is an important quantity. It describes the total radiant energy per unit time and is measured in Watts (W).

Photometry on the other hand is the measurement of light taking the human eye response into account. Similar to the radiant flux, the luminous flux Φ_{lum}, measured in lumen (lm), describes the total power perceived by the human eye, essentially, how bright a light source appears. It can be calculated from the radiant flux by taking the eye sensitivity function $V(\lambda)$ into account using the following equation:

$$\Phi_{lum}(\lambda) = 683 \frac{\text{lm}}{\text{W}} \int_{\lambda} V(\lambda)\Phi_{rad}(\lambda)\,d\lambda \tag{10.6}$$

For the characterization of white light generation, the *luminous efficacy* is an important parameter, which defines the efficiency of the energy conversion. The luminous efficacy is defined as the ratio of the luminous flux and power, with units in lumen per Watt (lm/W). For the *luminous efficacy of optical radiation*, the power is equivalent to the radiant flux of the device. More commonly, the *luminous efficacy of a light source* is used, where the power corresponds to the electrical input power of the light source.

The luminous efficacy is a general parameter describing the efficacies of white light sources. Recently, another parameter was introduced for characterizing white LEDs using blue LEDs [29]. The *blue-to-white (B-W) efficacy* is the ratio of the luminous flux of the white LED to the radiant flux of the underpinning blue LED. This efficacy essentially describes how much of the optical power of the blue LED is converted into perceived white light, that is, luminous flux, by the specific color converter. Commercial phosphor-based converters have values above 200 lm/W.

10.2.3 White Light Generation

The light emitted by LEDs is defined by the bandgap of the semiconductor. Commonly, there are two approaches used to achieve white light. Either a combination of inorganic LEDs emitting at different wavelengths is used or a single blue or UV LED is coated by a down-converting material. White light can be generated by different combinations of emission spectra as described below.

The trivial approach is using two or more LEDs on separate chips, each emitting nearly monochromatic light, which when combined emit white light. When using two LEDs (*dichromatic*), their wavelengths have to be complementary and operate at a certain power ratio in order to be perceived as white. Dichromatic white LEDs in general have a high luminous efficacy (ratio of luminous flux to power in units of lumen per watt), but a low CRI. The CRI can be improved by adding more LEDs of different emission wavelengths (e.g., *trichromatic*); however, its luminous efficiency will decrease [2,5]. This is a fundamental trade-off between the color rendering capabilities and the luminous efficiency of an LED, which cannot be eliminated. Examples would be a blue and yellow LED or a combination of three LEDs emitting in the RGB spectral regions.

Currently, the most common approach for inorganic white LEDs consists of a single LED, emitting in the UV or the visible spectrum, coated with one or more energy down-converting phosphors. The most commonly used combination is a blue LED, using InGaN/GaN MQWs as the active region, pumping a yellow-emitting phosphor, such as YAG:Ce, to provide the requisite white light.

The conversion efficiency using a converter is limited by two factors. Firstly, the external quantum efficiency (EQE) η_{EQE} of the phosphor describes the ratio of the numbers of photons emitted and absorbed by the converter. The EQE is also the product of the internal quantum efficiency (IQE) η_{IQE} and extraction efficiency $\eta_{extraction}$:

$$\eta_{EQE} = \eta_{IQE} \cdot \eta_{extraction} \tag{10.7}$$

The IQE is the intrinsic efficiency of the converter material. The second limitation is the Stokes shift, which is an inherent energy loss due to the conversion of the absorbed photon with wavelength λ_1 to the emitted photon with wavelength λ_2 ($\lambda_1 < \lambda_2$). This energy loss is fundamental, cannot be overcome and is given by:

$$\Delta E = E_1 - E_2 = hc\left(\frac{1}{\lambda_1} - \frac{1}{\lambda_2}\right) \tag{10.8}$$

The efficiency of the wavelength conversion can be expressed as:

$$\eta_{\text{Stokes}} = \frac{E_2}{E_1} = \frac{\lambda_1}{\lambda_2} \tag{10.9}$$

The total efficiency of the conversion process of the wavelength converter is the product of the EQE and efficiency of the wavelength conversion:

$$\eta_{\text{converter}} = \eta_{\text{EQE}} \cdot \eta_{\text{Stokes}} = \eta_{\text{EQE}} \cdot \frac{\lambda_1}{\lambda_2} \tag{10.10}$$

Due to the additional energy loss from the Stokes shift, white LEDs employing a phosphor for the conversion process show lower efficiencies than white LEDs using multiple single LEDs of different colors. The highest energy losses occur when UV light is converted into red light. Materials suited as wavelength converters include phosphors, organic dyes, and semiconductors. The EQE for each of these converters can be close to 100%.

An extensive introduction on the principle of LEDs, and their operation, design, packaging, and applications can be found in Refs. [2, 5].

10.2.4 Use of Phosphors in White LEDs

Phosphors possess a broad emission band making them very suitable for white light generation and are very stable. They are made of an inorganic host doped with an optically active element. Common hosts are garnets, such as yttrium aluminium garnet (YAG), $Y_3Al_5O_{12}$ [30,31]. The optically active element can be a rare earth element, rare earth oxide, or other rare earth compound. For white light generation cerium (Ce) is used, whereas neodymium (Ny) is utilized for lasers. By adding gadolinium (Gd) and/or gallium (Ga) to fabricate Ce-doped $(Y_{1-x}Gd_x)_3(Al_{1-y}Ga_y)_5O_{15}$, the emission of the phosphor can be shifted [32]. This makes it possible to change and adapt the chromaticity of the white light when used together with a blue LED. For the fabrication of white LEDs, the YAG phosphor is incorporated in epoxy resin. The epoxy is then deposited on the blue LED die. The fraction of the absorbed blue light by the phosphor depends on the thickness of the epoxy-phosphor layer and its concentration, which also defines the yellow emission intensity of the phosphor.

Issues of phosphors include challenging color tuning and rendering, self-absorption due to overlap of broad absorption and emission bands, lack of suitable red phosphors, and use of expensive rare-earth materials [6,7].

10.2.5 White Organic LEDs

White light can also be produced using entirely organic electroluminescent materials. Advantages of white organic LEDs (WOLEDs/OLEDs) include low-cost fabrication, ease of processability, fabrication of large area flexible sheets, and tuning of their emission properties through modification of the chemical structure [8,33,34]. In order to generate white light, emissive materials of varying color need to be combined, which can be achieved by several approaches [9]. One approach uses several individual emitters arranged in a multilayer structure, or by blending individual emitters into a single layer or by doping a host material with emissive materials at varying concentrations [35–37]. Alternatively, a single molecule can be used to generate white light by emission from its excited state and its aggregates (e.g., excimers or exciplexes) [38]. Similarly, electroluminescent polymers can be used as the emissive material [34]; however, their efficiencies are below that of multilayer WOLEDs.

10.3 CHEMISTRY

Perhaps the most attractive feature of organic materials is the wide variety of structures obtainable, and made possible through incorporation of different, readily accessible building blocks. This allows key properties to be controlled and manipulated to achieve the desired performance, including solubility, energy gap, and optoelectronic properties. There are significant seminal contributions on how to do this, and comprehensive overviews can be found in Refs. [39–41]. Here a brief discussion of key parameters as support for the remainder of this chapter will be given.

One feature common to all organic down-converting materials is that each generally contains a conjugated skeletal backbone, consisting of an alternating pattern of single and double bonds between carbon and/or heteroatoms. Often this takes the form of cyclic (and heterocyclic) rings connected either through C–C bonds, bridging double bonds or fusion of two rings together. The variety of structures available when designing organic materials for any application, including here as down-conversion units, is vast and ensures that fine control of structure and therefore application properties is possible on selection of the most appropriate candidate. Examples of selected synthetic building blocks are shown in Figure 10.5.

Often the most important parameter to control is the absorption and emission wavelengths to enable appropriate absorption of emitted LED light, as well as emission of the desired wavelength to achieve the sought-after hybrid light output [42–44]. Strategies to achieve this include incorporating structural units responsible for absorption and emission within a larger molecule or polymer, or introducing electron-rich and electron-poor components adjacent to each other to control the bandgap, and hence the optoelectronic properties. Hybrid white light emitting LEDs rely on yellow emissive organic materials in conjunction with an efficient blue inorganic LED for example. Therefore, combining yellow emissive organic materials with blue light absorbing materials in one structure, linked via an appropriate conjugated bridge, can produce the desired output.

A further key factor that needs to be considered is the solubility of the organic species, which can require consideration of the deposition method. For example, the

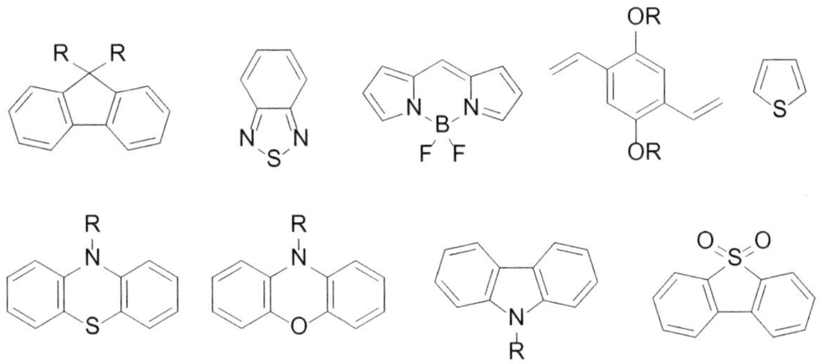

FIGURE 10.5 Examples of selected organic building blocks.

deposition of organic emissive active layers within OLED devices typically involves either solution processing, for example spin-coating or drop-casting, or vacuum deposition [45]. However, this either requires that a specific solvent is used to deposit the material, which can damage the existing LED structure (or underlying electronics), or the molecule can survive high temperatures and vacuum to be deposited via sublimation. An alternative method is encapsulation in a non-emissive matrix and offers several advantages. A low concentration of the bulk solution for deposition (i.e., 0.5–1% w/v) can be used, while matrices are often cured swiftly through UV light irradiation. Finally, often the existing solution-state optical properties are maintained, with minimal effects from aggregation or other morphological changes, ensuring a uniform and smooth down-conversion layer to boost the overall efficiency of the device [46].

An additional consideration is whether to employ polymers or small molecules as the organic converting layer within a hybrid device. Both types of material have their advantages. For example, organic small molecules are monodisperse allowing for absolute identification of structure and therefore synthesis is straightforward to replicate. They also benefit from being effectively tunable in terms of their properties via synthetic manipulation of their structure, although many materials require extensive, multi-step pathways to achieve the required product in high purity. On the other hand, polymers suffer from being polydisperse and batch-to-batch reproducibility is often challenging to achieve, although their production is more synthetically straightforward than monodisperse small molecules [47].

Overall, choice of material family is dependent on the factors outlined above and requires careful consideration prior to commencing the work.

10.4 LIGHT-EMITTING POLYMERS

10.4.1 Introduction

Light-emitting or luminescent polymers are one choice of organic color converters. They have several advantages over currently used phosphors [33,48,49]. Polymers with emission characteristics covering the entire visible spectrum are available. They can have low or negligible self-absorption to due large Stokes shift between their absorption and emission peaks and can exhibit high photoluminescence quantum yields (PLQYs). They are easily dissolved in solution for easy and cost-efficient processing and regular spin-coating techniques. When deposited on bendable substrates (e.g., plastics), polymers are also flexible. Issues of polymers include stability of their emission properties and lifetime. A review on polymers used for OLEDs can be found in Refs. [34, 50, 51]. For a polymer to be emissive and therefore useful in hybrid light-emitting devices, they must be conjugated, and therefore consist of a series of alternating single and double bonds. There are an unlimited number of ways to connect polymers together (see Section 10.3), but the polymer structure must enable crucial factors including solubility and high PLQY. Control of the synthetic method and therefore polymer structure also allows for close control of the energy (band) gap, either through incorporation of pendant functionalities and/or a donor-acceptor backbone and can provide fine control of the absorption and emission characteristics of the hybrid device.

This section shows different examples of combining a UV or blue-emitting nitride-based LED with either a combination of polymers or with polymers combined with other color converter components.

10.4.2 Polymers in Hybrid White LEDs

One of the first applications of an organic color converter optically-pumped by a blue inorganic nitride-based LED was reported by Hide *et al.* utilizing conjugated polymers [43]. In order to achieve white light, two conjugated polymers, both with an absorption maximum in the blue, were combined. The first emitted in the green wavelength region (BuEH-PPV) while the other emitted in the red spectral region (MEH-PPV). Their absorption and photoluminescence (PL) emission spectra can be seen in Figure 10.6a together with their chemical structures. The conjugated polymers were spin-coated either separately onto transparent glass slides or as a bilayer. Figure 10.6b shows the emission spectra of these hybrid LEDs. When separate slides of the conjugated polymers were placed above the LED, wave guiding effects occurred due to different refractive indices of the glass slide and the polymers leading to sideways emission from the glass slides reducing the forward emission intensity. Therefore, the polymers were applied as a bilayer. Varying the thickness of the MEH-PVV layer allowed the control of the red and green intensity ratio and hence the color of the hybrid LED. As the thickness of the MEH-PVV increased more red light contributed to the overall emission, the chromaticity coordinates shifted toward the right as seen in Figure 10.6c.

One of the issues of using a color conversion process for white light generation is the efficiency of the energy transfer from the blue LED to the color converter. In their study, Smith *et al.* [52] applied a commercial light-emitting polymer to a blue-emitting InGaN/GaN nanorod structure utilizing non-radiative resonant energy transfer (RET). This energy transfer, also called non-radiative Förster energy transfer (FRET), relies on close coupling between the donor (the InGaN/GaN nanorod structure) to the acceptor (the light-emitting polymer) [53,54]. The transfer rate is proportional to R^{-4} with R being the separation between a nanorod and the polymer [55]. In a regular, planer blue LED, the InGaN/GaN QW structure, which is the region where the blue

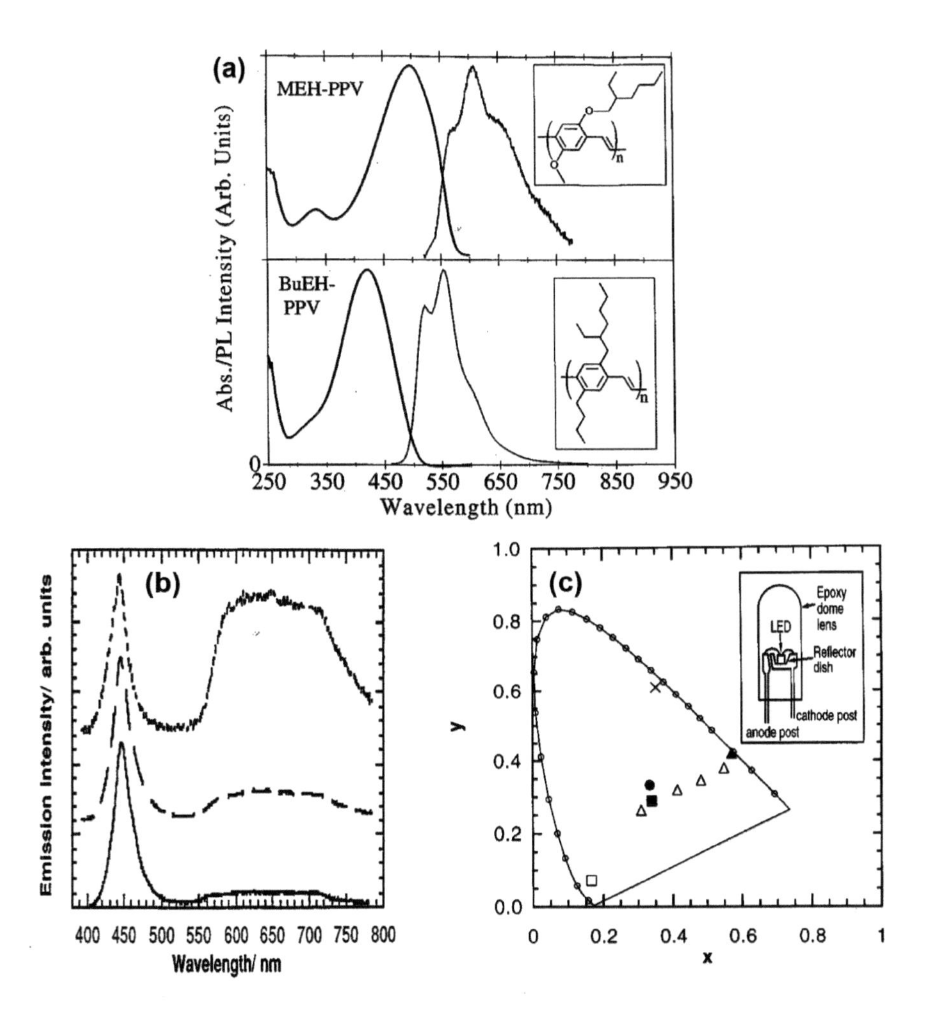

FIGURE 10.6 (a) Absorption and PL emission spectra of MEH-PVV and BuEH-PPV. Their chemical structure is shown in the insets. (b) Emission spectra of the hybrid LEDs using polymers: solid line: separate glass slides of BuEH-PPV and MEH-PVV films; dashed lines: bilayer of BuEH-PPV/MEH-PVV with two different MEH-PVV layer thicknesses. (c) CIE 1931 (x, y) chromaticity diagram: open square: blue LED; filled square and open triangles: emission from hybrid LED with BuEH-PPV/MEH-PVV bilayer with different MEH-PVV layer thicknesses; cross: PL from BuEH-PVV; solid triangle: PL from MEH-PVV. Reproduced from Ref. [43] with the permission of AIP Publishing.

light is emitted, is generally located over 100 nm below the surface. Therefore, the planar LED structure was etched into nanorods in order to expose the InGaN/GaN QWs. The polymer was then spin-coated to fill-in between the nanorods, as seen in Figure 10.7a, to be in direct contact with the QW structure for efficient FRET coupling as seen in Figure 10.7b. The polymer used was the commercially available F8BT (poly[(9,9-dioctylfluorenyl-2,7-diyl)-alt-co-(1,4-benzo-{2,1′,3})-thiadiazole)].

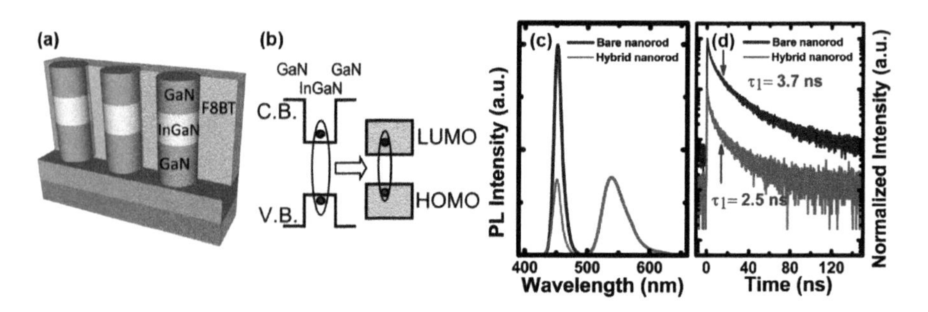

FIGURE 10.7 (a) Schematic of the hybrid structure with the polymer (F8BT) filled-in between the nanorods containing the InGaN/GaN quantum wells. (b) FRET coupling between the InGaN/GaN quantum wells and the polymer. (c) PL spectra of only the InGaN/GaN nanorods and the hybrid structure. (d) Time-resolved PL traces and decay lifetimes of the bare nanorod and hybrid structure. Adapted with permission from Ref. [52]. Copyright 2013 American Chemical Society.

PL emission spectra of only the InGaN/GaN nanorods and the hybrid structure, both measured under identical conditions using a using a 375 nm laser, are shown in Figure 10.7c. The decay lifetimes measured at the 450 nm nanorod peak were also determined for the bare nanorod and the hybrid structure as displayed in Figure 10.7d. The recombination lifetime of the hybrid structure is small due to the additional non-radiative FRET process between the nanorods and the polymer. A more detailed discussion of the lifetime measurements can be found in Ref. [52].

Regular planer nitride LEDs typically have a dimension of 500 μm × 500 μm or larger. An alternative are micro LEDs (or μ-LEDs) with dimensions in the tens of micron range [56]. The use of micro LEDs over planar LEDs has several advantages, including matrix addressable pixel for display applications and visible light communication [57,58]. In the study by Heliotis *et al.*, light from an array of UV-emitting micro LEDs was converted using three organic polyfluorene compounds emitting at different wavelengths [59]. The UV LED consisted of an array of 60 × 60 micro LEDs with a diameter of 20 μm emitting at about 370 nm that could be individually addressed, while the UV-absorbing polymers consisted of the blue-emitting F8DP (poly(9,9-dioctylfluorene-co-9,9-di(4-methoxy)phenylfluorene)), the green-emitting F8BT (poly(9,9-dioctylfluorene-co-benzothiadiazole)), and the red-emitting Dow Red F co-polymer. The emission from these polymer films, spin-coated on quartz substrates and placed above the micro LED array, can be seen in Figure 10.8a showing emission across the entire visible spectrum. To produce white light, the three compounds (in toluene solution) were blended at different concentrations. The emission spectra and chromaticity coordinates of four hybrid LEDs using four different mixing concentrations are shown in Figure 10.8b–c, respectively. Hybrid LEDs A, C, and D appeared white to the eye, whereas LED B was yellowish-white. The CIE 1931 chromaticity coordinates of LED D were (0.30, 0.34), making it close to pure white which has the coordinates (0.33, 0.33). The energy transfer of these systems is assisted by non-radiative FRET. It should be noted that F8DP was the host polymer due to its lowest emission wavelength (highest energy) and energy could be transferred to the

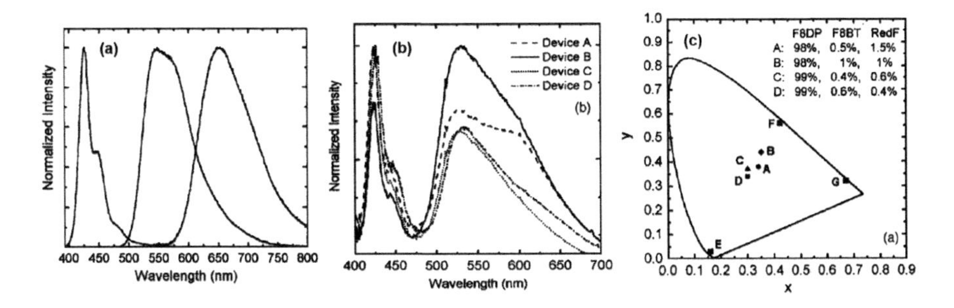

FIGURE 10.8 (a) Normalized emission spectra from F8DP, F8BT, and Dow Red F (left to right) placed on top of the micro LED array. (b) Emission spectra of four hybrid LEDs (A-D) using four different blending concentrations. The concentrations in wt % are shown in (c). (c) CIE 1931 (*x*, *y*) coordinates of these four hybrid LEDs. The CIE coordinates of the LEDs only using F8DP, F8BT, and Dow Red F (E, F and G) are also shown. Reproduced from Ref. [59] with the permission of AIP Publishing.

other two polymers (F8BT and Dow Red F), which possessed an absorption peak at the emission of F8DP. However, the energy transfer is more complicated than this since Dow Red F had an additional absorption peak at the emission of F8BT. Although challenging, this chain of energy transfers provides a successful route for controlling the white light emission.

The same group has demonstrated a much simpler system combining an array of blue-emitting micro LEDs (diameter of 50 μm) with a yellow-emitting conjugated co-polymer ("Super Yellow") [60], emulating the commonly used blue LED and yellow phosphor approach of commercial white LEDs. Another advantage of organic polymers over inorganic phosphors is their short luminescence lifetimes. The long lifetimes of phosphors (on the order of μs) pose a limit to achieve high data transmission rates for visible light communication [61].

An often encountered issue of white LEDs is the lack of red emission leading to blueish or cool appearing white light described by high CCTs. Although red-emitting phosphors are available they are less common compared with phosphors of other colors with lower wavelength. Chen *et al.* synthesized a novel red-emitting cationic iridium(III) coordination polymer and combined it with a standard yellow YAG:Ce phosphor in a silicone encapsulant coated on an inorganic blue LED [62]. The excitation spectra of the coordination polymer as a powder and in silicone are both quite broad, extending from the UV to the green spectral range with strong absorption in the blue spectral range as seen in Figure 10.9a. The emission spectrum when excited with blue light is displayed in Figure 10.9a with an emission maximum at 620 nm and 652 nm. Hybrid LEDs were prepared by blending different concentrations (0–0.3 wt %) of the coordination polymer with YAG:Ce at a fixed concentration of 7.0 wt % in silicone. The emission spectra and chromaticity coordinates of these hybrid LEDs are shown in Figure 10.9b–c, respectively. By adding up to 0.3 wt % of the coordination polymer the CCT could be shifted from 6157 K, with only YAG:Ce, to 3475 K with a slight change in CRI (from 72.7 to 75.6).

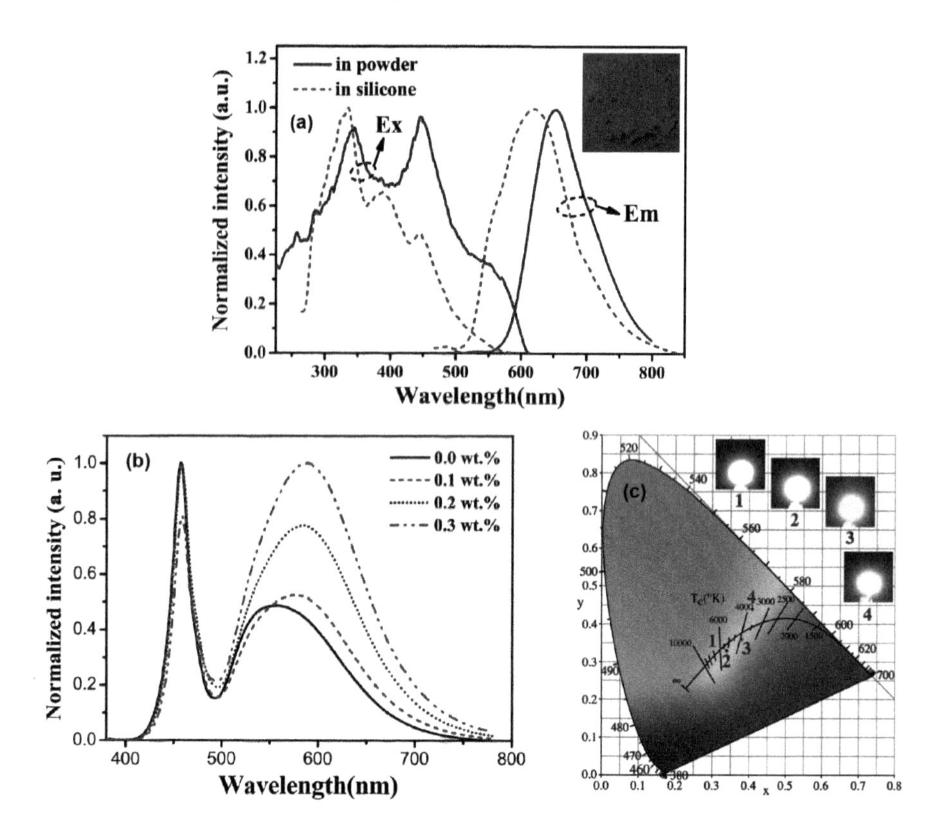

FIGURE 10.9 (a) Excitation (detection wavelength 620 nm or 653 nm) and emission (excitation wavelength 450 nm) spectra of the coordination polymer in powder form or in silicone. (b) Emission spectra and (c) chromaticity coordinates of the hybrid LEDs using different concentrations of the coordination polymer (0–0.3 wt %) and the YAG:Ce phosphor (7.0 wt %). This article was published in Ref. [62], Copyright Elsevier (2018).

The previous example combined the traditionally used phosphor and a polymer as color converters. Another possibility is combining an organic polymer and inorganic quantum dots (QDs). QDs have the advantage of easy tunability of their emission wavelength by changing their sizes and offer high efficiencies [63]. Compared with organic materials, the emission from QDs is also much narrower allowing better control of the color mixing with high color rendering [64]. Demir *et al.* combined a UV LED with a combination of a blue-emitting polyfluorene (9,9-*bis*(2-ethylhexyl) polyfluorene) and three QDs (or nanocrystals as they refer to them), emitting at different wavelengths for fine-tuning of the overall white light emission [65]. The QDs are commercial CdSe/ZnS core-shell structures of different diameters (2.4 nm, 3.2 nm, and 5.2 nm) emitting at 540 nm, 580 nm, and 620 nm, respectively. The emission spectra of the polyfluorene polymer and the QDs are displayed in Figure 10.10a. The converter material was applied to the LED using a layer-by-layer deposition in order of increasing emission wavelength. It was shown that the CRI increased from 53.4 for the polymer and yellow-emitting QD (580 nm) combination to 83.0

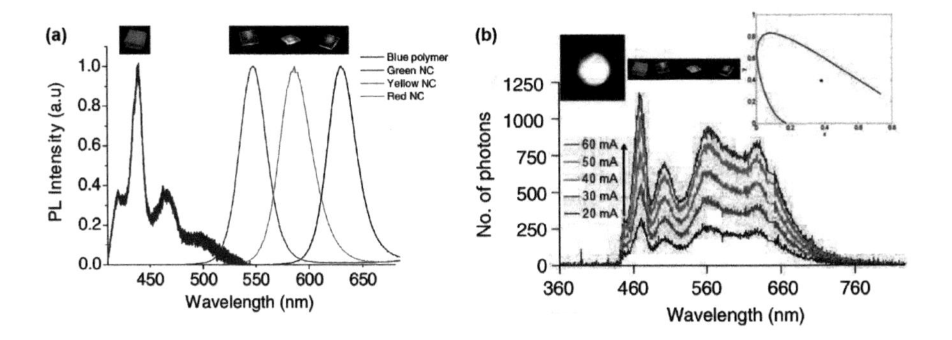

FIGURE 10.10 (a) Emission spectra of the polyfluorene polymer and the three quantum dot structures. (b) Emission spectra of the hybrid LED under different drive currents and chromaticity coordinates. Adapted and reprinted with permission under Creative Commons from Ref. [65].

when all three QDs where used. Simultaneously, the CCT was lowered from 6762 K to 3433 K, changing the white emission from cool to warm. The emission spectrum of the hybrid LED using the polymer and all three QDs under different drive currents and the chromaticity coordinates are shown in Figure 10.10b.

10.5 LUMINESCENT SMALL MOLECULES

10.5.1 Introduction

An alternative class of organic color converter is luminescent small molecules. Much like polymers, a huge variety of structural choice is available when designing luminescent small molecules (see Section 10.3 for details). This degree of choice allows for the full color spectrum to be achieved, while absolute control over the synthesis process ensures that the materials are monodisperse and therefore entirely reproducible. Additionally, this can lead to control of the energy gap and, through choice of appropriate sub-units, can minimize self-absorption and maximize the Stokes shift between the absorbing and emissive components. One issue that small molecule emitters can suffer from is aggregation, although often the attachment of solubilizing groups can render this as ineffective to the overall device characteristics. Perhaps the main issue that small molecules can suffer from is that they can become synthetically complex, leading to questions around their long-term commercial suitability. A review on small molecules used in OLED structures can be found in Ref. [34].

This section showcases selected examples of luminescent small molecules as color converters in hybrid LEDs, with a particular focus on how structural evolution can lead to enhanced device properties. It then will illustrate the influence of deposition method on device fabrication before applying the organic converter to blue LEDs. This section will also be used to show how to fully characterize and understand hybrid LED devices using the methodology described in Section 10.2.2. This will illustrate what properties of the color converter influence the performance of the hybrid LED, such as luminous and B-W efficacy and color rendering.

10.5.2 Manipulation of the Chemical Structure and Effect on Optical Properties

10.5.2.1 Introduction to BODIPY

The 4,4-difluoro-4-borata-3*a*-azonia-4*a*-aza-*s*-indacene unit, or BODIPY as it is commonly known, is an organic fluorescent dye and part of the difluoro-boraindacene family. Advantages of BODIPY are its thermal and photochemical stability, high fluorescence quantum yields, large absorption and emission profiles and good solubility in various organic solvents, such as acetone and toluene [66]. Due to its properties, it has found applications in biological labelling [67], luminescence [68], dye-sensitized solar cells [69] and as sensors for pH and ions [70,71]. Its tunable emission in the yellow spectral region makes it attractive for organic light-emitting devices [66]. BODIPY has a strong, but narrow absorption peak, which is confined to around 500 nm. Therefore, another absorbing partner unit in the UV or blue spectral region is needed for color conversion when combined with an inorganic LED.

For absorption in the blue region and emission in the yellow region, efficient energy transfer between the two partner units needs to occur. One type of energy transfer is the previously described FRET mechanism (see Section 10.4.2) where energy is non-radiatively transferred from a donor molecule to an acceptor molecule through dipole–dipole coupling [54]. The efficiency of the energy transfer strongly depends on the spatial separation of the involved molecules. Another possible energy transfer, called Dexter electron transfer, is a quantum mechanical mechanism based on the exchange of an electron between two molecules [72]. Strong overlap of the wave functions of the donor and acceptor molecule is necessary. Both Förster- and Dexter-type energy migrations processes were shown to contribute to the energy transfer in a star-shaped supramolecular system from the absorbing truxene core to three emitting BODIPY units [73].

10.5.2.2 Toward White Light: Yellow Emission from Oligofluorene-BODIPY Oligomers

The following set of small molecule compounds, four linear oligofluorene-BODIPY structures, is an example illustrating the impact of relatively simple modifications of the chemical structure on the absorption and emission properties [74]. The absorbing partner consists of a chain of either three or four fluorene units. This chain is then coupled with a BODIPY unit at either the *meso*- or *beta*-position as seen in Figure 10.11a–b, respectively. The unused positions of the BODIPY unit are blocked with a methyl (Me) or ethyl (Et) group, which also increases the stability of the final molecule. The end products are bright red-orange powders. In toluene or chloroform solution, they are fluorescent under UV irradiation. A full description of the synthesis can be found in Ref. [74].

UV–vis absorption spectra of all four compounds as a diluted solution in dichloromethane are shown in Figure 10.11c. The stronger and shorter wavelength absorption band around 354–370 nm corresponds to the absorption by the oligofluorene chain, whereas the longer wavelength and less intense band around 513–527 nm stems from the BODIPY unit. Extending the conjugation length from three to four fluorene units results in a redshift of the oligofluorene absorption band by about

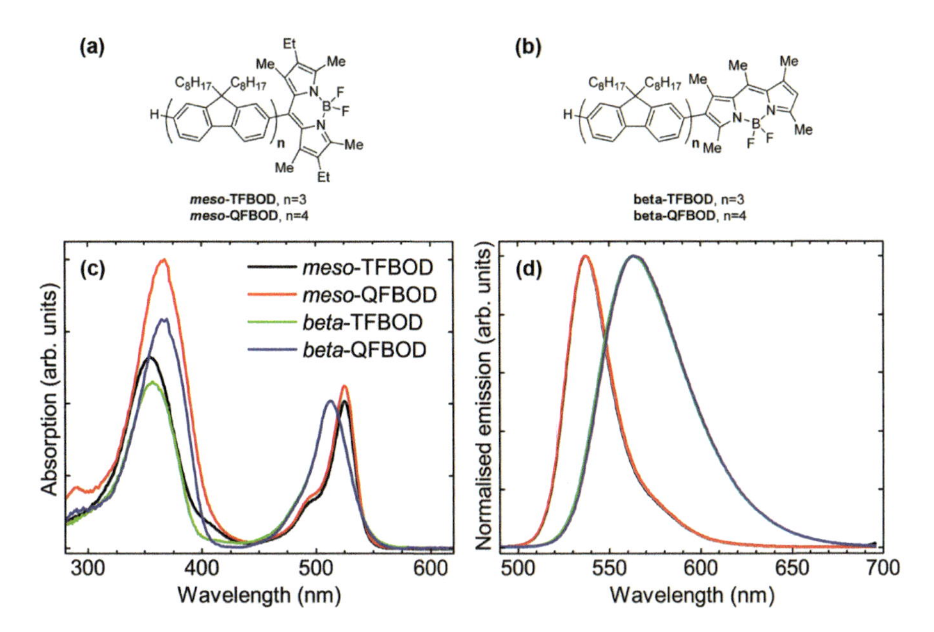

FIGURE 10.11 Chemical structure of the linear oligofluorene-BODIPY compounds with the fluorine chain coupled to the (a) *meso*- and (b) *beta*-position of BODIPY. (c) UV–vis absorption and (d) normalized PL spectra of the four linear oligofluorene-BODIPY compounds in dichloromethane solution. The excitation wavelength for the PL measurement corresponds to the absorption maximum around 350–370 nm of the oligofluorene chain. Adapted and reprinted with permission from Ref. [74] under Creative Commons License.

12–13 nm for both the *meso*- and *beta*-position. The absorption is also increased by extending the conjugation. Changing the substitution position does not seem to have a large effect on the absorption. The BODIPY absorption band redshifts by about 12 nm when changing the substitution position from *beta* to *meso* for either conjugation length. Figure 10.11d displays PL spectra measured for the four compounds in dilute dichloromethane solutions with an excitation wavelength corresponding to the absorption maximum of the oligofluorene chain. It shows that energy transfer between the oligofluorene chain and the BODIPY unit takes places since the oligofluorene arm is selectively excited while emission from the BODIPY was observed. Changing the substitution position from *meso* to *beta* broadens and redshifts the emission by about 26 nm regardless of the conjugation length of the fluorene chain.

In order to demonstrate the energy down-conversion process, *meso*-QFBOD was deposited from a toluene solution onto a UV LED emitting at around 365 nm. *meso*-QFBOD was chosen as its absorption maximum almost coincides with the emission peak of the UV LED and as it had the strongest absorption band of the four oligofluorene-BODIPY compounds. Optical microscope images of the LED with and without *meso*-QFBOD and under forward bias are displayed in Figure 10.12a–c, respectively. As seen in the images, a continuous film was not formed, but most of the material has accumulated around and between the two wire contacts of the LED. Figure 10.12d shows the EL spectra of the bare UV LED and the LED coated with *meso*-QFBOD

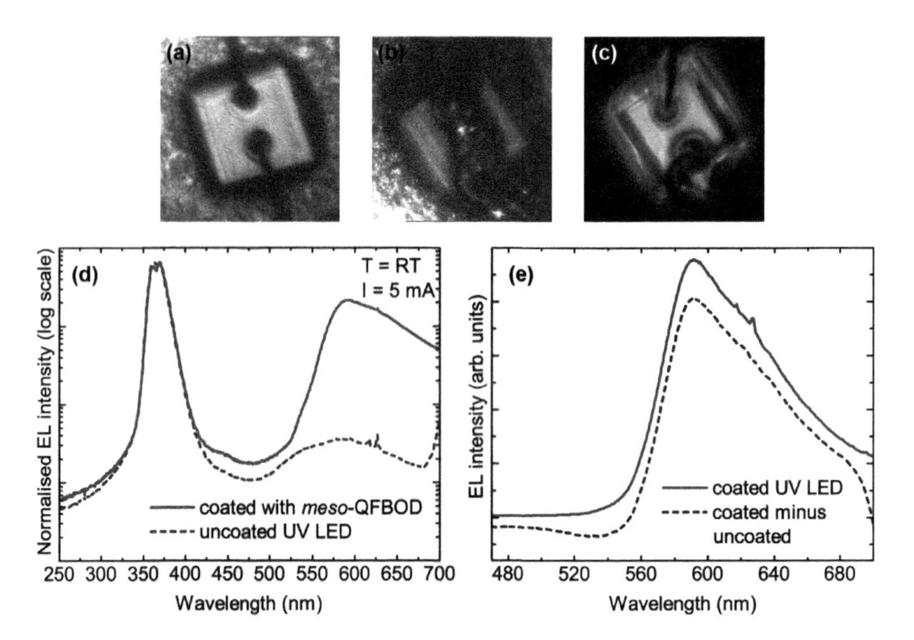

FIGURE 10.12 Optical microscope image of the (a) bare UV LED; of the UV LED coated with *meso*-QFBOD (b) switched off and (c) under forward current of 1mA. The dimension of the LED is 280 μm × 280 μm. (d) EL spectra of the uncoated and coated LED under forward current of 5 mA. Both spectra are normalized to the LED emission peak. (e) Spectrum of the uncoated UV LED subtracted from the spectrum of the *meso*-QFBOD coated LED compared with the spectrum of the coated LED in the region of the yellow emission. One 10 μL drop (30 mg in 1 mL of solvent) of *meso*-QFBOD was deposited on the LED die. Adapted and reprinted with permission from Ref. [74] under Creative Commons License.

under a forward current of 5 mA. The spectra are normalized to the emission peak of the UV LED around 365 nm in order to see the changes in the emission due to the organic material. Both spectra exhibit a broad emission band peaking just below 600 nm. In the spectrum of the bare UV LED, this weak peak around 585 nm is caused by impurities in the semiconductor material and referred to as yellow band [75]. In the spectrum of the coated LED, this second emission is much stronger and slightly redshifted to around 590 nm and is due to emission from *meso*-QFBOD. In order to show that the change in the yellow emission band is indeed caused by emission from the organic material, the spectrum of the uncoated LED is subtracted from the spectrum of the LED coated with *meso*-QFBOD. Almost no change in the subtracted spectrum is visible, indicating that the emission originates from *meso*-QFBOD with negligible contribution from the impurity band. The steeper drop on the shorter wavelength side is due to self-absorption by the second absorption peak of *meso*-QFBOD at 527 nm (see Figure 10.11c).

10.5.2.3 Toward White Light: Blue Light Absorption for White LEDs

For white light generation, blue light is needed in addition to yellow, which can be achieved by exchanging the absorbing component of the small molecule mentioned in Section 10.5.2.2.

Figure 10.13a displays the chemical structure of $(BODFluTh)_2FB$ and its synthesis is described in Ref. [46]. This compound has the absorbing unit placed in the center between BODIPY units on either side. The absorbing core consists of a fluorobenzene (FB) at its center, which is bordered by a thiophene (Th) and a fluorene (Flu) on either side. The fluorene units are substituted at the BODIPY *meso*-position. The hydrocarbon group (C_8H_{17}) on the fluorene units makes the compound soluble. The UV–vis absorption spectrum in Figure 10.13b exhibits an absorption peak around 403 nm corresponding to a charge transfer between the tetrafluorophenylene core and the neighbouring thiophene–fluorene unit and a second absorption peak associated with the absorption of the BODIPY unit. The PL spectrum in Figure 10.13b with a single emission peak at 550 nm shows energy transfer from the core unit to the BODIPY when excited at 440 nm, which is slightly higher than the absorption maximum, but closer to the blue emission needed for white light generation.

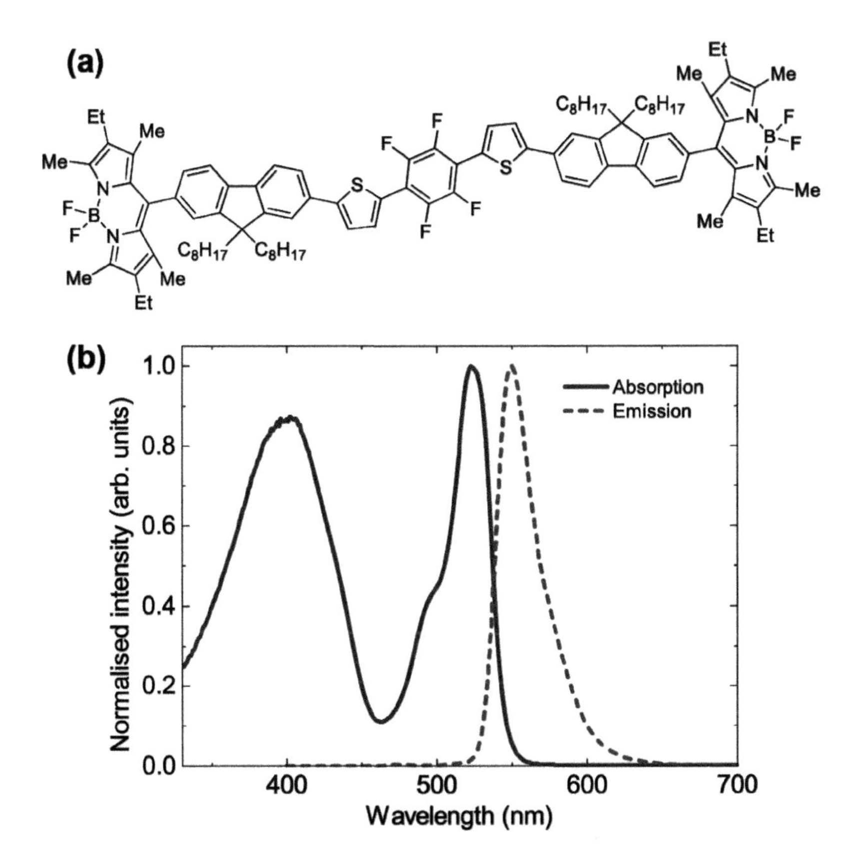

FIGURE 10.13 (a) Chemical structure of $(BODFluTh)_2FB$. (b) UV–vis absorption and emission spectra for $(BODFluTh)_2FB$ in dichloromethane solution. The excitation wavelength was 440 nm. Adapted and reprinted with permission from Refs. [29, 46] under Creative Commons License.

10.5.2.4 Toward White Light: Nanorod Encapsulation

The energy down-conversion process of $(BODFluTh)_2FB$ was studied by applying the organic compound to a GaN-based nanorod structure containing a blue-emitting InGaN/GaN multiple quantum well (MQW) structure ($\lambda \approx 445$ nm) [76]. Details on the used nanorod structure can be found in Ref. [77]. Direct contact of the organic material with the MQW structure should enhance the energy transfer between the blue-emitting nanorods and the converter through FRET mechanism. The converter was drop-casted in solution with toluene (10 mg/1 μL) on the nanorod structure. Figure 10.14a–d displays secondary electron (SE) images recorded using a variable pressure scanning electron microscope (SEM) from the center to the edge of the nanorod sample. The film coverage is very non-uniform. At the center the organic material filled in the gaps between the nanorods, visible as a semi-transparent haze as seen in Figure 10.14b, whereas toward the edge a thick film has formed on top of the nanorods.

The light conversion process was investigated through PL measurements. Since these samples are not complete LED structures (no p-n junction or metal contacts), the emission from the MQW structure embedded in the nanorods was selectively excited using a 405 nm laser. The excitation took place through the polished back surface in order to minimize absorption of the laser light by the organic material on top of the nanorods. PL spectra of the sample with different film thicknesses were recorded by subsequently adding further drops of converter material between the measurements (shown in Figure 10.14e). The emission peak around 446 nm originates from the MQW structure whereas the broad emission above 530 nm is due to down-conversion by the organic material. The steep drop on the shorter wavelength side is due to partial self-absorption of the BODIPY molecule. Overall, the emission intensity from the organic converter material increases with number of deposited drops. However, each additional drop likely leads to dissolution of the existing organic film leading to poor film formation and coverage of the LED with the organic material which impacts the overall color conversion and emission.

10.5.3 Toward White Light: Deposition and Encapsulation

The previous two examples use drop-casting as a method of deposition. In both cases the films suffered from strong non-uniformities. In the case of direct deposition on the LED, the small size and the wire bonds were problematic leading to the accumulation of material around the LED edge or wire bonds (Figure 10.12a–c). Similar issues were encountered for the nanorod sample with variable film thickness across the sample (Figure 10.14a–d). Alternative deposition methods are spin-coating or vacuum deposition, although either could still lead to issues with film formation around the wire bonds. Alternatively, the organic material can be incorporated in a transparent, non-emissive matrix to encapsulate the entire LED chip [78]. Advantages of this encapsulation method are that only small concentrations of the color converter are required, the solution emission properties are retained, and rapid curing can be achieved using UV light.

Therefore, the organic material, $(BODFluTh)_2FB$, was dissolved in 1,4-cyclohexanedimethanol divinyl ether (CHDV) containing trace amounts of the photoacid generator (PAG) 4-octyloxy diphenyliodonium hexafluoroantimonate at concentrations

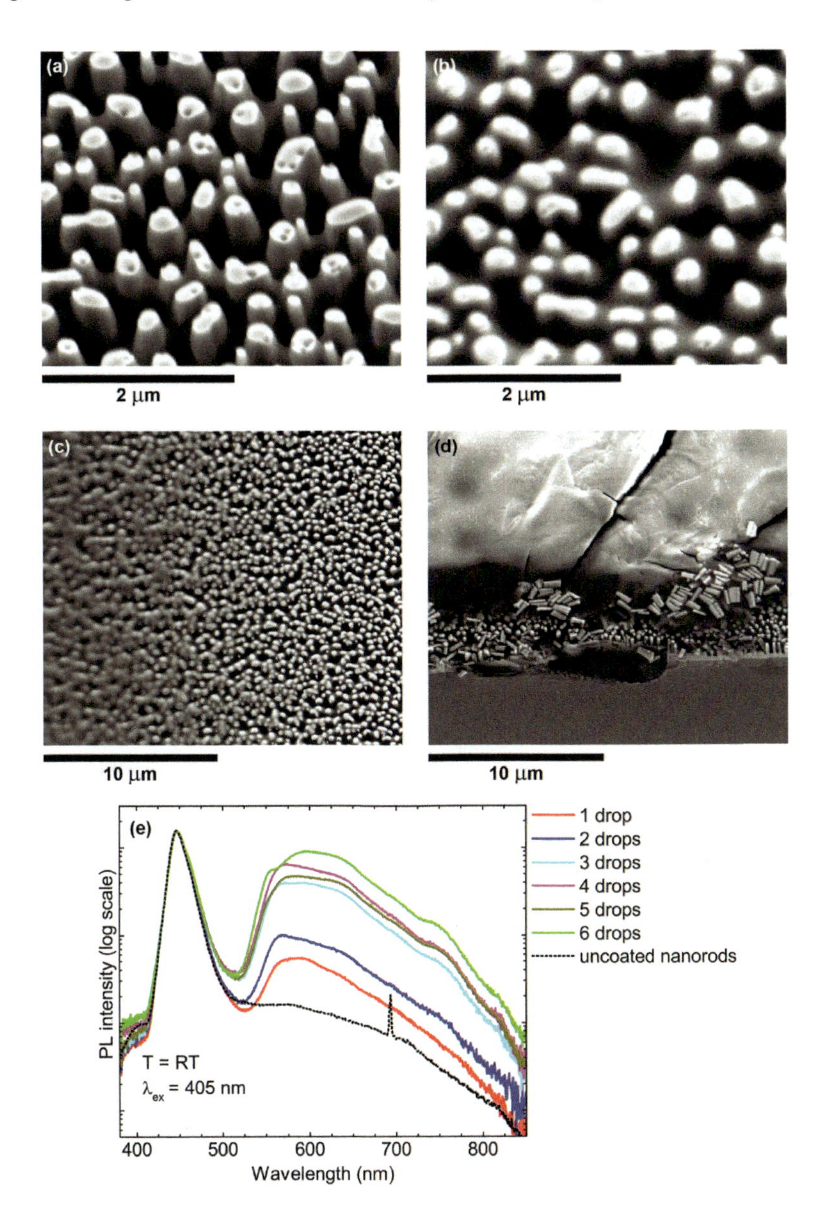

FIGURE 10.14 (a)–(d) SE images of nanorods coated with (BODFluTh)2-FB (in solution with toluene) using a variable pressure SEM (0.5 mbar). The deposited layer exhibits a gradient of increasing thickness toward the sample edge. (e) Normalized PL spectra (λ_{ex} = 405 nm) of InGaN/GaN MQW nanorod structures coated with $(BODFluTh)_2$-FB of various film thickness achieved by depositing one to six drops of the organic material. The spectra are normalized to the MQW emission peak of the nanorods. The excitation of the MQW emission using a 405 nm laser was carried out through the back of the sample to minimize absorption of the laser light by the organic material. The InGaN/GaN nanorod structures were fabricated by the group of Prof. Tao Wang at the University of Sheffield and more details on the structures can be found in Ref. [77]. Modified and reprinted with permission from Ref. [76].

of 0.25–4% weight per volume (w/v) [29]. The solution was drop-cast onto the blue LED chip filling the cup with the LED die and cured using UV light to harden the encapsulant. Optical microscope images of the hybrid LEDs are shown in Figure 10.15a where the organic material becomes orange from pale yellow with increasing concentration. The SE images in Figure 10.15b show that the LED is fully encapsulated and a dome has formed over the LED chip.

EL spectra of the hybrid LEDs under a continuous forward current of 25 mA using different concentrations are displayed in Figure 10.15c. With increasing

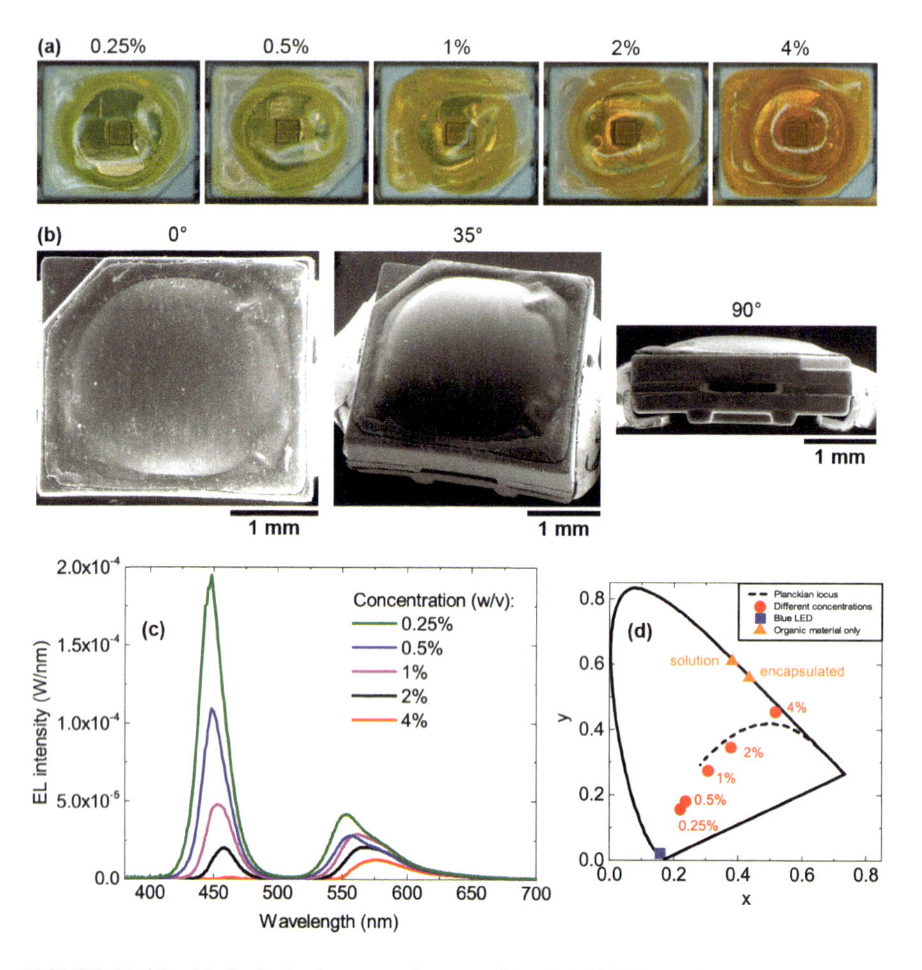

FIGURE 10.15 (a) Optical microscope images of the hybrid LEDs using concentrations of 0.25–4% w/v of the color converter $(BODFluTh)_2FB$. (b) SE images of one of the hybrid LEDs at different viewing angles. (c) EL spectra of the same hybrid LEDs operated under a continuous forward current of 25 mA. (d) CIE1931 chromaticity diagram showing the location of the hybrid LED with respect to the Planckian locus. The coordinates of the bare blue LED and the organic material in solution and encapsulated state are also shown. Adapted and reprinted with permission from Ref. [29] under Creative Commons License.

converter concentration, both the intensity of the blue LED and of the yellow organic emission decrease. Overall, however, the intensity of the yellow emission increases with respect to the blue intensity. This causes the emission color to shift from blueish to white and then into the yellow-orange as seen in the chromaticity diagram in Figure 10.15d by an almost linear shift with concentration. The simultaneous decrease of both emission peaks implies that with increasing concentration more blue light is getting absorbed but not re-emitted as yellow. This quenching effect is caused by aggregation of the organic molecules within the encapsulant with increasing concentration due to their close proximity leading to non-radiative recombination [46,79]. Similarly, this aggregation also causes a redshift of the organic emission with increasing concentration which leads to the slight deviation of the straight line at higher concentrations in the chromaticity diagram [80].

By controlling the concentration of the color converter, $(BODFluTh)_2FB$, in the encapsulant and the volume of the deposited drop, it is possible to produce warm and cool white light emission. The warm hybrid LED has a CCT of 2770 K, and CRI of 20 and chromaticity coordinates (0.45, 0.41), whereas the cool LED possesses a CCT of 7680 K, CRI of 46, and chromaticity coordinates (0.31, 0.27) [29]. As a reference, pure white is located at (0.33, 0.33). It should be noted that both LEDs exhibit poor color rendering due to the absence of emission in the green spectral range, where the human eye is most sensitive. The lack of green emission is further enhanced due to self-absorption of BODIPY.

10.5.4 White Light Device Efficiency and Efficacy

The "efficiency" of the device and the organic color converter can be assessed by different parameters, as defined in Section 10.2.2.

The efficiency of energy-down converting materials is commonly described by the PLQY, which is the ratio of emitted and absorbed photons. It is generally determined using an integrating sphere system and only describes the efficiency of the material itself. An example of a double integrating sphere system used to measure the PLQY is given in Ref. [81], which allows the differentiation of reflected and transmitted light through the sample which is placed between the two spheres.

For $(BODFluTh)_2FB$, the PLQY was determined in solution (dichloromethane) and in the encapsulated state and estimated to be 60% and 63%, respectively [46], confirming that the transparent matrix has a negligible impact on the optical properties.

The quantum conversion efficiency of $(BODFluTh)_2FB$ when applied to the blue LED in the encapsulant is also measured and fully described in Ref. [29] using a similar methodology as Ref. [81]. It is given by the ratio of the number of converted photons (yellow emission from the organic material) and the number of absorbed photons from the blue LED. It is only an estimation since only the entire emitted light is measured using a single integrating sphere. Neither transmitted nor reflected light can be differentiated, or light that is trapped in the LED package. However, it still is an estimation of the efficiency when combined directly with an LED. The conversion efficiency of different concentrations of $(BODFluTh)_2FB$ in the encapsulant applied to a blue LED is shown in Figure 10.16. It decreases with increasing concentration due to aggregation-induced luminescence quenching. The conversion efficiency of

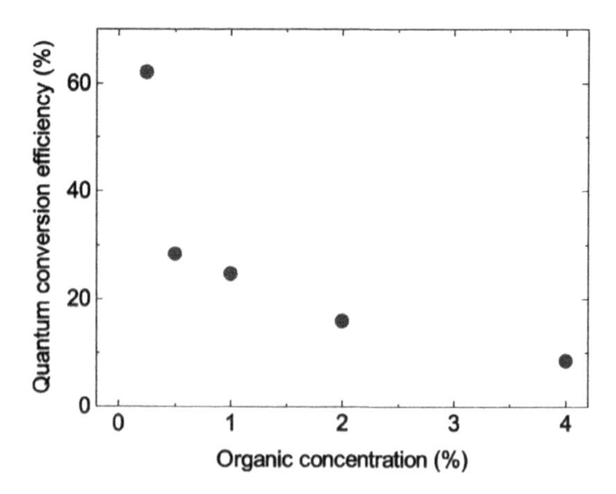

FIGURE 10.16 Quantum conversion efficiency of $(BODFluTh)_2FB$ applied to a blue LED as a function of concentration in the encapsulant. Reprinted with permission from Ref. [29] under Creative Commons License.

62% at the lowest concentration (0.25% w/v) is in agreement with the PLQY indicating low aggregation at that concentration.

Commonly, the luminous efficacy (of a light source) is given to provide a number for efficiency. The warm and cool white LEDs from the previous section have luminous efficacies of 3 lm/W and 11 lm/W, respectively. Alternatively, the efficiency of white LEDs can also be described by the B-W efficacy as defined in Section 10.2.2. The warm hybrid LED (4% w/v $(BODFluTh)_2FB$)) has a B-W efficacy of 29 lm/W, whereas the efficacy of the cool LED (1% w/v $(BODFluTh)_2FB$) is 96 lm/W. Phosphor-based white LEDs have B-W efficacies above 200 lm/W. The difference in efficacy between those two LEDs is due to the aggregation-induced luminescence quenching, while the low efficacies of the $(BODFluTh)_2FB$ hybrid LEDs are predominantly caused by the lack of green emission where the human eye is most sensitive (since the luminous efficacy is directly proportional to the luminous flux).

10.5.5 White Light Degradation and Lifetime

For real applications, the lifetime of the hybrid devices and hence organic color converters is crucial. For this purpose, degradation studies of the hybrid LEDs coated with $(BODFluTh)_2FB$ were performed [29,46]. In an initial experiment, the hybrid LED was measured periodically for a month, but only operated for the duration of the measurement (around 5 s). Negligible changes in emission characteristics, chromaticity coordinates, or CCT were observed. Therefore, the experiment was repeated, but with continuous operation of the LED. After 7 hours, the luminous efficacy dropped below 60% of its initial value as seen in Figure 10.17a. The emission from the organic material decreased drastically, while the blue LED emission remained constant indicating that the absorbed blue light is not reemitted in the spectral region. In order to investigate if the heat generated by the LED causes this degradation, the $(BODFluTh)_2FB$ encapsulant was deposited on a transparent glass slide and placed

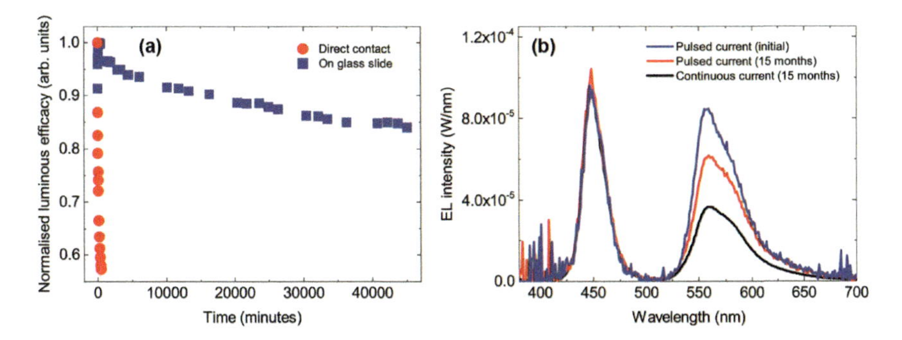

FIGURE 10.17 (a) Normalized luminous efficacy of a hybrid LED with the $(BODFluTh)_2FB$ encapsulant directly applied to the LED or on a glass slide 5 mm above the blue LED. The data shown here is the same as in Figures S6 and S8 of Ref. [46] except that the LED with the material applied on a glass slide was operated for longer since publication. (b) EL spectra of the same hybrid LED measured using pulsed and continuous current of 25 mA and remeasured after 15 months. Modified and reprinted with permission from Refs. [29, 46] under Creative Commons License.

5 mm above the LED. Again, the LED was continuously operated up to 750 hours (31 days) and spectra were recorded at different time intervals. The luminous efficacy is shown in Figure 10.17a together with the results from the LED where the material is directly applied to the LED. The values are normalized to the initial value for comparison. After about 200 hours, the luminous efficacy decreased by less than 10%, while almost no change was observed for the CCT or chromaticity coordinates. Even after 1 month of continuous operation, the luminous efficacy only decreased by 15%. The chromaticity coordinates shifted from (0.36, 0.32) to (0.32, 0.28) and the CCT from 4400 K to 6500 K. Similar continuous operation studies were carried out for the blue LED coated with only the transparent encapsulant (CHDV with PAG). The results showed that the transparent encapsulant was not affected by the LED or temperature over time, indicating it is optically stable during operation of the LED.

The same $(BODFluTh)_2FB$ encapsulated LED was also remeasured after 15 months using a pulsed current to minimize heating of the LED [29]. The initially recorded EL spectrum and spectrum after 15 months are displayed in Figure 10.17b. The decrease in intensity from the organic material in the transparent matrix indicates degradation due to oxygen and/or moisture in the environment. The LED emission was stable over time. Figure 10.17b also shows the spectra of the same LED measured using either continuous or pulsed current conditions. It can be observed that the yellow emission intensity is higher for pulsed current operation compared with continues current, while the blue LED emission remains the same. Time-resolved luminescence measurements showed that both the blue LED and the organic material have similar decay times in the order of 10s of nanoseconds ruling out different decays times as the cause for the observed differences in intensity. It is more likely that increased population of non-radiative triplet states during continuous current operation is the cause of a drop in intensity [82]. During pulsed current operation, the ground state can be populated again since the triplet lifetime (up to 100 μs) is shorter than the pulse period of the current pulse of 500 μs [83].

10.5.6 Next-Generation White-Emitting LEDs with Improved Efficacy

An alternative approach to small molecule color converters focused on removing the BODIPY emissive unit to eliminate the inherent self-absorption at green wavelengths, thereby increasing the color rendering and efficiency (see Section 10.2.2).

Two compounds bearing either one or two units of benzothiadiazole (BT) coupled to fluorene and peripheral triphenylamine donor units, $(TPA\text{-}Flu)_2BT$ and $(TPA\text{-}Flu)_2BTBT$, as shown in Figure 10.18a–b, respectively, were reported [84]. As seen in the UV–vis absorption and emission spectra in Figure 10.18c, the two absorption peaks at 345 nm and 430/443 nm of both compounds marginally overlap with their emission peaks at 570/584 nm. Introduction of the second BT unit redshifts both the second absorption peak and the emission by 13 nm and 14 nm, respectively.

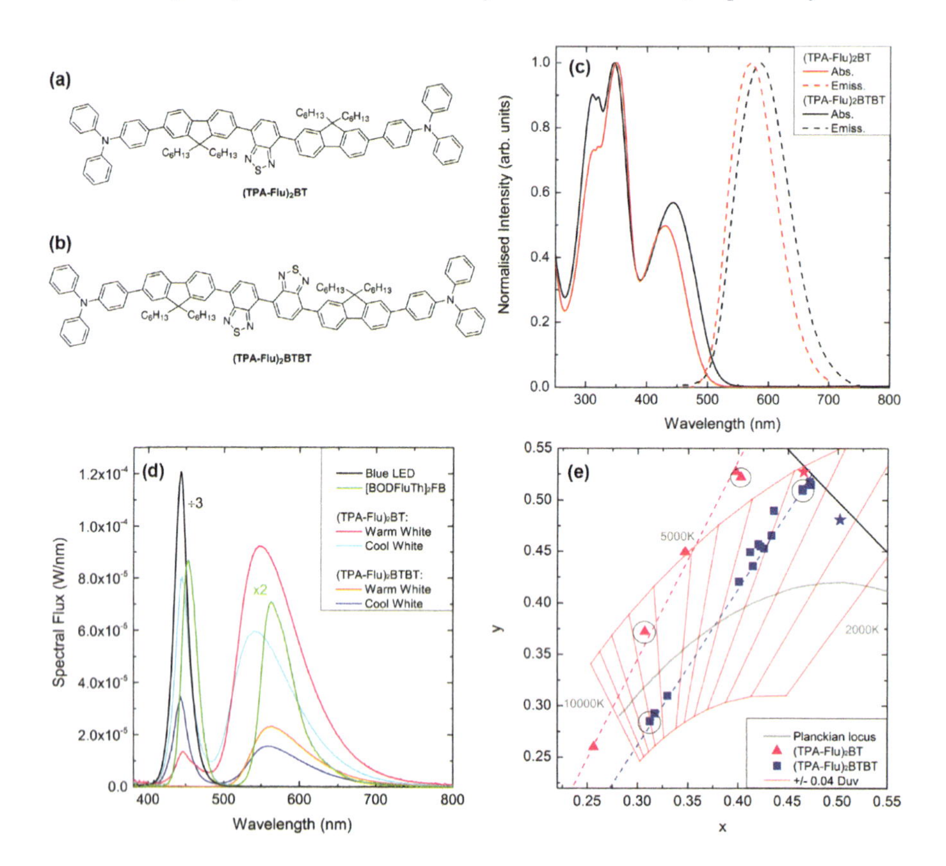

FIGURE 10.18 Chemical structure of (a) $(TPA\text{-}Flu)_2BT$ and (b) $(TPA\text{-}Flu)_2BTBT$. (c) Normalized UV–vis absorption and emission spectra of both compounds in dichloromethane dilute solution. (d) Example EL spectra of the $(TPA\text{-}Flu)_2BT$ and $(TPA\text{-}Flu)_2BTBT$ hybrid LEDs demonstrating warm and cool white light emission. For comparison, the spectrum for a $(BODFluTh)_2FB$ LED is shown. (e) 1931 CIE chromaticity diagram showing the chromaticity coordinates for both compounds as an evolution of concentration. The circled data points correspond to the warm and cool LEDs from (d). Adapted and reprinted with permission from Ref. [84] under Creative Commons License.

In order to investigate the down-conversion properties, the compounds were incorporated into a transparent encapsulant (CHDV with PAG) and deposited on blue LEDs. Example EL spectra of different concentrations of the two compounds yielding warm and cool white light are shown in Figure 10.18d. For comparison, the spectrum of a $(BODFluTh)_2FB$ hybrid LED is included as well showing that the two new compounds exhibit a much broader emission from the organic material in the green to yellow spectral range. This is reflected by the much higher CRI of 51–66 for the single BT compound and 40–61 for the double BT compound depending on the concentration showing that reasonable color rendering can be achieved. The chromaticity coordinates of the entire concentration series are shown in Figure 10.18e with the points corresponding to the warm and cool LEDs marked by circles. As before, changing the concentration shifts the chromaticity coordinates along a straight line from the blue to the yellow/green region in the chromaticity diagram.

Luminous efficacies of 41 lm/W and 10 lm/W, and B-W efficacies of 368 lm/W and 116 lm/W for $(TPA\text{-}Flu)_2BT$ and $(TPA\text{-}Flu)_2BTBT$, respectively, were reported. The much lower efficacies of $(TPA\text{-}Flu)_2BTBT$ are due to the low PLQY of 17% compared with the 63% for $(TPA\text{-}Flu)_2BT$. The much higher efficacy numbers for the $(TPA\text{-}Flu)_2BT$ LEDs compared with $(BODFluTh)_2FB$ (luminous and B-W efficacies: 13.6 lm/W and 100–120 lm/W [29,46]) are due to the enhanced green emission in the 490–520 spectral range. The B-W efficacy also exceeds the values of phosphor-based white LEDs (200–300 lm/W). It should be noted that these numbers are comparable because the blue LEDs used in these studies were identical for all measurements.

10.5.7 Metal–Organic Frameworks

One issue encountered with previous color converters is aggregation of the molecules with increasing concentration. This aggregation can impact optical properties by redshifting and quenching the luminescence due to non-radiative recombination [29,46,79].

One method to counteract this is to integrate small molecule emissive compounds into a metal–organic framework (MOF), which are well-defined rigid networks where the organic material, or ligand, is connected through metal ions to the network. This integration into a rigid network should prevent aggregation and preserve or improve the optical properties as schematically shown in Figure 10.19a [85,86].

The structure of the BT-based organic compound BTBMBA is shown in Figure 10.19b [87]. To synthesize the MOF structure MOF-MTBMBA, the ligand BTBMBA was linked by Zr ions to form a structure with UiO-68 topology (Figure 10.19c) [88,89]. The SE image of MOF-MTBMBA in Figure 10.19d shows the regular octahedral morphology of the crystals with sizes around 500–700 nm.

To investigate the optical properties, the material was incorporated into a commercially available polyurethane resin, OptiTEC™ 4200. The UV–vis absorption and emission spectra of the free ligand and the ligand incorporated into the MOF structure are shown in Figure 10.19e. Both have a single absorption peak at 408 and 412 nm and emission peak at 501 and 514 nm for MTBMBA and MOF-MTBMBA, respectively. The slight blueshift for the MOF structure for both absorption and

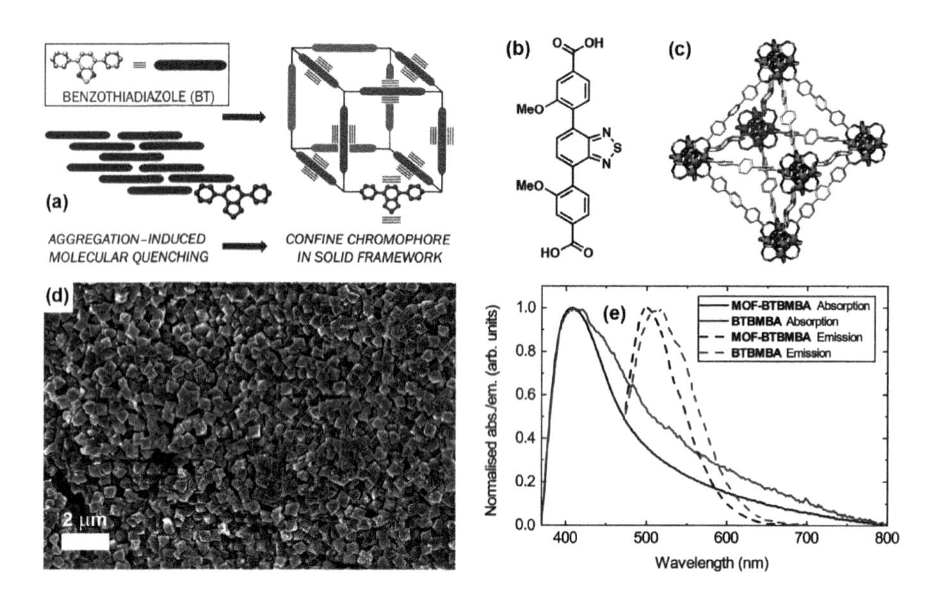

FIGURE 10.19 (a) Schematic of aggregation-induced luminescence quenching and integration into MOF structure. (b) Structure of BTBMBA. (c) Crystal structure of UiO-68 used for MOF-BTBMBA. (d) SE image of MOF-BTBMBA. (e) UV–vis absorption and emission of BTBMBA and MOF-BTBMBA encapsulated in the polyurethane resin. Adapted and reprinted with permission from Ref. [87] under Creative Commons License.

emission is associated with the steric confinement of the ligand in the MOF causing an altered environment of the ligand.

A series of hybrid LEDs were prepared using different concentrations (0.33–4%) of the ligand (BTBMA) or the MOF (MOF-BTBMA) incorporated into the resin. The EL spectra on identical logarithmic intensity scales of those two concentrations series are shown in Figure 10.20a–b. Although weak, both exhibit an emission peak around 550 nm, which is increasing with concentration, in addition to the emission from the blue LED. The spectra of the bare blue LEDs before they were coated with the converter materials do not have this emission peak, clearly indicating energy conversion into the yellow region by the organic material. Most noteworthy is the increased intensity for the MOF-BTBMA LEDs compared with the free ligand LEDs as displayed in Figure 10.20c, which shows the ratio of the integrated emission intensities of the coated LED and the bare blue LEDs in the wavelength range of 525–600 nm.

To further investigate this observation, the luminous efficacy was determined for both hybrid LED series and is shown in Figure 10.20d. It increases with increasing concentration for the LEDs coated with the MOF structure, whereas the free ligand versions show a decrease in efficacy. This is caused by the luminous flux which shows the same behavior. This indicates that with increasing concentration, the ligand keeps absorbing the blue light but does not transfer and re-emit it at longer wavelength. On the other hand, the MOF structure shows the opposite behavior. This is most likely caused by luminescence quenching due to aggregation of the free ligand.

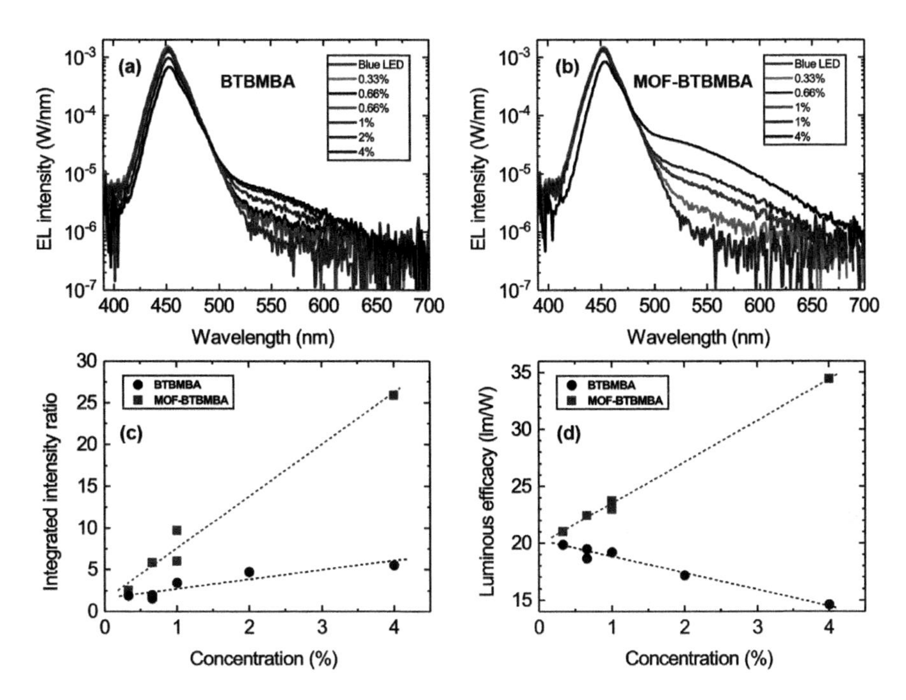

FIGURE 10.20 EL spectra of the blue LEDs coated with the (a) ligand BTBMBA and (b) MOF-BTBMBA using concentrations of 0.33–4% in the polyurethane resin. (c) Ratio of the integrated intensities of the coated LEDs and the bare blue LEDs in the wavelength range of 525–600 nm and (d) luminous efficacies of the two hybrid LED series. Adapted and reprinted with permission from Ref. [87] under Creative Commons License.

Incorporation into the MOF structure enhances the light emission by supressing aggregation. While the device properties are far from outstanding, this showcases the promise of this methodology as a means to fabricate hybrid LEDs using MOFs.

10.6 SUMMARY

This chapter has illustrated the functionality of hybrid inorganic–organic LEDs. This technology utilizes the high efficiency of blue inorganic LEDs and combines it with the versatility of organic color converters to provide an alternative method for white light generation.

The extended background was aimed to provide a sufficient introduction to nitride semiconductors and their LEDs, but also to familiarize the reader with the background to colorimetry and radiometry of light sources. For example, the terms efficiency and efficacy are related but can describe quite different properties. It is also important to consider what factors contribute to these color parameters, e.g. luminous efficacy and CRI, in order to know what properties of the color converter need to be adjusted or modified for the desired emission characteristics.

A brief introduction and background to the chemistry behind organic color converting compounds aimed to showcase the number of factors that can be controlled and modified when designing organic color converters. Although not comprehensive, it hopefully provided some context for the latter two sections.

Light-emitting polymers are one choice of organic color converter. They are readily commercially available and cost effective. The examples given demonstrate their color conversation capabilities using either one or multiple polymers or polymers in combination with other types of color converters.

Luminescent small molecules have the advantage of a precisely known molecular structure which allows straightforward replication. Their absorption and emission can also be easily tuned through the manipulation of their structure, such as exchanging units or changing the conjugation length, to achieve the desired color characteristics. This section presented examples that demonstrate the evolution of a series of novel small molecule-based converters to adjust and improve their color conversion properties when coupled with a blue inorganic LED. It highlighted the design considerations and changes made to the small molecule in order to achieve the desired white light emission. Before applying to the inorganic LED, the deposition method needs to be considered in order to achieve good coverage of the LED chip but also, for example, prevent detrimental aggregation-induced effects on the luminescence. Remaining challenges, such as lifetime and degradation, which are needed to make small molecules commercially viable, were also discussed.

ACKNOWLEDGMENTS

The research carried out by the authors was funded by the UK Engineering and Physical Sciences Research Council (EPSRC).

REFERENCES

1. Shchekin, O. and M.G. Craford, History of solid-state light sources, in *Handbook of advanced lighting technology*, ed. R. Karlicek, et al., 2017, Cham: Springer International Publishing, p. 1–30.
2. Zukauskas, A., M.S. Shur, and R. Gaska, *Introduction to solid-state lighting*. 2002: Wiley-Interscience.
3. *Introduction to nitride semiconductor blue lasers and light emitting diodes*, ed. S. Nakamura and S.F. Chichibu. 2000: CRC Press
4. Available from: https://www.nobelprize.org/prizes/physics/2014.
5. Schubert, E.F., *Light-emitting diodes*. 2006: Cambridge University Press.
6. McKittrick, J. and L.E. Shea-Rohwer, Review: down conversion materials for solid-state lighting. *J. Am. Ceram. Soc.*, 2014. 97(5): p. 1327–1352.
7. Massari, S. and M. Ruberti, Rare earth elements as critical raw materials: focus on international markets and future strategies. *Res. Policy*, 2013. 38(1): p. 36–43.
8. Kamtekar, K.T., A.P. Monkman, and M.R. Bryce, Recent advances in white organic light-emitting materials and devices (WOLEDs). *Adv. Mater.*, 2010. 22(5): p. 572–582.
9. Farinola, G.M. and R. Ragni, Organic emitters for solid state lighting. *J. Solid State Light.*, 2015. 2(1):9.
10. *III-nitride based light emitting diodes and applications*, ed. T.-Y. Seong, et al. 2013: Springer Netherlands.
11. *III-nitride ultraviolet emitters: technology and applications*, ed. M. Kneissl and J. Rass. 2016: Springer International Publishing.

12. Gorczyca, I., et al., Limitations to band gap tuning in nitride semiconductor alloys. *Appl. Phys. Lett.*, 2010. 96(10): p. 101907.
13. Morkoç, H., *Handbook of nitride semiconductors and devices: materials properties, physics and growth, Volume 1*. 2008: Wiley-VCH.
14. Neamen, D., *Semiconductor physics and devices*. 4th ed. 2011: McGraw-Hill Education.
15. Chichibu, S., K. Wada, and S. Nakamura, Spatially resolved cathodoluminescence spectra of InGaN quantum wells. *Appl. Phys. Lett.*, 1997. 71(16): p. 2346–2348.
16. Sugahara, T., et al., Direct evidence that dislocations are non-radiative recombination centers in GaN. *Jpn. J. Appl. Phys.*, 1998. 37(Part 2, No. 4A): p. L398–L400.
17. Casey, H.C., B.I. Miller, and E. Pinkas, Variation of minority-carrier diffusion length with carrier concentration in GaAs liquid-phase epitaxial layers. *J. Appl. Phys.*, 1973. 44(3): p. 1281–1287.
18. Ettenberg, M., H. Kressel, and S.L. Gilbert, Minority carrier diffusion length and recombination lifetime in GaAs:Ge prepared by liquid-phase epitaxy. *J. Appl. Phys.*, 1973. 44(2): p. 827–831.
19. *Commission internationale de l'Eclairage Proceedings*. 1931.
20. Vos, J.J., Colorimetric and photometric properties of a 2° fundamental observer. *Color. Res. Appl.*, 1978. 3(3): p. 125–128.
21. MacAdam, D.L., Specification of small chromaticity differences. *J. Opt. Soc. Am.*, 1943. 33(1): p. 18–26.
22. Wyszecki, G. and W.S. Stiles, *Color science: concepts and methods, quantitative data and formulae*. 2nd ed. 2000: Wiley-Blackwell.
23. Planck, M., Ueber eine verbesserung der Wien'schen spectralgleichung. *Verhandlungen der Deutschen physikalischen Gesellschaft*, 1900. 13: p. 202–202.
24. Planck, M., Ueber das gesetz der energieverteilung im normalspectrum. *Ann. Phys.*, 1901. 309(3): p. 553–563.
25. Wien, W., Ueber die energievertheilung im emissionsspectrum eines schwarzen Körpers. *Anal. Phys.*, 1896. 294: p. 662–662.
26. in *CIE No. 17.4: international lighting vocabulary*. 1987.
27. Taylor, E., P.R. Edwards, and R.W. Martin, Colorimetry and efficiency of white LEDs: spectral width dependence. *Phys. Status Solidi A*, 2012. 209(3): p. 461–464.
28. in *CIE No. 13.3: method of measuring and specifying colour rendering properties of light sources*. 1995.
29. Bruckbauer, J., et al., Colour tuning in white hybrid inorganic/organic light-emitting diodes. *J. Phys. D. Appl. Phys.*, 2016. 49(40): p. 405103–405103.
30. Bando, K., et al., Development of high-bright and Pure-white LED lamps. *J. Light. Vis. Environ.*, 1998. 22(1): p. 2–5.
31. Schlotter, P., et al., Fabrication and characterization of GaN/InGaN/AlGaN double heterostructure LEDs and their application in luminescence conversion LEDs. *J. Mater. Sci. Eng. B*, 1999. 59(1): p. 390–394.
32. Ronda, C.R., T. Jüstel, and H. Nikol, Rare earth phosphors: fundamentals and applications. *J. Alloys Compd.*, 1998. 275(0): p. 669–676.
33. *Organic light emitting devices: synthesis, properties and applications*, ed. K. Müllen and U. Scherf. 2006: Wiley-VCH.
34. Reineke, S., et al., White organic light-emitting diodes: status and perspective. *Rev. Mod. Phys.*, 2013. 85(3): p. 1245–1293.
35. Cho, Y.J., K.S. Yook, and J.Y. Lee, Cool and warm hybrid white organic light-emitting diode with blue delayed fluorescent emitter both as blue emitter and triplet host. *Sci. Rep.*, 2015. 5(1): p. 7859.
36. Yu, L., et al., Red, green, and Blue light-emitting polyfluorenes containing a dibenzothiophene-S,S-dioxide unit and efficient high-color-rendering-index white-light-emitting diodes made therefrom. *Adv. Funct. Mater.*, 2013. 23(35): p. 4366–4376.

37. Kim, T.H., et al., White-light-emitting diodes based on iridium complexes via efficient energy transfer from a conjugated polymer. *Adv. Funct. Mater.*, 2006. 16(5): p. 611–617.
38. Angioni, E., et al., A single emitting layer white OLED based on exciplex interface emission. *J. Mater. Chem. C*, 2016. 4: p. 3851–3856.
39. Roncali, J., Synthetic principles for bandgap control in linear π-conjugated systems. *Chem. Rev.*, 1997. 97(1): p. 173–206.
40. Roncali, J., Molecular engineering of the band gap of π-conjugated systems: facing technological applications. *Macromol. Rapid Commun.*, 2007. 28(17): p. 1761–1775.
41. Bredas, J.-L., Mind the gap! *Materi. Horizons*, 2014. 1(1): p. 17–19.
42. Schlotter, P., R. Schmidt, and J. Schneider, Luminescence conversion of blue light emitting diodes. *Appl. Phys. A Mater. Sci. Process.*, 1997. 64(4): p. 417–418.
43. Hide, F., et al., White light from InGaN/conjugated polymer hybrid light-emitting diodes. *Appl. Phys. Lett.*, 1997. 70(20): p. 2664–2666.
44. Zhang, C. and A.J. Heeger, Gallium nitride/conjugated polymer hybrid light emitting diodes: performance and lifetime. *J. Appl. Phys.*, 1998. 84(3): p. 1579–1582.
45. Gather, M.C., A. Köhnen, and K. Meerholz, White organic light-emitting diodes. *Adv. Mater.*, 2011. 23(2): p. 233–248.
46. Findlay, N.J., et al., An organic down-converting material for white-light emission from hybrid LEDs. *Adv. Mater.*, 2014. 26: p. 7290–7290.
47. Kanibolotsky, A.L., I.F. Perepichka, and P.J. Skabara, Star-shaped pi-conjugated oligomers and their applications in organic electronics and photonics. *Chem. Soc. Rev.*, 2010. 39: p. 2695–2728.
48. *Organic light-emitting materials and devices*, ed. Z.R. Li. 2015: CRC Press.
49. *Organic light-emitting devices*, ed. J. Shinar. Springer.
50. Bernius, M.T., et al., Progress with light-emitting polymers. *Adv. Mater.*, 2000. 12(23): p. 1737–1750.
51. Ying, L., et al., White polymer light-emitting devices for solid-state lighting: materials, devices, and recent progress. *Adv. Mater.*, 2014. 26(16): p. 2459–2473.
52. Smith, R., et al., Hybrid III-nitride/organic semiconductor nanostructure with high efficiency nonradiative energy transfer for white light emitters. *Nano Lett.*, 2013. 13(7): p. 3042–3047.
53. Baldo, M.A., et al., Highly efficient phosphorescent emission from organic electroluminescent devices. *Nature*, 1998. 395(6698): p. 151–154.
54. Förster, T., Zwischenmolekulare energiewanderung und fluoreszenz. *Ann. Phys.*, 1948. 437(1–2): p. 55–75.
55. Achermann, M., et al., Energy-transfer pumping of semiconductor nanocrystals using an epitaxial quantum well. *Nature*, 2004. 429: p. 642–642.
56. Jiang, H.X. and J.Y. Lin, Nitride micro-LEDs and beyond – a decade progress review. *Opt. Express*, 2013. 21(S3): p. A475–A484.
57. Rajbhandari, S., et al., A review of gallium nitride LEDs for multi-gigabit-per-second visible light data communications. *Semicond. Sci. Technol.*, 2017. 32(2): p. 023001.
58. Lin, J.Y. and H.X. Jiang, Development of microLED. *Appl. Phys. Lett.*, 2020. 116(10): p. 100502.
59. Heliotis, G., et al., Spectral conversion of InGaN ultraviolet microarray light-emitting diodes using fluorene-based red-, green-, blue-, and white-light-emitting polymer overlayer films. *Appl. Phys. Lett.*, 2005. 87(10): p. 103505.
60. Chun, H., et al., Visible light communication using a blue GaN μ LED and fluorescent polymer color converter. *IEEE Photon. Technol. Lett.*, 2014. 26(20): p. 2035–2038.
61. O'Brien, D.C., et al. *Visible light communications: challenges and possibilities*. in *2008 IEEE 19th International Symposium on Personal, Indoor and Mobile Radio Communcations*. 2008.

62. Chen, M., et al., Application of a novel red-emitting cationic iridium(III) coordination polymer in warm white light-emitting diodes. *Opt. Mater.*, 2018. 76: p. 141–146.
63. Jang, E., et al., White-light-emitting diodes with quantum dot color converters for display backlights. *Adv. Mater.*, 2010. 22(28): p. 3076–3080.
64. Mangum, B.D., et al., Exploring the bounds of narrow-band quantum dot downconverted LEDs. *Photonics Research*, 2017. 5(2): p. A13–A22.
65. Volkan Demir, H., et al., White light generation tuned by dual hybridization of nanocrystals and conjugated polymers. *New J. Phys.*, 2007. 9(10): p. 362–362.
66. Ulrich, G., R. Ziessel, and A. Harriman, The Chemistry of fluorescent bodipy dyes: versatility unsurpassed. *Angew. Chem. Int. Ed.*, 2008. 47(7): p. 1184–1201.
67. Ikawa, Y., S. Moriyama, and H. Furuta, Facile syntheses of BODIPY derivatives for fluorescent labeling of the 3′ and 5′ ends of RNAs. *Anal. Biochem.*, 2008. 378(2): p. 166–170.
68. Bonardi, L., et al., Fine-tuning of yellow or red photo- and electroluminescence of functional difluoro-boradiazaindacene films. *Adv. Funct. Mater.*, 2008. 18(3): p. 401–413.
69. Hattori, S., et al., Charge separation in a nonfluorescent donor-acceptor dyad derived from boron dipyrromethene dye, leading to photocurrent generation. *J. Phys. Chem. B*, 2005. 109(32): p. 15368–15375.
70. Boens, N., et al., Rational design, synthesis, and spectroscopic and photophysical properties of a visible-light-excitable, ratiometric, fluorescent near-neutral pH indicator based on BODIPY. *Chem.-Eur. J.*, 2011. 17(39): p. 10924–10934.
71. Dodani, S.C., Q. He, and C.J. Chang, A turn-on fluorescent sensor for detecting nickel in living cells. *J. Am. Chem. Soc.*, 2009. 131(50): p. 18020–18021.
72. Dexter, D.L., A theory of sensitized luminescence in solids. *J. Chem. Phys.*, 1953. 21(5): p. 836–850.
73. Diring, S., et al., Star-shaped multichromophoric arrays from bodipy dyes grafted on truxene core. *J. Am. Chem. Soc.*, 2009. 131(17): p. 6108–6110.
74. Findlay, N.J., et al., Linear oligofluorene-BODIPY structures for fluorescence applications. *J. Mater. Chem. C*, 2013. 1: p. 2249–2256.
75. Hofmann, D.M., et al., Properties of the yellow luminescence in undoped GaN epitaxial layers. *Phys. Rev. B*, 1995. 52(23): p. 16702–16706.
76. Bruckbauer, J., *Luminescence study of III-nitride semiconductor nanostructures and LEDs*. 2013, University of Strathclyde.
77. Bruckbauer, J., et al., Probing light emission from quantum wells within a single nanorod. *Nanotechnology*, 2013. 24(36): p. 365704–365704.
78. Kuehne, A.J.C., et al., Direct laser writing of nanosized oligofluorene truxenes in UV-transparent photoresist microstructures. *Adv. Mater.*, 2009. 21(7): p. 781–785.
79. Hong, Y., J.W.Y. Lam, and B.Z. Tang, Aggregation-induced emission: phenomenon, mechanism and applications. *Chem. Commun.*, 2009(29): p. 4332–4353.
80. Zhang, Z., et al., Color-tunable solid-state emission of 2,2′-biindenyl-based fluorophores. *Angew. Chem. Int. Ed.*, 2011. 50(49): p. 11654–11657.
81. Gorrotxategi, P., M. Consonni, and A. Gasse, Optical efficiency characterization of LED phosphors using a double integrating sphere system. *J. Solid State Light.*, 2015. 2:1.
82. Zhang, Y. and S.R. Forrest, Existence of continuous-wave threshold for organic semiconductor lasers. *Phys. Rev. B*, 2011. 84(24): p. 241301.
83. Ziessel, R. and A. Harriman, Artificial light-harvesting antennae: electronic energy transfer by way of molecular funnels. *Chem. Commun.*, 2011. 47(2): p. 611–631.
84. Taylor-Shaw, E., et al., Cool to warm white light emission from hybrid inorganic/organic light-emitting diodes. *J. Mater. Chem. C*, 2016. 4: p. 11499.
85. Nguyen, T.N., F.M. Ebrahim, and K.C. Stylianou, Photoluminescent, upconversion luminescent and nonlinear optical metal-organic frameworks: from fundamental photophysics to potential applications. *Coord. Chem. Rev.*, 2018. 377: p. 259–306.

86. Lustig, W.P. and J. Li, Luminescent metal–organic frameworks and coordination polymers as alternative phosphors for energy efficient lighting devices. *Coord. Chem. Rev.*, 2018. 373: p. 116–147.
87. Angioni, E., et al., Implementing fluorescent MOFs as down-converting layers in hybrid light-emitting diodes. *J. Mater. Chem. C*, 2019. 7: p. 2394–2394.
88. Bai, Y., et al., Zr-based metal–organic frameworks: design, synthesis, structure, and applications. *Chem. Soc. Rev.*, 2016. 45(8): p. 2327–2367.
89. Cavka, J.H., et al., A new zirconium inorganic building brick forming metal organic frameworks with exceptional stability. *J. Am. Chem. Soc.*, 2008. 130(42): p. 13850–13851.

11 Organic–Inorganic Semiconductor Heterojunctions for Resistive Switching Memories

Shuang Gao and Run-Wei Li

CONTENTS

11.1 INTRODUCTION

As an essential component for modern computers, memory devices have gained great success over the past half century. In particular, the rapid development of silicon-based flash memories is making consumer electronic devices like laptop computers and mobile phones to be faster and faster as well as more portable. However, such memories are now approaching severe downscaling constraints in terms of data fidelity, heat death, and unaffordable manufacturing costs for ultra-large-scale implementation (Gao et al. 2019). On the other hand, the current Big Data era is witnessing a striking explosion in global digital information, which is estimated to reach over 40 trillion gigabytes (about 5200 gigabytes for each person) until the end of 2020 but will still exhibit an exponential growing speed for a long period in future (Gantz and Reinsel 2012). The development of alternative memory devices based on novel structures, materials, and mechanisms is therefore of great importance for further promoting the progress of information technology.

Resistive switching memories, which store binary digital information through external electric field-induced resistance change phenomena in a simple "electrode/storage medium/electrode" sandwich-like structure, have aroused great interest among academic and industrial communities over the past two decades (Pan, Gao, et al. 2014; Gao et al. 2019). Unlike three-terminal transistor cell for flash memories, the simple two-terminal resistor cell enables resistive switching memories to be readily integrated into passive crossbar arrays (Figure 11.1a) with even three-dimensional stacking architecture for the highest possible storage density (Seok et al. 2014). Meanwhile, resistive switching usually occurs within a few nanoseconds (Yang et al. 2009; Choi et al. 2016), which promises an approximately one thousand times faster operation speed than flash memories. Together with the demonstrated low energy consumption of sub-pico joule (Tsai et al. 2013), high switching endurance over trillion cycles (Goswami et al. 2017), and nonvolatile data retention exceeding 10 years (Gao et al. 2013), resistive switching memories have been documented by the International Technology Roadmap for Semiconductors (ITRS) as one of the most promising candidates for next-generation nonvolatile information storage technology (ITRS 2015).

For a resistive switching memory cell under external electric fields, the two electrodes act mainly as current transporting paths, whereas the storage medium can

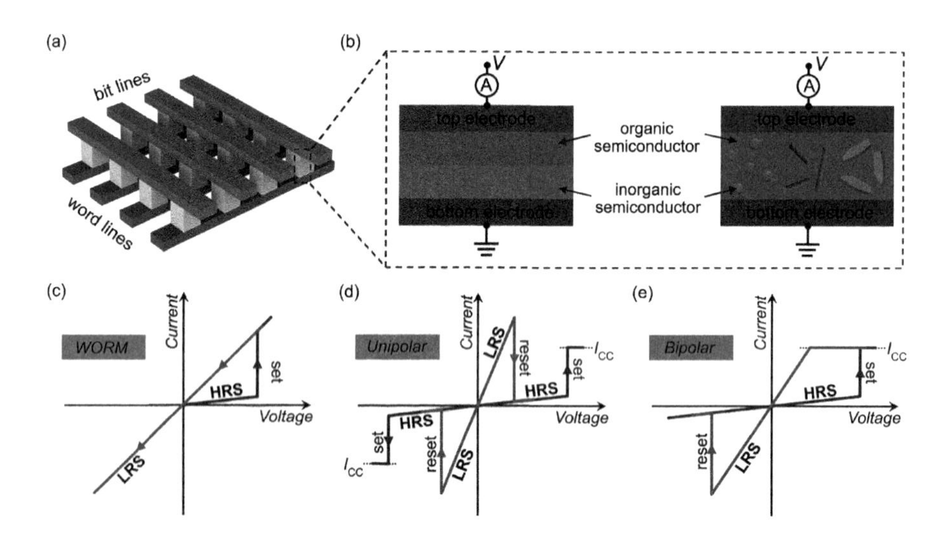

FIGURE 11.1 Schematic device structure and various current–voltage (*I–V*) characteristics of nonvolatile organic–inorganic semiconductor heterojunction resistive switching memories. (a) Passive crossbar array with one memory cell at each intersection. (b) Memory cell structure with either a bilayer structure or a single layer consisting of inorganic nanomaterials dispersed in an organic matrix. The inorganic nanomaterials can be zero-dimensional nanoparticles, one-dimensional nanowires, or two-dimensional nanosheets. (c–e) *I–V* characteristics of WORM, unipolar, and bipolar resistive switching behaviors. High resistance state (HRS) and low resistance state (LRS) are also frequently denoted as OFF and ON states, respectively. I_{CC} denotes the compliance current during set process to prevent the device from permanent breakdown. (Adapted with permission from (Gao et al. 2019), © 2019 The Royal Society of Chemistry)

undergo certain physical or chemical changes that account for cell resistance change. Storage media are therefore of critical importance for resistive switching device performance, and a great number of semiconductors and insulators have been explored thus far as the storage media for resistive switching memories, as documented in detail in some recent comprehensive review articles (Pan, Gao, et al. 2014; Gao et al. 2019). The focus herein is on organic–inorganic semiconductor heterojunctions for resistive switching applications, with either a bilayer structure (Smith and Forrest 2004; Khan et al. 2019) or a single layer consisting of inorganic nanomaterials dispersed in an organic matrix (Park et al. 2009; Wang et al. 2015), as shown in Figure 11.1b. Related devices can exhibit bistable write-once-read-many (WORM, Figure 11.1c), unipolar (Figure 11.1d), or bipolar (Figure 11.1e) current–voltage (I–V) hysteresis characteristics for nonvolatile information storage, depending on not only the compositions of adopted organic and inorganic semiconductors but also the ratio between them. For example, with the increase of Au@air@TiO_2-h yolk–shell microsphere content from 5 to 20 wt% in a poly(3-hexylthiophene) (P3HT) matrix, resistive switching characteristic of the Al/(P3HT + Au@air@TiO_2-h)/indium-tin oxide (ITO) device is found to change from WORM to bipolar (Wang et al. 2015). This actually provides us an additional, simple yet effective method to adjust resistive switching device performance for various applications. It is also demonstrated that a properly designed heterojunction can make related resistive switching devices more promising for practical applications. For example, the rectifying p-n junction between p-type poly(3,4-ethylenedioxythiophene):poly(styrenesulfonate) (PEDOT:PSS) and n-type ZnO makes the Ag/PEDOT:PSS/ZnO/ITO device to exhibit a highly asymmetric memory window (i.e., HRS/LRS resistance ratio or ON/OFF current ratio) under positive and negative reading voltages, which can greatly alleviate the sneak-path issue faced by resistive switching memory-based passive crossbar arrays (Khan et al. 2019). Besides, the incorporation of organic semiconductors into resistive switching media can make related devices to exhibit light weight, low-cost fabrication by solution processing, and also intrinsic softness for future flexible and even wearable electronics (Li, Sui, and Li 2017; Zhang et al. 2020).

In this chapter, the latest research progress of organic–inorganic semiconductor heterojunctions for resistive switching memory applications is comprehensively reviewed. The involved semiconducting materials for WORM, unipolar, and bipolar resistive switching memory devices, accompanied by working mechanisms and device performance, are summarized in detail in Sections 11.2–11.4, respectively. Finally, the current challenges and future prospects in this field are briefly discussed.

11.2 ORGANIC–INORGANIC SEMICONDUCTOR HETEROJUNCTIONS FOR WORM MEMORY DEVICES

The main characteristic of WORM memory devices is that they can be written only once but the written information can be permanently stored, as shown in Figure 11.1c. Accordingly, WORM memory devices are mainly applicable to store the information with high reliability but small capacity, such as the basic input/output system programs for computers and the electronic instructions for production and testing equipment. Compared with the commonly used erasable programable read-only memory that relies on three-terminal crystalline Si transistors, the newly proposed WORM resistive switching memory has huge advantages in terms of simple structure, easy fabrication, and high-density integration.

The first organic–inorganic semiconductor heterojunction for WORM resistive switching memory applications was proposed by Forrest et al. from Princeton University in 2003 (Möller et al. 2003). The memory cell at each intersection of the Au row and Al column electrodes was designed to consist of a spin-coated PEDOT:PSS layer on top of a non-crystalline Si p-n junction diode, as shown in Figure 11.2a. It is noted that PEDOT:PSS is a typical p-type organic semiconductor with good water solubility and highly tunable conductivity by changing the PEDOT/PSS component ratio. At initial state, the memory cell had a relatively high conductivity, with a rectifying ratio of ~50 at ±1.5 V for passive matrix memory addressing. Interestingly, a remarkable decrease in cell conductivity of ~10^4 at 1.5 V was observed after applying a voltage sweep from 0 to 10 V on it, as shown in Figure 11.2b. Besides, transient electrical analysis revealed that the time needed for switching the memory cell from conducing to insulting was ~2 μs.

To simplify memory cell structure, Forrest et al. subsequently suggested the direct using of PEDOT:PSS as one side of the rectifying junction (Smith and Forrest 2004). That is to say, the spin-coated PEDOT:PSS was simultaneously used as the active resistive switching layer and the p-type semiconductor for forming a p-n junction diode with the n-type Si substrate. With Au as the top electrode, the fabricated Au/PEDOT:PSS/n-Si memory cell was demonstrated unexpectedly to exhibit a larger rectifying ratio of ~10^3, a higher memory window of ~10^5, and also a faster switching speed of only ~300 ns. The simple structure, together with the excellent resistive switching performance, made the Au/PEDOT:PSS/n-Si device to be a highly promising WORM memory candidate. In addition, they had also conducted detailed x-ray photoelectron spectroscopy and temperature-dependent conductivity measurements to clarify the resistive switching mechanism. It was found that the aggregation of PSS^- component near anode in the presence of both strong electric field and high

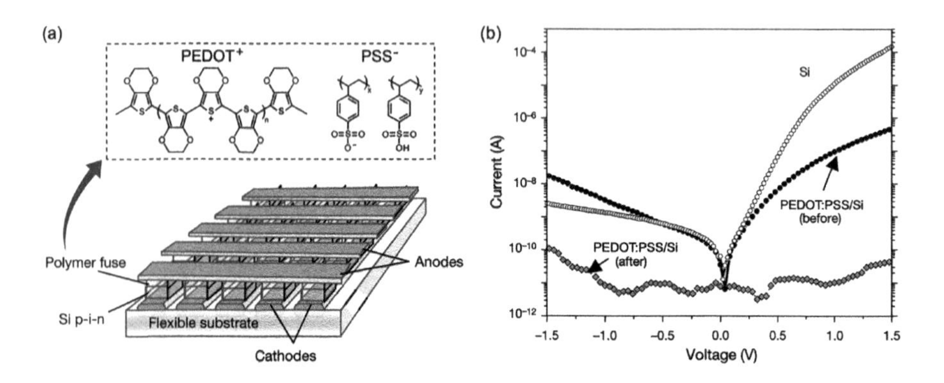

FIGURE 11.2 (a) Schematic device structure of the PEDOT:PSS WORM memory device. Inset: chemical structure of PEDOT:PSS. (b) *I–V* characteristics of the device before and after resistive switching. (Adapted with permission from (Möller et al. 2003), © 2003 Nature Publishing Group)

temperature should be responsible for the observed conducting-to-insulating transition phenomenon (Xu, Register, and Forrest 2006).

As the most famous way to modulate the performance of semiconductor devices, doping should also have a significant influence on memory device performance. Inspired by this, Sun et al. recently reported a careful study on the effect of single-wall carbon nanotube (SWCNT) doping level on the performance of PEDOT:PSS-based WORM-resistive switching memory (Sun et al. 2015). It is noted that, compared to the commonly used multi-wall CNTs, SWCNTs have a smaller diameter, fewer defects, and also a higher uniformity, which are beneficial to miniaturization and stability of memory devices. The composite solutions were obtained by dissolving a certain weight of SWCNTs into 1 mL isopropyl alcohol and then mixing it with 2 mL PEDOT:PSS (2.8 wt% dispersion in H_2O). The fabricated devices with 8, 5, 1, and 0.3 mg SWCNTs were denoted as samples A–D, respectively. With Al top and ITO bottom electrodes, all the as-fabricated samples exhibited a high conductance state, arising most likely from the formation of certain continuous electron pathways by close stacking of SWCNTs. However, under a negative voltage sweep of 0 to −8 V, the samples were able to be switched into a relatively low conductance state at about −6 V due to Joule heat-induced rupture of the existing continuous electron pathways, as shown by the representative *I*–*V* characteristic of sample A in Figure 11.3a. Further applying negative voltage sweeps could not recover the initial high-conduction state, indicating the so-called WORM resistive switching behavior. Also, an almost identical behavior was observed after changing the external voltage polarity, suggesting a bi-directional switchable characteristic of these samples. Further, almost no variation in conductance was detected for both states in more than 10^4 s, demonstrating a high reliability for information storage. More importantly, the memory window was found to increase significantly with the SWCNT doping level, reaching up to a very high value of ~5×10^4 for sample A with 8 mg SWCNTs, as shown in Figure 11.3b. This is possibly caused by the obvious decrease in the current of low conductance state and provides us a simple yet effective to design WORM memory devices with a desired memory window.

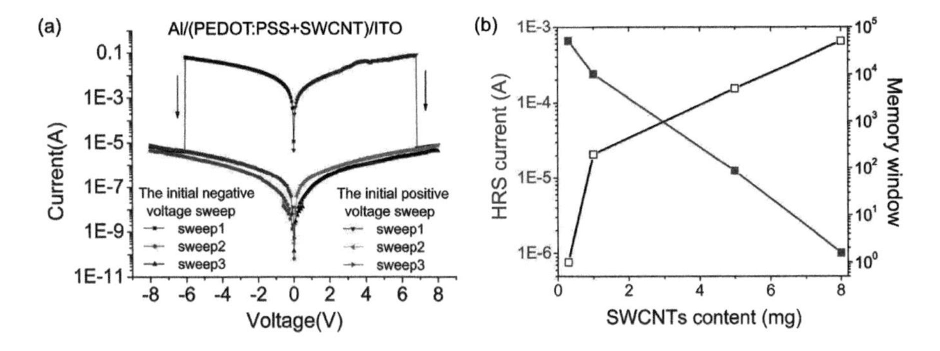

FIGURE 11.3 (a) Typical *I*–*V* characteristic of the PEDOT:PSS WORM memory device doped with 8 mg SWCNTs. (b) The dependence of memory window on SWCNT doping level. (Adapted with permission from (Sun et al. 2015), © 2015 the Owner Societies)

11.3 ORGANIC–INORGANIC SEMICONDUCTOR HETEROJUNCTIONS FOR UNIPOLAR MEMORY DEVICES

Unlike WORM memory devices that can be written only once, unipolar resistive switching memory devices are rewritable under the same voltage polarity but different threshold values, as shown in Figure 11.1d. Normally, the transition from high resistance state (HRS) to low resistance state (LRS) is denoted as set or write process, with its opposite as reset or erase process. The rewritable property under the same voltage polarity enables unipolar resistive switching memory devices to be readily integrated into crossbar arrays with two-terminal diodes (rather than there-terminal transistors) as selecting elements (Ji et al. 2013). In this case, the peripheral circuits needed for memory operation can also be simplified to some extent. As such, unipolar resistive switching memories have been acknowledged as a promising replacement for the currently used silicon-based Flash memories.

One representative organic–inorganic semiconductor heterojunction for unipolar resistive switching memory applications is the composite of aluminum tris-(8-hydroxyquinoline) (Alq_3) and Ni@NiO nanoparticles proposed by Park et al. from Hanyang University (Park et al. 2009), as shown in Figure 11.4a. The Alq_3 matrix was thermally evaporated, whereas the middle Ni@NiO nanoparticle layer was obtained by thermal evaporation of a 10 nm Ni layer followed by plasma oxidation. The selection of Ni as the metal nanocrystal was due to its high work function of 5.15 eV for electron trapping, and the surrounding NiO of 4 ~ 5 nm was expected to adhere well to the Alq_3 matrix. With Al as top and bottom electrodes, the obtained Al/(Alq_3 + Ni@NiO)/Al device was found to exhibit a stable unipolar resistive switching behavior with a high memory window of ~10^3 under either positive or negative voltage polarity, as shown in Figure 11.4b. The involved mechanism was explained by electron trapping into and de-trapping from Ni@NiO nanoparticles during set and reset processes, respectively. More interestingly, obvious negative

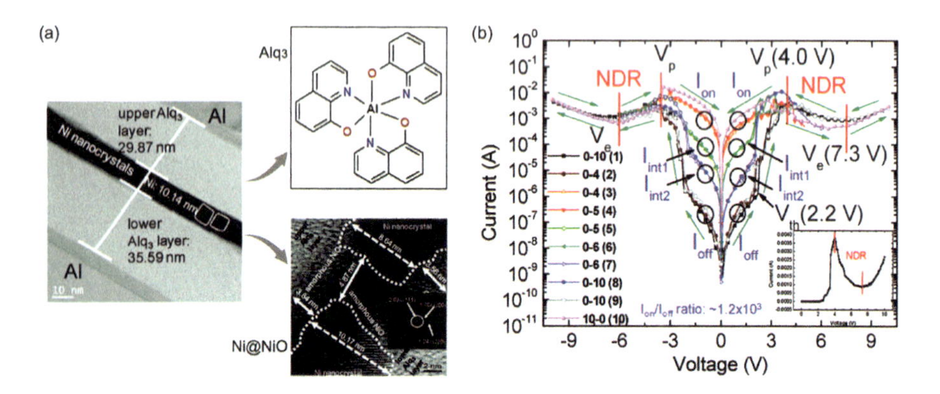

FIGURE 11.4 (a) Device structure of the Al/(Alq_3 + Ni@NiO)/Al unipolar resistive switching memory. Insets: chemical structure of Alq_3 and microstructure of Ni@NiO nanoparticles. (b) Typical *I–V* characteristic of the memory device. Inset: clear representation of the NDR region. (Adapted with permission from (Park et al. 2009), © 2009 American Chemical Society)

differential resistance (NDR) phenomenon was observed, which enabled the realization of multilevel storage by varying the stop voltage during reset process. As demonstrated in this figure, four-level storage was feasible by using 4, 5, 6, and 10 V as stop voltages, guaranteeing still a clear memory window of ~10 between any two neighboring resistance states. Furthermore, stable data retention of >10^5 s and good switching endurance of >200 cycles were also confirmed in this device. These results together make the Al/(Alq_3 + Ni@NiO)/Al device highly promising for high-density multilevel nonvolatile memory applications.

Besides small molecules, semiconducting polymers have also been embedded with transition metal oxide nanoparticles for unipolar resistive switching memory applications. For example, Lee et al. from Gwangju Institute of Science and Technology once reported a unipolar resistive switching memory based on the composite of poly(N-vinylcarbazole) (PVK) and TiO_2 nanoparticles (Cho et al. 2009), as shown in Figure 11.5a. A series of spin-coated composite films had been tried, with the solution volume ratios of PVK (5 mg/mL) and TiO_2 nanoparticles (~5 nm in diameter, 2 mg/mL) of 200:1, 150:1100:1,10:1, and 1:1. As demonstrated in Figure 11.5b, the solution volume ratio was found to have a huge effect on device memory window, arising from the obvious increase of HRS current with TiO_2 nanoparticle concentration and leading to a high memory of >10^3 in the solution volume range from ~150:1 to 100:1. Based on a detailed analysis of the charge transport mechanisms of different memory states, the observed unipolar resistive switching behavior was explained by the reversible formation and rupture of a conducting filament consisting of TiO_2 nanoparticles and some kind of defects within the PVK. Similar resistive switching behavior has also been recently reported by Zhao et al. in the composite of PVK and ZnO nanoparticles (Zhao et al. 2017). Strikingly, stable data retention of >10^5 s and excellent switching endurance of >10^4 cycles were demonstrated. These results suggest that semiconducting polymers embedded with inorganic semiconducting nanoparticles are also highly promising for unipolar resistive switching memory applications.

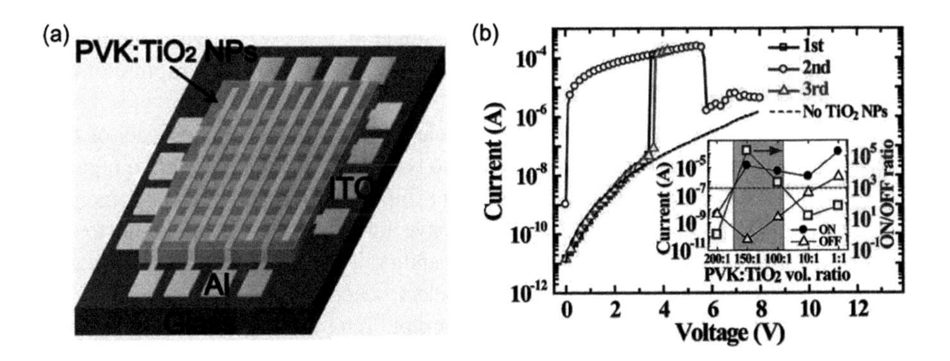

FIGURE 11.5 (a) Device structure of the Al/(PVK + TiO_2)/ITO unipolar resistive switching memory. (b) Typical *I–V* characteristic of the memory device. Inset: the effect of solution volume ratio on device memory window. (Adapted with permission from (Cho et al. 2009), © 2009 Elsevier B.V.)

11.4 ORGANIC–INORGANIC SEMICONDUCTOR HETEROJUNCTIONS FOR BIPOLAR MEMORY DEVICES

Compared to the unipolar counterparts, bipolar resistive switching memories are also rewritable but need to be written and erased under just opposite voltage polarities, as shown in Figure 11.1e. Normally, bipolar resistive switching memories are superior to the unipolar ones in terms of switching uniformity and endurance. For example, the record high endurance of $>10^{12}$ cycles was demonstrated exclusively in bipolar resistive switching memory devices (Lee et al. 2011; Goswami et al. 2017). As such, in addition to being a promising replacement for silicon-based Flash memories, bipolar resistive switching memories are also acknowledged as a potential alternative to the current main memory, that is, dynamic random access memory (DRAM). One may concern that the need of three-terminal transistors (rather than two-terminal diodes) as selecting elements in crossbar arrays will deteriorate the integration density of bipolar resistive switching memories to some degree (Lv et al. 2015). This issue could not be an obstacle in future, given the rapid development of two-terminal bipolar selecting devices in recent years (Xue et al. 2017; Sun et al. 2019).

Up to now, many organic–inorganic semiconductor heterojunctions have been explored for bipolar resistive switching memory applications, as summarized in Table 11.1. One can see that, compared to the organic–inorganic bilayer structure, that of organic matrix embedded with inorganic nanomaterials has received much more attentions, possibly due to its easy deposition through one-step solution processing like spin-coating. The adopted organic matrix materials are mainly Alq_3, P3HT, PEDOT:PSS, PVK, poly (9,9-di-octylfluorene-alt-benzothiadiazole) (PFBT), and poly-[2-methoxy-5-(2-ethyl-hexyloxy)-1,4-phenylenevinylene] (MEH-PPV). As for the embedded inorganic nanomaterials, not only zero-dimensional nanoparticles (e.g., TiO_2, ZnO, MoO_3, Cu_2S, Ag_2S, ZnS, and CdSe) but also one-dimensional nanowires like CNT and nitrogen-doped CNT (NCNT) have been frequently used. For example, both the composites of P3HT + Au@air@TiO_2-h and PEDOT:PSS + NCNT have been demonstrated to exhibit a notable bipolar resistive switching behavior with high memory window of $\geq 10^3$, good endurance of $>10^2$ cycles, and stable retention of $>10^4$ s (Wang et al. 2015; Sun et al. 2015). The switching mechanisms therein were both supposed to be the reversible formation and rupture of oxygen vacancy filaments.

In addition to commonly used nanoparticles and nanowires, nanosheets of two-dimensional layered semiconductors have also been recently introduced into organic–inorganic semiconductor heterojunctions for bipolar resistive switching memory applications. In particular, MoS_2 nanosheets have attracted much attention due to their unique properties, such as a small indirect bandgap (1.2 ~ 1.9 eV), a higher carrier mobility (200 ~ 500 $cm^2/V{\cdot}s$), a good photoelectrochemical stability, and larger surface areas (Fan et al. 2017). As an initial attempt, Teng et al. from Beijing Jiaotong University prepared MoS_2 nanosheets through a simple heating-up colloidal chemical approach and then dispersed them into a PVK matrix (Li, Tang, Li, Guan, et al. 2016; Li et al. 2018). With Al top electrode and PEDOT:PSS-modified ITO bottom electrode, the fabricated device could exhibit a repeatable bipolar resistive switching behavior with memory window of ~10 and endurance of $>10^2$ cycles. The involved switching

TABLE 11.1
Summary of organic–inorganic semiconductor heterojunctions for bipolar resistive switching memory applications. The symbol '—' means that no data concerning that performance item was provided

Semiconductor Heterojunction	Memory Window	Endurance (cycles)	Retention (s)	Reference
Organic–Inorganic Bilayer				
Alq_3/n-Si	$>10^4$	>4	$>10^3$	(Lee, Chang, and Chen 2008)
Alq_3/LSMO	$>10^2$	$>10^4$	—	(Prezioso et al. 2013)
PEDOT:PSS/ZnO	$>5 \times 10^2$	$>5 \times 10^2$	$>10^6$	(Khan et al. 2019)
Organic Matrix Embedded with Inorganic Nanomaterials				
$Alq_3 + MoO_3$	$\sim 10^5$	>8	$>4 \times 10^3$	(Chang, Cheng, and Lee 2010)
MEH-PPV + ZnO	$\sim 10^4$	—	$>3 \times 10^3$	(Ramana et al. 2013)
P3HT + CNT	$>10^2$	>50	$>10^3$	(Chaudhary et al. 2018)
P3HT + Au@air@TiO_2-h	$>10^4$	$>10^2$	$>10^4$	(Wang et al. 2015)
PEDOT:PSS + NCNT	$\sim 10^3$	$>5 \times 10^2$	$>10^4$	(Sun et al. 2015)
PFBT-TiO_2	$>10^3$	>30	—	(Zhang et al. 2020)
PFBT-ZnO	$>10^3$	>25	$>10^6$	(Sui et al. 2018)
PVK-TiO_2	$>10^3$	>10	—	(Li, Sui, and Li 2017)
PVK-Cu_2S	$\sim 10^4$	—	—	(Tang et al. 2011)
PVK-Ag_2S	$\sim 10^4$	—	$>10^4$	(Li et al. 2014)
PVK-ZnS	$\sim 10^4$	—	—	(Cao et al. 2015)
PVK-MoS_2	$>10^4$	$>2 \times 10^2$	$>10^4$	(Fan et al. 2017)
PVK-CdSe	$\sim 10^4$	>50	—	(Kaur, Singh, and Tripathi 2017)

mechanism was supposed to be the reversible formation and rupture of sulfur vacancy filaments. It is noted herein that the direct mixing of MoS_2 nanosheets and polymers usually gives rise to phase separation, especially at high nanosheet loading conditions, and thus could severely deteriorate the memory device performance. To address this issue, Chen et al. from East China University of Science and Technology recently proposed the synthesis of PVK chemically modified MoS_2 nanosheets through the reversible addition and fragmentation chain transfer (RAFT) polymerization reaction (Fan et al. 2017), as shown in Figure 11.6a. The DDAT in this figure is the abbreviation for the RAFT agent S-1-dodecyl-S′-(α,α′-dimethyl-α″-acetic acid)-trithiocarbonate. Although this synthesis procedure will cause some inevitable distortion into the crystal structure of MoS_2 nanosheets, such effect can be greatly relieved after annealing at only 80 °C, as demonstrated in Figure 11.6b. Also, the enhanced crystallization of PVK after annealing helps to the charge transfer process from PVK to MoS_2 that accounts for resistive switching. As expected, with Au and ITO as electrodes, the fabricated device exhibited a highly stable bipolar resistive switching behavior with memory window of $>10^4$, retention of $>10^4$ s, and endurance of >200 cycles, as shown in Figure 11.6c–e. These results indicate that two-dimensional layered semiconductors have a great potential for nonvolatile memory device applications.

For a give organic–inorganic semiconductor heterojunction bipolar resistive switching memory, many factors in addition to semiconductor types have been

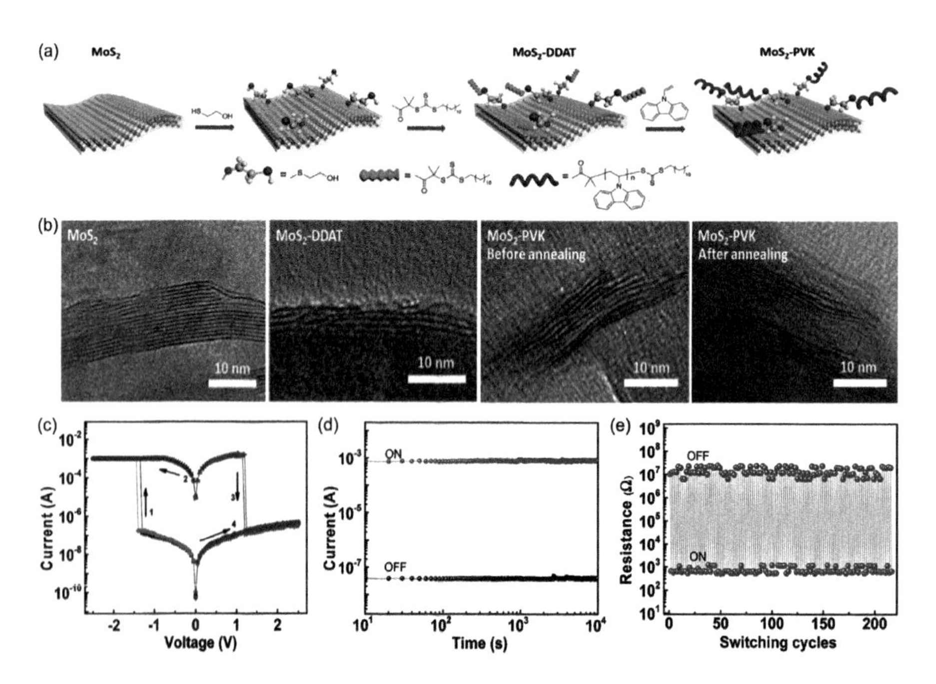

FIGURE 11.6 (a) Synthesis procedure of PVK covalently grafted MoS_2 nanosheets. (b) Microstructures of MoS_2 nanosheets at various periods. (c) Typical *I*–*V* characteristic of the Au/(annealed PVK + MoS_2)/ITO bipolar resistive switching memory. (d,e) Retention and endurance properties of the memory device. (Adapted with permission from (Fan et al. 2017), © 2017 The Royal Society of Chemistry)

demonstrated to greatly affect device performance. For example, in the Al/Alq_3/n-Si memory device, significant decreases in write voltage (from ~5 to ~2 V) and memory window (from ~10^4 to ~10^3) were observed after increasing the thermal deposition rate of Alq_3 from 0.05 to 0.3 nm/s (Lee, Chang, and Chen 2008). Careful morphology analyses suggested that a lower deposition rate can make the Alq_3 more rough and thus more defects exist at the interfaces, leading to a more pronounced interfacial charge trapping-based bipolar resistive switching behavior. In contrast, for the Al/(PVK + Cu_2S)/ITO memory device, a smoother active layer obtained by thermal annealing or covering the ITO electrode with a PEDOT:PSS buffer layer was demonstrated able to greatly enhance the memory window (Li, Lu, et al. 2016). This is because the mechanism of such device is charge transfer between Cu_2S and PVK, which process can be remarkably enhanced by high-quality electrode/active layer interfaces. More interestingly, Teng et al. from Beijing Jiaotong University observed a notable effect of measurement atmosphere on memory device performance (Li, Tang, Li, Wang, et al. 2016), as shown in Figure 11.7. One can see that, for the Al/(PVK + Ag_2S)/PEDOT:PSS/ITO memory device, the clear bipolar resistive switching behavior in air disappears almost completely in N_2 atmosphere (Figure 11.7b). As for the mechanism, it was suggested that molecular oxygen may react with sulfur

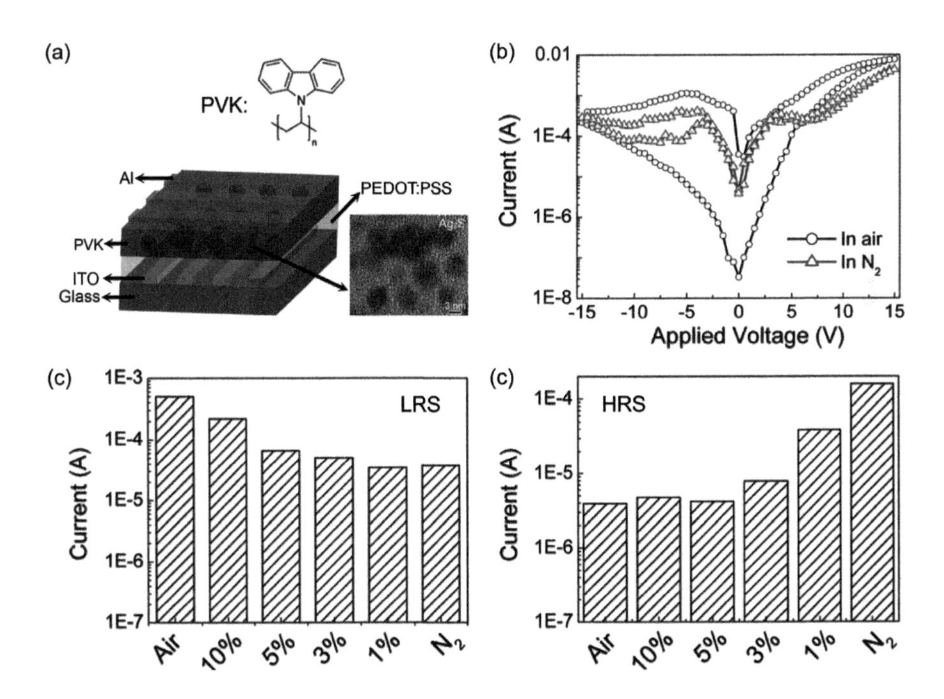

FIGURE 11.7 (a) Device structure of the Al/(PVK + Ag_2S)/PEDOT:PSS/ITO bipolar resistive switching memory. Insets: chemical structure of PVK and morphology of Ag_2S nanoparticles. (b) Typical *I–V* characteristics of the memory device under different atmospheres. (c,d) The currents of LRS and HRS states measured under different oxygen concentrations. (Adapted with permission from (Li, Tang, Li, Wang, et al. 2016), © 2016 American Chemical Society)

vacancies to form bridging oxide dimers to act as a potential barrier around the Ag_2S nanocrystals. Under external electric field, the barrier may help to trap more injected carriers and thus a more pronounced resistive behavior can be expected in air. Furthermore, detailed tests revealed that the current of LRS state increases rapidly with oxygen concentration of the atmosphere, whereas that of HRS state has a just opposite trend (Figure 11.7c–d). Such phenomenon makes the current memory device to be potentially used also for gas sensing.

Due to the rapid development of flexible and wearable electronics, the research on flexible memory has attracted an increasing amount of attentions in recent years. With simple structure and easy fabrication, resistive switching devices based on the composites of polymers and inorganic nanomaterials are highly promising for flexible memory applications. Concerning this point, Li et al. from Northeastern University have conducted a serious of detailed studies (Li, Sui, and Li 2017; Zhang, Yu, et al. 2020) It was found that, for the GaIn/(PVK + TiO_2)/ITO device, bending up to 500 times or down to 10 mm distance can notably decrease the memory window from as high as ~10^3 to only ~10 (Li, Sui, and Li 2017), as shown in Figure 11.8a–d. Based on quantum-chemical calculations and finite-element analyses, a possible explanation for such phenomenon was provided, as shown in Figure 11.8e. In brief, when the device is under bending, deformation should come mainly from the PVK matrix due to the rigid nature of TiO_2 nanoparticles, thus leading to the generation of many microscale cracks and voids near the composite surface by interfacial debonding or by cohesive failure of the matrix. After filled with the GaIn metal droplet top electrode, the microscale cracks can result in a thinner effective active switching layer. Meanwhile, the voids can act as blocking layers for charge trapping into TiO_2 nanoparticles. These two points together lead to a greatly decreased amount of effective charge traps in the active switching layer, thus severely deteriorate the memory window. To further improve device stability under bending, they recently conducted a comparative study on devices with PVK, P3HT, and PFBT as polymer matrix materials (Zhang, Yu, et al. 2020). The device based on PFBT was confirmed to exhibit the best memory performance under severe bending, which was attributed to the larger monomers of PFBT polymeric chains that can adsorb more TiO_2 nanoparticles through coordination bonds. These results could provide important guidance for future research on flexible resistive switching memory devices.

Besides the common applications for data storage, resistive switching memories have also been introduced into the field of data processing. The intrinsic nonvolatility of resistive switching memories enables the processed data to be stored therein, that is, the so-called processing-in-memory function that holds the promise of resolving the performance limits faced by current von Neumann computers (Borghetti et al. 2010; Gao et al. 2015). Since binary logic functions act as the basic algorithms for computers, their implementation based on resistive switching memories is certainly the most fundamental step for constructing future processing-in-memory computers. Concerning this respect, Prezioso et al. from the CNR-ISMN proposed an electric–magnetic coupled resistive switching memory based on the Co/Alq_3/$La_{0.7}Sr_{0.3}MnO_3$ (LSMO) sandwich structure (Prezioso et al. 2013), as shown in Figure 11.9a. Under electric stimuli, a typical bipolar resistive switching behavior with clear NDR was observed (Figure 11.9b), and multilevel storage was also possible by carefully controlling the stop

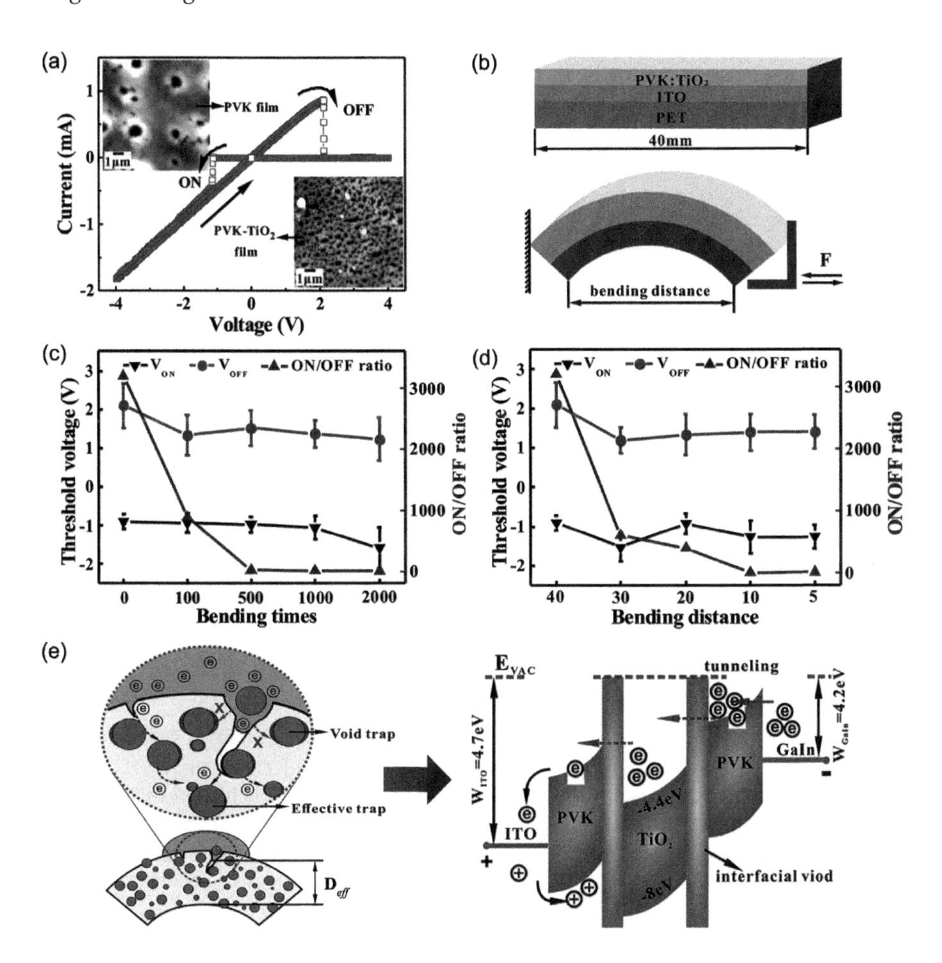

FIGURE 11.8 (a) Device structure of the GaIn/(PVK + TiO_2)/ITO bipolar resistive switching memory. Insets: morphologies of pure PVK and composite PVK + TiO_2 films. (b) Schematic illustration of bending test for the memory device. (c) Variations of threshold voltages and ON/OFF ratio with bending time. (d) Variations of threshold voltages and ON/OFF ratio with bending distance. (e) Schematic illustrations of bending effects on the microstructure and charge transport process. (Adapted with permission from (Li, Sui, and Li 2017), © 2017 American Chemical Society)

voltage during the NDR region (Figure 11.9c). In fact, the authors had realized 32 distinctive resistance states with an average difference of ~20% from each other, offering promising possibilities for five-bit memory applications. More importantly, clear giant magnetoresistance (GMR) effects with a higher MR ratio in lower resistance states were identified when the memory device was under magnetic stimuli, as indicated by the two resistance dips in each curve in Figure 11.9c. Herein, the two dips for a given curve correspond to the magnetic moments of Co and LSMO electrodes being arranged in antiparallel directions, while the high resistance background corresponds to the magnetic moments of electrodes being oriented parallel to each other. Based on

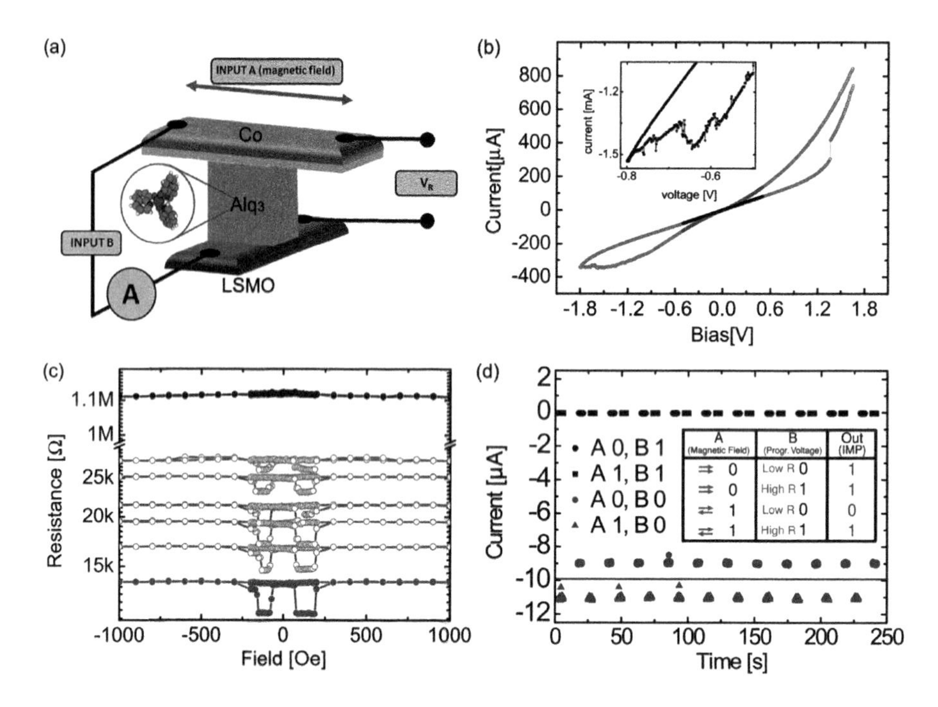

FIGURE 11.9 (a) Device structure of the Co/Alq_3/LSMO bipolar resistive switching memory with both electric and magnetic fields as inputs. (b) Typical *I–V* characteristic of the memory device. Inset: clear representation of the NDR behavior. (c) GMR behaviors of the memory device at various resistance states. (d) Implementation of IMP logic in the memory device with magnetic and electric fields as input A and input B, respectively. Inset: truth table of the IMP logic. (Adapted with permission from (Prezioso et al. 2013), © 2012 WILEY-VCH)

such electric–magnetic coupled behavior, the universal IMP logic gate was demonstrated able to be built with only a single memory device, as shown in Figure 11.9d. One can see that both electric and magnetic fields were set as the two inputs with the device resistance as the output. In detail, for the magnetic input A, '0' corresponds to the parallel orientation of the electrode magnetization, while '1' is assigned to the antiparallel configuration; for the electric input B, '0' and '1' correspond to the programing-bias needed to set the memory device into low and high resistance states, respectively. With such definitions, the lowest resistance state that assigned as output '0' was obtained only with the input combination of A = 1 and B = 0, that is, IMP logic. Although all the displayed data were collected at 100 K and the observed GMR effect would disappear after a few tens of operation cycles, this work definitely provides a new direction for future resistive switching memory research.

11.5 CHALLENGES AND PROSPECTS

Thanks to the great efforts from researchers all around the world in the past decade, significant advances have been made concerning organic–inorganic semiconductor heterojunctions for resistive switching memory applications, with either a bilayer structure or a single layer consisting of inorganic nanomaterials dispersed in an organic matrix. Both small molecules and polymers have been explored as the organic semiconductors, and inorganic nanomaterials including zero-dimensional nanoparticles, one-dimensional nanowires, and also two-dimensional nanosheets have been dispersed into the organic matrix layers. The performance of fabricated resistive switching devices has been carefully characterized, and its dependence on a series of factors like organic/inorganic component ratio and film deposition rate has been thoroughly discussed. After optimization, notable resistive switching performances of memory window of $>10^4$, endurance of $>10^2$ cycles, and retention of $>10^4$ s have been frequently reported. Moreover, novel applications beyond data storage, such as gas sensing and logic operation, have been preliminary demonstrated in some specially designed organic–inorganic semiconductor heterojunction devices.

Despite the above advances, the research of organic–inorganic semiconductor heterojunctions for resistive switching memory applications is still in the infancy stage, and much more efforts in the following aspects are urgently needed in future to realize the commercialization as well as widen the application scope of these memory devices. The first is to get a thorough understanding of the resistive switching mechanisms, which is not only a long-term desire of researchers but also highly beneficial to device performance optimization. Although the charge trapping/de-trapping process has been widely adopted to explain the observed resistive switching behaviors, it is highly speculative since no solid experimental evidence has been reported to date. Spectroscopic analysis and theoretical calculations could be powerful in this respect (Goswami et al. 2017; Zhang, Fan, et al. 2019). Second, the switching endurance performance must be significantly improved because the reported values are normally at least two orders of magnitude lower than that of currently used silicon-based flash memories (~10^6 cycles). The realization of close chemical bonding (rather than loose physical bonding) between the adopted organic and inorganic components and/or more uniform dispersion of inorganic nanomaterials in the organic matrix layers could be possible solutions for such issue (Fan et al. 2017). Also, a close bonding between the adopted organic and inorganic components should help to maintain device performance under bending conditions for flexible electronics applications. This can be corroborated by the excellent switching endurance of $>10^7$ cycles at small bending radius of 4.5 mm in metal–organic framework (MOF) films (Figure 11.10), which are typical organic–inorganic hybrid crystalline materials (Chui et al. 1999; Pan, Liu, et al. 2014; Pan et al. 2015). Third, given the high intrinsic photosensitivity of semiconductors, close research attention deserves to be paid to the interaction between light and organic–inorganic semiconductor heterojunction resistive switching memories. This could endow related memories with improved device performance or novel usable functions, such as low-voltage operation, low-power optical readout, optically modulated memory window, optoelectronic logic-in-memory operation, and merged property of optical sensing and storage (Gao et al. 2012; Tan et al. 2015; Tan et al. 2017; Mao et al. 2019). Last but not least, it is worthy of emulating biological synapses and even neurons using organic–inorganic semiconductor heterojunction resistive switching memories, which could contribute a lot to the development of future artificial intelligence technology (Xia and Yang 2019; Tang et al. 2019). After years of joint efforts from materials scientists, electronic

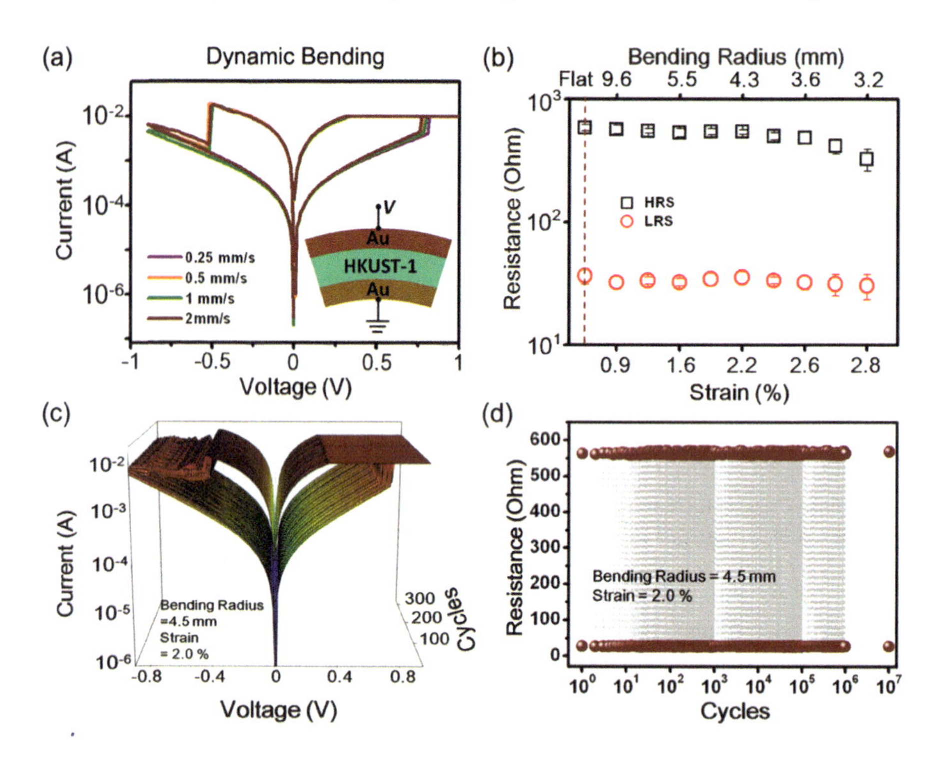

FIGURE 11.10 (a) Device structure and typical *I–V* characteristic of the Au/HKUST-1/Au bipolar resistive switching memory. (b) The dependence of memory device resistance on bending radius. (c,d) Switching uniformity and endurance of the memory device under bending with a small radius of 4.5 mm. (Adapted with permission from (Pan et al. 2015), © 2015 WILEY-VCH (b–d))

engineers, chemists, physicists, and even biologists, we can image that resistive switching memories based on organic–inorganic semiconductor heterojunctions could be at least an indispensable supplement in the modern information age.

ACKNOWLEDGMENTS

The authors acknowledge the financial supports from the National Key R&D Program of China (2017YFB0405604), National Natural Science Foundation of China (61704178, 61841404, 61974179, 61774161, 51525103, and 51931011), K. C. Wong Education Foundation (RCZX0800), Natural Science Foundation of Zhejiang Province (LR17E020001), and Ningbo Natural Science Foundation (2018A610020).

REFERENCES

Borghetti, Julien, Gregory S. Snider, Philip J. Kuekes, J. Joshua Yang, Duncan R. Stewart, and R. Stanley Williams. 2010. 'Memristive' switches enable 'stateful' logic operations via material implication. *Nature* 464 (7290):873–876.

Cao, Ya-Peng, Yu-Feng Hu, Jian-Tao Li, et al. 2015. Electrical bistable devices using composites of zinc sulfide nanoparticles and poly-(N-vinylcarbazole). *Chin. Phys. B* 24 (3):037201.

Chang, Tzu-Yueh, You-Wei Cheng, and Po-Tsung Lee. 2010. Electrical characteristics of an organic bistable device using an Al/Alq_3/nanostructured MoO_3/Alq_3/p^+-Si structure. *Appl. Phys. Lett.* 96 (4):043309.

Chaudhary, Deepti, Sandeep Munjal, Neeraj Khare, and V. D. Vankar. 2018. Bipolar resistive switching and nonvolatile memory effect in poly(3-hexylthiophene)–carbon nanotube composite films. *Carbon* 130:553–558.

Cho, Byungjin, Tae-Wook Kim, Minhyeok Choe, Gunuk Wang, Sunghoon Song, and Takhee Lee. 2009. Unipolar nonvolatile memory devices with composites of poly(9-vinylcarbazole) and titanium dioxide nanoparticles. *Org. Electron.* 10 (3):473–477.

Choi, Byung Joon, Antonio C. Torrezan, John Paul Strachan, et al. 2016. High-speed and low-energy nitride memristors. *Adv. Funct. Mater.* 26 (29):5290–5296.

Chui, Stephen S.-Y., Samuel M.-F. Lo, Jonathan P. H. Charmant, A. Guy Orpen, and Ian D. Williams. 1999. A chemically functionalizable nanoporous material $[Cu_3(TMA)_2(H_2O)_3]_n$. *Science* 283 (5405):1148–1150.

Fan, Fei, Bin Zhang, Yaming Cao, and Yu Chen. 2017. Solution-processable poly (N-vinylcarbazole)-covalently grafted MoS2 nanosheets for nonvolatile rewritable memory devices. *Nanoscale* 9 (7):2449–2456.

Gantz, John, and Davide Reinsel. 2012. https://www.emc.com/leadership/digital-universe/2012iview/index.htm (accessed February 23, 2020).

Gao, Shuang, Cheng Song, Chao Chen, Fei Zeng, and Feng Pan. 2012. Dynamic processes of resistive switching in metallic filament-based organic memory devices. *J. Phys. Chem. C* 116 (33):17955–17959.

Gao, Shuang, Xiaohui Yi, Jie Shang, Gang Liu, and Run-Wei Li. 2019. Organic and hybrid resistive switching materials and devices. *Chem. Soc. Rev.*48 (6):1531–1565.

Gao, Shuang, Fei Zeng, Chao Chen, et al. 2013. Conductance quantization in a Ag filament-based polymer resistive memory. *Nanotechnology* 24 (33):335201.

Gao, Shuang, Fei Zeng, Minjuan Wang, Guangyue Wang, Cheng Song, and Feng Pan. 2015. Implementation of complete boolean logic functions in single complementary resistive switch. *Sci. Rep.* 5:15467.

Goswami, Sreetosh, Adam J. Matula, Santi P. Rath, et al. 2017. Robust resistive memory devices using solution-processable metal-coordinated azo aromatics. *Nat. Mater.* 16 (12):1216–1224.

ITRS. 2015. https://www.semiconductors.org/resources/2015-international-technology-roadmap-for-semiconductors-itrs/ (accessed February 23, 2020).

Ji, Yongsung, David F. Zeigler, Dong Su Lee, et al. 2013. Flexible and twistable non-volatile memory cell array with all-organic one diode–one resistor architecture. *Nat. Commun.* 4:2707.

Kaur, Ramneek, Janpreet Singh, and S. K. Tripathi. 2017. Incorporation of inorganic nanoparticles into an organic polymer matrix for data storage application. *Curr. Appl. Phys.* 17 (5):756–762.

Khan, Muhammad Umair, Gul Hassan, Muhammad Asim Raza, Jinho Bae, and Nobuhiko P. Kobayashi. 2019. Schottky diode based resistive switching device based on ZnO/PEDOT:PSS heterojunction to reduce sneak current problem. *J. Mater. Sci.: Mater. Electron.* 30 (5):4607–4617.

Lee, Po-Tsung, Tzu-Yueh Chang, and Szu-Yuan Chen. 2008. Tuning of the electrical characteristics of organic bistable devices by varying the deposition rate of Alq_3 thin film. *Org. Electron.* 9 (5):916–920.

Lee, Myoung-Jae, Chang Bum Lee, Dongsoo Lee, et al. 2011. A fast, high-endurance and scalable non-volatile memory device made from asymmetric Ta_2O_{5-x}/TaO_{2-x} bilayer structures. *Nat. Mater.* 10 (8):625–630.

Li, Xu, Yue Lu, Li Guan, et al. 2016. Effects of buffer layer and thermal annealing on the performance of hybrid Cu_2S/PVK electrically bistable devices. *Solid-State Electron.* 123:101–105.

Li, Yue, Wen Sui, and Jian-Chang Li. 2017. Interfacial effects on resistive switching of flexible polymer thin films embedded with TiO_2 nanoparticles. *J. Phys. Chem. C* 121 (14): 7944–7950.

Li, Guan, Fenxue Tan, Bokun Lv, et al. 2018. Thiol-modified MoS_2 nanosheets as a functional layer for electrical bistable devices. *Opt. Commun.* 406:112–117.

Li, Xu, Aiwei Tang, Jiantao Li, Li Guan, Guoyi Dong, and Feng Teng. 2016. Heating-up synthesis of MoS_2 nanosheets and their electrical bistability performance. *Nanoscale Res. Lett.* 11 (1).

Li, Jiantao, Aiwei Tang, Xu Li, et al. 2014. Negative differential resistance and carrier transport of electrically bistable devices based on poly(N-vinylcarbazole)-silver sulfide composites. *Nanoscale Res. Lett.* 9 (1):128.

Li, Jiantao, Aiwei Tang, Xu Li, et al. 2016. Oxygen effects on performance of electrically bistable devices based on hybrid silver sulfide poly(N-vinylcarbazole) nanocomposites. *Nanoscale Res. Lett.* 11 (1).

Lv, Hangbing, Xiaoxin Xu, Hongtao Liu, et al. 2015. Evolution of conductive filament and its impact on reliability issues in oxide-electrolyte based resistive random access memory. *Sci. Rep.* 5:7764.

Mao, Jing-Yu, Li Zhou, Xiaojian Zhu, Ye Zhou, and Su-Ting Han. 2019. Photonic memristor for future computing: A perspective. *Adv. Opt. Mat.* 7 (22):1900766.

Möller, Sven, Craig Perlov, Warren Jackson, Carl Taussig, and Stephen R. Forrest. 2003. A polymer/semiconductor write-once read-many-times memory. *Nature* 426 (6963): 166–169.

Pan, F., S. Gao, C. Chen, C. Song, and F. Zeng. 2014. Recent progress in resistive random access memories: Materials, switching mechanisms, and performance. *Mat. Sci. Eng.: R: Rep.* 83:1–59.

Pan, Liang, Zhenghui Ji, Xiaohui Yi, et al. 2015. Metal-organic framework nanofilm for mechanically flexible information storage applications. *Adv. Funct. Mater.* 25 (18): 2677–2685.

Pan, Liang, Gang Liu, Hui Li, et al. 2014. A resistance-switchable and ferroelectric metal–organic framework. *J. Am. Chem. Soc.* 136 (50):17477–17483.

Park, Jea-Gun, Woo-Sik Nam, Sung-Ho Seo, et al. 2009. Multilevel nonvolatile small-molecule memory cell embedded with Ni nanocrystals surrounded by a NiO tunneling barrier. *Nano Lett.* 9 (4):1713–1719.

Prezioso, Mirko, Alberto Riminucci, Patrizio Graziosi, et al. 2013. A single-device universal logic gate based on a magnetically enhanced memristor. *Adv. Mater.* 25 (4):534–538.

Ramana, Ch V. V., M. K. Moodley, V. Kannan, and A. Maity. 2013. Solution based-spin cast processed organic bistable memory device. *Solid-State Electron.* 81:45–50.

Seok, Jun Yeong, Seul Ji Song, Jung Ho Yoon, et al. 2014. A review of three-dimensional resistive switching cross-bar array memories from the integration and materials property points of view. *Adv. Funct. Mater.* 24 (34):5316–5339.

Smith, Shawn, and Stephen R. Forrest. 2004. A low switching voltage organic-on-inorganic heterojunction memory element utilizing a conductive polymer fuse on a doped silicon substrate. *Appl. Phys. Lett.* 84 (24):5019–5021.

Sui, Wen, Chi Zhang, He-Yuan Xu, and Jian-Chang Li. 2018. Mechanical strain effects on resistive switching of flexible polymer thin films embedded with ZnO nanoparticles. *Mater. Res. Exp.* 5 (6):066425.

Sun, Yanmei, Lei Li, Dianzhong Wen, Xuduo Bai, and Gang Li. 2015. Bistable electrical switching and nonvolatile memory effect in carbon nanotube–poly(3,4-ethylenedioxythiophene):poly(styrenesulfonate) composite films. *Phys. Chem. Chem. Phys.*17 (26): 17150–17158.

Sun, Yiming, Xiaolong Zhao, Cheng Song, et al. 2019. Performance-enhancing selector via symmetrical multilayer design. *Adv. Funct. Mater.* 29 (13):1808376.

Tan, Hongwei, Gang Liu, Huali Yang, et al. 2017. Light-gated memristor with integrated logic and memory functions. *ACS Nano* 11 (11):11298–11305.

Tan, Hongwei, Gang Liu, Xiaojian Zhu, et al. 2015. An optoelectronic resistive switching memory with integrated demodulating and arithmetic functions. *Adv. Mater.* 27 (17): 2797–2803

Tang, Aiwei, Feng Teng, Jie Liu, et al. 2011. Electrical bistability and charge-transport mechanisms in cuprous sulfide nanosphere-poly(N-vinylcarbazole) composite films. *J. Nanopart. Res.* 13 (12):7263–7269.

Tang, Jianshi, Fang Yuan, Xinke Shen, et al. 2019. Bridging biological and artificial neural networks with emerging neuromorphic devices: Fundamentals, progress, and challenges. *Adv. Mater.* 31 (49):1902761.

Tsai, Cheng-Lin, Feng Xiong, Eric Pop, and Moonsub Shim. 2013. Resistive random access memory enabled by carbon nanotube crossbar electrodes. *ACS Nano* 7 (6):5360–5366.

Wang, Peng, Quan Liu, Chun-Yu Zhang, et al. 2015. Preparing non-volatile resistive switching memories by tuning the content of Au@air@TiO_2-h yolk–shell microspheres in a poly(3-hexylthiophene) layer. *Nanoscale* 7 (46):19579–19585.

Xia, Qiangfei, and J. Joshua Yang. 2019. Memristive crossbar arrays for brain-inspired computing. *Nat. Mater.* 18 (4):309–323.

Xu, Xin, Richard A. Register, and Stephen R. Forrest. 2006. Mechanisms for current-induced conductivity changes in a conducting polymer. *Appl. Phys. Lett.* 89 (14):142109.

Xue, Wuhong, Gang Liu, Zhicheng Zhong, et al. 2017. A 1D vanadium dioxide nanochannel constructed via electric-field-induced ion transport and its superior metal–insulator transition. *Adv. Mater.* 29 (39):1702162.

Yang, Yu Chao, Feng Pan, Qi Liu, Ming Liu, and Fei Zeng. 2009. Fully room-remperature-fabricated nonvolatile resistive memory for ultrafast and high-density memory application. *Nano Lett.* 9 (4):1636–1643.

Zhang, Bin, Fei Fan, Wuhong Xue, et al. 2019. Redox gated polymer memristive processing memory unit. *Nat. Commun.* 10:736.

Zhang, Chi, Pei-Lun Yu, Yue Li, and Jian-Chang Li. 2020. Polymer/TiO_2 nanoparticles interfacial effects on resistive switching under mechanical strain. *Org. Electron.* 77:105528

Zhao, Enming, Diyou Liu, Lu Liu, Xinghua Yang, Wei Kan, and Yanmei Sun. 2017. Unipolar nonvolatile memory devices based on the composites of poly(9-vinylcarbazole) and zinc oxide nanoparticles. *J. Mater. Sci.: Mater. Electron.* 28 (16):11749–11754.

12 Optoelectronic Sensors for Health Monitoring

Zheng Li

CONTENTS

12.1 INTRODUCTION

Optoelectronic sensors typically use inorganic semiconductors, organic dyes or electronics, or polymeric electrolytes as fundamental building blocks to probe subtle changes in relevant physical or chemical properties of the target into an analytically useful readout.[1] Those sensing materials are typically supported by flexible substrates or templates that can be readily mounted on living systems as wearable or stretchable sensors or actuators to monitor the environment and interconnect in real time with a multitude of physical or chemical parameters including pressure, strain, torsion, temperature, humidity, as well as various molecular chemicals or biomarkers.[2] Optoelectronic devices have recently found a wide range of advanced applications in clinical and biomedical research, industrial production, and commercialization, including soft robotics and automation,[3] flexible microchips,[4] electronic skins,[5–6] and implantable healthcare monitoring.[7] To meet the increasing demand for these applications, a growing number of requirements and criteria related to the construction of effective optoelectronic sensory systems have been continuously implemented. As such, the recent decade has witnessed the advent of a variety of state-of-the-art sensing devices, which utilizes functional materials with micro- or nanostructures and optical, mechanical, or electric outputs for signal transduction.[8–9] Those sensors usually take advantages of high degrees of durability and selectivity, enhanced sensitivity, long-term stability, and exceptional optical transparency.[10] In an attempt to provide a quantitative assessment of the scientific significance and impact of this topic, a citation report from the Thomson-Reuters Web of Knowledge has been generated using the search term "optoelectronic sensors." As shown in Figure 12.1, this field is growing rapidly both in the number of publications per year (more than 550 publications per year as of 2018) and even exponentially in the number of citations per year (more than 14,000 citations per year in the past 20 years); both are strong indicators of the health of the field.

Optoelectronic sensors can be fabricated from a myriad of composite materials, that is, active sensing materials supported by rigid or more useful flexible templates or substrates. Particularly, flexible sensors are more adaptable to a variety of field applications and therefore more commonly employed in clinical settings.[11] The deformable nature and physicochemical properties render those devices intriguing mechanical and photoelectric characteristics for biomedical sensing applications.[12] Generally speaking, flexible sensors are composed of organically polymeric substrates such as polypropylene, polycarbonate and polyethylene terephthalate (PET), which offer high mechanical strength and optical transparency for clinical analysis.[13–14] Of special interest is another class of flexible materials, organosilicon elastomers, which includes polydimethylsiloxane (PDMS), silicone rubbers, and resins. Owing to their unique advantages such as high elasticity, stretchability, ease of surface modification as well as low toxicity, silicone elastomers have been broadly used as alternatives to traditional organic polymers for healthcare purposes.[15]

In general, any sensing device may operate in a way that relative variations in the physical or chemical properties (i.e., capacitance, resistance, hydrophobicity/hydrophilicity, dielectricity, and molecular polarity) of sensing elements are captured and

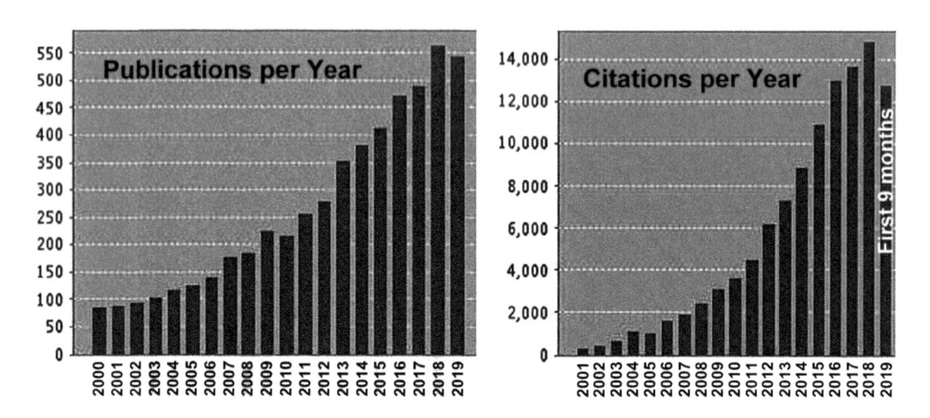

FIGURE 12.1 Thomson-Reuters Web of Science Citation Report on "optoelectronic sensors" in the past two decades, according to the number of publications and citations per year. Data statistics are as of November, 2019. Both the number of publications and citations are growing rapidly from year to year.

quantified by the central data processing system, and interpreted as digitalized outputs. A wide variety of optoelectronic sensors have demonstrated the efficiency of recognizing electrical signals created by the sensing elements through the introduction of the simplest mechanical deformations (i.e., touching, pressing, or stretching) as induced by stimuli in the solid, gaseous, or liquid state. This simple and effective transduction mechanism[16–17] has led to a numerous number of flexible and stretchable functional sensors being developed and optimized in recent years for various healthcare and medical engineering applications, notably for artificial electronic skins, physiological monitoring and assessment systems. Depending on the categories of sensing elements that experience such changes, sensors can be largely classified into solid- and liquid-state devices. As the names suggest, active sensing elements made of solids are solid-state sensors. Common examples are a wide range of nanomaterials or their composites, including conductive polymers, metal oxide semiconductors, carbon-based materials (carbon dots, graphene, and carbon nanotubes), and metallic nanoparticles. On the other hand, few examples of reported sensors that employ liquid sensing components, such as ionic liquids and liquid metals, are classified as liquid-state sensors.[18–20]

In the book chapter, we will summarize various kinds of optoelectronic sensing platforms for healthcare monitoring developed in the recent decade. We will begin with the introduction of representative flexible substrates, fabrication methods, and materials with remarkable optical or electrical properties used as functional sensor elements. The present and emerging optoelectronic sensing platforms based on single or integrated sensing components will be discussed, and the involved operating principles or mechanisms of the flexible or stretchable sensors will be highlighted. In the next part, particular focuses will be placed on the healthcare and biomedical applications of those flexible and wearable optoelectronic sensory systems, specifically for in situ and easy-to-use physiological diagnosis. In the end, we will give some deep insights into the limitations, challenges, and opportunities facing the continuous development and improvement of optoelectronic sensing platforms.

12.2 SENSING MECHANISMS AND MATERIALS

The most common optoelectronic sensors use electrochemical outputs for signal transduction. Electrochemical sensors are made of resistors or capacitors as electronic units that keep track of changes in conductivity or capacitance during interactions between analyte molecules and the electrically active surface.[21] Optical sensors use infrared, visible, or ultraviolet light to probe chemical interactions at liquid or solid interfaces, and exist as the other main class of optoelectronic sensors.[22] Many other classification standards are used to categorize a wide variety of optoelectronic sensors, including the sensor configuration (i.e., single versus differential), the detection basis (i.e., material versus geometry change), the size of the sensor, the resolution of sensor response, and the type of readout circuit.[23]

Among various electrical sensors, chemiresistive or chemicapacitive devices are easy to manufacture: a pair of parallel electrodes placed on a substrate and separated by a dielectric material, where the sensing component is coated or drop-casted to tune the electrochemical reactivity.[24] An interdigitated electrode array is often used to adjust overall resistance or capacitance. In terms of active sensing components, a broad range of different materials have been explored and designed. The most widely used sensing materials include metal oxide chemiresistive semiconductors (e.g., tin oxide),[25] organic field-effect transistors (OFETs),[26–27] conducting polymers of both inherently conductive (e.g., polythiophenes), and those integrated with conductive particles (e.g., graphene- or carbon nanotube-polymer composite).[28–29] This chapter will describe several major categories of active sensing components and supporting substrates, as well as their fabrication methods.

12.2.1 Active Sensing Components

Various solid- or liquid-state sensing components have been employed and embedded in flexible substrates in the construction of powerful optoelectronic sensors for a multitude of clinical applications.[30–31] Solid-state sensing elements with superior conductivity, stretchability, and durability are considered excellent candidates for the use. So far, a majority of solid-state sensing elements consist of conductive or semiconductive carbon, metallic conductor, or organically polymer nanomaterials, such as polymer nanofibers, silver or gold nanoparticles and nanowires, carbon nanotubes, and graphene.[32] In addition to individual sensing components, hybrids of these conductive constituents are also actively explored as the composite sensing elements of wearable sensors. One common example of these hybrid structures is the elastomeric composites integrated with highly dispersed 3D carbon nanofillers forming a reciprocal architecture, which greatly enhance the electrical conductivity.[33]

Surface modification of electrochemical sensor surfaces can offer improved selectivity, especially for bioanalytical applications, for example with molecular imprinted membranes or aptamers,[34] but at the expense of added complexity and susceptibility to drift from irreversible chemical reactions. Surface modification of electrochemical sensors has also had success using various nanostructures; the enhanced sensitivity is largely due to the extremely high surface area to volume ratio inherent to nanoparticles.[35] Nanowires, nanotubes, nanofibers, nanospheres, graphene, and other two-dimensional layered materials have thus far shown great promise, although specificity among similar physical signals or chemical/biological analytes can be problematic. Hybrid heterostructures of these nanomaterials likewise act as excellent candidates for chemical sensors and have drawn significant attention in the past decade.[36] Owing to the low dimensionality of the data resulting from most present manifestations of limited number of sensor elements, engineering chemical selectivity into these systems remains the key to the future success for applications in chemical sensing or biosensing associated with human health monitoring. One of the possible solutions is to multiplexing the detection channels that can recognize a wide range of targets, for example using array-based sensors.[37]

12.2.2 Supporting Substrates and Fabrication Methods

Stretchable solid-state sensors typically use highly pliable, conductive, and lightweight materials as supporting substrates or templates. The proper fabrication of sensor devices that can provide optimal geometry and configuration of active sensing elements for signal transduction remains a great challenge.[38–39] New innovations in the selection of core sensing materials have resulted in the generation of a wide variety of fabrication approaches that can endow different sensing materials with desired structural morphologies. Among different sophisticated and effective means of sensor fabrication, photolithography, printing, electrospinning, spray coating, molecular imprinting, and solution drop-casting are commonly employed by researchers.[40–42] In this section we will briefly introduce the principle and current progress of two mainstream methods: photolithography and functional printing.

12.2.2.1 Photolithography

Photolithography is a gold standard process for the fabrication of many sensor devices with well-defined spatial patterns and architectures.[43] Photolithography aims to form the masking layer with desired geometry for etching and ion doping using photosensitive materials called photoresist in the patterning process.[44–45] Using photolithography, the pattern of the photomask can be transferred into a desired film that is coated on the substrate.[46] The general process for photolithography is shown in Figure 12.2a. For high volume fabricating, the photolithography equipment usually adopts a design with an in-line configuration where all the stations for pre-baking and UV light exposure processes including cleaning, photoresist coating, exposure, and development are connected together in a photoresist track system to allow high throughput in-line motion of glass substrates. One end of the photoresist-track system is connected to the exposure system, and the loader or unloader station is installed at the other end. The equipment for etching and stripping equipment is set up in separated regions, to which the glass substrates are automatically attached.

Photolithography makes it possible to engineer different sensing components into small or microdevices to enhance their performance and reduce the cost. Trends in functionalization and miniaturization of sensor design enabled by lithography require in situ and accurate deposition of engineered active materials on a specified area of a substrate.[47–49] Yuan and co-workers reported an in situ patterning method for instant preparation of porous poly(ionic liquid)s (PILs) microstructures using maskless photolithography.[50] The porous PILs shows fast response time (<18 s) and great sensitivity (45 mM) toward H_2O_2, and is expected to be incorporated in various clinical applications, including soft robotics and wearable healthcare devices for in vivo peroxide detection (Figure 12.2b). Javey and co-workers used traditional photolithography method to fabricate a fully integrated wearable sensor for real-time monitoring of human sweat electrolytes, metabolites, and temperature during physical activities (Figure 12.2c).[51] The platform is particularly valuable for monitoring the perspiration system and could be exploited for in situ analyses of biologically important biomarkers within sweat and other human fluids to facilitate personalized physiological investigations. Nguyen et al. described a robust and nanoscale SiC membrane microstructure with a large aspect ratio (up to 1:3200).[52] The method eliminates the sticking effects on free-standing architectures, enhancing mechanical performance which is favorable for applications in micro-electro-mechanical system (MEMS). The SiC sensor exhibits excellent linear response to the applied pressure with impressive repeatability. Standard photolithographic methods are also employed in the construction of microresonators, waveguides based on silicon carbide or silica-based photonic crystals for biomolecular sensing.[53–55]

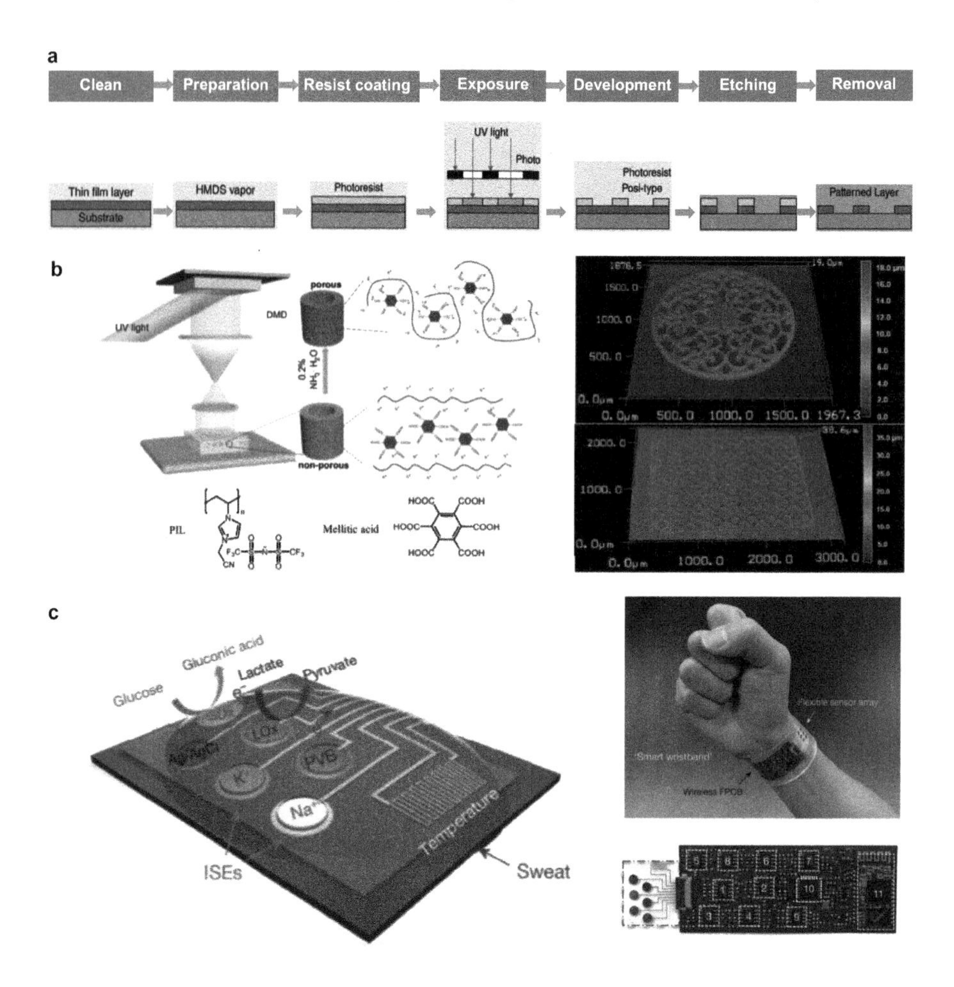

FIGURE 12.2 Photolithography and its applications to wearable sensors. (a) General fabrication workflow of photolithography, including pre-baking (clean, preparation, and resist coating), photo exposure and post-baking (development, etching, and template removal) processes. (b) Scheme illustration of the fabrication procedure of porous PIL microstructures by maskless photolithography (left panel) and the resultant confocal images of the patterns (right panel). Reproduced with permission from Ref. 50. Copyright 2018 American Chemical Society. (c) Photograph of a wearable sensor array on a subject's wrist (top right panel). The red dashed box (bottom right panel) indicates the location of the sensor array and the white dashed boxes indicate the locations of the integrated circuit components. The sensor array includes glucose, lactate, sodium, potassium, and temperature-sensitive elements and used for multiplexed perspiration analysis (left panel). Reproduced with permission from Ref. 51. Copyright 2016 Springer Nature.

However, photolithography sometimes is incompatible with flexible substrates with poor thermal or etching resistance, which may result in poor pattern uniformity.[56] Alternative fabrication routes are required to be incorporated for the production of flexible sensing platforms. Low-cost, efficient, and photolithography-free fabrication of stretchable electronics such as mechanical cutter- or laser printer-patterned methods has been actively investigated.[57–59] Among various non-lithographic approaches, functional printing techniques as will be discussed in the next sub-chapter are the prevailing manufacturing methods. Printed electronics represent an attractive alternative for the patterning of insulating and piezoelectric polymeric substrates, providing the excellent electrical conductivity necessary to become the fundamental building block of functional devices.

12.2.2.2 Functional Printing of Electronic Sensing Devices

Functional printing techniques are a class of predominant methods for the preparation of various printed electronics, which represents an alternative manner for surface patterning of conductive or piezoelectric films on flexible polymeric templates.[60–61] The core component of printed electronics is conductive ink, which originates from a liquid suspension of highly dispersed metallic nanoparticles or inorganic solids. The regularity in the structure and composition ensures that the inks can be uniformly deposited on the substrates and subsequently sintered to form conformal and compact conductive layers at high curing temperatures or under laser. However, due to the demand for high sintering temperature, the technique was once limited to only a few thermoplastics such as polyimide (PI).[62] Recent progress in material synthesis and fabrication has sufficiently moderated the prerequisite for high temperatures such that more accessible and less heat-sensitive thermoplastic alternatives such as polyethylene naphthalate (PEN) could be employed for the fabrication as well.[63]

Conductive ink printing is usually performed on different platforms, which includes screen printing, roll-to-roll printing, pin printing, and inkjet printing.[64] These techniques normally involve masks, nozzles, or patterned templates for the deposition of the conductive inks at the desired locations. The transition of traditional single-point printing techniques to roll-to-roll printing makes it possible for the fast preparation of a considerable number of microarray sensors over the size of a few meters length by a few meters width, thus facilitating large-scale deployment of wearable sensors.[65]

12.2.2.3 Fabrication of Optical Sensor Arrays

Array-based sensing technologies, or "electronic noses," typically use several to many different cross-reactive sensors to interact with analytes, most commonly through physical adsorption, and generate an electrical response (e.g., changes in resistance or capacitance).[66] The pattern of the sensor array, due to the cross-reactivity of individual sensor elements, enables molecular recognition of response patterns by comparison to a predetermined library of responses. The first example of an artificial nose was reported by Persaud et al. in 1982,[67] who attempted to mimic the biological olfactory system using semiconductor transducers that were semi-specific to analytes: three different metal oxides were employed to sense similar mixtures that achieved excellent discrimination results. In recent years, an increasing number of cross-reactive analytical devices have emerged and have been applied to environmental monitoring, security screening, biomedical diagnosis, and food inspection.[68]

The vigorous development of novel techniques in chemical sensing has brought about great accessibility to many useful sensors not simply based on electrical responses as alternatives to electronic noses. Among those, optical sensors are especially noteworthy. The most common optical sensors are dependent on colorimetric or fluorescent changes originating from intermolecular interactions between the chromophore or fluorophore with the analytes.[69] Combining array-based techniques that employ a chemically diverse set of cross-reactive sensor elements with novel digital imaging methods, one can produce a composite pattern of response as a unique optical "fingerprint" for any given analyte. By analogy, such optical sensor arrays are often referred to as optoelectronic noses or tongues.

Current version of optoelectronic noses mostly uses organically chemoresponsive dyes and primarily measure the chemical reactivity of analytes, rather than their physical properties such as the physisorption. This provides a high dimensionality to chemical sensing that permits high sensitivity (often as low as parts per billion (ppb) or even parts per trillion (ppt) levels), impressive discrimination among very similar analytes, and fingerprinting of extremely similar mixtures over a wide range of analyte categories, in both gaseous and liquid phases. Colorimetric and fluorometric sensor arrays therefore effectively overcome the limitations of traditional array-based sensors that solely depend on physisorption or nonspecific chemical interactions. Such optical array sensing has shown excellent performance in the detection and identification of diverse analytes, ranging from chemical hazards to energetic explosives, to medical biomarkers, and to food additives.

One of the common drawbacks in many optoelectronic sensors is their sensitivity to changes in humidity. For any field applications, humidity changes over a single day could be as high as tens of thousands parts per million in water vapor concentration. False responses to humidity prove lethal to the detection of parts per million levels of any gaseous analytes (e.g., volatile organic chemicals (VOCs)). One solution to the reduction of such sensor response to humidity is to use optoelectronic noses as described above and employ hydrophobic materials as both matrices and substrates of sensor inks. Sufficiently humidity-unresponsive arrays can sometimes even be applied to sensing in aqueous phase. For optical sensors, while the choice of chromophores or fluorophores will predominate the response, there are other factors that play important roles in the improvement of sensing efficacy: most notably, the formulation of the dye and the substrate upon which the dye formulations are printed. Sensitivity, response time, reproducibility, selectivity, susceptibility to interferents, and shelf life of the sensor array will be heavily influenced by the choice of substrates and matrices of the sensing pigments.[70]

Particularly for applications of optoelectronic noses, a large number of solid supports have been used for the construction of colorimetric and fluorometric arrays. The desired properties of such substrates include good chemical resistance (e.g., stability in the presence of acids or bases, gases, or solvents), high surface area (for sufficient contact with analytes), suitable permeability, optical transparency or high reflectivity, and ease of fabrication. A simple method for production of colorimetric sensor arrays was developed by the Suslick group, involving robotic dip-pen printing dye formulations on the surface of reverse-phase silica gel plates, acid-free paper, or porous membranes made of polymeric materials including PET, polyvinylidene difluoride (PVDF), and PP.[71–72] Various methods have been used to print colorimetric sensor arrays, including spin-coating, ink- or aerosol-jet printing, and an adapted microarray robotic pin printer to print multiple sensor elements with controllable shapes and geometry.

Despite the above advantages, most of the paper- or membrane-based substrates employed by the current version of optoelectronic noses are not stretchable on any dimensions of the 3D contour, thus not being quite useful for field applications. In comparison, some soft elastomers as discussed in the last section, whose mechanical properties are similar to biological skins, do not have those drawbacks. Examples such as PDMS chips have been widely used for environmental monitoring of biologically important analytes.[73]

12.3 CLINICAL APPLICATIONS OF OPTOELECTRONIC SENSORS

As a useful analytical technique, the discrimination among highly similar complex mixtures often remains critical even using the most sophisticated equipment. A great number of clinical samples are known for their complexity in chemical composition, and a component-by-component analysis is generally impractical, given the thousands of different compounds present in biological specimens or gas phase.[74–75] On the other hand, early and effective diagnosis of important human biomarkers has become increasingly imperative for the control and treatment of relevant diseases, and the ability to discriminate infected samples from the healthy ones is highly desirable. For the above two reasons, pattern-based fingerprinting that array-based sensors are able to perform have proven extremely desirable.[37,76]

Wearable and stretchable sensor devices based on optoelectronic materials have shown extensive applications in healthcare diagnosis. Those sensors can be easily mounted on clothes or attached to the human body with variable textures, shapes, and configurations, for real-time monitoring of health conditions. Others can be embedded into wearing apparel or accessories such as gloves, watches, hats, socks, skin patches, earrings, etc.[77–78] In recent years, wearable sensors have exhibited tremendous potentials in a wide range of clinical applications. In this section, specific applications of wearable sensors in the monitoring of health parameters, disease biomarkers, as well as human body motion will be mainly summarized.

12.3.1 Physical Index Monitoring

Extensive studies regarding the combination of wearable devices with optoelectronic sensors have been carried out particularly for monitoring important physical indices of great relevance to human health.[79–80] Those integrated sensory systems can constantly detect, acquire, analyze, and transmit data of great relevance to human physiological factors including blood pressure, heart rate, body temperature, respiratory rate, and sweat excretion rate, which provides long-term and useful information about personal health conditions for users. Those sensors are required to be stretchable, flexible, light weight, biocompatible, and naturally conformal to the surface of the skins with negligible user awareness, and sensors made of wearable materials can be directly attached to flat or curved surfaces of human skin, and therefore being well-suited for in situ monitoring of health parameters.

Body temperature is an important parameter to evaluate human health condition.[81] Hattori et al. reported an epidermal electronics system (EES) that uses an electronic sensor reversibly coated at wounds. The device enables precise quantification of surgical wound healing by mapping of temperature and thermal conductivity of the skin near the wounds, as shown in Figure 12.3a.[82] A novel graphite-filled polyethylene oxide and PVDF composite exhibits a high accuracy in temperature measurement with deviation of ±0.1 °C and high repeatability over 2000 cycles.[83] Heart beat rate is also a common physiological parameter, which is a critical index that indicates the health and disease of human. Carbon-based optoelectronic sensors or hybrid heterojunctions have been intensively studied for the signal collection and quantification of human heart beats.[84–86] Graphene strain sensors attached to the chest is used to measure relative resistance change of the respiratory during normal and exercising states, as shown in Figure 12.3b.[87] A PVDF plastic wrap transparent capacitive pressure sensor displays an accurate measurement of the pressure changes and conformal detection of the cyclic change in pressure (Figure 12.3c).[88] The cost-effectiveness for the fabrication of such thin-film sensor and its excellent signal reproducibility is highly desirable. Single-walled carbon nanotube has also shown promising applications as a wearable pressure sensor to be incorporated into electronic skins, for the monitoring of human dynamic physiological signals, including radial arterial pulse and muscle activity at multiple body locations.[89–90]

12.3.2 Human Movement Monitoring

Wearable sensors integrated with optoelectronic materials have been extensively applied to monitor human motions in the healthcare settings.[91–92] Sensing mechanisms are generally based on electrical signal transduction, which basically

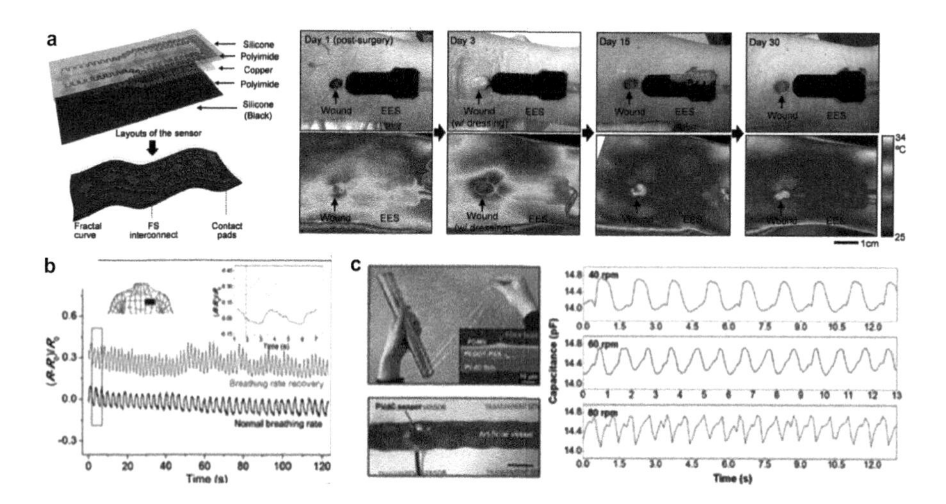

FIGURE 12.3 Wearable sensors for physical index monitoring. (a) Schematic illustration of an epidermal electronics system (left panel) and photos of the wound with an epidermal electronic system (EES) and corresponding IR images of the temperature distribution (right panel) during wound healing. Reproduced with permission from Ref. 82. Copyright 2014 Wiley. (b) Relative resistance change of the respiratory during normal and exercise as measured by a graphene strain sensor attached to the chest. Reproduced with permission from Ref. 87. Copyright 2014 Wiley. (c) Flexible and transparent capacitive sensor for monitoring pressure of an artificial blood vessel (left panel) and repetitive capacitive curves of real-time pulse waves of the artificial vessel (right panel). Reproduced with permission from Ref. 88. Copyright 2018 Wiley.

originates from changes in strain, pressure, or bending caused by body motion or vibration. Detected movements can be classified into two categories: monitoring of long-range movements such as bending of the knees and hands that are associated with walking or other exercising events, and monitoring of short-range, subtle motions such as many physiological features including pulse, heart beat, gesture, and facial expression.[93]

Highly stretchable and sensitive wearable sensors that can detect long-range human movements are essential to many medical applications, including personalized healthcare monitoring, continuous acquisition of human daily physical activities, rehabilitation, and soft robotics.[94–96] High-performance wearable sensors must be capable of integrating with any curved and moving textiles or soft skins to accurately detect physical stimuli for health monitoring. In addition, other characteristics are required as well: (1) high stretchability for monitoring human activities with large dynamic ranges; (2) high sensitivity for detection of tiny-scale motions; (3) fast response and recovery rates for real-time monitoring. Conventional electrical sensing materials, mainly composed of metallic semiconductors, possess brittle and rigid features with limited stretchability and therefore cannot be ideally applied to in situ monitoring at the interfaces of human skin and clothing.[97] To solve those issues, the combination of various nanomaterials and nanocomposites with flexible substrates has been widely employed to construct outstanding wearable and stretchable sensors.[98]

Resistive sensors use piezoresistive-type electronics to record a resistance change in the case when conductive materials are subjected to an applied mechanical deformation.[99–100] One of the most common fabrication methods for nanomaterial-based resistive sensors is to coat functional nanomaterials on the surface or dope them into an elastomeric substrate, such as polyurethane (PU), PDMS, or polyester (PE).[101–102] Applied mechanical forces on resistive sensor can result in the translation, disconnection, or sliding between layers of nanomaterials, hence changing the contact area between nanomaterials, consequently altering the electrical resistance. Among different nanomaterials used in this sort of sensors, graphene and nanowire sensors have been extensively studied with wide dynamic ranges and adjustable sensitivity.[103–106] Figure 12.4a describes a unique structural design for the construction of a fish-scale-like, graphene-based strain sensor. This microstructure based on the disconnection mechanism greatly enhances sensitivity of the sensor by largely increasing its stretchability up to 82% strain. The flexible strain sensor could be readily attached onto human skin to monitor human motions such as phonation recognition and wrist pulse (Figure 12.4b-d).[94]

Capacitive sensors also stand for a common class of detection tools to which flexible electronics can be readily incorporated. The capacitive electrical sensors respond to the applied strain by recognizing the capacitance variations. A typical capacitive sensor consists of a dielectric layer sandwiched between two electrodes, whose thickness fluctuates upon stretching.[107] The parallel-plate structure is the most common configuration employed in most part of capacitive sensors since it is straightforward to fabricate and easy to modify with functional materials. In order to obtain conformal skin attachment, the electrodes of the sensor must be bendable and stretchable.[108] Therefore, stretchable nanostructured semiconductors are regularly used as the electrodes of capacitive sensors.[109] In the recent years, a variety of silver

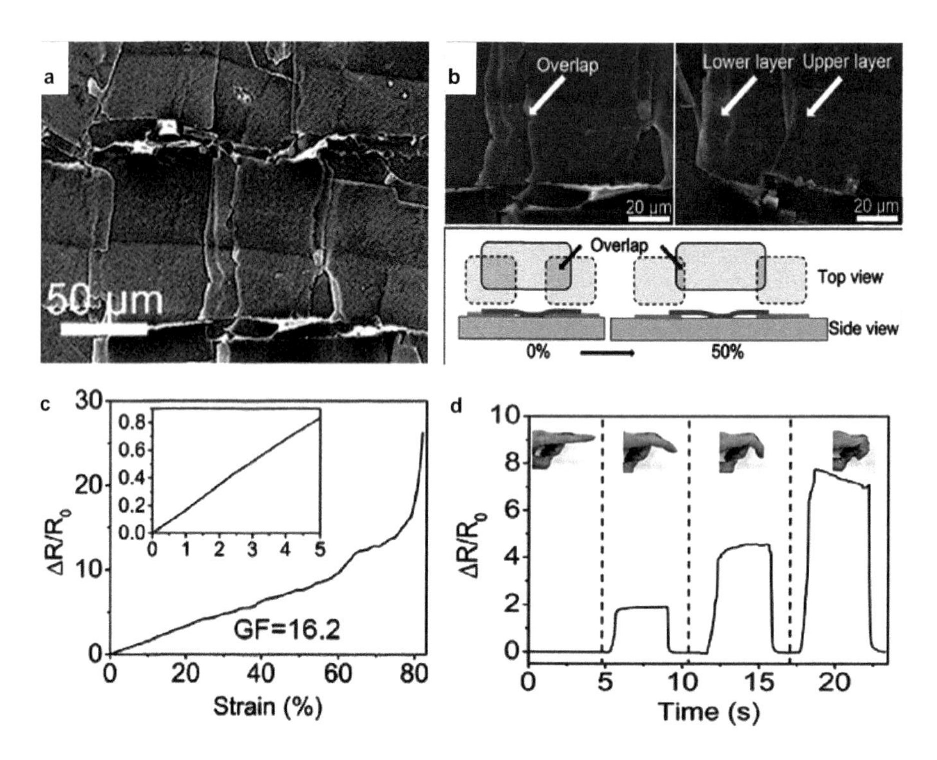

FIGURE 12.4 Characterization and performance of nanomaterial-based wearable strain sensors. (a) Scanning electron microscopy (SEM) image of fish-scale-like strain sensor with overlapped rGO sheets. b) SEM image of the strain sensor at 0% strain (left) and 50% strain (right), respectively; and schematic illustration of the sensing mechanism of FSG strain sensor upon stretching. c) Typical resistance change as a function of strain percentage. d) Resistance changes of the strain sensor in detecting finger bending. All figures reproduced with permission from Ref. 94. Copyright 2016 American Chemical Society.

nanowire composites with advanced architectures are extensively developed for building stretchable capacitive sensors in terms of their cost-effectiveness, high conductivity, perfect stretchability, and ideal optical transparency.[110–111] In order to obtain high mechanical sensitivity, highly stretchable dielectric layer materials integrated with small mechanical modules such as PDMS, PU, and PE are desirable. As wearable strain sensors, the applied strain usually causes the deformation of the elastic materials, thereby the dielectric layer thickness and the overlapped area between two electrodes alter and the overall capacitance changes accordingly.

12.3.3 Chemical Index Monitoring

12.3.3.1 Glucose

The increasingly growing interest in mobile health sensing has led to new innovations in non-invasive detection methods for blood sugar contents, especially for glucose.[112] Continuous monitoring of glucose levels addresses the limitations in finger-stick blood testing such as long analyzing time, complicated sample disposal, and potential wound infection, hence becoming essential for timely and proper management of relevant diseases such as diabetes.[113] Many reports have been focused on electrochemical biosensors or optical sensors for epidermal glucose monitoring.[114–115] However, realizing the potential clinical function of these novel sensing techniques requires extensive efforts, including addressing key technological challenges and establishing a reliable correlation to present diagnostic approaches and any improvements from gold standard glucometers.

The first commercial non-invasive glucose monitor approved by the U.S. Food and Drug Administration (FDA) is the GlucoWatch biographer.[116] This wrist-worn system electrochemically measured glucose concentrations in skin interstitial fluids extracted by reverse iontophoresis, which is carried out by applying a moderate current with two skin-worn electrodes to induce ion migration across the skin.[117] The positively charged Na ions trigger an electronic flow toward the cathode due to the negatively charged skin surface, which leads to the movement of the neutral glucose to the same electrode. The GluoWatch quantifies extracted glucose concentration via the enzymatic glucose oxidation reaction at the interface of skin-worn sensing electrodes coated with glucose oxidase. The entire electrode configuration, enzymatic components, and display are housed in a wrist-attached watch device (Figure 12.5a) as a commercialized product. Glucose concentrations in extracted interstitial fluids are expected to be 1000 times more diluted than blood glucose. Therefore, a highly sensitive glucose sensing system is demanded for this particular use. It turns out that clinical trials of the GlucoWatch have shown sufficient precision for blood-glucose monitoring in the hospital or at home. However, the more advanced GlucoWatch version was retracted from the market in the late 2000s due to various reasons, including reported skin irritation due to reverse iontophoresis, long necessary warm up time (>2 h), and the need for repetitive calibration using standard glucose strips. To address these drawbacks, recent research has focused on developing reliable non-invasive glucose monitoring platforms (Figure 12.5b).[118]

Strip-type glucose sensor is a common device for quantification of glucose level in the vascular system.[119] However, collecting the venous plasma with a needle is invasive and time consuming. Alternatively, the glucose content is typically measured with capillary blood drawn from fingertips by a blood test strip.[120] For users' convenience, the strips have a multilayer capillary channel to take up the blood drop effectively and deliver the fluid to the sensors, which takes a small amount of blood sample for analysis. After sufficient amount of blood is drawn by the strip inserted in the glucometer, the device automatically analyzes the collected blood and quantifies the glucose level (Figure 12.5c).[118,121–122] By measuring the alternating current impedance using the three-electrode system (Figure 12.5d), the glucometer estimates the hematocrit ratio (Figure 12.5e) and transform it into the glucose level in the plasma. However, although the conversion factor increases the accuracy of the reported glucose level, glucose monitoring via a blood strip is still prone to errors, and therefore the conversion from invasive to non-invasive detection methods for implantable and wearable sensing have been highly recommended.[123–124] Various kinds of body fluids such as tear, saliva, and sweat can be readily collected for rapid analysis, an indirect reflection of overall glucose level in the human body.[116,125]

To monitor blood glucose continuously, implantable sensors and microdialysis-type devices have been developed. The implantable glucose sensor has suffered from

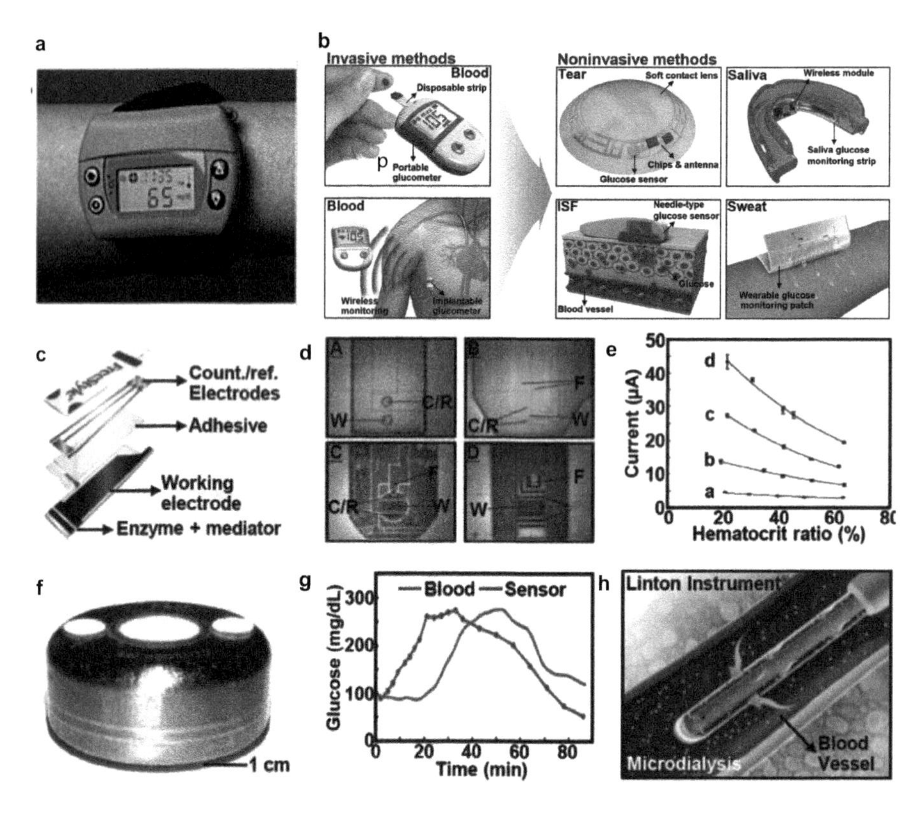

FIGURE 12.5 Various wearable glucose-sensing devices. (a) Photo of GlucoWatch biographer device. Reproduced with permission from Ref. 116. Copyright 2001 Elsevier. (b) Evolution of glucose sensors from invasive to non-invasive electrochemical glucose monitoring system. Reprinted with permission from Ref. 118. Copyright 2018 Wiley. (c) Schematic illustration of exploded view of the blood glucose test strip. Reproduced with permission from Ref. 120. Copyright 2018 Elsevier. (d) Schematic illustration of electrode configuration of blood glucose monitoring strip, with W, C/R, and F referring to working, counter/reference, and fill detection electrodes, respectively. Reproduced with permission from Ref. 120. Copyright 2018 Elsevier. (e) Amperometric responses of glucose sensor to glucose concentrations of 75.7 a), 170.4 b), 318.0 c), and 462.0 mg dL^{-1} d) at different hematocrit ratio. Reproduced with permission from Ref. 120. Copyright 2018 Elsevier. (f) An optical image of implantable PDMS glucose sensor. Reproduced with permission from Ref. 118. Copyright 2018 Wiley. (g) The delay response to an intravascular glucose tolerance test. Reproduced with permission from Ref. 118. Copyright 2018 Wiley. (h) Schematic illustration of microdialysis tube for glucose sensing. Reproduced with permission from Ref. 118. Copyright 2018 Wiley.

several major drawbacks, including a short lifetime and poor biocompatibility. Gough and co-workers addressed these issues by using a nonporous PDMS layer with the co-immobilization of excess catalase, which demonstrated that the implanted device could be used in mammalians over 1 year (Figure 12.5f).[126] However, the performance of the implanted detector is subject to oxygen level since the sensing mechanism is dependent on oxymetry.[127] To resolve this potential issue, a new design that regenerated half of the oxygen consumed by GO_x with catalase and limited the radial diffusion of glucose using a 2D membrane was proposed, which aided to create an environment with more concentrated oxygen. However, the sensor exhibited an inherent 10-min lag between the actual concentration and the reported value (Figure 12.5g) as the circulatory transport to and mass diffusion of glucose within local tissues are slow.[128] As another useful technique for in situ glucose-sensing, intravenous microdialysis is a sampling technique that uses a microdialysis tube with a semipermeable membrane to penetrate into the vein.[129] The perfusate is perfused continuously through the tube, and the glucose level in the dialysate is detected by the connected glucose sensor (Figure 12.5h). However, the dialysis-type device requires that a probe be inserted into a vein and connected to a pump and a glucose sensor, which is unsuitable for users during physical activity. Therefore, intravenous microdialysis has limited applications to patients in the hospital equipped with sufficient facility.[130]

12.3.3.2 Pathogens

Effective methods for detection of bacteria related to human health are highly desirable in medicine and industry. Bacterial infections are widely involved in food poisoning, hospital-acquired infections, and many other areas that cause diseases and that are of great importance to the health of the general population.[131] In industry, a numerous number of products have to be screened to avoid potential bacterial contamination before they may be released, and as a consequence regulation of the food industry must be particularly strict.[132] Challenges for the identification of pathogens include the necessity of long culturing times, the need for highly trained laboratory personnel, the demand for expensive and high-maintenance equipment, and the ineffectiveness due to antibiotic resistance.[133–134]

Bacteria stink, that is, the VOCs produced from bacteria metabolism, is a class of characteristic markers from pathogens to which the mammalian olfactory system is highly responsive. Consequently, an experienced microbiologist can readily identify many bacteria by smell. Applications of prior electronic nose technology, however, have been limited by the low dimensionality of traditional electrical sensor arrays (e.g., conductive polymers or metal oxides) and have achieved only modest success, even when attempting to classify small numbers of bacterial variants.[135–136] Disposable and wearable optical sensor arrays have shown numerous successful applications in the identification of microbial species, especially for bacteria. Carey et al. used the 6 × 6 array to recognize strains of human pathogenic bacteria grown on standard agar based on specific VOCs produced from different bacterial cultures.[137] Through monitoring bacterial growth during 10 h of incubation time, 10 strains of bacteria including *E. faecalis* and *S. aureus* as well as their antibiotic-resistant strains were identified with 98.8% accuracy out of 164 trials, as assessed by PCA, HCA, and leave-one-out LDA using time-stacked data.

Lim and co-workers have developed an 80-element optical sensor array for screening of other specific classes of bacteria,[138] including *Yersinia pestis* and *Bacillus anthracis* that feature on the Center for Disease Control and Prevention's list of potential biothreats.[139] Through headspace gas analysis of bacteria incubated in Petri dishes, the sensor array was capable of distinguishing four different bacterial species and five strains of *Y. pestis* and *B. anthracis*, with the detection limit as low as ~10 CFU/plate for several strains (Figure 12.6). In addition to the promising results demonstrated in the work, the authors also suggested several aspects of considerations for the future improvement of array-based sensing techniques, in terms of the array sensitivity, selectivity, incubation volume, data analysis, and portability of the imaging device.

Rotello, Bunz, and co-workers employed a displacement assay consisting of hydrophobically functionalized AuNPs and a conjugated polymer, poly(p-phenyleneethynylene) (PPE), to identify several different bacteria in liquid growth media based on the fluorescence quenching of fluorophores attached to gold nanoparticles.[140] As described here, the negatively charged conjugated polymer fluorophore is replaced by negatively charged bacteria on the surface of positively charged AuNPs, generating a differential fluorescence probe whose response is dictated by the binding strength of each specific bacterium. By diversifying the surface chemistry of AuNPs and the structure of the conjugated polymer, they generated array-like data that provided high dimensionality. Nine bacterial species and three strains of *E. coli* were examined in 64 trials with a 95% accuracy of classification using linear discriminant analysis (LDA).

Bunz and co-workers have very recently reported a similar fluorescent array by combining PPE with four variants of antimicrobial peptides, for the identification of 14 different bacteria based on their Gram status and genetic relationship (Figure 12.7)[141]; the array showed 100% discrimination accuracy using LDA with leave-one-out cross-validation. This chemical sensor was also successfully applied to target relevant microbes in urine and serum samples, which enabled the discrimination of all bacteria at the upper ranges of their clinically relevant concentrations. Detection of infectious bacteria on biofilms is a great challenge due to the complexity and heterogeneity of biofilm matrices. Rotello and co-workers used a AuNP-based multichannel sensor for the in situ, real-time monitoring of different biofilms based on their physicochemical properties, as an important complement to the currently

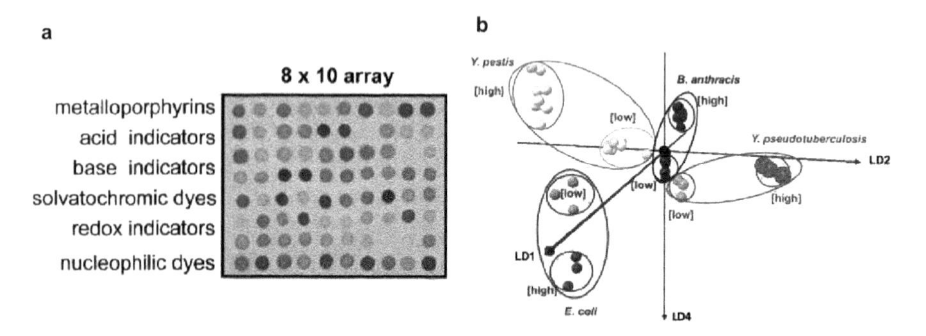

FIGURE 12.6 Optical array-based sensing technique for bacterial detection and discrimination. (a) An 8 × 10 colorimetric sensor array comprising 80 different chemically responsive nanoporous dyes. (b) Species concentration trajectories in LDA multidimensional space. All species can be identified independent of inoculum concentration. Six control trials (black) are located at the origin. Reproduced with permission from Ref. 139. Copyright 2013 Lonsdale et al.

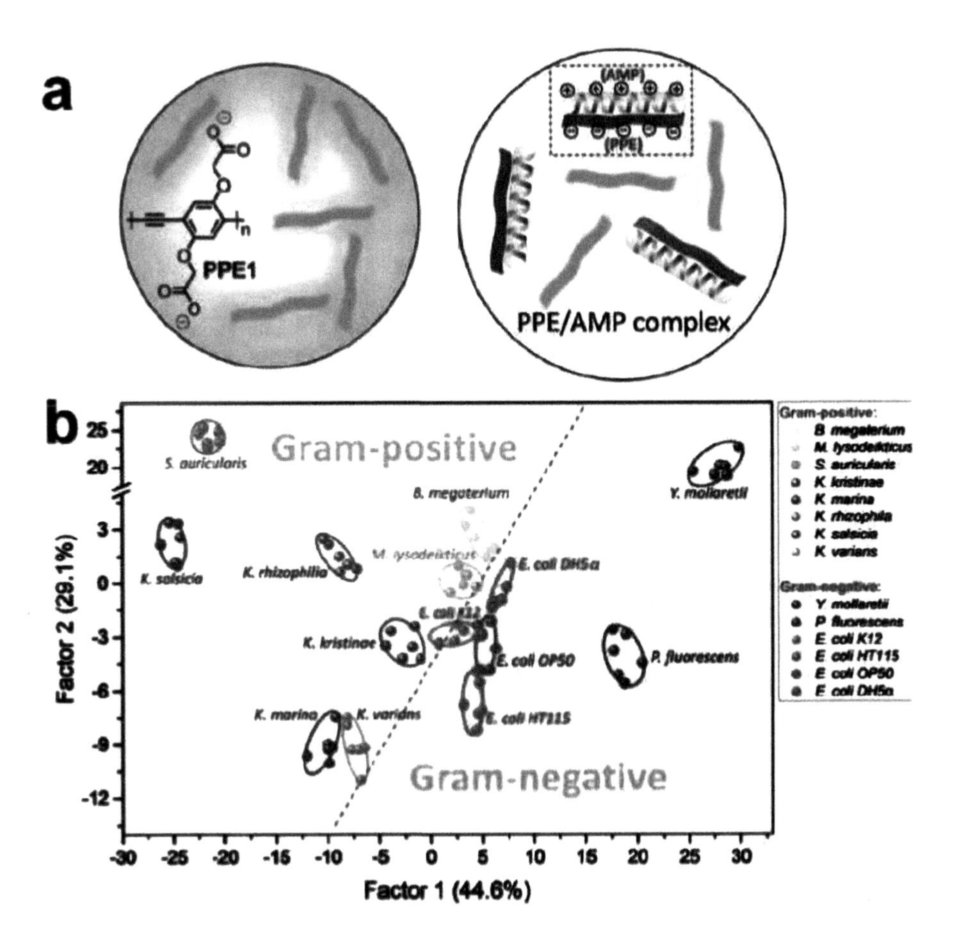

FIGURE 12.7 Polymeric sensor array for pathogen differentiation. (a) Structure of PPE and electrostatic complex formed between negatively charged PPE and positively charged AMP. (b) Two-dimensional canonical score plot obtained from fluorescence response patterns. 95% confidence ellipses were described for the individual bacteria. Both figures reproduced with permission from Ref. 141. Copyright 2017 Wiley.

available methods for early infection detection and antibiotic treatment.[142] Through the selective binding between the AuNP–fluorescent protein conjugates and biofilms, the authors were able to discriminate six types of bacterial biofilms including two uropathogenic bacteria. The sensor array was further proven effective *in vitro*, in which it worked perfectly in a mixed bacteria–mammalian cell co-culture wound model.

The immobilized fluorescent bead strategy that Walt and coworkers developed for use with microbead optical fiber bundles has also been recently applied to bacterial identification. Fixed arrays of this sort are difficult to reuse multiple times, partially due to photobleaching. Walt and co-workers have attempted to resolve this problem by using time-shared optical tweezers to dynamically create sensing materials.[143] The quantum-dot-encoded microbead arrays were able to collect fluorescence signals from biological samples in the real time.

Especially for human immuno-compromised patients, fungal infections are a serious problem worldwide and have gained recent notoriety following contamination of pharmaceuticals in the compounding process. With the growing of fungal cells, fungi produce gas-phase and carbon-based volatile organic compounds as characteristic metabolites, and these represent an alternative diagnostic approach for identification of different even highly similar fungal strains. Relevant studies have proven the utility of this technique for identification of clinically important pathogenic yeasts in standard blood cultures.[144] This proof-of-principle study has shown that seven types of clinically important yeasts growing in blood culture have distinctive signatures in their profile of volatile organic compounds, which allows for accurate identification using a simple and disposable optical sensor array and is faster than the conventional detection protocol.

12.3.3.3 Cancer and Other Disease Biomarkers

Optical array sensing has begun to find applications in the area of cancer diagnosis of biomarker detection.[145] Different cell lines produce different types of VOCs as featured metabolites; this applies to any rapidly growing cells, and in principle the breath composition is consistent with the volatiles produced in the body. Breath analysis has a long history as an underutilized diagnostic approach. Limitations in traditional analytical tools that are lack of sensitivity, selectivity, and cost-effectiveness have restricted their uses in clinical settings.[146–147] Electronic noses have certainly been evaluated for breath analysis, especially for diagnosis of lung cancer and of respiratory infections, albeit with inadequate success. The prototype optical sensor arrays developed by the Suslick group for fungal and bacterial identification have shown some preliminary clinical success for breath diagnosis. Lim et al.[148] designed a tuberculosis testing tool that incorporated a 73-indicator colorimetric sensor array for fingerprinting VOC signatures of human urine samples. Sensor array responses to 22 tuberculosis urine samples and the other 41 symptomatic controls were collected under different sample treatment conditions, which proved that basified condition was able to provide the best accuracy of >85% sensitivity and > 79% specificity of all samples. The urine assay using a colorimetric sensor array offers a powerful tool for a quick and simple diagnosis of tuberculosis in low resource settings.

Lung cancer detection via breath analysis using colorimetric sensing techniques has also been elucidated by Mazzone et al. at the Cleveland Clinic with promising classification results.[149] The test revealed that, based on exhaled breath screening of 229 study subjects (92 with lung cancer and 137 as the controls), lung cancer patients can be distinguished from control subjects with high accuracy (C-statistic of 0.889 for adenocarcinoma versus squamous cell carcinoma), and the accuracy for identification of lung cancer can be further improved by adjusting clinical and breath predictors.[150]

The direct analysis of nonvolatile cancer cells or disease biomarkers in aqueous biological samples is also highly desirable. In an entirely different approach to cell differentiation, it is not surprising that different cell lines interact differently with different nanoparticles and that those interactions are strongly affected by the chemical nature of the nanoparticle surfaces, particularly as biomolecules adsorb on the surface of nanoparticles forming a "protein corona." [151]

Taking advantage of interactions between polymers/AuNPs and various biomolecules, the multiplexed soluble fluorescent displacement assays for protein detection

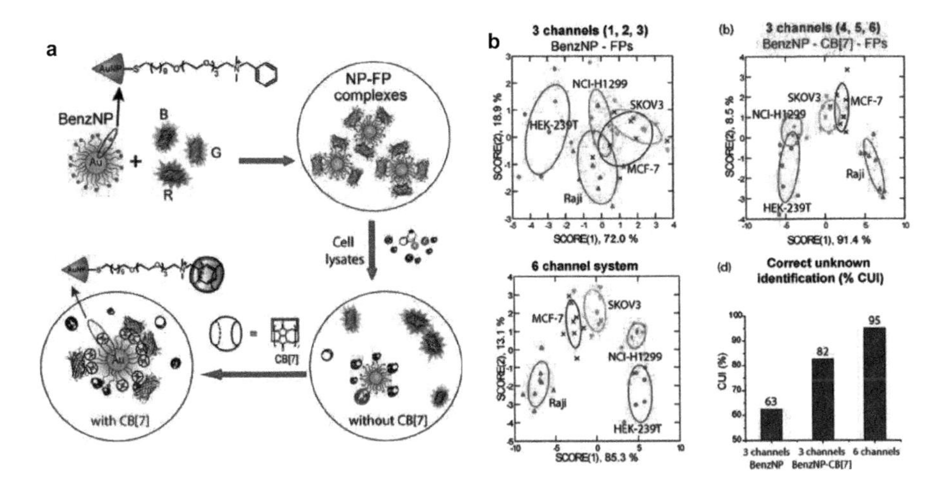

FIGURE 12.8 Fluorescent displacement assays for protein detection. (a) Construction of a six-channel sensor in a single well. Quenched AuNP–fluorescent protein (BenzNP–FP) conjugates serve as differential probes for the detection of cell lysates based on fluorescence signal changes in three emission channels. CB[7] is then added to the same well to gain three additional fluorescent channels based on the interactions between the analytes and newly formed composites incorporating CB[7]. (b) LDA classification of five human cancerous cell lines based on only three channels of BenzNP–FPs (top left panel), only three channels of BenzNP–CB[7] (top right panel), and the combinatory classification with all six channels (bottom left panel), and overall classification accuracy of three sensing systems (bottom right panel). Both figures reproduced from Ref. 154. Copyright 2017 American Chemical Society.

have also been extended to identification of cancer cells.[152–153] Rotello et al. reported a nanosensor using cell lysates to rapidly profile the tumorigenicity of cancer cells.[154] In addition to the aforementioned approach involving differential probes, this sensing platform used a host–guest interaction between the macrocyclic cucurbit[7]uril (CB[7]) and the ammonium head of gold nanoparticles to build the second recognition receptor that extends the number of dimensions within each sensor element from three to six (Figure 12.8a). The overall accuracy in classification accordingly increased from 63% of using only three channels of the composite without the supramolecular assembly to 95% when taking all six channels into account (Figure 12.8b). It is worth noting that his method required the minimal sample quantity of ~200 ng (~1000 cells) for analysis and could become an ideal tool for microbiopsy-based cancer diagnosis.

12.4 CONCLUSIONS

The thriving development of electronics or bioelectronics is among one of the most important commercial and industrial objects for a large number of useful applications, especially those associated with healthcare and medicine. Having provided a comprehensive overview of the emerging optical or electrical sensing platforms for healthcare and biomedical monitoring, we put forward a few challenges that current sensing technologies are facing and perspectives toward the future development of advanced detection devices. Despite the success of a wide range of sensing techniques achieved in controlled laboratory environments, there are still challenges to overcome before these methods become sufficiently mature to be put into practice.

Admittedly, the fast progress in optoelectronic sensors and devices in the last decade has addressed the great power and precious merits of this particular sort of monitoring tools. Advances in sensing technologies have brought out new concepts and methodologies in device operation, deployment, and implementation. However, as is the case for most novel techniques, there still remain a great number of challenges and opportunities for this field, both in the fundamental studies and for applied purposes. These challenges mostly include (1) fabrication and integration of the flexible sensors and devices to enhance the application-oriented sensing capabilities, such as sensing modality, functionality, and directionality; (2) design and synthesis of sensor components whose key sensing parameters, particularly sensitivity and dynamic range, fall in the ranges required for the diagnosis of specific diseases; (3) enhancement of the sensors' specificity in order to create useful optoelectronic devices that can accurately identify different physical signals or chemical analytes; and (4) elimination of disturbances from possible interferents. In a word, developing portable, low noise, and accurate sensing instrumentation with the capability of onboard analysis has shown significant progress and begins to match the increasingly urgent demand for a large number of clinical applications. This consideration generally requires sophisticated fabrication processes and therefore necessitates the rational selection and optimization of sensing elements and detector configurations. As a last step toward real-world applications, device integration is always of primary importance for the complete realization of multimodal, multifunctional, and multidirectional optoelectronic sensing platforms.

Overall, despite all the ongoing challenges for the practical deployment of optoelectronic sensing platforms, there are promising positive signs from the continuous progress of this field that signal future opportunities for fundamental research and commercial use. The precise detection of physical signals or chemical species from both the human body and the surrounding environment is the key to study personal health because of the abundant information that those factors can provide when determining a person's health condition. In this respect, sensor applications based on flexible optoelectronic platforms have various advantages, including signal amplification, molecular design capability, low cost, and mechanical robustness (e.g., flexibility and stretchability). With deeper investigations into the essence of functional materials and fabrication of electronics, we expect to see more exciting accomplishments in the near future in the realization of functional sensing platforms for a variety of biomedical and healthcare applications.

REFERENCES

1. Xu, H.; Yin, L.; Liu, C.; Sheng, X.; Zhao, N., Recent advances in biointegrated optoelectronic devices. *Adv. Mater.* 2018, 30 (33), 1800156.
2. Matsuhisa, N.; Chen, X.; Bao, Z.; Someya, T., Materials and structural designs of stretchable conductors. *Chem. Soc. Rev.* 2019, 48 (11), 2946–2966.
3. Cianchetti, M.; Laschi, C.; Menciassi, A.; Dario, P., Biomedical applications of soft robotics. *Nat. Rev. Mater.* 2018, 3 (6), 143–153.

4. Huang, Y.; Li, F.; Qin, M.; Jiang, L.; Song, Y., A multi-stopband photonic-crystal microchip for high-performance metal-ion recognition based on fluorescent detection. *Angew. Chem. Int. Ed.* 2013, 52 (28), 7296–7299.
5. Ha, M.; Lim, S.; Park, J.; Um, D.-S.; Lee, Y.; Ko, H., Bioinspired interlocked and hierarchical design of ZnO nanowire arrays for static and dynamic pressure-sensitive electronic skins. *Adv. Funct. Mater.* 2015, 25 (19), 2841–2849.
6. Bauer, S., Sophisticated skin. *Nat. Mater.* 2013, 12 (10), 871–872.
7. Ma, Y.; Zheng, Q.; Liu, Y.; Shi, B.; Xue, X.; Ji, W.; Liu, Z.; Jin, Y.; Zou, Y.; An, Z.; Zhang, W.; Wang, X.; Jiang, W.; Xu, Z.; Wang, Z. L.; Li, Z.; Zhang, H., Self-powered, one-stop, and multifunctional implantable Triboelectric active sensor for real-time biomedical monitoring. *Nano Lett.* 2016, 16 (10), 6042–6051.
8. Chen, D.; Pei, Q., Electronic muscles and skins: A review of soft sensors and actuators. *Chem. Rev.* 2017, 117 (17), 11239–11268.
9. Amjadi, M.; Kyung, K.-U.; Park, I.; Sitti, M., Stretchable, skin-mountable, and wearable strain sensors and their potential applications: A review. *Adv. Funct. Mater.* 2016, 26 (11), 1678–1698.
10. Bansal, A. K.; Hou, S.; Kulyk, O.; Bowman, E. M.; Samuel, I. D. W., Wearable organic optoelectronic sensors for medicine. *Adv. Mater.* 2015, 27 (46), 7638–7644.
11. Bao, Z.; Chen, X., Flexible and stretchable devices. *Adv. Mater.* 2016, 28 (22), 4177–4179.
12. Vilela, D.; Romeo, A.; Sánchez, S., Flexible sensors for biomedical technology. *Lab Chip* 2016, 16 (3), 402–408.
13. Wang, L.; Luo, J.; Yin, J.; Zhang, H.; Wu, J.; Shi, X.; Crew, E.; Xu, Z.; Rendeng, Q.; Lu, S.; Poliks, M.; Sammakia, B.; Zhong, C.-J., Flexible chemiresistor sensors: Thin film assemblies of nanoparticles on a polyethylene terephthalate substrate. *J. Mater. Chem.* 2010, 20 (5), 907–915.
14. Zhang, M.; Wang, M.; Zhang, M.; Maimaitiming, A.; Pang, L.; Liang, Y.; Hu, J.; Wu, G., Fe3O4 nanowire arrays on flexible polypropylene substrates for UV and magnetic sensing. *ACS Appl. Nano Mater.* 2018, 1 (10), 5742–5752.
15. Song, Y.; Chen, H.; Su, Z.; Chen, X.; Miao, L.; Zhang, J.; Cheng, X.; Zhang, H., Highly compressible integrated Supercapacitor–Piezoresistance-sensor system with CNT–PDMS sponge for health monitoring. *Small* 2017, 13 (39), 1702091.
16. Heylman, K. D.; Knapper, K. A.; Horak, E. H.; Rea, M. T.; Vanga, S. K.; Goldsmith, R. H., Optical microresonators for sensing and transduction: A materials perspective. *Adv. Mater.* 2017, 29 (30), 1700037.
17. Abriata, L. A.; Albanesi, D.; Dal Peraro, M.; de Mendoza, D., Signal sensing and transduction by Histidine kinases as unveiled through studies on a temperature sensor. *Acc. Chem. Res.* 2017, 50 (6), 1359–1366.
18. Clark, K. D.; Nacham, O.; Purslow, J. A.; Pierson, S. A.; Anderson, J. L., Magnetic ionic liquids in analytical chemistry: A review. *Anal. Chim. Acta* 2016, 934, 9–21.
19. Hirsch, A.; Dejace, L.; Michaud, H. O.; Lacour, S. P., Harnessing the rheological properties of liquid metals to shape soft electronic conductors for wearable applications. *Acc. Chem. Res.* 2019, 52 (3), 534–544.
20. Dickey, M. D., Stretchable and soft electronics using liquid metals. *Adv. Mater.* 2017, 29 (27), 1606425.
21. He, Q.; Wu, S.; Yin, Z.; Zhang, H., Graphene-based electronic sensors. *Chem. Sci.* 2012, 3 (6), 1764–1772.
22. Tong, L., Micro/Nanofibre optical sensors: Challenges and prospects. *Sensors* 2018, 18 (3), 903.
23. Carminati, M., Advances in high-resolution microscale impedance sensors. *J. Sens.* 2017, 15, 7638389.
24. Fennell Jr., J. F.; Liu, S. F.; Azzarelli, J. M.; Weis, J. G.; Rochat, S.; Mirica, K. A.; Ravnsbæk, J. B.; Swager, T. M., Nanowire chemical/biological sensors: Status and a roadmap for the future. *Angew. Chem. Int. Ed.* 2016, 55 (4), 1266–1281.

25. Yoon, J.-W.; Lee, J.-H., Toward breath analysis on a chip for disease diagnosis using semiconductor-based chemiresistors: Recent progress and future perspectives. *Lab Chip* 2017, 17 (21), 3537–3557.
26. Surya, S. G.; Raval, H. N.; Ahmad, R.; Sonar, P.; Salama, K. N.; Rao, V. R., Organic field effect transistors (OFETs) in environmental sensing and health monitoring: A review. *TrAC, Trends Anal. Chem.* 2019, 111, 27–36.
27. Lee, M. Y.; Lee, H. R.; Park, C. H.; Han, S. G.; Oh, J. H., Organic transistor-based chemical sensors for wearable bioelectronics. *Acc. Chem. Res.* 2018, 51 (11), 2829–2838.
28. Liu, H.; Li, Q.; Zhang, S.; Yin, R.; Liu, X.; He, Y.; Dai, K.; Shan, C.; Guo, J.; Liu, C.; Shen, C.; Wang, X.; Wang, N.; Wang, Z.; Wei, R.; Guo, Z., Electrically conductive polymer composites for smart flexible strain sensors: A critical review. *J Mater. Chem. C* 2018, 6 (45), 12121–12141.
29. Jiang, H.; Lee, E.-C., Highly selective, reusable electrochemical impedimetric DNA sensors based on carbon nanotube/polymer composite electrode without surface modification. *Biosens. Bioelectron.* 2018, 118, 16–22.
30. Huang, L.; Wang, Z.; Zhu, X.; Chi, L., Electrical gas sensors based on structured organic ultra-thin films and nanocrystals on solid state substrates. *Nanoscale Horizons* 2016, 1 (5), 383–393.
31. Ota, H.; Chen, K.; Lin, Y.; Kiriya, D.; Shiraki, H.; Yu, Z.; Ha, T.-J.; Javey, A., Highly deformable liquid-state heterojunction sensors. *Nat. Commun.* 2014, 5 (1), 5032.
32. Jiang, C.; Yao, Y.; Cai, Y.; Ping, J., All-solid-state potentiometric sensor using single-walled carbon nanohorns as transducer. *Sens. Actuators B* 2019, 283, 284–289.
33. Chen, M.; Zhang, L.; Duan, S.; Jing, S.; Jiang, H.; Li, C., Highly stretchable conductors integrated with a conductive carbon nanotube/Graphene network and 3D porous poly(dimethylsiloxane). *Adv. Funct. Mater.* 2014, 24 (47), 7548–7556.
34. Jo, N.; Kim, B.; Lee, S.-M.; Oh, J.; Park, I. H.; Jin Lim, K.; Shin, J.-S.; Yoo, K.-H., Aptamer-functionalized capacitance sensors for real-time monitoring of bacterial growth and antibiotic susceptibility. *Biosens. Bioelectron.* 2018, 102, 164–170.
35. Asadian, E.; Ghalkhani, M.; Shahrokhian, S., Electrochemical sensing based on carbon nanoparticles: A review. *Sens. Actuators B* 2019, 293, 183–209.
36. Bae, G.; Jeon, I. S.; Jang, M.; Song, W.; Myung, S.; Lim, J.; Lee, S. S.; Jung, H.-K.; Park, C.-Y.; An, K.-S., Complementary Dual-Channel gas sensor devices based on a role-allocated ZnO/Graphene hybrid Heterostructure. *ACS Appl. Mater. Interfaces* 2019, 11 (18), 16830–16837.
37. Li, Z.; Askim, J. R.; Suslick, K. S., The optoelectronic nose: Colorimetric and Fluorometric sensor arrays. *Chem. Rev.* 2019, 119 (1), 231–292.
38. Heikenfeld, J.; Jajack, A.; Rogers, J.; Gutruf, P.; Tian, L.; Pan, T.; Li, R.; Khine, M.; Kim, J.; Wang, J.; Kim, J., Wearable sensors: Modalities, challenges, and prospects. *Lab Chip* 2018, 18 (2), 217–248.
39. Wang, H.; Totaro, M.; Beccai, L., Toward perceptive soft robots: Progress and challenges. *Adv. Sci.* 2018, 5 (9), 1800541.
40. Mercante, L. A.; Scagion, V. P.; Migliorini, F. L.; Mattoso, L. H. C.; Correa, D. S., Electrospinning-based (bio)sensors for food and agricultural applications: A review. *TrAC, Trends Anal. Chem.* 2017, 91, 91–103.
41. Ahmad, O. S.; Bedwell, T. S.; Esen, C.; Garcia-Cruz, A.; Piletsky, S. A., Molecularly imprinted polymers in electrochemical and optical sensors. *Trends Biotechnol.* 2019, 37 (3), 294–309.
42. Luo, S.; Wang, Y.; Wang, G.; Liu, F.; Zhai, Y.; Luo, Y., Hybrid spray-coating, laser-scribing and ink-dispensing of graphene sensors/arrays with tunable piezoresistivity for in situ monitoring of composites. *Carbon* 2018, 139, 437–444.
43. Leggett, G. J., Scanning near-field photolithography—Surface photochemistry with nanoscale spatial resolution. *Chem. Soc. Rev.* 2006, 35 (11), 1150–1161.
44. Sumaru, K.; Takagi, T.; Morishita, K.; Satoh, T.; Kanamori, T., Fabrication of pocket-like hydrogel microstructures through photolithography. *Soft Matter* 2018, 14 (28), 5710–5714.

45. Paik, S.; Kim, G.; Chang, S.; Lee, S.; Jin, D.; Jeong, K.-Y.; Lee, I. S.; Lee, J.; Moon, H.; Lee, J.; Chang, K.; Choi, S. S.; Moon, J.; Jung, S.; Kang, S.; Lee, W.; Choi, H.-J.; Choi, H.; Kim, H. J.; Lee, J.-H.; Cheon, J.; Kim, M.; Myoung, J.; Park, H.-G.; Shim, W., Near-field sub-diffraction photolithography with an elastomeric photomask. *Nat. Commun.* 2020, **11** (1), 805.
46. Nishimura, Y.; Yano, K.; Itoh, M.; Ito, M., Photolithography. In *Flat panel display manufacturing*, pp. 287–310.
47. Zhang, Y.; Zhang, F.; Yan, Z.; Ma, Q.; Li, X.; Huang, Y.; Rogers, J. A., Printing, folding and assembly methods for forming 3D mesostructures in advanced materials. *Nat. Rev. Mater.* 2017, **2** (4), 17019.
48. Yin, M.-J.; Yao, M.; Gao, S.; Zhang, A. P.; Tam, H.-Y.; Wai, P.-K. A., Rapid 3D patterning of poly(acrylic acid) ionic hydrogel for miniature pH sensors. *Adv. Mater.* 2016, **28** (7), 1394–1399.
49. Ganter, P.; Lotsch, B. V., Photocatalytic Nanosheet lithography: Photolithography based on organically modified photoactive 2D Nanosheets. *Angew. Chem. Int. Ed.* 2017, **56** (29), 8389–8392.
50. Yin, M.-J.; Zhao, Q.; Wu, J.; Seefeldt, K.; Yuan, J., Precise micropatterning of a porous poly(ionic liquid) via Maskless photolithography for high-performance nonenzymatic H2O2 sensing. *ACS Nano* 2018, **12** (12), 12551–12557.
51. Gao, W.; Emaminejad, S.; Nyein, H. Y. Y.; Challa, S.; Chen, K.; Peck, A.; Fahad, H. M.; Ota, H.; Shiraki, H.; Kiriya, D.; Lien, D.-H.; Brooks, G. A.; Davis, R. W.; Javey, A., Fully integrated wearable sensor arrays for multiplexed in situ perspiration analysis. *Nature* 2016, **529** (7587), 509–514.
52. Phan, H.-P.; Nguyen, T.-K.; Dinh, T.; Iacopi, A.; Hold, L.; Shiddiky, M. J. A.; Dao, D. V.; Nguyen, N.-T., Robust free-standing Nano-thin SiC membranes enable direct photolithography for MEMS sensing applications. *Adv. Eng. Mater.* 2018, **20** (1), 1700858.
53. Girault, P.; Azuelos, P.; Lorrain, N.; Poffo, L.; Lemaitre, J.; Pirasteh, P.; Hardy, I.; Thual, M.; Guendouz, M.; Charrier, J., Porous silicon micro-resonator implemented by standard photolithography process for sensing application. *Opt. Mater.* 2017, **72**, 596–601.
54. Gajos, K.; Budkowski, A.; Petrou, P.; Awsiuk, K.; Misiakos, K.; Raptis, I.; Kakabakos, S., Spatially selective biomolecules immobilization on silicon nitride waveguides through contact printing onto plasma treated photolithographic micropattern: Step-by-step analysis with TOF-SIMS chemical imaging. *Appl. Surf. Sci.* 2020, **506**, 145002.
55. Ooka, Y.; Tetsumoto, T.; Daud, N. A. B.; Tanabe, T., Ultrasmall in-plane photonic crystal demultiplexers fabricated with photolithography. *Opt. Express* 2017, **25** (2), 1521–1528.
56. Song, J.; Lu, H.; Foreman, K.; Li, S.; Tan, L.; Adenwalla, S.; Gruverman, A.; Ducharme, S., Ferroelectric polymer nanopillar arrays on flexible substrates by reverse nanoimprint lithography. *J Mater. Chem. C* 2016, **4** (25), 5914–5921.
57. Xu, R.; Huang, Y.; Lee, P.; Yeh, P.; Ren, L.; Togelang, A.; Liang, J.; Wu, Y.; Irie, L.; Rodgers, A.; He, W.; Ding, G.; Sanghadasa, M.; Lin, L. In *Low-Cost, Efficient, Photolithography-Free Fabrication of Stretchable Electronics Systems on a Vinyl Cutter, 2019 IEEE 32nd International Conference on Micro Electro Mechanical Systems (MEMS)*, 27–31 Jan. 2019; 2019; pp. 343–346.
58. Liu, Q.; Aroonyadet, N.; Song, Y.; Wang, X.; Cao, X.; Liu, Y.; Cong, S.; Wu, F.; Thompson, M. E.; Zhou, C., Highly sensitive and quick detection of acute myocardial infarction biomarkers using In2O3 Nanoribbon biosensors fabricated using shadow masks. *ACS Nano* 2016, **10** (11), 10117–10125.
59. Huang, K.-C.; Lin, C.-H.; Anuratha, K.S.; Huang, T.-Y.; Lin, J.-Y.; Tseng, F.-G.; Hsieh, C.-K., Laser printer patterned sacrificed layer for arbitrary design and scalable fabrication of the all-solid-state interdigitated in-planar hydrous ruthenium oxide flexible micro supercapacitors. *J. Power Sources* 2019, **417**, 108–116.
60. Sirringhaus, H.; Shimoda, T., Inkjet printing of functional materials. *MRS Bull.* 2011, **28** (11), 802–806.

61. Nie, Z.; Kumacheva, E., Patterning surfaces with functional polymers. *Nat. Mater.* 2008, 7 (4), 277–290.
62. Tan, H. W.; Saengchairat, N.; Goh, G. L.; An, J.; Chua, C. K.; Tran, T., Induction sintering of silver nanoparticle inks on polyimide substrates. *Adv. Mater. Technol.* 2020, 5 (1), 1900897.
63. Farraj, Y.; Bielmann, M.; Magdassi, S., Inkjet printing and rapid ebeam sintering enable formation of highly conductive patterns in roll to roll process. *RSC Adv.* 2017, 7 (25), 15463–15467.
64. Kim, J.; Kumar, R.; Bandodkar, A. J.; Wang, J., Advanced materials for printed wearable electrochemical devices: A review. *Adv. Electron. Mater.* 2017, 3 (1), 1600260.
65. Molina-Lopez, F.; Gao, T. Z.; Kraft, U.; Zhu, C.; Öhlund, T.; Pfattner, R.; Feig, V. R.; Kim, Y.; Wang, S.; Yun, Y.; Bao, Z., Inkjet-printed stretchable and low voltage synaptic transistor array. *Nat. Commun.* 2019, 10 (1), 2676.
66. Turner, A. P. F.; Magan, N., Electronic noses and disease diagnostics. *Nat. Rev. Microbiol.* 2004, 2 (2), 161–166.
67. Persaud, K.; Dodd, G., Analysis of discrimination mechanisms in the mammalian olfactory system using a model nose. *Nature* 1982, 299 (5881), 352–355.
68. Sanaeifar, A.; ZakiDizaji, H.; Jafari, A.; Guardia, M. D. L., Early detection of contamination and defect in foodstuffs by electronic nose: A review. *TrAC, Trends Anal. Chem.* 2017, 97, 257–271.
69. Li, Z.; Bassett, W. P.; Askim, J. R.; Suslick, K. S., Differentiation among peroxide explosives with an optoelectronic nose. *Chem. Commun.* 2015, 51 (83), 15312–15315.
70. Kemling, J. W.; Suslick, K. S., Nanoscale porosity in pigments for chemical sensing. *Nanoscale* 2011, 3 (5), 1971–1973.
71. Li, Z.; Suslick, K. S., Colorimetric sensor Array for monitoring CO and ethylene. *Anal. Chem.* 2019, 91 (1), 797–802.
72. Li, Z.; Fang, M.; LaGasse, M. K.; Askim, J. R.; Suslick, K. S., Colorimetric recognition of aldehydes and ketones. *Angew. Chem. Int. Ed.* 2017, 56 (33), 9860–9863.
73. Li, Z.; Zhang, S.; Yu, T.; Dai, Z.; Wei, Q., Aptamer-based fluorescent sensor Array for multiplexed detection of cyanotoxins on a smartphone. *Anal. Chem.* 2019, 91 (16), 10448–10457.
74. Assen, A. H.; Yassine, O.; Shekhah, O.; Eddaoudi, M.; Salama, K. N., MOFs for the sensitive detection of ammonia: Deployment of fcu-MOF thin films as effective chemical capacitive sensors. *ACS Sens.* 2017, 2 (9), 1294–1301.
75. Mao, S.; Chang, J.; Pu, H.; Lu, G.; He, Q.; Zhang, H.; Chen, J., Two-dimensional nanomaterial-based field-effect transistors for chemical and biological sensing. *Chem. Soc. Rev.* 2017, 46 (22), 6872–6904.
76. Li, Z.; Li, H.; LaGasse, M. K.; Suslick, K. S., Rapid quantification of Trimethylamine. *Anal. Chem.* 2016, 88 (11), 5615–5620.
77. Koydemir, H. C.; Ozcan, A., Wearable and implantable sensors for biomedical applications. *Annu. Rev. Anal. Chem.* 2018, 11 (1), 127–146.
78. Jayathilaka, W. A. D. M.; Qi, K.; Qin, Y.; Chinnappan, A.; Serrano-García, W.; Baskar, C.; Wang, H.; He, J.; Cui, S.; Thomas, S. W.; Ramakrishna, S., Significance of Nanomaterials in Wearables: A review on wearable actuators and sensors. *Adv. Mater.* 2019, 31 (7), 1805921.
79. Liu, Y.; Pharr, M.; Salvatore, G. A., Lab-on-skin: A review of flexible and stretchable electronics for wearable health monitoring. *ACS Nano* 2017, 11 (10), 9614–9635.
80. Wang, X.; Liu, Z.; Zhang, T., Flexible sensing electronics for wearable/attachable health monitoring. *Small* 2017, 13 (25), 1602790.
81. Harada, S.; Kanao, K.; Yamamoto, Y.; Arie, T.; Akita, S.; Takei, K., Fully printed flexible fingerprint-like three-Axis tactile and slip force and temperature sensors for artificial skin. *ACS Nano* 2014, 8 (12), 12851–12857.
82. Hattori, Y.; Falgout, L.; Lee, W.; Jung, S.-Y.; Poon, E.; Lee, J. W.; Na, I.; Geisler, A.; Sadhwani, D.; Zhang, Y.; Su, Y.; Wang, X.; Liu, Z.; Xia, J.; Cheng, H.; Webb, R. C.;

Bonifas, A. P.; Won, P.; Jeong, J.-W.; Jang, K.-I.; Song, Y. M.; Nardone, B.; Nodzenski, M.; Fan, J. A.; Huang, Y.; West, D. P.; Paller, A. S.; Alam, M.; Yeo, W.-H.; Rogers, J. A., Multifunctional skin-like electronics for quantitative, clinical monitoring of cutaneous wound healing. *Adv. Healthcare Mater.* 2014, 3 (10), 1597–1607.

83. Huang, Y.; Zeng, X.; Wang, W.; Guo, X.; Hao, C.; Pan, W.; Liu, P.; Liu, C.; Ma, Y.; Zhang, Y.; Yang, X., High-resolution flexible temperature sensor based graphite-filled polyethylene oxide and polyvinylidene fluoride composites for body temperature monitoring. *Sens. Actuators A* 2018, 278, 1–10.
84. Kwak, Y. H.; Kim, W.; Park, K. B.; Kim, K.; Seo, S., Flexible heartbeat sensor for wearable device. *Biosens. Bioelectron.* 2017, 94, 250–255.
85. Ramírez, J.; Rodriquez, D.; Urbina, A. D.; Cardenas, A. M.; Lipomi, D. J., Combining high sensitivity and dynamic range: Wearable thin-film composite strain sensors of Graphene, ultrathin palladium, and PEDOT:PSS. *ACS Appl. Nano Mater.* 2019, 2 (4), 2222–2229.
86. Shin, K.-Y.; Lee, J. S.; Jang, J., Highly sensitive, wearable and wireless pressure sensor using free-standing ZnO nanoneedle/PVDF hybrid thin film for heart rate monitoring. *Nano Energy* 2016, 22, 95–104.
87. Wang, Y.; Wang, L.; Yang, T.; Li, X.; Zang, X.; Zhu, M.; Wang, K.; Wu, D.; Zhu, H., Wearable and highly sensitive Graphene strain sensors for human motion monitoring. *Adv. Funct. Mater.* 2014, 24 (29), 4666–4670.
88. Kim, H.; Kim, G.; Kim, T.; Lee, S.; Kang, D.; Hwang, M.-S.; Chae, Y.; Kang, S.; Lee, H.; Park, H.-G.; Shim, W., Transparent, flexible, conformal capacitive pressure sensors with nanoparticles. *Small* 2018, 14 (8), 1703432.
89. Zhan, Z.; Lin, R.; Tran, V.-T.; An, J.; Wei, Y.; Du, H.; Tran, T.; Lu, W., Paper/carbon nanotube-based wearable pressure sensor for physiological signal acquisition and soft robotic skin. *ACS Appl. Mater. Interfaces* 2017, 9 (43), 37921–37928.
90. Nela, L.; Tang, J.; Cao, Q.; Tulevski, G.; Han, S.-J., Large-area high-performance flexible pressure sensor with carbon nanotube active matrix for electronic skin. *Nano Lett.* 2018, 18 (3), 2054–2059.
91. Ryu, S.; Lee, P.; Chou, J. B.; Xu, R.; Zhao, R.; Hart, A. J.; Kim, S.-G., Extremely elastic wearable carbon nanotube Fiber strain sensor for monitoring of human motion. *ACS Nano* 2015, 9 (6), 5929–5936.
92. Park, J. J.; Hyun, W. J.; Mun, S. C.; Park, Y. T.; Park, O. O., Highly stretchable and wearable Graphene strain sensors with controllable sensitivity for human motion monitoring. *ACS Appl. Mater. Interfaces* 2015, 7 (11), 6317–6324.
93. Kenry; Yeo, J. C.; Lim, C. T., Emerging flexible and wearable physical sensing platforms for healthcare and biomedical applications. *Microsyst. Nanoeng.* 2016, 2 (1), 16043.
94. Liu, Q.; Chen, J.; Li, Y.; Shi, G., High-performance strain sensors with fish-scale-like Graphene-sensing layers for full-range detection of human motions. *ACS Nano* 2016, 10 (8), 7901–7906.
95. Yeo, J. C.; Yap, H. K.; Xi, W.; Wang, Z.; Yeow, C.-H.; Lim, C. T., Flexible and stretchable strain sensing actuator for wearable soft robotic applications. *Adv. Mater. Technol.* 2016, 1 (3), 1600018.
96. Lu, N.; Kim, D.-H., Flexible and stretchable electronics paving the way for soft robotics. *Soft Robotics* 2014, 1 (1), 53–62.
97. Wagner, S.; Bauer, S., Materials for stretchable electronics. *MRS Bull.* 2012, 37 (3), 207–213.
98. Takei, K.; Honda, W.; Harada, S.; Arie, T.; Akita, S., Toward flexible and wearable human-interactive health-monitoring devices. *Adv. Healthcare Mater.* 2015, 4 (4), 487–500.
99. Fiorillo, A. S.; Critello, C. D.; Pullano, S. A., Theory, technology and applications of piezoresistive sensors: A review. *Sens. Actuators A* 2018, 281, 156–175.
100. Ma, Y.; Yue, Y.; Zhang, H.; Cheng, F.; Zhao, W.; Rao, J.; Luo, S.; Wang, J.; Jiang, X.; Liu, Z.; Liu, N.; Gao, Y., 3D Synergistical MXene/reduced Graphene oxide aerogel for a Piezoresistive sensor. *ACS Nano* 2018, 12 (4), 3209–3216.

101. Weng, W.; Chen, P.; He, S.; Sun, X.; Peng, H., Smart Electronic Textiles. *Angew. Chem. Int. Ed.* 2016, 55 (21), 6140–6169.
102. Zhang, F.; Zang, Y.; Huang, D.; Di, C.-A.; Zhu, D., Flexible and self-powered temperature–pressure dual-parameter sensors using microstructure-frame-supported organic thermoelectric materials. *Nat. Commun.* 2015, 6 (1), 8356.
103. Bae, S.-H.; Lee, Y.; Sharma, B. K.; Lee, H.-J.; Kim, J.-H.; Ahn, J.-H., Graphene-based transparent strain sensor. *Carbon* 2013, 51, 236–242.
104. Amjadi, M.; Pichitpajongkit, A.; Lee, S.; Ryu, S.; Park, I., Highly stretchable and sensitive strain sensor based on silver nanowire–elastomer Nanocomposite. *ACS Nano* 2014, 8 (5), 5154–5163.
105. Shin, S.-H.; Park, D. H.; Jung, J.-Y.; Lee, M. H.; Nah, J., Ferroelectric zinc oxide nanowire embedded flexible sensor for motion and temperature sensing. *ACS Appl. Mater. Interfaces* 2017, 9 (11), 9233–9238.
106. Jia, J.; Huang, G.; Deng, J.; Pan, K., Skin-inspired flexible and high-sensitivity pressure sensors based on rGO films with continuous-gradient wrinkles. *Nanoscale* 2019, 11 (10), 4258–4266.
107. Hammock, M. L.; Chortos, A.; Tee, B. C.-K.; Tok, J. B.-H.; Bao, Z., 25th anniversary article: The evolution of electronic skin (E-skin): A brief history, design considerations, and recent Progress. *Adv. Mater.* 2013, 25 (42), 5997–6038.
108. Frutiger, A.; Muth, J. T.; Vogt, D. M.; Mengüç, Y.; Campo, A.; Valentine, A. D.; Walsh, C. J.; Lewis, J. A., Capacitive soft strain sensors via multicore–Shell Fiber printing. *Adv. Mater.* 2015, 27 (15), 2440–2446.
109. Yao, S.; Zhu, Y., Nanomaterial-enabled stretchable conductors: Strategies, Materials and Devices. *Adv. Mater.* 2015, 27 (9), 1480–1511.
110. Choi, T. Y.; Hwang, B.-U.; Kim, B.-Y.; Trung, T. Q.; Nam, Y. H.; Kim, D.-N.; Eom, K.; Lee, N.-E., Stretchable, transparent, and stretch-unresponsive capacitive touch sensor Array with selectively patterned silver nanowires/reduced Graphene oxide electrodes. *ACS Appl. Mater. Interfaces* 2017, 9 (21), 18022–18030.
111. Wang, J.; Lou, H.; Meng, J.; Peng, Z.; Wang, B.; Wan, J., Stretchable energy storage E-skin supercapacitors and body movement sensors. *Sens. Actuators B* 2020, 305, 127529.
112. Hong, Y. J.; Lee, H.; Kim, J.; Lee, M.; Choi, H. J.; Hyeon, T.; Kim, D.-H., Blood sugar monitoring: Multifunctional wearable system that integrates sweat-based sensing and vital-sign monitoring to estimate pre-/post-exercise glucose levels. *Adv. Funct. Mater.* 2018, 28 (47), 1870336.
113. Isensee, K.; Müller, N.; Pucci, A.; Petrich, W., Towards a quantum cascade laser-based implant for the continuous monitoring of glucose. *Analyst* 2018, 143 (24), 6025–6036.
114. Martín, A.; Kim, J.; Kurniawan, J. F.; Sempionatto, J. R.; Moreto, J. R.; Tang, G.; Campbell, A. S.; Shin, A.; Lee, M. Y.; Liu, X.; Wang, J., Epidermal microfluidic electrochemical detection system: Enhanced sweat sampling and metabolite detection. *ACS Sens.* 2017, 2 (12), 1860–1868.
115. Kim, J.; Campbell, A. S.; de Ávila, B. E.-F.; Wang, J., Wearable biosensors for healthcare monitoring. *Nat. Biotechnol.* 2019, 37 (4), 389–406.
116. Tierney, M. J.; Tamada, J. A.; Potts, R. O.; Jovanovic, L.; Garg, S., Clinical evaluation of the GlucoWatch® biographer: A continual, non-invasive glucose monitor for patients with diabetes. *Biosens. Bioelectron.* 2001, 16 (9), 621–629.
117. Giri, T. K.; Chakrabarty, S.; Ghosh, B., Transdermal reverse iontophoresis: A novel technique for therapeutic drug monitoring. *J. Contr. Rel.* 2017, 246, 30–38.
118. Lee, H.; Hong, Y. J.; Baik, S.; Hyeon, T.; Kim, D.-H., Enzyme-based glucose sensor: From invasive to wearable device. *Adv. Healthcare Mater.* 2018, 7 (8), 1701150.
119. Liu, C.; Xu, T.; Wang, D.; Zhang, X., The role of sampling in wearable sweat sensors. *Talanta* 2020, 212, 120801.
120. Yang, J., Blood glucose monitoring with smartphone as glucometer. *Electrophoresis* 2019, 40 (8), 1144–1147.

121. Kim, J.; Campbell, A. S.; Wang, J., Wearable non-invasive epidermal glucose sensors: A review. *Talanta* 2018, 177, 163–170.
122. Heller, A.; Feldman, B., Electrochemical glucose sensors and their applications in diabetes management. *Chem. Rev.* 2008, 108 (7), 2482–2505.
123. Cha, K. H.; Meyerhoff, M. E., Compatibility of nitric oxide release with implantable enzymatic glucose sensors based on osmium (III/II) mediated electrochemistry. *ACS Sens.* 2017, 2 (9), 1262–1266.
124. Dautta, M.; Alshetaiwi, M.; Escobar, J.; Tseng, P., Passive and wireless, implantable glucose sensing with phenylboronic acid hydrogel-interlayer RF resonators. *Biosens. Bioelectron.* 2020, 151, 112004.
125. La Belle, J. T.; Adams, A.; Lin, C.-E.; Engelschall, E.; Pratt, B.; Cook, C. B., Self-monitoring of tear glucose: The development of a tear based glucose sensor as an alternative to self-monitoring of blood glucose. *Chem. Commun.* 2016, 52 (59), 9197–9204.
126. Lucisano, J. Y.; Routh, T. L.; Lin, J. T.; Gough, D. A., Glucose monitoring in individuals with diabetes using a long-term implanted sensor/telemetry system and model. *IEEE T. Bio-Med. Eng.* 2017, 64 (9), 1982–1993.
127. Chang Ming, L.; Hua, D.; Xiaodong, C.; John, H. T. L.; Xueji, Z., Implantable electrochemical sensors for biomedical and clinical applications: Progress, problems, and future possibilities. *Curr. Med. Chem.* 2007, 14 (8), 937–951.
128. Gough, D. A.; Kumosa, L. S.; Routh, T. L.; Lin, J. T.; Lucisano, J. Y., Function of an implanted tissue glucose sensor for more than 1 year in animals. *Sci. Transl. Med.* 2010, 2 (42), 42ra53-42ra53.
129. Rooyackers, O.; Blixt, C.; Mattsson, P.; Wernerman, J., Continuous glucose monitoring by intravenous microdialysis: Influence of membrane length and dialysis flow rate. *Acta Anaesthesiol. Scand.* 2013, 57 (2), 214–219.
130. Krinsley, J. S.; Chase, J. G.; Gunst, J.; Martensson, J.; Schultz, M. J.; Taccone, F. S.; Wernerman, J.; Bohe, J.; De Block, C.; Desaive, T.; Kalfon, P.; Preiser, J.-C., Continuous glucose monitoring in the ICU: Clinical considerations and consensus. *Crit. Care* 2017, 21 (1), 197.
131. Peleg, A. Y.; Hooper, D. C., Hospital-acquired infections due to gram-negative bacteria. *N. Engl. J. Med.* 2010, 362 (19), 1804–1813.
132. Capita, R.; Alonso-Calleja, C., Antibiotic-resistant bacteria: A challenge for the food industry. *Crit. Rev. Food Sci. Nutr.* 2013, 53 (1), 11–48.
133. Váradi, L.; Luo, J. L.; Hibbs, D. E.; Perry, J. D.; Anderson, R. J.; Orenga, S.; Groundwater, P. W., Methods for the detection and identification of pathogenic bacteria: Past, present, and future. *Chem. Soc. Rev.* 2017, 46 (16), 4818–4832.
134. Sauer, S.; Kliem, M., Mass spectrometry tools for the classification and identification of bacteria. *Nat. Rev. Microbiol.* 2010, 8 (1), 74–82.
135. Zulkifli, S. N.; Rahim, H. A.; Lau, W.-J., Detection of contaminants in water supply: A review on state-of-the-art monitoring technologies and their applications. *Sens. Actuators B* 2018, 255, 2657–2689.
136. Mortari, A.; Lorenzelli, L., Recent sensing technologies for pathogen detection in milk: A review. *Biosens. Bioelectron.* 2014, 60, 8–21.
137. Carey, J. R.; Suslick, K. S.; Hulkower, K. I.; Imlay, J. A.; Imlay, K. R. C.; Ingison, C. K.; Ponder, J. B.; Sen, A.; Wittrig, A. E., Rapid identification of bacteria with a disposable colorimetric sensing Array. *J. Am. Chem. Soc.* 2011, 133 (19), 7571–7576.
138. Lim, S. H.; Martino, R.; Anikst, V.; Xu, Z.; Mix, S.; Benjamin, R.; Schub, H.; Eiden, M.; Rhodes, P. A.; Banaei, N., Rapid diagnosis of tuberculosis from analysis of urine volatile organic compounds. *ACS Sens.* 2016, 1 (7), 852–856.
139. Lonsdale, C. L.; Taba, B.; Queralto, N.; Lukaszewski, R. A.; Martino, R. A.; Rhodes, P. A.; Lim, S. H., The use of colorimetric sensor arrays to discriminate between pathogenic bacteria. *PLoS One* 2013, 8 (5), e62726.

140. Phillips, R. L.; Miranda, O. R.; You, C.-C.; Rotello, V. M.; Bunz, U. H. F., Rapid and efficient identification of bacteria using gold-nanoparticle–poly(Para-phenyleneethynylene) constructs. *Angew. Chem. Int. Ed.* 2008, 47 (14), 2590–2594.
141. Han, J.; Cheng, H.; Wang, B.; Braun, M. S.; Fan, X.; Bender, M.; Huang, W.; Domhan, C.; Mier, W.; Lindner, T.; Seehafer, K.; Wink, M.; Bunz, U. H. F., A polymer/peptide complex-based sensor Array that discriminates bacteria in urine. *Angew. Chem. Int. Ed.* 2017, 56 (48), 15246–15251.
142. Li, X.; Kong, H.; Mout, R.; Saha, K.; Moyano, D. F.; Robinson, S. M.; Rana, S.; Zhang, X.; Riley, M. A.; Rotello, V. M., Rapid identification of bacterial biofilms and biofilm wound models using a multichannel Nanosensor. *ACS Nano* 2014, 8 (12), 12014–12019.
143. Manesse, M.; Phillips, A. F.; LaFratta, C. N.; Palacios, M. A.; Hayman, R. B.; Walt, D. R., Dynamic microbead arrays for biosensing applications. *Lab Chip* 2013, 13 (11), 2153–2160.
144. Shrestha, N. K.; Lim, S. H.; Wilson, D. A.; SalasVargas, A. V.; Churi, Y. S.; Rhodes, P. A.; Mazzone, P. J.; Procop, G. W., The combined rapid detection and species-level identification of yeasts in simulated blood culture using a colorimetric sensor Array. *PLoS One* 2017, 12, e0173130.
145. Wu, L.; Qu, X., Cancer biomarker detection: Recent achievements and challenges. *Chem. Soc. Rev.* 2015, 44 (10), 2963–2997.
146. Montuschi, P.; Mores, N.; Trové, A.; Mondino, C.; Barnes, P. J., The electronic nose in respiratory medicine. *Respiration* 2013, 85, 72–84.
147. Fitzgerald, J. E.; Bui, E. T. H.; Simon, N. M.; Fenniri, H., Artificial nose technology: Status and prospects in diagnostics. *Trends Biotechnol.* 2017, 35, 33–42.
148. Queralto, N.; Berliner, A. N.; Goldsmith, B.; Martino, R.; Rhodes, P.; Lim, S. H., Detecting cancer by breath volatile organic compound analysis: A review of Array-based sensors. *J. Breath Res.* 2014, 8, 027112.
149. Mazzone, P. J., Analysis of volatile organic compounds in the exhaled breath for the diagnosis of lung cancer. *J. Thorac. Oncol.* 2008, 3, 774–780.
150. Mazzone, P. J.; Wang, X.-F.; Xu, Y.; Mekhail, T.; Beukemann, M. C.; Na, J.; Kemling, J. W.; Suslick, K. S.; Sasidhar, M., Exhaled breath analysis with a colorimetric sensor Array for the identification and characterization of lung cancer. *J. Thorac. Oncol.* 2012, 7 (1), 137–142.
151. Buchman, J. T.; Gallagher, M. J.; Yang, C.-T.; Zhang, X.; Krause, M. O. P.; Hernandez, R.; Orr, G., Research highlights: Examining the effect of shape on nanoparticle interactions with organisms. *Environ. Sci.: Nano* 2016, 3 (4), 696–700.
152. Jiang, Z.; Le, N. D. B.; Gupta, A.; Rotello, V. M., Cell surface-based sensing with metallic nanoparticles. *Chem. Soc. Rev.* 2015, 44 (13), 4264–4274.
153. Rana, S.; Elci, S. G.; Mout, R.; Singla, A. K.; Yazdani, M.; Bender, M.; Bajaj, A.; Saha, K.; Bunz, U. H. F.; Jirik, F. R.; Rotello, V. M., Ratiometric Array of conjugated polymers–fluorescent protein provides a robust mammalian cell sensor. *J. Am. Chem. Soc.* 2016, 138 (13), 4522–4529.
154. Le, N. D. B.; Yesilbag Tonga, G.; Mout, R.; Kim, S.-T.; Wille, M. E.; Rana, S.; Dunphy, K. A.; Jerry, D. J.; Yazdani, M.; Ramanathan, R.; Rotello, C. M.; Rotello, V. M., Cancer cell discrimination using host–guest "doubled" arrays. *J. Am. Chem. Soc.* 2017, 139 (23), 8008–8012.

13 Organic–Inorganic Semiconductor Heterojunction Photocatalysts

Tao Lv, Zhengyuan Jin, Luhong Zhang, and Yu-Jia Zeng

CONTENTS

13.1 INTRODUCTION

With the development of society and the improvement of the economic level, human demand for green energy is increasing day by day. As a kind of abundant renewable energy, solar energy is gradually incorporated into the future energy map [1,2]. Based on the big fever on solar energy and materials technology, there are bright prospects for photocatalysis in the energy chemical industry. The onset of photocatalysis which has attracted great attention from worldwide researchers can be traced back to 1970s. In 1972, Fujishima and Honda discovered that a TiO_2 photoanode could photoelectrochemically split water into H_2 and O_2 [3]. As a clean technology, it has become one of the key technologies for the conversion of solar energy to chemical energy [4,5]. Particularly, in the last half-century, the application of semiconductor photocatalysts in the environment, energy, nanotechnology, and other fields has been widely and in-depth researched [6–8]. Presently, photocatalytic materials including organic and inorganic semiconductor materials have been the focus of photocatalytic research.

Semiconductor photocatalytic reaction is a process in which the properties of photocatalytic materials do not change with the participation of photons but can promote the redox reaction. The mechanism of photocatalysis is usually explained by the unique energy band structures of semiconductors which are quite different from other materials. Inside the semiconductors, the energy levels of the atoms interact with each other, expanding the atomic energy level to a band. Electrons in an atom comply with the Pauli Exclusion Principle, filling from the ground state with low energy until the Fermi level (E_f). The band above the Fermi level is called the conduction band (CB) which is unfilled with electrons. On the other hand, the band filled with electrons is called the valence band (VB). Electrons in VB are not free to move and the interval between CB and VB is called the bandgap (E_g). Based on above band theory, the photocatalytic reaction can be divided into three processes (Figure 13.1): (i) electrons in the VB absorb photons with the energy greater than or equal to the E_g, then move to the CB and become excited leaving the same amount of holes in the VB. The excited electrons can move freely, thus forming photogenerated electron–hole pairs with very strong redox ability; (ii) the electron–hole pairs are in an unstable state with the lifetime of nanoseconds. Therefore, only when the migration of the

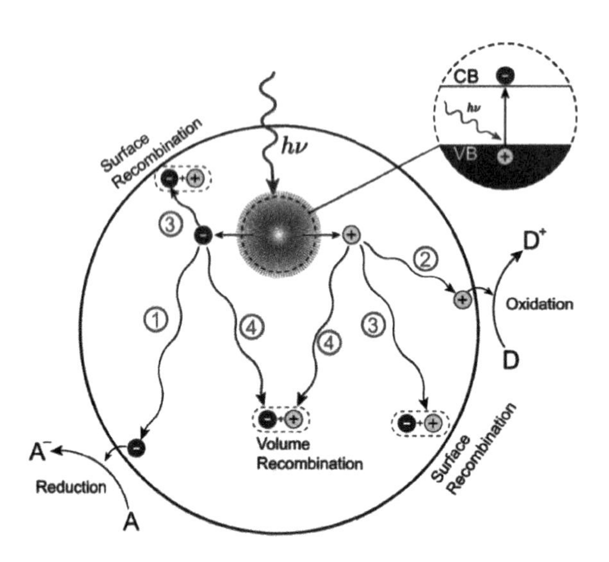

FIGURE 13.1 Schematic illustration of the principle of semiconductor photocatalysis [9].

charge carriers to the surface of the material is finished in the same time range, the charges can take place in the photocatalytic reaction instead of recombination. However, the presence of internal and surface defects or the same carriers in the material will cause internal or surface recombination of the photogenerated carriers during the migration process; (iii) after the second process, only a few photogenerated carriers migrate to the surface of the material, and redox reactions will occur with the pre-adsorbed substance [9]. The CB and VB positions of semiconductor determine the reduction and oxidation ability of photogenerated electrons and holes. Theoretically, it is necessary for the semiconductor to have a suitable energy band position, so as to ensure that photogenerated carriers have sufficient redox capacity to achieve catalytic reactions.

After about half a century of development, photocatalytic technology has been widely used in many fields such as (1) photolysis water to yield hydrogen (H_2) or oxygen (O_2); (2) photocatalytic organic synthesis; (3) photocatalytic degradation of pollutants and purification of waste gas; (4) photocatalytic reduction of carbon dioxide (CO_2) and fixation of nitrogen; (5) recycling of heavy metals; (6) sterilization and anti-toxin; (7) self-cleaning decontamination; (8) health care. In the following sections, we will focus on three main applications.

The first application is photocatalytic H_2 production. H_2 is considered as one of the most ideal alternatives to fossil fuels. H_2 production in the modern industry mainly depends on electrolytic water, steam reforming, and synthetic gas from petroleum cracking. However, these methods have serious problems of environmental pollution and energy consumption. Photocatalytic H_2 production is a clean and pollution-free technique which is worth looking forward to replace traditional technologies. It takes 273 kJ energy to split one mole of water. Therefore, the bandgap of the semiconductor needs to be greater than 1.23 eV so that it can be used for photocatalytic water splitting [10]. In addition to the requirement on the bandgap of the semiconductor, the CB and VB potential also determine whether the reaction can occur. To completely split water, the CB potential of the semiconductor needs more negative than the reduction potential of H_2 (0 eV vs. NHE), and the VB potential needs more positive than the oxidation potential of O_2 (1.23 eV vs. NHE). Figure 13.2 shows the relationship between the CB and VB potential of some semiconductors and the overall water splitting potential [11]. A suitable bandgap can satisfy the thermodynamics demand of the semiconductor water splitting, but the H_2 evolution activity is affected by the generation, migration, and recombination of the photogenerated carriers.

The second application is the photocatalytic degradation of pollutants. With the acceleration of industrialization and the continuous growth of population, environmental pollution caused by the organic pollutants has become a worldwide problem. At present, the main methods to treat organic pollutants in wastewater are traditional biological technology and physical technology, such as adsorption, ultrafiltration, and coagulation. However, these methods have the disadvantages of low removal rate, high operating cost, and secondary pollution. Photocatalytic degradation of pollutants has the characteristics of environmental friendliness and high efficiency, so it has great application value in environmental pollution control. It is generally believed that the active substances involved in photocatalytic degradation of pollutants are

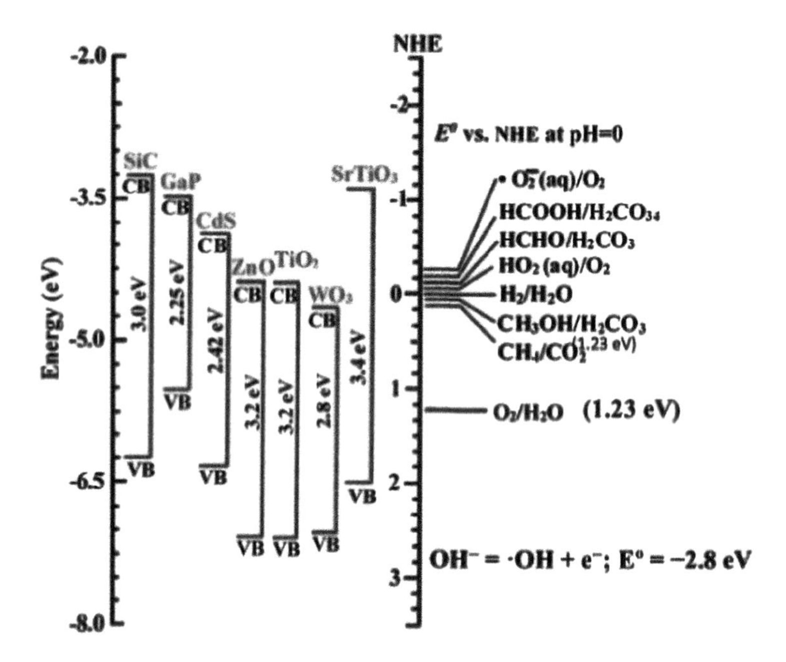

FIGURE 13.2 Band-edge positions of semiconductor photocatalysts relative to the energy levels of various redox couples in water [11].

$\cdot O_2^-$, OH, and H_2O_2. The photogenerated electrons can reduce the O_2 absorbed on the surface of a semiconductor to O_2^-, then O_2^- will react with H^+ to produce H_2O_2. At the same time, photogenerated holes oxidize H_2O and OH^- to highly active $\cdot OH$. O_2^-, OH, and H_2O_2 are all strong oxidizing free radicals, thus they can degrade organic pollutants adsorbed on the surface of semiconductors to CO_2 and H_2O, etc. [12].

The third application is photocatalytic CO_2 reduction. Inoue's group first found that a photocatalytic reaction was occurred at the suspended TiO_2 in the solution, which resulted in the reduction of CO_2 with H_2O to form formic acid, methanal, methanol, and a small amount of methane [13]. This reaction converts CO_2 into organic matters, which can effectively reduce the content of greenhouse gas in nature. CO_2 has a natural chemical inertness and is also very difficult to adsorb on the surface of common catalysts, resulting in a very low conversion rate and selectivity. The ground-breaking discovery of photocatalytic CO_2 reduction points another way for researchers to convert CO_2 into other chemicals, which is of great significance to the improvement of the global environment.

Although photocatalytic technology has made some progress in continuous research in recent years, photocatalytic efficiency is still relatively low, which greatly restricts the application of photocatalytic technology in many aspects. Therefore, improving the efficiency of the photocatalytic reaction is the most important target. Generally speaking, there are two main factors limiting the photocatalytic activity. The first factor is the narrow light-absorption range of semiconductor photocatalysts. In essence, semiconductor photocatalytic technology is to convert solar energy into chemical energy, and the more solar energy the material absorbs, the more chemical energy conversion is obtained. In the early stage of research, the mainstream photocatalytic materials, such as titanium dioxide (TiO_2) and zinc oxide (ZnO), all have a large bandgap (3.0–3.2 eV), so they can only absorb the ultraviolet (UV) light which makes up 5% of the total sunlight. However, most of the solar energy is concentrated in the visible and infrared bands. Therefore, widening the range of optical response range is very important in improving the utilization of solar energy. The second factor is the low separation efficiency of photogenerated carriers. Photogenerated carriers need to migrate to the surface of the material from generation to reaction. This process usually takes 10^{-3} s, but the electrons and holes recombination only takes 10^{-9} s [9]. At the same time, defects and impurity levels in the material accelerate the recombination probability of carriers, which greatly reduces the number of carriers migrating to the surface of the material. Therefore, some materials do not exhibit a high photocatalytic activity even though they have a good absorption property.

The apparent efficiency of the photocatalytic reaction is determined by photon absorption efficiency, charge transfer efficiency, and catalytic reaction efficiency. It can be expressed by the following formula [14]:

$$\eta_{photocatalysis} = \eta_{Abs.} * \eta_{CT} * \eta_{cat.} \tag{13.1}$$

Therefore, the photocatalytic efficiency can be improved from these three perspectives.

i) Enhance light-absorption intensity or extend light-absorption range: stronger light-absorption ability generally means that higher concentration of photogenerated electron–hole pairs can be excited in the photocatalyst bulk phase. Without altering the electronic structure of semiconductors, the photocatalyst with hierarchical architecture shows higher photoexcited carriers concentration and photocurrent output due to the stronger absorption capacity caused by multiple reflections [15]. Similarly, reducing the size of the photocatalyst or improving pore structure can increase the specific surface area, which is also a way to enhance the light absorption and mass transfer. For the traditional thin-film catalysts, the catalytic efficiency is much lower than that of the powder photocatalysts because of its greatly reduced specific surface area relative to the powder catalysts. Therefore, the preparation of porous/mesoporous films is one of the effective ways to improve their photocatalytic performance. The hierarchical architecture or porous structure can only increase the light-absorption intensity but cannot expand the light-absorption region. In the photovoltaic field, in order to enhance the utilization of visible light, a lot of works are to use the sensitization ability of dyes to absorb visible light and then transfer the electrons of the transition through the semiconductor to form the photovoltaic effect [16]. This method is introduced into the field of photochemistry and photocatalysis. However, because of the strong redox ability of photogenerated electrons and holes, the dyes can be destroyed during the reaction. Therefore, it is necessary to find a light-stable dye to prevent the phenomenon of self-destruction while photosensitization [17]. Some commonly used in sensitizing dyes are phthalocyanine, rose red, bipyridine ruthenium, etc. In addition, many researchers try to expand the absorption of sunlight by changing the electronic structure of semiconductors and reducing the bandgap. The most common method is the doping of metallic or non-metallic elements. The impurity level generated by the shallow doping is used to achieve the in-band transition to reduce the apparent bandgap width of the semiconductor photocatalyst [18]. Selective fabrication of defects within or on the surface of semiconductors is also a common strategy for expanding light absorption. Other methods, such as localized surface plasmon resonance (LSPR) of noble metals and multiple exciton generation of quantum dots, can extend the range of light absorption of semiconductors. In principle, the enhancement of semiconductor photocatalytic capacity brought by the regulation of semiconductor energy band structure from energy band engineering is limited. Therefore, to realize the large-scale application of photocatalytic technology, semiconductor photocatalysts with a narrow bandgap for visible-light absorption or near-infrared absorption must be developed. At this stage, researchers have reported a variety of visible-light and near-infrared light-responsive photocatalysts, such as bismuth ferrite ($BiFeO_3$), bismuth molybdate (Bi_2MoO_6), halogen oxide, silver phosphate (Ag_3PO_4), carbon nitride (C_3N_4), and so on [19–24]. Most of them have good photocatalytic activity, but it is difficult to achieve practical application due to the limited specific surface areas and the instability caused by acid corrosion and photocorrosion.

ii) Improve space separation of electron–hole pairs and accelerate carriers' migration to the surface of photocatalyst: one of the simplest ways to reduce the carriers' recombination probability is to decrease the dimension and scale of the photocatalyst and increase the specific surface area. The diffusion capacity of different carriers is limited, and larger size of the photocatalyst particles, will make more recombination of the carriers before diffusion to the catalyst surface. Compared with the semiconductor at three-dimensional scale, the carriers transmit along the one/two-dimensional direction in the one/two-dimensional semiconductor material, which greatly reduces the carriers' transmission distance and recombination probability in the transmission process. In addition, the lower-dimensional semiconductor has more contact with the environment solution, which can better enable the carriers to be captured by polar molecule H_2O and non-polar molecule O_2 in the transmission process, thus producing hydroxyl active radicals (OH) and superoxide radicals (O^{2-}) triggering the radical chain reaction. Noble metals have higher work functions and are often used as co-catalysts in photocatalytic processes. The Mott-Schottky barrier at the interface of semiconductor–metal can effectively inhibit the recombination of free electrons and holes and therefore increase the photon quantum efficiency of the photocatalytic reaction. Based on the recombination tendency of the free charge at high-energy state in the bulk phase, the construction of semiconductor heterojunction with matching energy band structure is also an effective means to promote electron–hole pair separation. Because of the existence of the depletion layer, a fixed built-in electric field is produced at the heterojunction interface. Driven by this built-in electric field, the high-energy free electrons generated by the light excitation are transferred to the appropriate VB of semiconductor under the energy gradient. Meanwhile, the holes are transferred to the CB of another semiconductor. Therefore, the electron–hole pair separation rate within the semiconductor bulk phase can be significantly improved with achieving accelerated photocatalytic reaction rate and enhanced photocatalytic activity.

iii) Promote chemical reaction on catalyst surface: in addition to the above-mentioned two methods, photocatalytic reaction efficiency can be also achieved by promoting the chemical reaction on the catalyst surface. For example, Bian et al. introduced Ti^{3+} defects on the surface of the TiO_2 crystal to make the surface positively charged [25]. The desorption of Cr (III) on the photocatalyst surface was promoted by the introduced Ti^{3+} to improve the photocatalytic reduction rate of Cr (VI). Most of the photocatalysts suffer from the sluggish surface reactions, which result in the accumulation of photogenerated carriers on the surface and cause charge recombination and photocorrosion. The presence of a co-catalyst could significantly accelerate the surface chemical reaction and improve the photocatalytic activity.

To sum up, the current hot issues in photocatalytic technology mainly focus on broadening the light-absorption range, improving the photon conversion efficiency and the photocatalytic activity of photocatalysts. The construction of heterogeneous structures, which can take these issues into account, has been widely studied by researchers all over the world. Organic semiconductor photocatalyst has become a research hotspot in recent years because of its adjustable bandgap structure and excellent photocatalytic performance. The researches on jointly built organic–inorganic heterojunction photocatalysts based on traditional inorganic and organic semiconductor photocatalysts have also gradually risen. The content of this chapter aims to briefly introduce and summarize the development of the existing inorganic and organic semiconductor photocatalysts and organic–inorganic heterojunction semiconductor photocatalysis in recent years. A prospect for the future development of organic–inorganic heterojunction semiconductor photocatalysis will also be presented at the end of this chapter.

13.2 PHOTOCATALYSTS

13.2.1 Inorganic Photocatalyst

13.2.1.1 Metal Oxide

Metal oxides are the most used and the earliest studied photocatalysts. At present, the commonly used metal oxide semiconductor photocatalysts mainly include TiO_2, ZnO, ZrO_2, WO_3, SnO_2, MoO_3, and so on [26–31]. Among them, TiO_2 has been the most in-depth studied photocatalyst. In the following sections, we take the TiO_2 as the representative. In the natural state, TiO_2 usually has four co-existing crystal structures: anatase, rutile, brookite, and TiO_2 (B) [32]. Rutile is the most stable crystal form of TiO_2 even in nanoscale size. Both anatase, brookite, and TiO_2 (B) can be

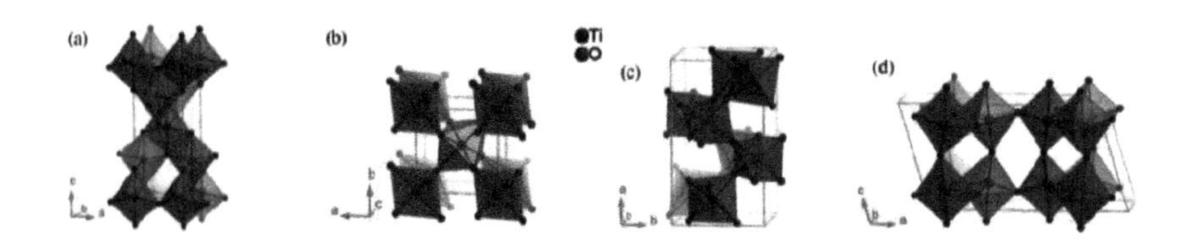

FIGURE 13.3 Crystalline structures of four forms of TiO_2: (a) anatase, (b) rutile, (c) brookite, and (d) TiO_2 [33].

transformed into rutile at high temperatures. Different crystal forms of TiO_2 usually present different morphologies and properties. Therefore, the preparation methods and conditions required for the synthesis of TiO_2 nanomaterials with different crystal forms are also different. For example, anatase TiO_2 nanomaterials are usually obtained by solution synthesis or low-temperature vapor deposition method, while high-temperature deposition and heating reactions normally produce rutile type of TiO_2. Figure 13.3 is the crystal structures for four different forms of TiO_2, which can be seen as they have different crystal structures and symmetries. Anatase has a tetragonal system and its {011} and {100} lattice plane have the lowest crystal facet energy. The thermodynamic stable morphology of anatase is a truncated octahedral structure. The rutile TiO_2 has a tetragonal system which is different from that of anatase and the lattice plane with lower energy is the same as anatase. Therefore, rutile TiO_2 also possesses a truncated octahedral structure. The brookite TiO_2 has an oblique square crystalline system, and its structural units are larger than that of anatase and rutile. Similarly, TiO_2 (B) has larger structural units too.

However, because of the large bandgap, TiO_2 can only absorb UV light, which seriously limits its photocatalytic performance under sunlight irradiation and practical application. The simplest strategy to improve the photocatalytic efficiency of TiO_2 is to regulate the pH of the solution. The pH value of the solution has a great influence on the activity of the photocatalyst. Most photocatalytic reactions occur in the neutral environments, but some photocatalysts perform better in acidic or alkaline environments [34,35]. The effect of pH value on the photocatalytic system can be summarized as follows: (i) changing the band structure of semiconductor materials, including bandwidth and energy-level position [36]; (ii) affecting the surface charge, thus changing the adsorption capacity of the target degradation product; (iii) changing the production rate of active substances [37]. For example, in the Pt/TiO_2 (anatase)-TiO_2 (rutile)-IO_3^-/I^- system, as the pH value of the reaction system increases, the rate of photocatalytic water splitting also increases [35]. When $pH < 3$, the I^- ions on the Pt/TiO_2 (anatase) surface could be oxidized to I_3^-, so that H_2O cannot be oxidized to O_2. But in the alkaline environment, the oxidation product of I^- was only IO_3^-, thus the photocatalytic activity was improved.

The doping of metal elements can broaden the absorption spectrum to a longer wavelength for TiO_2 and improve the surface adsorption capacity, conductivity, and separation efficiency of photoelectron–hole pairs. Therefore, various metallic elements such as Pt, Au, Ag, Ni, and Pd have been widely used in the TiO_2 modification

[38–42]. In addition, the improvement of the photocatalytic activity of the TiO_2 can also be achieved by the LSPR effect of the doped metals, such as Pt, Au, and Ag [43–45]. When the metals are in contact with the semiconductor, the generated thermal electrons by the LSPR effect can be transferred directly to the CB of the semiconductor, thus realizing the separation of electron–hole pairs.

In addition to metal doping, non-metal doping is another important means to improve the catalytic activity and visible light absorption capacity of TiO_2. Doping of non-metal elements (e.g., N, F, C, S, etc.) can narrow the energy band of TiO_2, resulting in a hybrid 2p energy level between CB and VB [46–48]. Under visible light irradiation, electrons transition from the VB to the middle 2p level, and then transition to the CB after absorbing a photon, so that the electrons can achieve the transition process while consuming lower energy, which also prevents the recombination of electron–hole pairs. Wang et al. reported that when the doping content of N in TiO_2 nanoparticles reached 4.91%, the bandgap width of TiO_2 decreased from 3.2 eV to 2.65 eV [49]. Different from metal doping, non-metal doping produces fewer recombination centers for the electron–hole pairs.

Combining TiO_2 with other semiconductor materials can broaden the light-absorption range and prolong the recombination time of electron–hole pairs, thus improving the photocatalytic performance. The combined semiconductors should have suitable bandgaps and energy-level positions which can form heterostructures at the contact interface. When the Fermi level of the two semiconductors is coupled with each other, the photogenerated charge carriers are easier transported and separated between the VB and CB of semiconductors. For example, commercial homologous semiconductor composites of TiO_2 (P25: 80% anatase phase and 20% rutile phase) exhibit better photocatalytic performance compared with monocrystalline TiO_2 (anatase, rutile, brookite). In addition, the semiconductor heterojunction formed by TiO_2 and other semiconductor materials such as ZnO, CuO, CdS, etc. can expand the spectral response range of the composites and promote the absorption of light by the semiconductor energy band overlap [50–52].

There are many modification strategies that can redshift the absorption edge of TiO_2, but seldom can these modification strategies enhance the optical response of TiO_2 over the whole solar wavelength range. Not until Chen et al. reported a black TiO_2 material that can absorb a wide range of sunlight wavelengths [53]. The black TiO_2 nanoparticles were prepared by annealing TiO_2 in an H_2 atmosphere of 2 MPa at about 200 °C for 5 days. After structural modification, the presence of Ti^{3+} self-doping, surface hydroxyl, oxygen vacancy, and hydrogenation in black TiO_2 are responsible for the dark color of the material as well as the excellent optical and electronic properties.

13.2.1.2 Sulfide

Relatively speaking, the electronic structures of metal sulfide compounds are more suitable for energy conversion and photocatalytic processes in the visible light region. Because the metal sulfides have a wider solar spectral response range, which also makes sulfides easily being deactivated by photooxidation during the catalytic reaction, almost all the sulfide photocatalysts are composed of metal cations with d10 configuration. From the perspective of thermodynamics, the S 2p orbitals contain the VB, sulfides are not able to split water. Therefore, a sacrificial agent is often necessary to obtain H_2 via photocatalytic water splitting with sulfides as the photocatalysis. Sulfide photocatalysts can be divided into following groups according to their elemental composition and material combination: IIB-VIA, IIB-IIIAVIA, IB-IIIA-VIA and IB-IIB-IVA-VIA. ZnS, CdS, and $Cd_{1-x}Zn_xS$ solid solution all belong to the IIB-VIA group, which is the dominant type of sulfide photocatalyst reported to date. Wurtzite and zinc-blende are two basic crystal structures for CdS, ZnS, and their solid solution, which are composed of tetrahedral metal coordination structures arranged in different ways (Figure 13.4) [54].

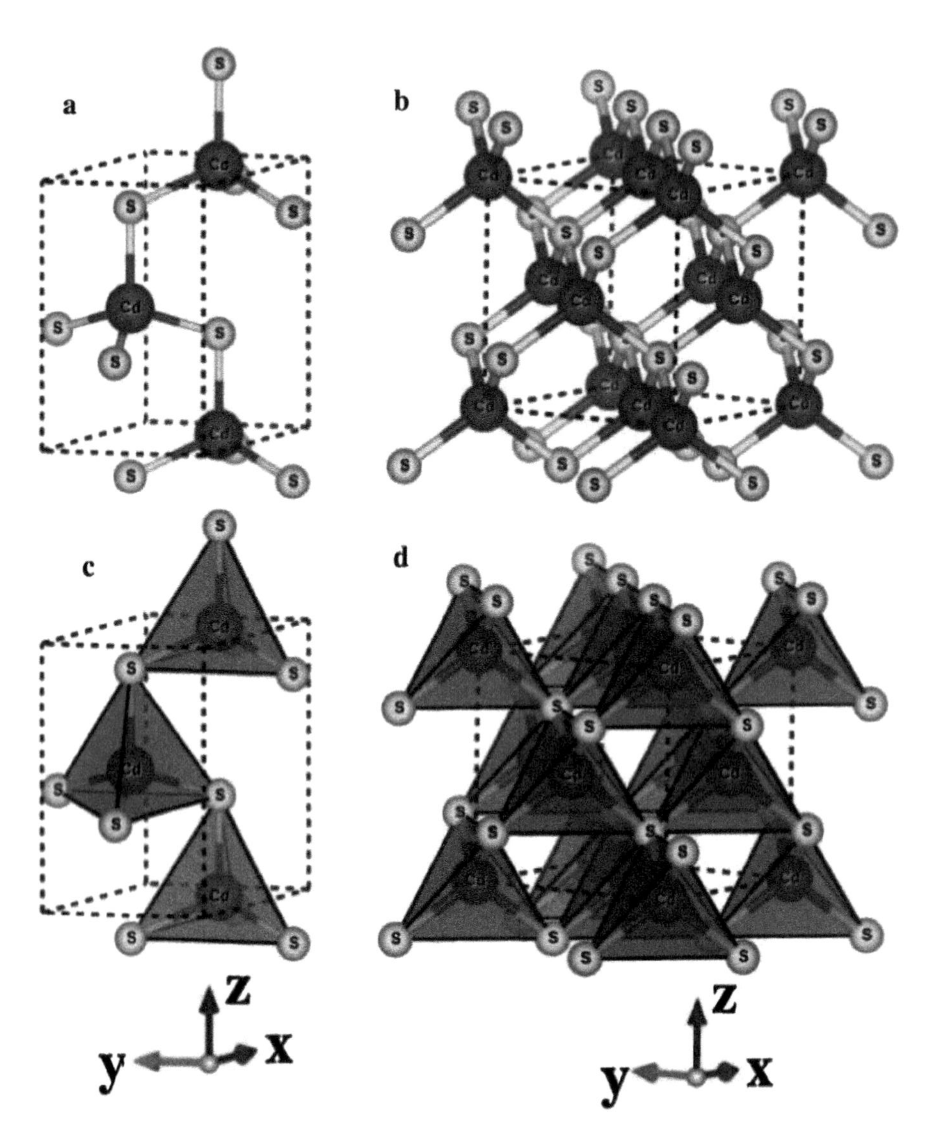

FIGURE 13.4 Schematic diagram of the wurtzite and zinc-blende CdS crystal structures (a and b are ball and stick models; c and d are the polyhedron models) [54].

ZnS is a typical IIB-VIA sulfide photocatalyst with a high CB position. The photogenerated electrons not only can reduce H^+ to H_2 by water splitting but also can achieve CO_2 reduction. However, due to the large bandgap (3.6 eV), ZnS can only utilize solar energy in the UV region. Therefore, on the basis of retaining the high-conductivity band position of ZnS, researchers tried a number of techniques to extend its spectral response. The most effective method is doping metal ions (Cu, Bi, Pb, Ni, Mn, etc.) to form a new donor energy level within the bandgap of ZnS, which can not only maintain the strong reducibility but also absorb visible light energy [55–59]. In addition, Muruganandham et al. successfully synthesized C, N-co-doped porous ZnS photocatalyst, which was a pioneer study on non-metallic-doped sulfide semiconductor [60]. The C, N-co-doped porous ZnS exhibited excellent photocatalytic activity under visible light irradiation, thus initiating the route to prepare non-metallic element-doped ZnS photocatalysts. The control on the morphology and size of synthetic ZnS nanocrystals can also enhance photocatalytic activity. Xie's research group systematically synthesized ZnS hierarchical structures of different sizes [61]. The study showed that ZnS with a three-dimensional flower-like structure had a better photocatalytic degradation performance toward RhB compared with the spherical ZnS attributing to larger surface area. Moreover, its photocatalytic activity decreased along with the decrease of particle size. Furthermore, a number of studies have confirmed that the self-assembled three-dimensional hierarchical structures can increase the specific surface area of the catalyst, while the heterostructure can effectively promote the photogenerated electron transfer. Combining these two concepts to improve photocatalytic efficiency, Chen's group prepared ZnS microspheres co-doped by In and Cd [62]; this three-dimensional hierarchical structure was formed by self-assembly of cubic and hexagonal mixed-phase nanosheets. The positively charged zinc ions and negatively charged sulfur ions are located on the surface of ZnS, spontaneously producing polarization. The electric field on the nanosheets promotes the separation of photogenerated electrons and holes and facilitates the migration of charge carriers to the surface. Therefore, in this study, the H_2 evolution efficiency under visible light irradiation was obviously improved.

Unlike ZnS, CdS has a direct bandgap of 2.4 eV and can absorb visible light radiation. Therefore, CdS has become one of the most widely studied semiconductor photocatalytic material except TiO_2. However, the photocatalytic activity of CdS is unsatisfactory because the photogenerated electrons and holes cannot be separated and transferred effectively. Moreover, it is easily dissociated by self-generated holes to generate Cd^{2+} and S, resulting in poor stability. In order to reduce photocorrosion, sacrificial agents such as lactic acid, triethanolamine, and NaS/Na_2SO_4 are generally used. The photocatalytic activity of CdS is related to the particle size, morphology, and crystalline structure. Researchers successfully fabricated CdS into hollow spheres, nanorods, lines, capsules, quantum dots, and other structures with different morphologies. And the structural changes effectively improve the specific surface area, light utilization, and stability of CdS. In addition, the photocatalytic activity can be significantly improved by doping noble metals such as Pt, Pd, RuO_2, etc., mainly ascribing to noble metals' effective transfer in photogenerated electrons and acting as the active sites, which effectively accelerate the photocatalytic reaction [63,64]. Recent studies have shown that transition metal compounds, especially transition metal sulfides (PdS, NiS), can be used as excellent co-catalysts for CdS photocatalytic H_2 production. As reported by Li's group, the CdS nanoparticles co-deposited by Pt and PdS achieved a very high quantum efficiency with nearly 93% which is quite close to the level of natural photosynthesis under 420 nm visible light irradiation. In this work, Pt and PdS were used as co-catalysts for reduction and oxidation reactions, respectively [65]. Similarly, modification of CdS can also be achieved by combination with other semiconductor photocatalysts. Semiconductor materials used in fabricating heterogeneous structures generally should possess a wide bandgap and suitable band structure, such as TiO_2, ZnO, ZnS, WO_3, and g-C_3N_4 [66–70].

13.2.1.3 Solid Solution

Doping is an effective way to expand the light-response range of materials. However, the overall photocatalytic activity is still difficult to be significantly improved. This is because the local impurity level introduced by doping tends to become the recombination center of carriers, which dams the quantum efficiency and photocatalytic activity of the material. In addition, while doping reduces the bandgap, the redox ability of photogenerated carriers has also been degraded. In order to realize the precise regulation of bandgap and band edge position, fabrication of solid solution is used to improve the photocatalytic activity of semiconductor photocatalytic materials. The method can continuously regulate the light-response range and redox potential of the material. The photocatalytic performance of the material can be improved to the greatest extent by balancing the effective light absorption and the redox capacity of the carriers. The so-called solid solution refers to the circumstance when the solute atoms are dissolved in the solvent lattice, solvent lattice structure still keeps unchanged, which means the retaining of crystalline structure. Briefly, a solid solution is a semiconductor material with high concentration of isoelectronic doping. The conditions of forming solid solution are as follows: ionic radius, polarization property, and electronegativity are relatively close, and the shape and size of lattice are similar. When the two semiconductors are combined to form a solid solution, the structure will not change. However due to the change of constituent elements, the electronic structure will change accordingly. From the perspective of electronic structure, the CB of metal oxide semiconductor materials is generally contributed by the vacant d or s orbital of metal elements, while the VB is generally contributed by the p orbital of oxygen or the occupied d orbital of metallic elements. Therefore, using two semiconductors with different CB positions to construct solid solution can regulate the CB position, while using two semiconductors with different VB positions to construct solid solution can regulate the VB position. Two semiconductors with different VB and CB positions can also be used to construct solid solution, which can realize the simultaneous regulation of VB and CB position.

The solid solution structure can also regulate the mobility of carriers. Generally speaking, the orbital of the introduced element is more delocalized than the original element, and the migration ability of electrons or holes in the solid solution can be effectively improved. Recently, Hong et al. prepared $SnNb_{2-x}Ta_xO_6$ solid solution by solid-state reaction method [71]. Compared with the pure $SnNb_2O_6$, the photocatalytic decomposition activity of the solid solution was significantly improved. The hybridization between Sn 5 s and O 2p elevated the VB position of $SnNb_2O_6$, narrowing down the bandgap. However, the Nb 4d orbital is localized so the photocatalytic activity is decreased due to its weak electron migration ability. Since the 5d rail of Ta is higher than the 4d rail of Nb, the bandgap of $SnNb_{2-x}Ta_xO_6$ solid solution broadens with the increase of Ta concentration. The upshift of the CB position of solid solution enhances the reduction ability of the photogenerated electrons. Meanwhile, the 5d orbital of Ta is more delocalized than the 4d orbital of Nb, and the electron migration ability is also improved. These are all favorable factors to improve the photocatalytic activity.

13.2.1.4 Perovskite

Perovskite compounds were found to have excellent catalytic activity as photocatalysts by Voorhoeve et al. in the 1970s [72]. Since then, perovskite photocatalysts have been widely studied by researchers worldwide and applied in many fields such as photocatalytic water splitting, degradation of organic pollutants, purification of air, and so on [73–75]. Perovskites have excellent catalytic, superconductivity, piezoelectric, magnetoresistance properties, among which the perovskite materials in the field of photocatalysis are more widely studied. Compared with other kinds of photocatalysts, perovskite photocatalysts have more advantages: (1) the crystal configuration is easy to adjust; (2) low cost and abundant resources; (3) the stability is high and can be reused multiple times, etc.

Normally perovskites have the same unit cell as calcium titanate ($CaTiO_3$) [78]. The perovskite structure can usually be expressed as ABX_3, where B and X represent a metal ion and an anion (O^{2-}, Br^-, I^-, etc.) [79]. As shown in Figure 13.5a, B and X form the BX_6 octahedra, where X is located on the corner around the center B of the octahedral. These BX_6 octahedra form a three-dimensional extension system by sharing each corner (Figure 13.5b). A represents metal cations, which are located in vacancies within the octahedra, maintaining the charge neutrality of the entire perovskite network system. Cation A such as Ca^{2+}, Na^+, Pb^{2+}, etc. is located at 12 coordination points between octahedra. The cations A, B, and the anion X in the cubic cell are located in each corner, body center, and surface center position, respectively.

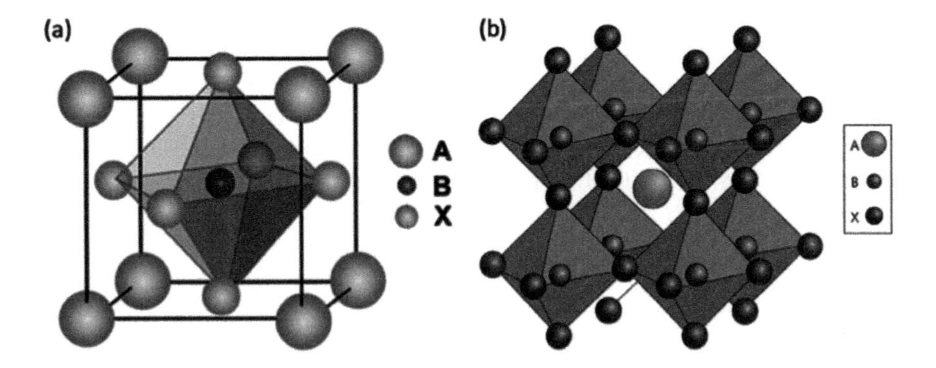

FIGURE 13.5 (a) ABX_3-type perovskite cell and (b) angular octahedral connected perovskite extended network structure [76,77].

One of the most attractive advantages of perovskites in photocatalysis is the flexibility of its composition and structure. Two classes of metal elements include alkali metals (Li, Na, K, etc.) and alkaline earth metals (Mg, Ca, Ba, etc.) can occupy the A site, while the B site can accommodate a wide range of transition metal elements (Ti, Nb, Fe, etc.). Moreover, both A and B sites can be doped with other metal ions to form $AA'BO_3$ or $ABB'O_3$ structures. Similar to other semiconductor materials, the position of O can also be partially substituted by other non-metallic anions to narrow the bandgap, which could enhance the absorption of visible light. The energy band structure, charge transfer, and adsorption of perovskites can be regulated and optimized by changing the composition of cations at A and B sites. For perovskite photocatalysts, the CB is generally composed of the outer d orbitals of B-site transition metals (e.g., Ti 3d and Nb 4d), while the top of the VB is composed of O 2p orbitals. In general, for the B-site cation, the CB edge shifts to a more negative potential and the bandgap becomes larger when the elements are transformed to the next cycle like $NaNbO_3$ (Eg = 3.3 eV) and $NaTaO_3$ (Eg = 4.0 eV). Whereas for the A-site cation, the CB edge will be transferred to a more positive potential to narrow the bandgap, such as $LiTaO_3$ (Eg = 4.8 eV) and $NaTaO_3$ (Eg = 4.0 eV) [80,81].

Several common perovskite photocatalysts are briefly described below. The application of titanate perovskites in photocatalysis has been studied for a long time. The bandgap of most titanate perovskites is greater than 3.0 eV, thus exhibiting excellent photocatalytic performance under UV irradiation. It is a very common strategy to use titanate perovskites as basic materials to change their optical properties and induce their response to visible light by doping method. And titanate perovskites $MTiO_3$ (M = Sr, Ni, Ba, Mn, Ca, etc.) have good optical stability and corrosion resistance in aqueous solution. $NaTaO_3$ and $KTaO_3$ are the most common types of tantalate perovskites. This kind of perovskite usually has a suitable CB and VB position, so it has a good photocatalytic performance. Compared with other types of perovskite photocatalysts, $NaTaO_3$ has a stable layered structure and excellent charge separation efficiency, which improves the activity of photoelectric catalysis and photocatalytic system [82]. Ferrite perovskites usually have a narrow bandgap and good visible light response. Hematite and other ferric oxide compounds as photocatalysts are limited by the defects of short exciton diffusion length, low electronic conductivity, and low CB edge potential. But some ferrite perovskites overcome these disadvantages of binary iron oxides and exhibit better photocatalytic activity. Some researchers have reported the degradation of pollutants and H_2 evolution by $LaFeO_3$ under visible light irradiation [83,84]. In other perovskite photocatalysts, the bandgaps have a response in the visible region when bismuth (Bi), cobalt (Co), nickel (Ni), and antimony (Sb) elements occupy the B site of the perovskites.

13.2.2 Organic Photocatalyst

Compared to inorganic photocatalyst, organic photocatalysts generally have a wide spectral response range up to the visible light even the infrared light, which greatly broadens the utilization of the solar spectrum, and the band structure of organic photocatalysts can be simply adjusted by such modification of molecular functional groups at an atomic level and molecular level. Since the firstly graphitic carbon nitride had been utilized in water splitting under visible light irradiation in 2009, the organic photocatalysts have gradually attracted interest [85]. Currently, the most studied organic photocatalysis materials mainly involve the C_3N_4 materials, metal-organic framework (MOF) materials, and Perylene diimide (PDI) materials.

13.2.2.1 C_3N_4-Based Materials

The name of graphitic carbon nitride (g-C_3N_4) comes from its molecular structure, which has the two-dimensional morphology as graphite and the s-triazine or tri-s-triazine ring with the molar ratio of carbon to nitride of 0.75 in the basic tectonic units (Figure 13.6) [85,86]. C_3N_4 has various advantages as one of the most studied photocatalysis materials, such as medium bandgap (2.7 eV) for visible light-responsive, cost-effective, high reduction ability (with CB locating in the high position), non-toxic, high stability, and easy to fabricate. However, it also has disadvantages like small specific surface area, high recombination rate, low oxidation ability, and the light-absorption range up to ca. 450 nm [87,88].

In order to obtain a larger specific surface area, C_3N_4 with various morphologies has been developed, such as nanorods, nanotubes, flowers, nanochips, hollow structures, and so forth. It seems easier to control the structure of C_3N_4 than that of inorganic photocatalysts, in addition to the hard and soft templating methods, template-free preorganization assembly and exfoliation approaches are effective for designing C_3N_4. Xiao et al. reported that porous nanosheet C_3N_4 exhibited almost 26 times higher photocatalytic activity for H_2 generation in half water splitting (with the presenting of sacrificial agent) than pure C_3N_4 [89].

According to recent research, the high recombination rate of C_3N_4 is caused by internal defects, which play the role of recombination centers to inhibit the transmission of photogenerated electrons from inside to outside of photocatalysts. Therefore, the preparation of highly crystalline C_3N_4 has become the focus of research. Wang et al. employed ionothermal synthesis approach to obtain high crystal C_3N_4, which showed the world-record high apparent quantum efficiency of 60% at H_2 generation from half water splitting in "Sea Water" [90].

Yu et al. used a different approach to obtained C_3N_4 with a reduced recombination rate [91]. The heterogeneous structure was generated among the nanosheet C_3N_4 by forming the C/N concentration gradient from the inside to the surface. The nanosheet

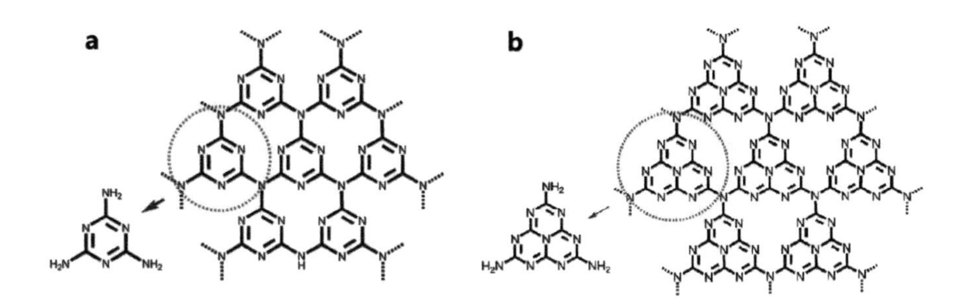

FIGURE 13.6 (a) s-Triazine and (b) tri-s-triazine as tectons of graphite C_3N_4 [87].

morphology provides shorter transmission distance for photogenerated electrons. Moreover, the concentration gradient provides energy-level gradient, which makes photogenerated electrons get a driving force for efficient transmission from inside to the outside surface of photocatalysts. The photocatalytic activity is 21 times higher than pure C_3N_4.

Latest, the most attractive approach for designing C_3N_4 is bandgap engineering, which is beneficial for not only broaden the absorption light range of materials but also adjusting the redox ability of materials. Traditionally the bandgap engineering realized through element doping [92,93]. Zhang et al. used barbituric acid and melamine as the precursors, N-doped C_3N_4 was successfully realized [94]. With the increased mass ratio of barbituric acid in precursor, the obtained C_3N_4 showed reduced bandgap. Ohno et al. designed and synthesized B-doped C_3N_4. They found the B-doped C_3N_4 turned out to be p-type semiconductor with exhibiting high oxidation ability [95].

The molecular functional group modification of C_3N_4 is easy to achieve, which means facile to control the bandgap by introducing defects (vacancy) in frameworks of C_3N_4. Yu et al. found the introduction of nitrogen defects redshifted the C_3N_4 absorption edge, which can be easily controlled by the ratio of precursor (KOH and urea) [96]. Similarly, the C_3N_4 with cyanamide defect was obtained by Yuan et al. The ratio of the cyanamide defect (N vacancy) could be facilely adjusted by KCl dosage, accompanying tunable bandgap of C_3N_4 [97]. On the other hand, Yang et al. employed steam-etching strategy to obtained C_3N_4 with carbon vacancy, which showed a 45-fold improvement in photocatalytic activity upon CO_2-to-CO conversion compared with that of pure C_3N_4. In the work, the carbon vacancy was confirmed playing key roles in CO evolution rate [98].

As C_3N_4 has high reduction ability in contrast to most inorganic photocatalysts which have high oxidation ability; therefore, there are more reports on C_3N_4 combination with some inorganic materials to form organic–inorganic heterojunction photocatalysts, which will be discussed in the next section.

13.2.2.2 MOF

Metal Organic Frameworks (MOFs) are framework structures composed of organic ligands (bridges) with different connection numbers and metal ion nodes. At present, the common expressions of MOFs materials in the research literature also include porous coordination polymers, organic–inorganic hybrid materials, etc. [99,100]. They are attractive materials in photocatalysis due to their channels for the diffusion of substrates and the ultrahigh porosities for accessible active sites. MOFs are especially widely used in decomposition of hazardous organic compounds, and also in other application of photocatalysis, including water splitting, CO_2 reduction, etc. The typical structure of MOFs (ZIF-67) and some merits for photocatalytic H_2 evolution are shown in Figure 13.7.

Due to the different types of connection ways between metal ions and ligands, the MOFs materials have various classifications. At present, the representative MOFs mainly consist of the following types: (1) IRMOF (Isoreticular Metal-Organic Frameworks), which are prepared from Zn^{2+} and terephthalic acid ligands; (2) MIL (Material Institute Lavoisier Frameworks), which are built from Cr, Fe, V, and other

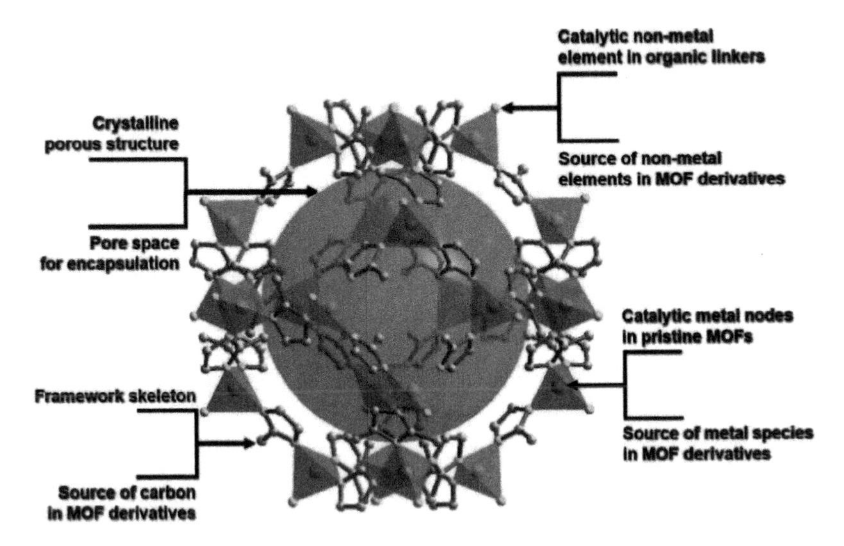

FIGURE 13.7 Key merits of MOFs (ZIF-67 as an example) as catalysts for hydrogen evolution reaction. Green ball: pore space; purple: Co; yellow: nitrogen; gray: carbon [101].

transition metals or lanthanide metals with dicarboxylic acid ligands; (3) ZIF (zeoliticim idazolate frameworks), which are different from IRMOF and MIL, the organic ligands of ZIF are imidazole or imidazole derivatives containing nitrogen instead of carboxyl group of IRMOF and MIL; (4) UiO (University of Oslo), which are two-dimensional porous materials constructed by Zr^{4+} and dicarboxylic acid ligands [101–104].

Back in 2010, Garcia et al. firstly reported UiO66 (Zr) for photocatalytic H_2 evolution reaction under UV light irradiation. MIL-100 (Fe) and MIL-101 (Fe) exhibit extraordinary photocatalytic activity in decomposition of chemical dyes even under 550 ± 50 nm visible light irradiation. As photocatalytic materials for CO_2 reduction applications, MOF-based catalysts (metal cluster of Ti-O, Zr-O, and Fe-O with functional organic linkers of amino modified, photosensitizer-functionalized, and electron-rich conjugated linkers) showed excellent photocatalytic activity [105]. The unlimited potential of MOFs can be confirmed from the fact that the development of a new generation of MOFs is still in progress.

Covalent organic frameworks (COFs) and hydrogen-bonded linked organic frameworks (HOFs) are other kinds of special MOFs [106]. Although the development of COFs and HOFs are subsequent to that of MOFs, the characteristics of these materials have advantages like metal-free and high potential in cyclic utilization. Therefore, they have attracted more and more attention in photocatalysis.

13.2.2.3 PDI

Perylene diimide (PDI) is a kind of special n-type organic semiconductors, which has excellent chemical stability and strong absorption range of light (nearly full-spectrum). As can be seen from Figure 13.8, PDI can be considered to consist of two half units

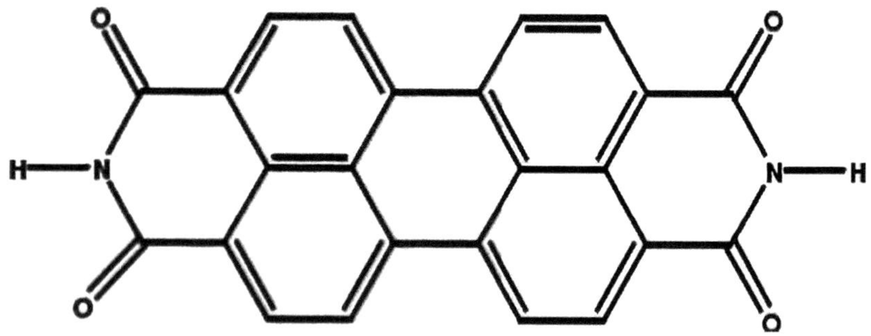

FIGURE 13.8 Molecular structures of PDI.

of naphthalene, each connected to the imide unit and connected to another naphthalene unit through two C sp2-C sp2 single bonds. PDI can form supramolecular structure through π–π stacking interaction which can regulate the band position. Strong π–π interaction is generally considered to be beneficial to the transportation of photogenerated carriers. It has been widely used in solar cell, biological fluorescence probe, light-emitting diodes, and so on. Zang et al. achieved enhanced hydrogen generation activity under visible light irradiation through decorate TiO_2 layers and/or a co-catalyst (Pt) on 1D self-assembled PDI derivatives [107]. It gradually becomes a research hotspot in photocatalysis due to its excellent optical trapping and carrier transport properties. Liu et al. successfully fabricated PDI nanoassemblies with excellent photocatalytic activity for degradation of phenol and splitting water for O_2 evolution under visible light by a rapid and simple solution dispersion method. This work provided a simple and economic method for controlled formation of PDI self-assemblies, which precipitates more research to this material in photocatalysis [108]. Recently, due to high $^1O_2$ quantum yields of PDI under red light irradiation, this material exhibits a strong potential as a material of phototherapy for tumors [109].

13.2.3 Organic–Inorganic Heterojunction Photocatalyst

The synergistic reaction between the substrate materials and co-catalysts is crucial to the improvement of photocatalytic activity. Recently, inorganic–organic heterojunction structures have attracted more attention. They usually possess excellent electronic, magnetic, rigidity and thermal stability, and optical properties of inorganic frameworks and processability, flexibility, and good geometric controllability of organic molecules at the same time. Meanwhile, the hybrids also bring forth new features that are different from those of single component.

13.2.3.1 C_3N_4-Inorganic Heterojunction

The construction of C_3N_4-based heterojunctions has great advantages including the following: (1) more efficient charge separation and transfer through Schottky junctions; (2) the presence of co-catalyst reduces the redox overpotential of the active site; (3) improvement in the utilization rate upon visible light; (4) the stability of the catalyst is improved by protecting active sites and functional groups on the semiconductor surface through appropriate surface passivation.

Over the past few years, many research groups have reported different types of heterojunction photocatalysts consisting of C_3N_4 and metal oxides or sulfides. The heterojunctions have improved efficiency relative to their respective separate components. Among abundant non-noble metal oxides, TiO_2 (3.2 eV) is one of the most widely used photocatalyst and is considered as a promising material for constructing heterojunction with C_3N_4. Natarajan et al. synthesized a direct Z-scheme C_3N_4/TiO_2 heterojunction photocatalyst via a facile wetness impregnation method [110]. The influence of the TiO_2 morphologies on the photocatalytic efficiency was investigated in detail. Among several different morphologies, the direct Z-scheme 3%-C_3N_4/TiO_2 nanotubes showed the best photocatalytic degradation efficiency of isoniazid. The Z-scheme-type transfer mechanism of photogenerated carriers plays a decisive role in the improved photocatalytic activity (Figure 13.9a). Recently, Li et al. [111] reported a traditional type-II heterojunction of Ti^{3+} self-doping B-TiO_2/C_3N_4 hollow core-shell nanostructure for photocatalytic hydrogen production (Figure 13.9b). The photocatalytic activity of the heterojunction is nearly 18 times and 65 times compared with that of normal TiO_2 and C_3N_4, respectively. The self-doping of Ti^{3+} induced to form a self-hydrogenated shell which could reduce the activation barrier of the H_2. The core-shell heterojunction could drive the transfer of photogenerated carriers to reduce the recombination. Moreover, composite of C_3N_4 with different

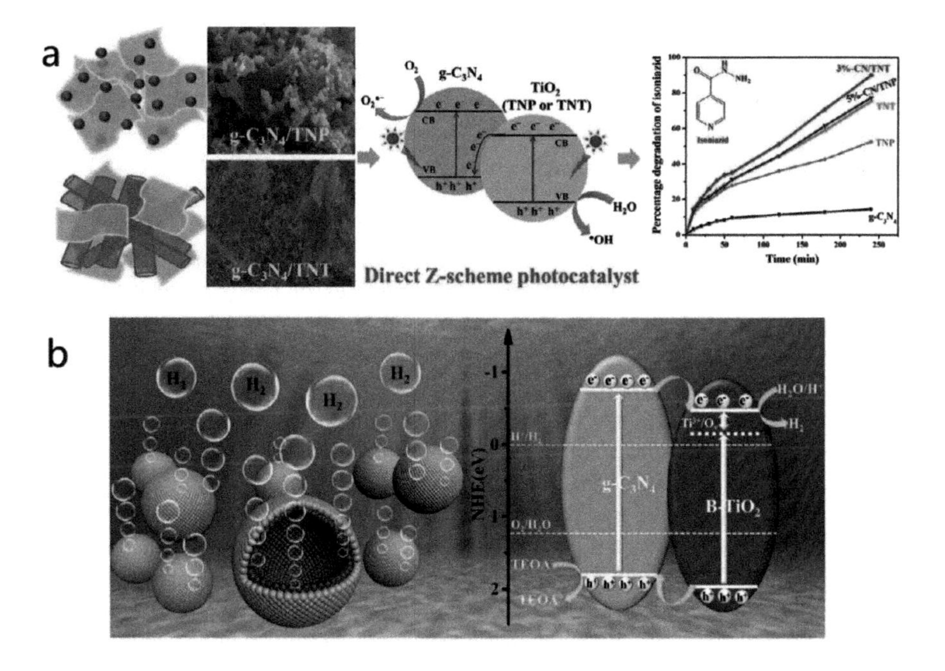

FIGURE 13.9 The photocatalytic mechanism of (a) Z-scheme C_3N_4/TiO_2 and (b) Ti^{3+} self-doping B-TiO_2/C_3N_4 hollow core-shell nano-heterojunctions [111].

morphologies (0D nanoparticles, 1D nanowires, 2D nanosheets, 3D mesoporous materials) and TiO_2 exposed different crystal faces have been widely reported. In addition to the TiO_2/C_3N_4 composite, ZnO/C_3N_4 heterostructures have been also reported. The ZnO/C_3N_4 composites are often used to construct Z-scheme heterojunctions for the improvement of photocatalytic activity. For example, Xiang et al. prepared a C_3N_4/OD-ZnO Z-scheme heterojunction photocatalyst with a core-shell structure using C_3N_4 nanosheets coupled with oxygen-defective ZnO nanorods (OD-ZnO) [112]. The rich oxygen vacancies in OD-ZnO play dual-function roles of improving visible light absorption and mediating the efficient Z-scheme charge separation of the heterojunction, which dramatically enhanced the photocatalytic efficiency of 4-chlorophenol degradation and H_2 evolution.

Unlike TiO_2 and ZnO, the bandgap of WO_3 (2.6–2.8 eV) is relatively narrow and locating in the visible light region which is attractive for the construction of heterojunctions with C_3N_4 [113]. Therefore, the WO_3/C_3N_4 composite usually exhibits type-II or Z-scheme heterojunction, which generally has superior absorption and catalytic efficiency in the visible light region. Yu et al. developed a direct Z-scheme WO_3/C_3N_4 heterojunction photocatalyst which exhibits dramatically enhanced activity for photocatalytic H_2 production [114]. Besides, SnO_2, MnO_2, and Fe_3O_4 are also used as co-catalysts to construct heterojunctions with C_3N_4 [115,116]. Most of them are non-toxic or hypotoxic, satisfying the requirements of green chemistry. In addition to the suitable bandgap matching with C_3N_4, the specific surface area of the composite is significantly increased too. Therefore, more organics can be absorbed on the surface of catalysts and photodegraded. Zhang et al. prepared C_3N_4/SnO_2 heterojunction with ultrasonic stirring and high-temperature calcination methods, with the material prepared by mechanical stirring as a contrast [117]. The characterization results showed that SnO_2 was scattered on the layered C_3N_4 surface in a globular form and well bonded with each other. The degradation rate of methyl orange by C_3N_4/SnO_2 (with 47.5 wt% of SnO_2) reached 73% within 3 h. In addition, they applied the composite for water splitting, and the H_2 evolution rate of C_3N_4/SnO_2 heterojunction in 3 h was 1.5 times higher that of pure C_3N_4.

In addition to metal oxides, metal sulfides (ZnS, CdS, MoS_2, SnS_2, etc.) are also good choices [118–121]. The energy band structures of C_3N_4 and CdS are well matched, which is beneficial to construct heterojunction. A core-shell structured CdS@C_3N_4 photocatalyst was prepared using self-assembly method [122]. The heterojunction exhibited a high visible light photocatalytic activity for oxidative coupling of amines to produce corresponding imines with robust more than 99% selectivity. As a two-dimensional transition metal sulfide with more outstanding physicochemical properties than other materials, MoS_2 is one of the most popular photocatalyst materials in recent years. The bandgap can be changed between 1.3 eV and 1.8 eV when the morphology changes from bulk to monolayer. When the lamellar MoS_2 is combined with C_3N_4 sheets, MoS_2 can grow on the edge of C_3N_4 and form an in-plane heterostructure. The commonly used methods to construct MoS_2/C_3N_4 heterojunction are ultrasonic-assisted hydrothermal method, impregnation method, and so on. Recently in 2019, Yan et al. prepared a 2D-2D MoS_2/C_3N_4 photocatalyst by a simple probe sonication assisted liquid exfoliation method [119]. The photocatalytic H_2 evolution rate of the heterojunction could reach 1155 $\mu mol \cdot h^{-1} \cdot g^{-1}$

with an apparent quantum yield of 6.8% at 420 nm monochromatic light, which is much higher than that of the optimized 0D-2D Pt/C_3N_4 photocatalyst. The large surface area and 2D interfaces between the two components contribute to the high photocatalytic H_2 production activity.

In recent years, some ternary compounds with good optical properties (Perovskites, Bi_2WO_6, BiOX (x = Cl, I, Br), Ag_2WO_4, $SmVO_4$, Ag_3PO_4, etc.) are also applied in the field of photocatalysis [123–127]. After forming a heterogeneous structure with C_3N_4, the photocatalytic efficiency is greatly improved, but the application is limited to a certain extent due to the higher cost of these materials. BiOX, in which X is halogen, has been used as catalyst, photoluminescent material, and pigment. BiOX has attracted much attention with its indirect transition bandgap and hierarchical structure properties. Its indirect transition bandgap would cause the excited electrons to travel a certain k space distance to arrive VB, which greatly reduces the recombination of excited electrons and holes [128].

13.2.3.2 MOFs-Inorganic Heterojunction

Pure MOFs photocatalysts are usually unable to show satisfactory photocatalytic efficiency. Therefore, the heterojunctions of MOFs with other inorganic materials are widely studied in recent years. The composites combine with their respective advantages and complement each other's shortcomings. The porous structures of MOFs are beneficial for the loading of semiconductors and the adjustable pore size structure is helpful to control the size and loading amount of loaded semiconductor particles.

After years of development, researchers have developed different fabrication methods to control the structure and morphology of MOFs composites, thereby regulating and optimizing their physicochemical properties. The commonly used methods include hydrothermal/solvothermal method, ultrasonic chemical method, gel crystallization method, microwave synthesis method, rapid precipitation method, electrochemical synthesis method, etc. To sum up, as shown in Figure 13.10, these

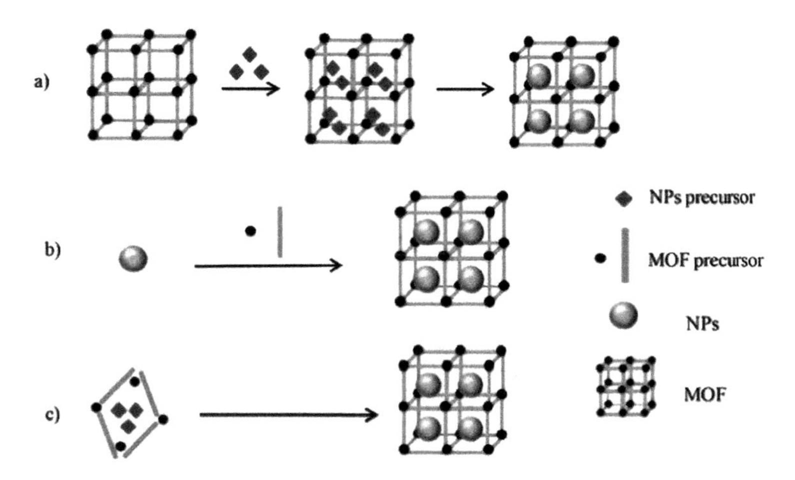

FIGURE 13.10 Main strategies for preparation of MOFs-semiconductor composites: (a) ship-in-bottle, (b) bottle-around-ship, and (c) one-pot [130].

methods can come down to three preparation strategies: "ship-in-a-bottle," "bottle-around-ship," and "one-pot" [129,130].

The "ship-in-a-bottle" method is to prepare the MOFs material firstly and then embed the semiconductor nanoparticles into the MOFs material so as to prepare the semiconductor/MOFs composite. However, several problems should be considered in this preparation method: (1) the precursor should be compatible with the structure of MOFs; (2) the MOFs should be stable under the condition of forming nanoparticles; (3) it should be clear whether the nanoparticles are formed in the framework or on the outer surface. The "bottle-around-ship" method is to first prepare semiconductor nanoparticles and then obtain semiconductor/MOFs composites through secondary synthesis. The method can control the size and morphology of semiconductor nanoparticles. But because of the high interfacial energy barrier between the two materials, the growth of MOFs may be more difficult in some cases. The "one-pot" method is for the synthesis of semiconductor/MOFs composites by one-step preparation of two components simultaneously. This synthesis strategy is simple and direct, but self-nucleation and synchronous control on the growth of MOFs and nanoparticles are more difficult.

Although the application of MOFs-inorganic heterojunctions for photocatalysis is still in its infancy, the experimental results have shown that the combination of inorganic semiconductors and MOFs has a good application prospect. Jiang et al. synthesized $Cu_3(BTC)_2@TiO_2$ composite, and the experimental results showed that photogenerated electrons could transfer from the TiO_2 to the MOFs effectively, which not only facilitated charge separation on the inorganic semiconductor but also provided high-energy electrons to the gas molecules adsorbed in the MOFs [131]. In addition, CO_2 could easily penetrate through the shell and become the core of the capture, increasing the activity and selectivity of photocatalytic reduction. Wang et al. prepared a ZnO@ZIF-8 core-shell structure rapidly (within 60 min) by in situ crystal-growth method [132]. The study found that the rate of surface-bound Zn^{2+} and 2-methylazole (Hmim) were related to the molar ratio of Hmim/Zn^{2+}. By controlling the concentration of Hmim, an ideal ZnO@ZIF-8 core-shell heterostructure can be obtained with a thinner shell (Figure 13.11). The synthesized ZnO@ZIF-8 can

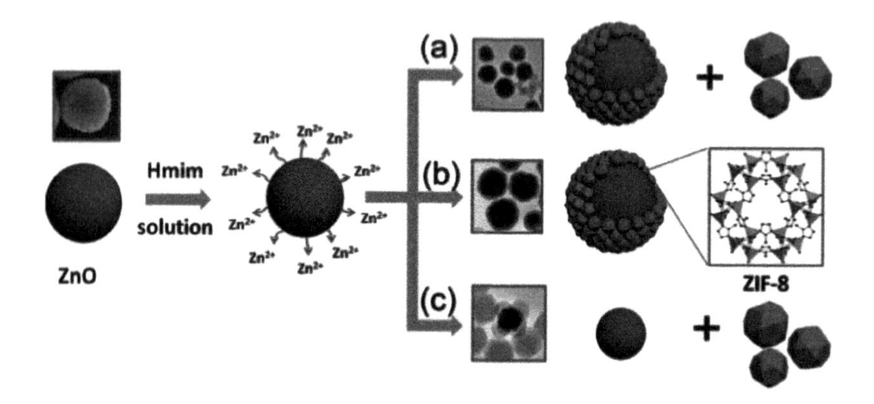

FIGURE 13.11 Schematic diagram of ZnO@ZIF-8 structure obtained at different concentrations of Hmim: (a) 4.51 M, (b) 3.66 M, and (c) 1.83 M [132].

selectively decrease the Cr (VI) concentration, which is caused by the selective permeation of ZIF-8 shell.

In practical applications, because of the high dispersibility of MOFs photocatalysts, it is difficult to separate them from the reaction solution for recovery. To solve this problem, the method of coating MOFs photocatalyst with magnetic particles can be used to modify it. Khataee et al. reported a Fe_3O_4@MOF-2 nanocomposite. As a heterojunction photocatalyst, the Fe_3O_4@MOF-2 was applied to activate persulfate (PS) in the attendance of ultrasonic (US) irradiation (Fe_3O_4@MOF-2/US/PS) for degradation of diazinon [133]. The evaluation results showed that, in the acidic solution (pH = 3), the reaction rate constant of Fe_3O_4@MOF-2/US/PS system (0.0546 min^{-1}) was more than six folds than that of the US/PS system (0.0079 min^{-1}). At the same time, the Fe_3O_4@MOF-2 nanocomposite showed an admirable recyclability.

In visible-light-driven semiconductor photocatalysts, CdS has attracted much attention for its excellent electronic optical properties and suitable bandgap. Zeng et al. prepared CdS@ZIF-8 nanocomposites by two-step method [134]. During the preparation process, the concentration of Zn^{2+} was over-changed to prepare multi-core and mononuclear CdS@ZIF-8 core-shell structures. The photocatalytic H_2 evolution activity of CdS@ZIF-8 was higher than that of pure CdS, while the CO production rate decreased from 12.5% to 5.7%. This is because the pore size (0.34 nm) of ZIF-8 is smaller than the kinetic diameter of CO (0.38 nm); therefore, the generated CO cannot pass through the pores of ZIF-8.

Recently, some halide perovskite materials have received widespread concern in optoelectronic applications, due to their wide absorption ranges, high extinction coefficients, and long electron–hole diffusion lengths. Zhong et al. fabricated a $CsPbBr_3$ QDs/UiO-66(NH_2) nanocomposite for photocatalytic CO_2 reduction under visible light irradiation [135]. The significantly enhanced photocatalytic activity of the heterojunction compared with each single component could attribute to the fast charge separation, large accessible specific surface area and enhanced visible light-absorption capacity. Meanwhile, the composite also possesses excellent recyclability during the photocatalytic reaction.

13.2.3.3 PDI-Inorganic Heterojunction

PDI aggregates have excellent photocatalytic activity and also possess some unique advantages in structure and property when combined with inorganic semiconductors in forming heterojunctions. However, it is often difficult for PDI aggregates to form effective coverage automatically on the surface of inorganic semiconductors. Instead, they can be arranged regularly on the surface in the form of monolayer. The preparation methods include evaporative coating, Langmuir—Blodgett technology, impregnation-rotating coating technology, metal fit, electrostatic stacking, and covalent bonding, which can achieve a firm combination between PDI and inorganic semiconductors.

PDI and derivatives (PTCDI) are n-type organic semiconductor materials with wide light-absorption range and good air stability. Because one PDI molecule can continuously absorb two photons and accumulate sufficient energy, so some reactions that single-photon process energy is not sufficient to trigger can proceed smoothly with PDI and derivatives as catalysts. However, because of the large conjugated benzene ring in the PDI structure, its solubility in most solvents is relatively low, especially in the aqueous phase, which limits its application. Therefore, PDI and PTCDI can be self-assembled or combined with other semiconductors to demonstrate excellent photocatalytic performance.

The research on PDI for photocatalysis is still in its infancy. Therefore, the studies on PDI-inorganic heterojunctions are also relatively rare. But in the view of its excellent physicochemical properties, the research field of constructing PDI-inorganic heterojunctions has great potential. Zhu et al. prepared a novel effective p-Ag_2S/n-PDI self-assembled supramolecular heterojunction for phenol degradation and O_2 evolution via an in situ precipitation method [136]. As shown in Figure 13.12, the heterojunction exhibits superior UV-light, visible light and full-spectrum photocatalytic performance. Ag_2S quantum dots were tightly loaded onto the surface of PDI which could promote the π–π stacking degree of PDI, and accelerate the migration of photogenerated electrons along the quasi-one-dimensional π–π stacking. Simultaneously, Ag_2S also can improve the light absorption to facilitate the light-chemical energy conversion. The constructed heterojunction with a built-in electrical potential is conducive for the more efficient transportation of photogenerated carriers. Moreover, the p-Ag_2S/n-PDI heterojunction produces more active species than pure PDI, resulting in much stronger oxidation ability.

A semi-core-shell TiO_2@PDI full-spectrum heterojunction photocatalyst was prepared by Zhu et al. [137]. It was constructed with self-assembly of PDI as the core and TiO_2 nanoparticles as the shell via the simple sol mixing method. And thereafter, it was used for photocatalytic degradation of phenol. The synergistic interaction between TiO_2 and PDI enables the catalyst to improve its UV, visible light, and full-spectrum performance, with benefit from the formation of new stacking states along the π–π stacking direction. Therefore, the photogenerated electron transfer from PDI to TiO_2 is greatly improved. Most recently, Liu et al. fabricated an organic–inorganic Z-scheme WO_3@Cu@PDI heterojunction photocatalyst [138]. Under visible light irradiation, the degradation rate of tetracycline hydrochloride (TC) is 40 and 5 times

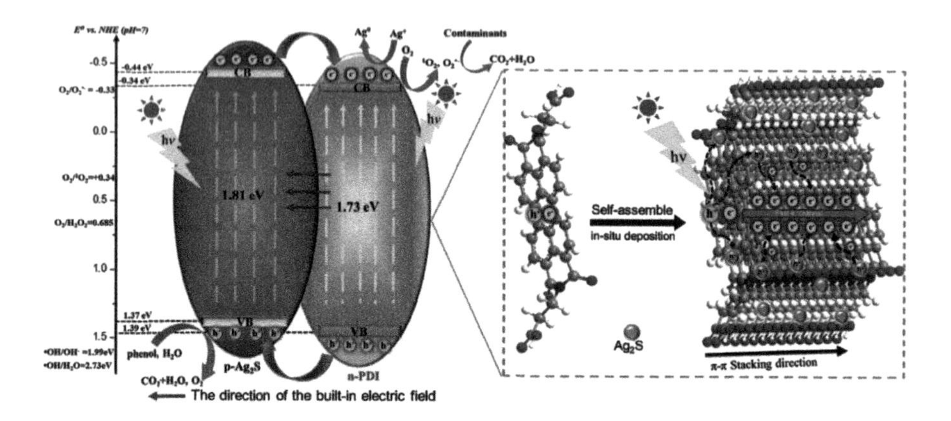

FIGURE 13.12 The separation process of photogenerated carriers and the photocatalytic process of p-Ag_2S/n-PDI heterojunction [136].

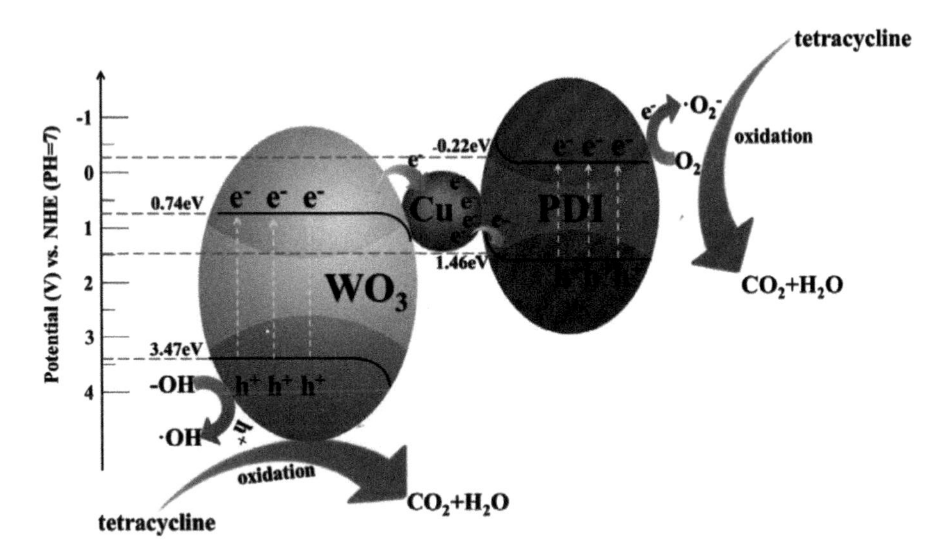

FIGURE 13.13 Z-scheme electron-transfer mechanism of WO_3@Cu@PDI heterojunction photocatalyst [138].

higher than that of pure WO_3 and PDI, respectively. The heterojunction also has great stability, and the photocatalytic activity toward degradation of TC could still keep 85% after three times of reuse. As shown in Figure 13.13, Z-scheme accelerate the photogenerated electrons on WO_3 to move toward the donor level of self-assemble PDI molecule through Cu atoms and then recombine with h^+ which enables a stronger redox active of e^- and h^+ on the CB of PDI and VB of WO_3, respectively.

13.3 MECHANISM OF THE PHOTOCATALYSIS FOR HETEROJUNCTIONS

In summary, in the design and modification of photocatalysts, the construction of heterojunctions has been proved to be one of the most promising methods for the fabrication of photocatalysts with high activity and visible light response due to their feasibility and effectiveness in the spatial separation of electron–hole pairs. In general, heterojunctions are considered to be the interface between two semiconductors with different band structures. The semiconductor–semiconductor heterojunctions can be divided into p-n heterojunctions and non-p-n heterojunctions. The p-n heterojunction refers to the formation of junctions at the interface and space charge region, resulting in a built-in potential that can guide the electrons and holes to move in opposite directions. Thus a good separation of electrons from holes can be achieved. The non-p-n heterojunction is based on the difference of CB and VB level of the two materials to realize the transfer of electrons and holes. According to the way of the transfer of electrons and holes in the semiconductors, non-p-n heterojunctions can be subdivided into type I, type II, type III, and Z-scheme heterojunction. The following are the detailed discussions of their mechanisms.

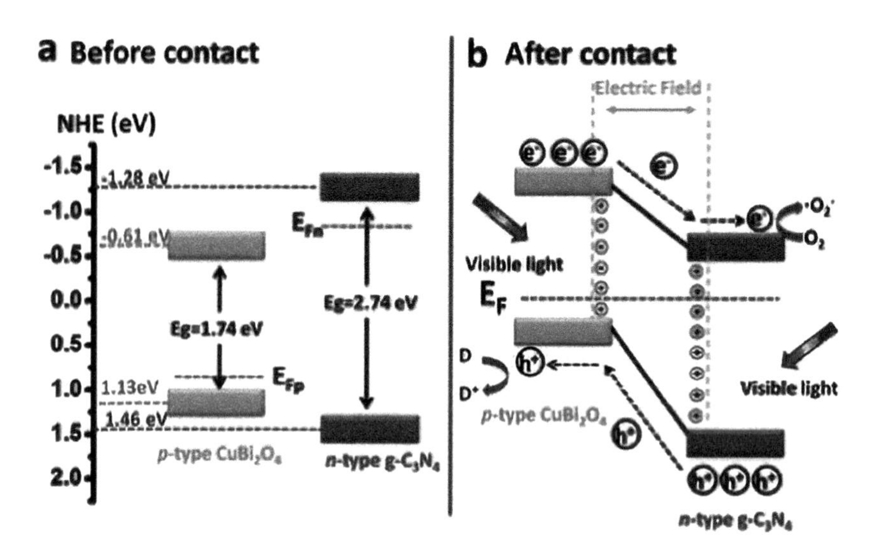

FIGURE 13.14 (a) Band alignment of p-type ($CuBi_2O_4$) and n-type ($g\text{-}C_3N_4$) semiconductors before contact and (b) transportation of the charge carries in p-n type heterojunction [139].

13.3.1 P-N Heterojunction

When two different types of semiconductors (p- or n- type) contacting each other, their energy band positions would be changed, thus altering the electron transfer pathway. As can be seen in Figure 13.14, the Fermi energy level of n-type semiconductor is near the CB, and the Fermi energy level of p-type semiconductor is near its VB. After contact, the energy band of p-type semiconductor moves up while the energy band of n-type semiconductor moves down with the diffusion of electron and hole. When the drift current is off set with the diffusion current, the space charge region formed near the interface and a built-in electric field is generated. The two types of semiconductors are excited by illumination at the same time, the electrons are transferred to the n-type semiconductor quickly under the action of built-in electric field. And the holes are transferred to the p-type semiconductor rapidly, which promotes the separation of photogenerated carriers effectively. In addition, in the p-n heterojunctions, the photogenerated electrons are enriched on the CB of n-type semiconductor with higher energy-level position (before the contact of two semiconductors), and the photogenerated holes are enriched on the VB of p-type semiconductor lower energy-level position, which can not only provide more photogenerated electrons but also make the photogenerated electrons have strong reducibility.

13.3.2 Type I and Type II Heterojunctions

For type I heterojunction photocatalysts (Figure 13.15a), the VB of semiconductor A is lower than that of semiconductor B, while its CB is higher than that of semiconductor B. Therefore, electrons and holes will be concentrated in the CB and VB of

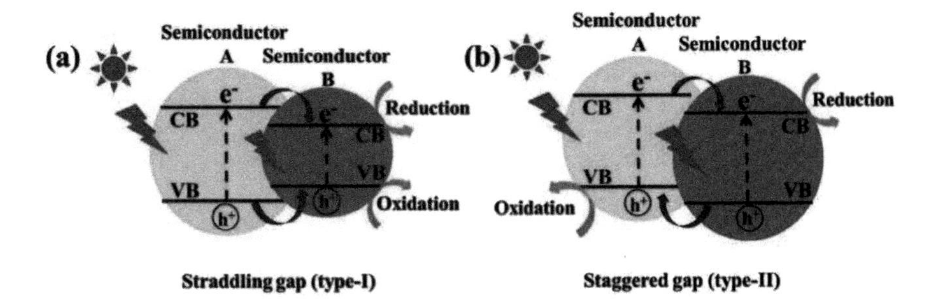

FIGURE 13.15 Band alignment and transportation of the charge carries in type I (a) and type II (b) heterojunction [140].

semiconductor B under the action of potential difference, respectively. The different transfer rates of photogenerated electrons and holes make the electrons have longer lifetimes and thus possess higher activity. However, the photogenerated electrons and holes are enriched on the same semiconductor, so there is still a large recombination possibility. Moreover, since both electrons and holes are transferred to the semiconductor B with lower redox potentials, the redox ability of heterojunction photocatalysts is significantly reduced.

Type II heterojunctions are widely used in photocatalysis. For type II heterojunction photocatalysts (Figure 13.15b), the CB and VB positions of semiconductor B are both higher than the corresponding band positions of semiconductor A. Therefore, the photogenerated electrons of semiconductor B will be transferred to semiconductor A, and the photogenerated holes will migrate to semiconductor B under light irradiation, which leads to the spatial separation of electron–hole pairs. Similar to the type I heterojunction, the redox ability of the type II heterojunction photocatalyst will also be reduced because the reduction and oxidation occur on the semiconductor A with lower reduction potential and the semiconductor B with lower oxidation potential, respectively.

13.3.3 Z-Scheme Heterojunction

In order to overcome the drawbacks existing in the traditional type I, type II, and p-n heterojunctions, researchers have proposed a Z-scheme heterostructure with strong redox ability and excellent spatial separation property of charge carriers. The Z-scheme photocatalytic system simulating the natural biological photosynthesis system has been widely used in many efficient photocatalytic reactions. As shown in Figure 13.16a, photosynthesis system can be divided into two parts: PC I and PC II. The fixation of CO_2 and the release of O_2 are carried out in these two systems, respectively. Under solar irradiation, the excited electrons of PC II transfer from HOMO (highest occupied molecular orbital) to LUMO (lowest unoccupied molecular orbital) and then pass through cytochrome to the HOMO of PC I that are excited to the LUMO of PC I to form a "Z" shape path. It is an important research direction to combine two kinds of semiconductors to form all-solid-state Z-scheme heterojunction. As shown

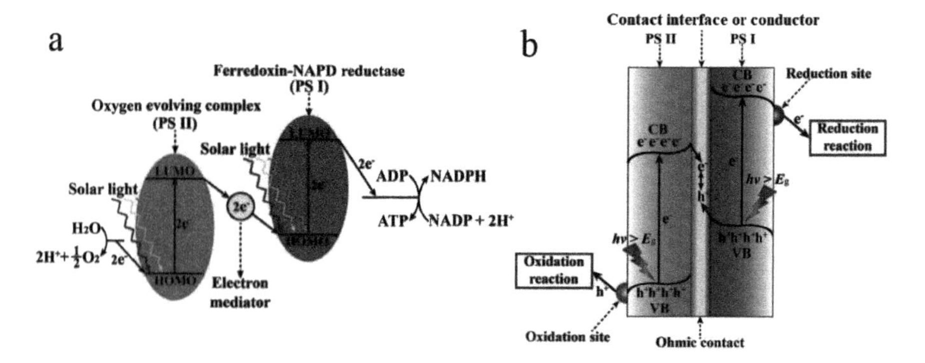

FIGURE 13.16 Transportation of the electrons in (a) photosynthesis system and (b) all-solid-state Z-scheme heterojunction [141].

in Figure 13.16b, in the Z-scheme heterojunction, photogenerated electrons on PC II photogenic and holes on PC I are transferred by interfacial phase or conductive medium and then recombined. Although the number of carries involved in the reaction is reduced, the rest electrons and holes are maintained in the higher CB and lower VB, thus possessing stronger redox ability.

13.4 CONCLUSION AND OUTLOOK

In summary, in this chapter, we introduce the basic principle of photocatalysis, the main applications and the modification methods of photocatalysts. The related research progresses of mainstream organic and inorganic photocatalysts are discussed in detail. Inorganic semiconductor materials are the first developed materials for the utilization of light energy. A series of superior photocatalytic materials have been developed and applied in the various photocatalytic reactions. Organic semiconductor materials are the focus of current research because of their advantages compared with inorganic materials, such as structural diversity, modifiability, and function tunability. Based on the unique strengths and weaknesses of organic and inorganic semiconductor materials, the organic–inorganic heterojunctions prepared by effective synthesis methods are considered as a new generation of photocatalytic materials with great application prospects. Therefore, the research on the design, synthesis, and application of new organic–inorganic heterojunction photocatalysts has become a hot topic at present. Finally, we discuss in detail the classification of organic–inorganic heterojunction photocatalysts and the corresponding charge transfer modes. This research field is still in the initial stage, which is of great theoretical and practical value to the green energy applications.

REFERENCES

1. J. Kou, C. Lu, J. Wang, Y. Chen, Z. Xu, R.S. Varma, Selectivity enhancement in heterogeneous photocatalytic transformations, *Chem. Rev.*, 117 (2017) 1445–1514.

2. X. Chen, N. Li, Z. Kong, W.-J. Ong, X. Zhao, Photocatalytic fixation of nitrogen to ammonia: state-of-the-art advancements and future prospects, *Mater. Horiz.*, 5 (2018) 9–27.
3. A. Fujishima, K. Honda, Electrochemical photolysis of water at a semiconductor electrode, *Nature*, 238 (1972) 37–38.
4. K. Maeda, K. Teramura, D. Lu, T. Takata, N. Saito, Y. Inoue, K. Domen, Photocatalyst releasing hydrogen from water, *Nature*, 440 (2006) 295–295.
5. T.P. Yoon, M.A. Ischay, J. Du, Visible light photocatalysis as a greener approach to photochemical synthesis, *Nat. Chem.*, 2 (2010) 527.
6. J.-M. Lei, Q.-X. Peng, S.-P. Luo, Y. Liu, S.-Z. Zhan, C.-L. Ni, A nickel complex, an efficient cocatalyst for both electrochemical and photochemical driven hydrogen production from water, *Mol. Catal.*, 448 (2018) 10–17.
7. Z. Wang, Z. Jin, H. Yuan, G. Wang, B. Ma, Orderly-designed Ni_2P nanoparticles on g-C_3N_4 and UiO-66 for efficient solar water splitting, *J. Colloid Interface Sci.*, 532 (2018) 287–299.
8. V. Vaiano, M.A. Lara, G. Iervolino, M. Matarangolo, J.A. Navío, M. Hidalgo, Photocatalytic H_2 production from glycerol aqueous solutions over fluorinated Pt-TiO_2 with high {001} facet exposure, *J. Photoch. Photobio. A*, 365 (2018) 52–59.
9. A.L. Linsebigler, G. Lu, J.T. Yates Jr, Photocatalysis on TiO_2 surfaces: principles, mechanisms, and selected results, *Chem. Rev.*, 95 (1995) 735–758.
10. X. Zong, J. Han, B. Seger, H. Chen, G. Lu, C. Li, L. Wang, An integrated photoelectrochemical-chemical loop for solar-driven overall splitting of hydrogen sulfide, *Angew. Chem. Int. Edit.*, 53 (2014) 4399–4403.
11. J.M. Coronado, F. Fresno, M.D. Hernández-Alonso, R. Portela, *Design of advanced photocatalytic materials for energy and environmental applications*, Springer, 2013.
12. X. Wang, J.C. Yu, H.Y. Yip, L. Wu, P.K. Wong, S.Y. Lai, A mesoporous Pt/TiO_2 nanoarchitecture with catalytic and photocatalytic functions, *Chem.-Eur. J.*, 11 (2005) 2997–3004.
13. T. Inoue, A. Fujishima, S. Konishi, K. Honda, Photoelectrocatalytic reduction of carbon dioxide in aqueous suspensions of semiconductor powders, *Nature*, 277 (1979) 637–638.
14. Z. Li, W. Luo, M. Zhang, J. Feng, Z. Zou, Photoelectrochemical cells for solar hydrogen production: current state of promising photoelectrodes, methods to improve their properties, and outlook, *Energ. Environ. Sci.*, 6 (2013) 347–370.
15. I.S. Cho, Z. Chen, A.J. Forman, D.R. Kim, P.M. Rao, T.F. Jaramillo, X. Zheng, Branched TiO_2 nanorods for photoelectrochemical hydrogen production, *Nano Lett.*, 11 (2011) 4978–4984.
16. M. Freitag, J. Teuscher, Y. Saygili, X. Zhang, F. Giordano, P. Liska, J. Hua, S.M. Zakeeruddin, J.-E. Moser, M. Grätzel, Dye-sensitized solar cells for efficient power generation under ambient lighting, *Nat. Photonics*, 11 (2017) 372.
17. J. Zhi, H. Wang, A. Fujishima, Development of photocatalytic coating agents with indicator dyes, *Ind. Eng. Chem. Res.*, 41 (2002) 726–731.
18. W. Wang, T. Ai, Q. Yu, Electrical and photocatalytic properties of boron-doped ZnO nanostructure grown on PET-ITO flexible substrates by hydrothermal method, *Sci. Rep.*, 7 (2017) 42615.
19. L.W. Zhang, Y.J. Wang, H.Y. Cheng, W.Q. Yao, Y.F. Zhu, Synthesis of porous Bi_2WO_6 thin films as efficient visible-light-active photocatalysts, *Adv. Mater.*, 21 (2009) 1286–1290.
20. B. Jiang, P. Zhang, Y. Zhang, L. Wu, H. Li, D. Zhang, G. Li, Self-assembled 3D architectures of $Bi_2TiO_4F_2$ as a new durable visible-light photocatalyst, *Nanoscale*, 4 (2012) 455–460.

21. Z. Bian, J. Ren, J. Zhu, S. Wang, Y. Lu, H. Li, Self-assembly of $Bi_xTi_{1-x}O_2$ visible photocatalyst with core-shell structure and enhanced activity, *Appl. Catal. B: Environ.*, 89 (2009) 577–582.
22. Y. Huo, M. Miao, Y. Zhang, J. Zhu, H. Li, Aerosol-spraying preparation of a mesoporous hollow spherical $BiFeO_3$ visible photocatalyst with enhanced activity and durability, *Chem. Commun.*, 47 (2011) 2089–2091.
[23. Z. Jin, Q. Zhang, J. Chen, S. Huang, L. Hu, Y.-J. Zeng, H. Zhang, S. Ruan, T. Ohno, Hydrogen bonds in heterojunction photocatalysts for efficient charge transfer, *Appl. Catal. B: Environ.*, 234 (2018) 198–205.
24. Z. Jin, J. Chen, S. Huang, J. Wu, Q. Zhang, W. Zhang, Y.-J. Zeng, S. Ruan, T. Ohno, A facile approach to fabricating carbonaceous material/g-C_3N_4 composites with superior photocatalytic activity, *Catal. Today*, 315 (2018) 149–154.
25. H. Qin, Y. Bian, Y. Zhang, L. Liu, Z. Bian, Effect of Ti (III) surface defects on the process of photocatalytic reduction of hexavalent chromium, *Chin. J. Chem.*, 35 (2017) 203–208.
26. T. Lv, H. Wang, W. Hong, P. Wang, L. Jia, In situ self-assembly synthesis of sandwich-like TiO_2/reduced graphene oxide/$LaFeO_3$ Z-scheme ternary heterostructure towards enhanced photocatalytic hydrogen production, *Mol. Catal.*, 475 (2019) 110497.
27. S.G. Kumar, K.K. Rao, Comparison of modification strategies towards enhanced charge carrier separation and photocatalytic degradation activity of metal oxide semiconductors (TiO_2, WO_3 and ZnO), *Appl. Surf. Sci.*, 391 (2017) 124–148.
28. K. Kaviyarasu, L. Kotsedi, A. Simo, X. Fuku, G.T. Mola, J. Kennedy, M. Maaza, Photocatalytic activity of ZrO_2 doped lead dioxide nanocomposites: investigation of structural and optical microscopy of RhB organic dye, *Appl. Surf. Sci.*, 421 (2017) 234–239.
29. Z. Jin, Q. Zhang, L. Hu, J. Chen, X. Cheng, Y.-J. Zeng, S. Ruan, T. Ohno, Constructing hydrogen bond based melam/WO_3 heterojunction with enhanced visible-light photocatalytic activity, *Appl. Catal. B: Environ.*, 205 (2017) 569–575.
30. I. Ullah, A. Munir, S. Muhammad, S. Ali, N. Khalid, M. Zubair, M. Sirajuddin, S.Z. Hussain, S. Ahmed, Y. Khan, Influence of W-doping on the optical and electrical properties of SnO_2 towards photocatalytic detoxification and electrocatalytic water splitting, *J. Alloys Compd.*, 827 (2020) 154247.
31. P.M. Shafi, R. Dhanabal, A. Chithambararaj, S. Velmathi, A.C. Bose, α-MnO_2/h-MoO_3 hybrid material for high performance supercapacitor electrode and photocatalyst, *ACS Sustain. Chem. Eng.*, 5 (2017) 4757–4770.
32. J.F. Banfield, D.R. Veblen, Conversion of perovskite to anatase and TiO_2 (B): a TEM study and the use of fundamental building blocks for understanding relationships among the TiO_2 minerals, *Am. Mineral.*, 77 (1992) 545–557.
33. Y. Ma, X. Wang, Y. Jia, X. Chen, H. Han, C. Li, Titanium dioxide-based nanomaterials for photocatalytic fuel generations, *Chem. Rev.*, 114 (2014) 9987–10043.
34. Y. Sasaki, H. Nemoto, K. Saito, A. Kudo, Solar water splitting using powdered photocatalysts driven by Z-schematic interparticle electron transfer without an electron mediator, *J. Phys. Chem. C*, 113 (2009) 17536–17542.
35. R. Abe, K. Sayama, H. Sugihara, Development of new photocatalytic water splitting into H_2 and O_2 using two different semiconductor photocatalysts and a shuttle redox mediator IO^{3-}/I, *J. Phys. Chem. B*, 109 (2005) 16052–16061.
36. Y. Xu, M.A. Schoonen, The absolute energy positions of conduction and valence bands of selected semiconducting minerals, *Am. Mineral.*, 85 (2000) 543–556.
37. Z.-F. Huang, L. Pan, J.-J. Zou, X. Zhang, L. Wang, Nanostructured bismuth vanadate-based materials for solar-energy-driven water oxidation: a review on recent progress, *Nanoscale*, 6 (2014) 14044–14063.

38. R. Saravanan, D. Manoj, J. Qin, M. Naushad, F. Gracia, A.F. Lee, M.M. Khan, M. Gracia-Pinilla, Mechanothermal synthesis of Ag/TiO_2 for photocatalytic methyl orange degradation and hydrogen production, *Process Saf. Environ. Prot.*, 120 (2018) 339–347.
39. J. Zhang, X. Jin, P.I. Morales-Guzman, X. Yu, H. Liu, H. Zhang, L. Razzari, J.P. Claverie, Engineering the absorption and field enhancement properties of Au-TiO_2 nanohybrids via whispering gallery mode resonances for photocatalytic water splitting, *ACS Nano*, 10 (2016) 4496–4503.
40. A. Ofiarska, A. Pieczyńska, A.F. Borzyszkowska, P. Stepnowski, E.M. Siedlecka, Pt-TiO_2-assisted photocatalytic degradation of the cytostatic drugs ifosfamide and cyclophosphamide under artificial sunlight, *Chem. Eng. J.*, 285 (2016) 417–427.
41. M.R. Elahifard, R.V. Meidanshahi, Photo-deposition of Ag metal particles on Ni-doped TiO_2 for photocatalytic application, *Prog. React. Kinet. Mech.*, 42 (2017) 244–250.
42. H. Safajou, H. Khojasteh, M. Salavati-Niasari, S. Mortazavi-Derazkola, Enhanced photocatalytic degradation of dyes over graphene/Pd/TiO_2 nanocomposites: TiO_2 nanowires versus TiO_2 nanoparticles, *J. Colloid Interface Sci.*, 498 (2017) 423–432.
43. S. Bera, J.E. Lee, S.B. Rawal, W.I. Lee, Size-dependent plasmonic effects of Au and Au@SiO_2 nanoparticles in photocatalytic CO_2 conversion reaction of Pt/TiO_2, *Appl. Catal. B: Environ.*, 199 (2016) 55–63.
44. M.S. Rodrigues, D. Costa, R. Domingues, M. Apreutesei, P. Pedrosa, N. Martin, V. Correlo, R. Reis, E. Alves, N. Barradas, Optimization of nanocomposite Au/TiO_2 thin films towards LSPR optical-sensing, *Appl. Surf. Sci.*, 438 (2018) 74–83.
45. H. Ran, J. Fan, X. Zhang, J. Mao, G. Shao, Enhanced performances of dye-sensitized solar cells based on Au-TiO_2 and Ag-TiO_2 plasmonic hybrid nanocomposites, *Appl. Surf. Sci.*, 430 (2018) 415–423.
46. I. Horovitz, D. Avisar, M.A. Baker, R. Grilli, L. Lozzi, D. Di Camillo, H. Mamane, Carbamazepine degradation using a N-doped TiO_2 coated photocatalytic membrane reactor: influence of physical parameters, *J. Hazard. Mater.*, 310 (2016) 98–107.
47. S. Komatsuda, Y. Asakura, J.J.M. Vequizo, A. Yamakata, S. Yin, Enhanced photocatalytic NO_x decomposition of visible-light responsive F-TiO_2/(N, C)-TiO_2 by charge transfer between F-TiO_2 and (N, C)-TiO_2 through their doping levels, *Appl. Catal. B: Environ.*, 238 (2018) 358–364.
48. H.R. Kim, A. Razzaq, C.A. Grimes, S.-I. In, Heterojunction pnp Cu_2O/S-TiO_2/CuO: synthesis and application to photocatalytic conversion of CO_2 to methane, *J. CO2 Util.*, 20 (2017) 91–96.
49. X. Yan, Z. Jia, H. Che, S. Chen, P. Hu, J. Wang, L. Wang, A selective ion replacement strategy for the synthesis of copper doped carbon nitride nanotubes with improved photocatalytic hydrogen evolution, *Appl. Catal. B: Environ.*, 234 (2018) 19–25.
50. S. Liu, N. Zhang, Z.-R. Tang, Y.-J. Xu, Synthesis of one-dimensional CdS@TiO_2 core-shell nanocomposites photocatalyst for selective redox: the dual role of TiO_2 shell, *ACS Appl. Mater. Interfaces*, 4 (2012) 6378–6385.
51. Z. Jin, X. Zhang, Y. Li, S. Li, G. Lu, 5.1% Apparent quantum efficiency for stable hydrogen generation over eosin-sensitized CuO/TiO_2 photocatalyst under visible light irradiation, *Catal. Commun.*, 8 (2007) 1267–1273.
52. F.-X. Xiao, Construction of highly ordered ZnO-TiO_2 nanotube arrays (ZnO/TNTs) heterostructure for photocatalytic application, *ACS Appl. Mater. Interfaces*, 4 (2012) 7055–7063.
53. X. Chen, L. Liu, Y.Y. Peter, S.S. Mao, Increasing solar absorption for photocatalysis with black hydrogenated titanium dioxide nanocrystals, *Science*, 331 (2011) 746–750.
54. J. Zhang, S. Wageh, A. Al-Ghamdi, J. Yu, New understanding on the different photocatalytic activity of wurtzite and zinc-blende CdS, *Appl. Catal. B: Environ.*, 192 (2016) 101–107.

55. G.-J. Lee, S. Anandan, S.J. Masten, J.J. Wu, Photocatalytic hydrogen evolution from water splitting using Cu doped ZnS microspheres under visible light irradiation, *Renew. Energ.*, 89 (2016) 18–26.
56. J. Zhang, S. Liu, J. Yu, M. Jaroniec, A simple cation exchange approach to Bi-doped ZnS hollow spheres with enhanced UV and visible-light photocatalytic H_2-production activity, *J. Mater. Chem.*, 21 (2011) 14655–14662.
57. J. Chen, F. Xin, S. Qin, X. Yin, Photocatalytically reducing CO_2 to methyl formate in methanol over ZnS and Ni-doped ZnS photocatalysts, *Chem. Eng. J.*, 230 (2013) 506–512.
58. I. Tsuji, A. Kudo, H_2 evolution from aqueous sulfite solutions under visible-light irradiation over Pb and halogen-codoped ZnS photocatalysts, *J. Photoch. Photobio. A*, 156 (2003) 249–252.
59. S. Joicy, R. Saravanan, D. Prabhu, N. Ponpandian, P. Thangadurai, Mn^{2+} ion influenced optical and photocatalytic behaviour of Mn-ZnS quantum dots prepared by a microwave assisted technique, *RSC Adv.*, 4 (2014) 44592–44599.
60. M. Muruganandham, Y. Kusumoto, Synthesis of N, C codoped hierarchical porous microsphere ZnS as a visible light-responsive photocatalyst, *J. Phys. Chem. C*, 113 (2009) 16144–16150.
61. Q. Zhao, Y. Xie, Z. Zhang, X. Bai, Size-selective synthesis of zinc sulfide hierarchical structures and their photocatalytic activity, *Cryst. Growth Des.*, 7 (2007) 153–158.
62. Y. Yu, G. Chen, Q. Wang, Y. Li, Hierarchical architectures of porous ZnS-based microspheres by assembly of heterostructure nanoflakes: lateral oriented attachment mechanism and enhanced photocatalytic activity, *Energ. Environ. Sci.*, 4 (2011) 3652–3660.
63. N. Zhang, S. Liu, X. Fu, Y.-J. Xu, Fabrication of coenocytic Pd@CdS nanocomposite as a visible light photocatalyst for selective transformation under mild conditions, *J. Mater. Chem.*, 22 (2012) 5042–5052.
64. J.M. Lee, E.K. Mok, S. Lee, N.S. Lee, L. Debbichi, H. Kim, S.J. Hwang, A conductive hybridization matrix of RuO_2 two-dimensional nanosheets: a hybrid-type photocatalyst, *Angew. Chem. Int. Edit.*, 55 (2016) 8546–8550.
65. H. Yan, J. Yang, G. Ma, G. Wu, X. Zong, Z. Lei, J. Shi, C. Li, Visible-light-driven hydrogen production with extremely high quantum efficiency on Pt-PdS/CdS photocatalyst, *J. Catal.*, 266 (2009) 165–168.
66. M. Zubair, I.-H. Svenum, M. Rønning, J. Yang, Facile synthesis approach for core-shell TiO_2-CdS nanoparticles for enhanced photocatalytic H_2 generation from water, *Catal. Today*, 328 (2019) 15–20.
67. S. Wang, B. Zhu, M. Liu, L. Zhang, J. Yu, M. Zhou, Direct Z-scheme ZnO/CdS hierarchical photocatalyst for enhanced photocatalytic H_2-production activity, *Appl. Catal. B: Environ.*, 243 (2019) 19–26.
68. Z. Wang, H. Zhang, H. Cao, L. Wang, Z. Wan, Y. Hao, X. Wang, Facile preparation of ZnS/CdS core/shell nanotubes and their enhanced photocatalytic performance, *Int. J. Hydrogen Energy*, 42 (2017) 17394–17402.
69. T. Hu, P. Li, J. Zhang, C. Liang, K. Dai, Highly efficient direct Z-scheme WO_3/CdS-diethylenetriamine photocatalyst and its enhanced photocatalytic H_2 evolution under visible light irradiation, *Appl. Surf. Sci.*, 442 (2018) 20–29.
70. F. Cheng, H. Yin, Q. Xiang, Low-temperature solid-state preparation of ternary CdS/g-C_3N_4/CuS nanocomposites for enhanced visible-light photocatalytic H2-production activity, *Appl. Surf. Sci.*, 391 (2017) 432–439.
71. C.W. Lee, H.K. Park, S. Park, H.S. Han, S.W. Seo, H.J. Song, S. Shin, D.-W. Kim, K.S. Hong, Ta-substituted $SnNb_{2-x}Ta_xO_6$ photocatalysts for hydrogen evolution under visible light irradiation, *J. Mater. Chem. A.*, 3 (2015) 825–831.
72. R. Voorhoeve, D. Johnson, J. Remeika, P. Gallagher, Perovskite oxides: materials science in catalysis, *Science*, 195 (1977) 827–833.

73. J. Hwang, K. Akkiraju, J. Corchado-García, Y. Shao-Horn, A perovskite electronic structure descriptor for electrochemical CO_2 reduction and the competing H_2 evolution reaction, *J. Phys. Chem. C*, 123 (2019) 24469–24476.
74. T. Lv, M. Wu, M. Guo, Q. Liu, L. Jia, Self-assembly photocatalytic reduction synthesis of graphene-encapusulated $LaNiO_3$ nanoreactor with high efficiency and stability for photocatalytic water splitting to hydrogen, *Chem. Eng. J.*, 356 (2019) 580–591.
75. S. Wu, Y. Lin, C. Yang, C. Du, Q. Teng, Y. Ma, D. Zhang, L. Nie, Y. Zhong, Enhanced activation of peroxymonosulfte by $LaFeO_3$ perovskite supported on Al_2O_3 for degradation of organic pollutants, *Chemosphere*, 237 (2019) 124478.
76. N.-G. Park, Crystal growth engineering for high efficiency perovskite solar cells, *CrystEngComm*, 18 (2016) 5977–5985.
77. M.A. Green, A. Ho-Baillie, H.J. Snaith, The emergence of perovskite solar cells, *Nat. Photonics*, 8 (2014) 506.
78. M. Moniruddin, B. Ilyassov, X. Zhao, E. Smith, T. Serikov, N. Ibrayev, R. Asmatulu, N. Nuraje, Recent progress on perovskite materials in photovoltaic and water splitting applications, *Mater. Today Energy*, 7 (2018) 246–259.
79. Z. Cheng, J. Lin, Layered organic-inorganic hybrid perovskites: structure, optical properties, film preparation, patterning and templating engineering, *CrystEngComm*, 12 (2010) 2646–2662.
80. F. Yoshitomi, K. Sekizawa, K. Maeda, O. Ishitani, Selective formic acid production via CO_2 reduction with visible light using a hybrid of a perovskite tantalum oxynitride and a binuclear ruthenium (II) complex, *ACS Appl. Mater. Interfaces*, 7 (2015) 13092–13097.
81. H. Shi, Z. Zou, Photophysical and photocatalytic properties of $ANbO_3$ (A= Na, K) photocatalysts, *J. Phys. Chem. Solids*, 73 (2012) 788–792.
82. H. Kato, K. Asakura, A. Kudo, Highly efficient water splitting into H_2 and O_2 over lanthanum-doped $NaTaO_3$ photocatalysts with high crystallinity and surface nanostructure, *J. Am. Chem. Soc.*, 125 (2003) 3082–3089.
83. K. Peng, L. Fu, H. Yang, J. Ouyang, Perovskite $LaFeO_3$/montmorillonite nanocomposites: synthesis, interface characteristics and enhanced photocatalytic activity, *Sci. Rep.*, 6 (2016) 1–10.
84. S.N. Tijare, M.V. Joshi, P.S. Padole, P.A. Mangrulkar, S.S. Rayalu, N.K. Labhsetwar, Photocatalytic hydrogen generation through water splitting on nano-crystalline $LaFeO_3$ perovskite, *Int. J. Hydrogen Energy*, 37 (2012) 10451–10456.
85. X. Wang, K. Maeda, A. Thomas, K. Takanabe, G. Xin, J.M. Carlsson, K. Domen, M. Antonietti, A metal-free polymeric photocatalyst for hydrogen production from water under visible light, *Nat. Mater.*, 8 (2009) 76–80.
86. Y. Wang, Y. Di, M. Antonietti, H. Li, X. Chen, X. Wang, Excellent visible-light photocatalysis of fluorinated polymeric carbon nitride solids, *Chem. Mater.*, 22 (2010) 5119–5121.
87. Y. Wang, X. Wang, M. Antonietti, Polymeric graphitic carbon nitride as a heterogeneous organocatalyst: from photochemistry to multipurpose catalysis to sustainable chemistry, *Angew. Chem. Int. Edit.*, 51 (2012) 68–89.
88. J. Zhu, P. Xiao, H. Li, S.A. Carabineiro, Graphitic carbon nitride: synthesis, properties, and applications in catalysis, *ACS Appl. Mater. Interfaces*, 6 (2014) 16449–16465.
89. Y. Xiao, G. Tian, W. Li, Y. Xie, B. Jiang, C. Tian, D. Zhao, H. Fu, Molecule self-assembly synthesis of porous few-layer carbon nitride for highly efficient photoredox catalysis, *J. Am. Chem. Soc.*, 141 (2019) 2508–2515.
90. G. Zhang, L. Lin, G. Li, Y. Zhang, A. Savateev, S. Zafeiratos, X. Wang, M. Antonietti, Ionothermal synthesis of triazine-heptazine-based copolymers with apparent quantum yields of 60% at 420 nm for solar hydrogen production from "sea water", *Angew. Chem. Int. Edit.*, 57 (2018) 9372–9376.

91. Y. Yu, W. Yan, X. Wang, P. Li, W. Gao, H. Zou, S. Wu, K. Ding, Surface engineering for extremely enhanced charge separation and photocatalytic hydrogen evolution on g-C_3N_4, *Adv. Mater.*, 30 (2018) 1705060.
92. L. Zhang, Z. Jin, H. Lu, T. Lin, S. Ruan, X.S. Zhao, Y.-J. Zeng, Improving the visible-light photocatalytic activity of graphitic carbon nitride by carbon black doping, *ACS Omega*, 3 (2018) 15009–15017.
93. L. Zhang, Z. Jin, S. Huang, X. Huang, B. Xu, L. Hu, H. Cui, S. Ruan, Y.-J. Zeng, Bio-inspired carbon doped graphitic carbon nitride with booming photocatalytic hydrogen evolution, *Appl. Catal. B: Environ.*, 246 (2019) 61–71.
94. J. Zhang, X. Chen, K. Takanabe, K. Maeda, K. Domen, J.D. Epping, X. Fu, M. Antonietti, X. Wang, Synthesis of a carbon nitride structure for visible-light catalysis by copolymerization, *Angew. Chem. Int. Edit.*, 49 (2010) 441–444.
95. N. Sagara, S. Kamimura, T. Tsubota, T. Ohno, Photoelectrochemical CO_2 reduction by a p-type boron-doped g-C_3N_4 electrode under visible light, *Appl. Catal. B: Environ.*, 192 (2016) 193–198.
96. H. Yu, R. Shi, Y. Zhao, T. Bian, Y. Zhao, C. Zhou, G.I. Waterhouse, L.Z. Wu, C.H. Tung, T. Zhang, Alkali-assisted synthesis of nitrogen deficient graphitic carbon nitride with tunable band structures for efficient visible-light-driven hydrogen evolution, *Adv. Mater.*, 29 (2017) 1605148.
97. J. Yuan, X. Liu, Y. Tang, Y. Zeng, L. Wang, S. Zhang, T. Cai, Y. Liu, S. Luo, Y. Pei, Positioning cyanamide defects in g-C_3N_4: engineering energy levels and active sites for superior photocatalytic hydrogen evolution, *Appl. Catal. B: Environ.*, 237 (2018) 24–31.
98. P. Yang, H. Zhuzhang, R. Wang, W. Lin, X. Wang, Carbon vacancies in a melon polymeric matrix promote photocatalytic carbon dioxide conversion, *Angew. Chem. Int. Edit.*, 58 (2019) 1134–1137.
99. A. Mahmood, W. Guo, H. Tabassum, R. Zou, Metal-organic framework-based nanomaterials for electrocatalysis, *Adv. Energy. Mater.*, 6 (2016) 1600423.
100. H. Wang, Q.-L. Zhu, R. Zou, Q. Xu, Metal-organic frameworks for energy applications, *Chem*, 2 (2017) 52–80.
101. B. Zhu, R. Zou, Q. Xu, Metal-organic framework based catalysts for hydrogen evolution, *Adv. Energy. Mater.*, 8 (2018) 1801193.
102. C.G. Silva, A. Corma, H. García, Metal-organic frameworks as semiconductors, *J. Mater. Chem.*, 20 (2010) 3141–3156.
103. Y. Bai, Y. Dou, L.-H. Xie, W. Rutledge, J.-R. Li, H.-C. Zhou, Zr-based metal-organic frameworks: design, synthesis, structure, and applications, *Chem. Soc. Rev.*, 45 (2016) 2327–2367.
104. K. Meyer, M. Ranocchiari, J.A. van Bokhoven, Metal organic frameworks for photocatalytic water splitting, *Energ. Environ. Sci.*, 8 (2015) 1923–1937.
105. R. Li, W. Zhang, K. Zhou, Metal-organic-framework-based catalysts for photoreduction of CO_2, *Adv. Mater.*, 30 (2018) 1705512.
106. Y. Lin, X. Jiang, S.T. Kim, S.B. Alahakoon, X. Hou, Z. Zhang, C.M. Thompson, R.A. Smaldone, C. Ke, An elastic hydrogen-bonded cross-linked organic framework for effective iodine capture in water, *J. Am. Chem. Soc.*, 139 (2017) 7172–7175.
107. S. Chen, D.L. Jacobs, J. Xu, Y. Li, C. Wang, L. Zang, 1D nanofiber composites of perylene diimides for visible-light-driven hydrogen evolution from water, *RSC Adv.*, 4 (2014) 48486–48491.
108. D. Liu, J. Wang, X. Bai, R. Zong, Y. Zhu, Self-assembled PDINH supramolecular system for photocatalysis under visible light, *Adv. Mater.*, 28 (2016) 7284–7290.
109. J. Wang, D. Liu, Y. Zhu, S. Zhou, S. Guan, Supramolecular packing dominant photocatalytic oxidation and anticancer performance of PDI, *Appl. Catal. B: Environ.*, 231 (2018) 251–261.

110. W.-K. Jo, T.S. Natarajan, Influence of TiO_2 morphology on the photocatalytic efficiency of direct Z-scheme g-C_3N_4/TiO_2 photocatalysts for isoniazid degradation, *Chem. Eng. J.*, 281 (2015) 549–565.
111. J. Pan, Z. Dong, B. Wang, Z. Jiang, C. Zhao, J. Wang, C. Song, Y. Zheng, C. Li, The enhancement of photocatalytic hydrogen production via Ti^{3+} self-doping black TiO_2/g-C_3N_4 hollow core-shell nano-heterojunction, *Appl. Catal. B: Environ.*, 242 (2019) 92–99.
112. J. Wang, Y. Xia, H. Zhao, G. Wang, L. Xiang, J. Xu, S. Komarneni, Oxygen defects-mediated Z-scheme charge separation in g-C_3N_4/ZnO photocatalysts for enhanced visible-light degradation of 4-chlorophenol and hydrogen evolution, *Appl. Catal. B: Environ.*, 206 (2017) 406–416.
113. S. Huang, Y. Long, S. Ruan, Y.-J. Zeng, Enhanced photocatalytic CO_2 reduction in defect-engineered Z-scheme WO_{3-x}/g-C_3N_4 heterostructures, *ACS Omega*, 4 (2019) 15593–15599.
114. W. Yu, J. Chen, T. Shang, L. Chen, L. Gu, T. Peng, Direct Z-scheme g-C_3N_4/WO_3 photocatalyst with atomically defined junction for H_2 production, *Appl. Catal. B: Environ.*, 219 (2017) 693–704.
115. P. Xia, B. Zhu, B. Cheng, J. Yu, J. Xu, 2D/2D g-C_3N_4/MnO_2 nanocomposite as a direct Z-scheme photocatalyst for enhanced photocatalytic activity, *ACS Sustain. Chem. Eng.*, 6 (2018) 965–973.
116. A. Habibi-Yangjeh, M. Mousavi, K. Nakata, Boosting visible-light photocatalytic performance of g-C_3N_4/Fe_3O_4 anchored with $CoMoO_4$ nanoparticles: novel magnetically recoverable photocatalysts, *J. Photoch. Photobio. A*, 368 (2019) 120–136.
117. Y. Zang, L. Li, X. Li, R. Lin, G. Li, Synergistic collaboration of g-C_3N_4/SnO_2 composites for enhanced visible-light photocatalytic activity, *Chem. Eng. J.*, 246 (2014) 277–286.
118. T. Di, B. Zhu, B. Cheng, J. Yu, J. Xu, A direct Z-scheme g-C_3N_4/SnS_2 photocatalyst with superior visible-light CO_2 reduction performance, *J. Catal.*, 352 (2017) 532–541.
119. Y.-J. Yuan, Z. Shen, S. Wu, Y. Su, L. Pei, Z. Ji, M. Ding, W. Bai, Y. Chen, Z.-T. Yu, Liquid exfoliation of g-C_3N_4 nanosheets to construct 2D-2D MoS_2/g-C_3N_4 photocatalyst for enhanced photocatalytic H_2 production activity, *Appl. Catal. B: Environ.*, 246 (2019) 120–128.
120. W.-K. Jo, N.C.S. Selvam, Z-scheme CdS/g-C_3N_4 composites with RGO as an electron mediator for efficient photocatalytic H_2 production and pollutant degradation, *Chem. Eng. J.*, 317 (2017) 913–924.
121. X. Hao, J. Zhou, Z. Cui, Y. Wang, Y. Wang, Z. Zou, Zn-vacancy mediated electron-hole separation in ZnS/g-C_3N_4 heterojunction for efficient visible-light photocatalytic hydrogen production, *Appl. Catal. B: Environ.*, 229 (2018) 41–51.
122. Y. Xu, Y. Chen, W.-F. Fu, Visible-light driven oxidative coupling of amines to imines with high selectivity in air over core-shell structured CdS@C_3N_4, *Appl. Catal. B: Environ.*, 236 (2018) 176–183.
123. Y. Wu, H. Wang, W. Tu, Y. Liu, Y.Z. Tan, X. Yuan, J.W. Chew, Quasi-polymeric construction of stable perovskite-type $LaFeO_3$/g-C_3N_4 heterostructured photocatalyst for improved Z-scheme photocatalytic activity via solid pn heterojunction interfacial effect, *J. Hazard. Mater.*, 347 (2018) 412–422.
124. J. Wang, L. Tang, G. Zeng, Y. Deng, Y. Liu, L. Wang, Y. Zhou, Z. Guo, J. Wang, C. Zhang, Atomic scale g-C_3N_4/Bi_2WO_6 2D/2D heterojunction with enhanced photocatalytic degradation of ibuprofen under visible light irradiation, *Appl. Catal. B: Environ.*, 209 (2017) 285–294.
125. Y. Bai, L. Ye, L. Wang, X. Shi, P. Wang, W. Bai, P.K. Wong, g-C_3N_4/$Bi_4O_5I_2$ heterojunction with I^{3-}/I^- redox mediator for enhanced photocatalytic CO_2 conversion, *Appl. Catal. B: Environ.*, 194 (2016) 98–104.

126 B. Zhu, P. Xia, Y. Li, W. Ho, J. Yu, Fabrication and photocatalytic activity enhanced mechanism of direct Z-scheme g-C_3N_4/Ag_2WO_4 photocatalyst, *Appl. Surf. Sci.*, 391 (2017) 175–183.
127. P. Shandilya, D. Mittal, M. Soni, P. Raizada, A. Hosseini-Bandegharaei, A.K. Saini, P. Singh, Fabrication of fluorine doped graphene and $SmVO_4$ based dispersed and adsorptive photocatalyst for abatement of phenolic compounds from water and bacterial disinfection, *J. Clean. Prod.*, 203 (2018) 386–399.
128. L. Zhao, X. Zhang, C. Fan, Z. Liang, P. Han, First-principles study on the structural, electronic and optical properties of BiOX (X = Cl, Br, I) crystals, *Physica B*, 407 (2012) 3364–3370.
129. G. Li, S. Zhao, Y. Zhang, Z. Tang, Metal-organic frameworks encapsulating active nanoparticles as emerging composites for catalysis: recent progress and perspectives, *Adv. Mater.*, 30 (2018) 1800702.
130. W. Xiang, Y. Zhang, H. Lin, C.-j. Liu, Nanoparticle/metal-organic framework composites for catalytic applications: current status and perspective, *Molecules*, 22 (2017) 2103.
131. X. Wang, J. Liu, S. Leong, X. Lin, J. Wei, B. Kong, Y. Xu, Z.-X. Low, J. Yao, H. Wang, Rapid construction of ZnO@ZIF-8 heterostructures with size-selective photocatalysis properties, *ACS Appl. Mater. Interfaces*, 8 (2016) 9080–9087.
132. R. Li, J. Hu, M. Deng, H. Wang, X. Wang, Y. Hu, H.L. Jiang, J. Jiang, Q. Zhang, Y. Xie, Integration of an inorganic semiconductor with a metal-organic framework: a platform for enhanced gaseous photocatalytic reactions, *Adv. Mater.*, 26 (2014) 4783–4788.
133. S. Sajjadi, A. Khataee, N. Bagheri, M. Kobya, A. Şenocak, E. Demirbas, A.G. Karaoğlu, Degradation of diazinon pesticide using catalyzed persulfate with Fe_3O_4@MOF-2 nanocomposite under ultrasound irradiation, *J. Ind. Eng. Chem.*, 77 (2019) 280–290.
134. M. Zeng, Z. Chai, X. Deng, Q. Li, S. Feng, J. Wang, D. Xu, Core-shell CdS@ZIF-8 structures for improved selectivity in photocatalytic H_2 generation from formic acid, *Nano Res.*, 9 (2016) 2729–2734.
135. S. Wan, M. Ou, Q. Zhong, X. Wang, Perovskite-type $CsPbBr_3$ quantum dots/UiO-66 (NH_2) nanojunction as efficient visible-light-driven photocatalyst for CO_2 reduction, *Chem. Eng. J.*, 358 (2019) 1287–1295.
136. J. Yang, H. Miao, W. Li, H. Li, Y. Zhu, Designed synthesis of a p-Ag_2S/n-PDI self-assembled supramolecular heterojunction for enhanced full-spectrum photocatalytic activity, *J. Mater. Chem. A*, 7 (2019) 6482–6490.
137. W. Wei, Y. Zhu, TiO_2@perylene diimide full-spectrum photocatalysts via semi-core-shell structure, *Small*, 15 (2019) 1903933.
138. W. Zeng, T. Cai, Y. Liu, L. Wang, W. Dong, H. Chen, X. Xia, An artificial organic-inorganic Z-scheme photocatalyst WO_3@Cu@PDI supramolecular with excellent visible light absorption and photocatalytic activity, *Chem. Eng. J.*, 381 (2020) 122691.
139. F. Guo, W. Shi, H. Wang, H. Huang, Y. Liu, Z. Kang, Fabrication of a $CuBi_2O_4$/gC_3N_4 p-n heterojunction with enhanced visible light photocatalytic efficiency toward tetracycline degradation, *Inorg. Chem. Front.*, 4 (2017) 1714–1720.
140. J. Low, J. Yu, M. Jaroniec, S. Wageh, A.A. Al-Ghamdi, Heterojunction photocatalysts, *Adv. Mater.*, 29 (2017) 1601694.
141. P. Zhou, J. Yu, M. Jaroniec, All-solid-state Z-scheme photocatalytic systems, *Adv. Mater.*, 26 (2014) 4920–4935.

Index

Page numbers in *italic* indicate figures. Page numbers in **bold** indicate tables.

D

I

M

P

Q

R

S

T

U

V

9780367342128